[4.2]	P_1P_2, or d	the distance between two points P_1 and P_2
	m	the slope of a line
[4.5]	$[x]$	the greatest integer not greater than x
[6.1]	b^x	the power with base $b > 0$ and exponent x
	e	an irrational number, approximately equal to 2.718281828
[6.2]	$\log_b x$	the logarithm to the base b of x, or the logarithm of x to the base b
	$\ln x$	$\log_e x$
[7.2]	(x, y, z)	the ordered triple of numbers whose first component is x, second component is y, and third component is z
[8.1]	$\begin{bmatrix} a_1 & b_1 & c_1 \\ a_2 & b_2 & c_2 \end{bmatrix}$, etc.	matrix
	$A_{m \times n}$	m by n matrix
	a_{ij}	the element in the ith row and jth column of the matrix A
	A^t	the transpose of the matrix A
	$0_{m \times n}$	the m by n zero matrix
	$-A_{m \times n}$	the negative of $A_{m \times n}$
[8.2]	$I_{n \times n}$	the identity matrix for all n by n matrices
[8.3]	$A \sim B$	A is row-equivalent to B (for matrices)

SEVENTH EDITION
College Algebra

SEVENTH EDITION
College Algebra

MICHAEL D. GRADY
Loyola Marymount University

IRVING DROOYAN
Los Angeles Pierce College *Emeritus*

EDWIN F. BECKENBACH

Wadsworth Publishing Company
Belmont, California ■ A Division of Wadsworth, Inc.

Mathematics Editor: Kevin Howat
Editorial Assistant: Sally Uchizono
Production Editor: Michael G. Oates
Managing Designer: Andrew H. Ogus
Print Buyer: Barbara Britton
Art Editor and Interior Designer: Wendy Calmenson
Copy Editor: Margaret Moore
Technical Illustrator: Victor Royer
Compositor: Polyglot Ltd., Singapore
Cover: Carolyn Deacey

Printed in the United States of America   34

1 2 3 4 5 6 7 8 9 10—92 91 90 89 88

Library of Congress Cataloging-in-Publication Data

Grady, Michael D., 1946–
   College algebra.

   Beckenbach's name appears first on earlier editions.
   Includes index.
   1. Algebra.   I. Drooyan, Irving.   II. Beckenbach,
Edwin F.   III. Title.

QA154.2.G72   1988   512.9   87–24268
ISBN 0–534–08556–3

Contents

Preface

The seventh edition of *College Algebra,* like its predecessors, is designed to provide a contemporary mathematics course for college students who have one to two years of high school algebra or its community college equivalent. In this edition, as in previous editions, close attention has been given to pedagogical detail and the organization provides for maximum flexibility.

Review and Reference Material

The *Preliminary Concepts* section provides a review of basic facts from prerequisite courses and serves as a reference throughout the text. In addition, the first three chapters provide an extensive review for students who need it. If students are prepared in the prerequisite algebra, these chapters can be omitted entirely, or a review can be limited to the chapter reviews which consist of a list of the important terms and concepts as well as review problems. These reviews help students focus on concepts as well as on mechanical techniques.

The *Reference Outline* in the appendix provides an overview of all the important ideas that are introduced in the text. This provides an excellent reference for the student while studying for chapter tests and the final examination.

Use of Calculators

Procedures for approximating solutions to polynomial equations using a calculator are introduced along with the traditional results in the theory of equations, thus giving instructors the flexibility to choose the approach that is best suited for their objectives.

In addition, decimal approximations to powers and logarithms are obtained using a calculator, but the methods for obtaining these results from the tables are provided in the appendix in case this approach better meets the needs of the instructor.

Improved Problem Sets

The exercises in most sections have been expanded to provide more variety and more drill exercises on which the student can practice. This includes many more applied problems, particularly in Chapter 6 on logarithms and exponentials.

The exercises have been carefully graded beginning with the A exercises, which are designed to give the students confidence, and proceeding through the B exercises, which are designed to stretch the students' abilities.

ix

A set of B exercises which are not keyed to specific sections has been added to most chapter reviews. These exercises will provide students the opportunity to test and improve their skills while studying for chapter tests.

Organization and Format

Examples that were included in the exercise sets in the sixth edition are now included in the textual discussions and are numbered. The exercises for each section are keyed to these examples. This organization allows students to develop their problem solving skills with less dependence on examples, while at the same time providing easy access to help should they need it.

Chapters 6 through 11 are sufficiently independent to allow one or more to be omitted in the interest of time. This organization of the material allows instructors to design the course according to their particular curriculum needs.

The format of the seventh edition is one which will make it easier for the student to focus on the important facts and procedures and make the text easier to read.

Supplemental Material

A *Student Manual* that includes solutions to the even-numbered exercises and other helpful study aids is available for student use. Additional ancillary materials that relate to the text are available to instructors.

Acknowledgments

We offer our thanks to the reviewers of this edition. Their comments and suggestions were appreciated: Dale R. Bedgood, East Texas State University; Barbara H. Briggs, Tennessee Technical University; Douglas G. Burkholder, Wichita State University; Jugal Ghorai, University of Wisconsin, Milwaukee; James Miller, West Virginia University; William H. Price, Middle Tennessee State University; Gloria Rivkin, Lawrence Institute of Technology; Marvin Roof, Charles S. Mott Community College; J. Bryan Sperry, Pittsburgh State University; Karen Walls, Northeastern Oklahoma A & M College; and David S. Weinstein, California State University, Northridge.

This edition is dedicated to the memory of Professor Edwin F. Beckenbach, an eminent mathematician, an outstanding teacher, and an inspirational coauthor.

We sincerely thank Mary-Margaret M. Grady for her excellent work in preparing the manuscript for publication, and Pamela Christopherson for her excellent help in proofreading the examples and exercises.

Michael D. Grady
Irving Drooyan

Preliminary Concepts

The following definitions, notations, algebraic properties, formulas, and geometric facts have been introduced in previous courses. They are presented here to provide you with a convenient reference for your work in this text.

Lowercase letters, such as a, b, x, and y, will be used to denote real numbers both in the following lists and in the text.

GLOSSARY

Prime and Composite Numbers

If a is an element of the set N of positive integers, and $a \neq 1$, then a is a **prime number** if and only if a has no factor in N other than itself and 1; otherwise, a is a **composite number**. For example, 2 and 3 are prime numbers; $6 = 2 \cdot 3$ is composite. The number 1 is considered to be neither prime nor composite.

Two integers a and b are said to be **relatively prime** if and only if they have no prime factors in common. For example, 6 and 35 are relatively prime, since $6 = 2 \cdot 3$ and $35 = 5 \cdot 7$; but 6 and 8 are not, since they have the prime factor 2 in common.

Set

A **set** is a collection of objects. Any one of the objects of a collection is called a **member** or an **element** of the set. For example, the collection of numbers 2, 4, 6, 8 is a set denoted by $\{2, 4, 6, 8\}$ and the number 2 is a member or an element of that set. We denote this by

$$2 \in \{2, 4, 6, 8\}.$$

Set Equality

Two sets A and B are **equal**, $A = B$, if and only if they consist of exactly the same elements. For example,

$$\{3, 1\} = \{3, 2 - 1\}.$$

Subset

If every element of a set A is an element of a set B, then A is a **subset** of B (denoted $A \subset B$). For example,

$$\{1, 2, 3\} \subset \{1, 2, 3, 4\}.$$

Union of Two Sets

The **union** of two sets A and B (denoted $A \cup B$) is the set of all elements that belong either to A or B or to both. For example, for sets $A = \{2, 4, 6\}$ and $B = \{1, 3, 4, 5\}$, "A union B" is

$$A \cup B = \{1, 2, 3, 4, 5, 6\}.$$

Intersection of Two Sets

The **intersection** of two sets A and B (denoted $A \cap B$) is the set of all elements common to both A and B. For example, for sets $A = \{1, 2, 3, 4\}$ and $B = \{2, 3, 4, 5\}$, "A intersect B" is

$$A \cap B = \{2, 3, 4\}.$$

NOTATION

$\{a, b\}$	the set whose elements are a and b
$A \subset B$	A is a subset of B
$\varnothing$	the null, or empty, set
$A \cup B$	the union of sets A and B
$A \cap B$	the intersection of sets A and B
$\in$	is an element of
$\{x \vert \ldots\}$	the set of all x such that . . .
J	the set of integers; . . . , $-2, -1, 0, 1, 2, \ldots$
N	the set of positive integers; $1, 2, 3, \ldots$
Q	the set of rational numbers
H	the set of irrational numbers
R	the set of real numbers
C	the set of complex numbers
$a + b$	the sum of a and b
$a \cdot b$	the product of a and b (also written ab)
$-a$	the negative of a; $(-1)a = -a$
$a - b$	the difference b subtracted from a; $a - b = a + (-b)$
$\dfrac{a}{b}$	the quotient a divided by b; $\dfrac{a}{b} = a \cdot \dfrac{1}{b}$
$a < b$	a is less than b
$a > b$	a is greater than b
$a \leqslant b$	a is less than or equal to b
$a \geqslant b$	a is greater than or equal to b
$\vert a \vert$	the absolute value of a;
	$\vert a \vert = \begin{cases} a & \text{if } a \geqslant 0 \\ -a & \text{if } a < 0 \end{cases}$
$P(x)$	algebraic expression in the variable x

ALGEBRAIC AXIOMS FOR THE REAL NUMBERS

1. $a + b$ is a unique real number. — *Closure law for addition.*
2. $(a + b) + c = a + (b + c)$ — *Associative law for addition.*
3. For each $a \in R$, — *Additive-identity law.*

$$a + 0 = a.$$

   0 is called the *additive identity for R*.

4. For each $a \in R$, — *Additive-inverse law.*

$$a + (-a) = 0.$$

   $-a$ is called the *additive inverse of a*.

5. $a + b = b + a$ — *Commutative law for addition.*
6. $a \cdot b$ is a unique real number. — *Closure law for multiplication.*
7. $(a \cdot b) \cdot c = a \cdot (b \cdot c)$ — *Associative law for multiplication.*

8. $a \cdot (b + c) = (a \cdot b) + (a \cdot c)$ — *Distributive law.*
9. For each $a \in R$, — *Multiplicative-identity law.*

$$a \cdot 1 = a.$$

   1 is called the *multiplicative identity in R*.

10. $a \cdot b = b \cdot a$ — *Commutative law for multiplication.*

11. For each $a \in R$, $a \neq 0$, — *Multiplicative-inverse law.*

$$a \cdot \frac{1}{a} = 1.$$

   $1/a$ is called the *multiplicative inverse of a*.

ORDER AXIOMS OF REAL NUMBERS

O-1. If a is a real number, then exactly one of the following statements is true: a is positive, a is zero, or a is negative. — *Trichotomy law.*

O-2. If a and b are positive real numbers, then $a + b$ is positive and ab is positive. — *Closure law for positive numbers.*

O-3. There is a one-to-one correspondence between the real numbers and the points on a geometric line. — *Completeness property.*

GEOMETRIC INTERPRETATION OF ORDER

Certain algebraic statements concerning the order of real numbers can be interpreted geometrically. We summarize some of the more common correspondences in the table, where in each case $a, b, c \in R$.

Algebraic Statement	Geometric Statement	Graph
1. a is positive	1. The graph of a lies to the right of the origin.	1. (number line with 0 and a to the right)
2. a is negative	2. The graph of a lies to the left of the origin.	2. (number line with a to the left, 0)
3. $a > b$	3. The graph of a lies to the right of the graph of b.	3. (number line with b then a)
4. $a < b$	4. The graph of a lies to the left of the graph of b.	4. (number line with a then b)
5. $a < c < b$	5. The graph of c is to the right of the graph of a and to the left of the graph of b.	5. (number line with a, c, b)
6. $|a| < c$	6. The graph of a is less than c units from the origin.	6. (number line with $-c$, 0 a, c)
7. $|a - b| < c$	7. The graphs of a and b are less than c units from each other.	7. $|a - b| < c$ (bracket over a and b)
8. $|a| < |b|$	8. The graph of a is closer to the origin than the graph of b.	8. (number line with $-|b|$, a 0, $|b|$)

Notice that Statement 5, $a < c < b$, is a contraction of the two inequalities $a < c$ and $c < b$ and is read "a is less than c and c is less than b" or "c is between a and b." Similarly, $a < c$ and $c \leq b$ can be written as $a < c \leq b$.

PROPERTIES OF REAL NUMBERS

1. If $a + b = 0$, then $b = -a$ and $a = -b$.

2. If $a \cdot b = 1$, then $a = 1/b$ and $b = 1/a$.

3. $a \cdot b = 0$ if and only if either $a = 0$ or $b = 0$, or both.

4. $-(-a) = a$

5. $\dfrac{-a}{b} = \dfrac{a}{-b} = -\dfrac{a}{b} = -\dfrac{-a}{-b}$, $b \neq 0$

6. $\dfrac{a}{b} = \dfrac{-a}{-b} = -\dfrac{a}{-b} = -\dfrac{-a}{b}$, $b \neq 0$

7. $\dfrac{a}{b} = \dfrac{c}{d}$ if and only if $ad = bc$, $b, d \neq 0$

8. $\dfrac{ac}{bc} = \dfrac{a}{b}$, $b, c \neq 0$

9. $\dfrac{1}{a} \cdot \dfrac{1}{b} = \dfrac{1}{ab}$, $a, b \neq 0$

10. $\dfrac{a}{b} \cdot \dfrac{c}{d} = \dfrac{ac}{bd}$, $b, d \neq 0$

11. $\dfrac{a}{c} + \dfrac{b}{c} = \dfrac{a + b}{c}$, $c \neq 0$

12. $\dfrac{a}{b} + \dfrac{c}{d} = \dfrac{ad + bc}{bd}$, $b, d \neq 0$

13. $\dfrac{a}{c} - \dfrac{b}{c} = \dfrac{a - b}{c}$, $c \neq 0$

14. $\dfrac{a}{b} - \dfrac{c}{d} = \dfrac{ad - bc}{bd}$, $b, d \neq 0$

15. $\dfrac{1}{a/b} = \dfrac{b}{a}$, $a, b \neq 0$

16. $\dfrac{a/b}{c/d} = \dfrac{a}{b} \div \dfrac{c}{d} = \dfrac{ad}{bc}$, $b, c, d \neq 0$

17. If $a = b$, then $a + c = b + c$.

18. If $a = b$, then $ac = bc$.

19. If $a < b$, then $a + c < b + c$.

20. If $a < b$, then $a - c < b - c$.

21. If $a < b$ and $c > 0$, then $a \cdot c < b \cdot c$.

22. If $a < b$ and $c < 0$, then $a \cdot c > b \cdot c$.

FACTS FROM GEOMETRY

Similar Triangles

Two triangles are said to be **similar** if corresponding angles are equal. If two triangles are similar, then the ratios of corresponding sides are equal. For example, the two triangles below are similar.

 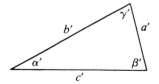

In this case $\alpha = \alpha'$, $\beta = \beta'$, and $\gamma = \gamma'$; further,

$$\frac{a'}{a} = \frac{b'}{b} = \frac{c'}{c}.$$

Pythagorean Theorem

If a, b, and c denote the lengths of the sides of a right triangle, and c is the length of the side opposite the right angle, then

$$a^2 + b^2 = c^2.$$

PERIMETER AND AREA FORMULAS

Circle

Circumference: $C = 2\pi r$
Area: $A = \pi r^2$

Triangle

Perimeter: $P = a + b + c$
Area: $A = \frac{1}{2}bh$

Rectangle

Perimeter: $P = 2l + 2w$
Area: $A = lw$

College Algebra

1

Polynomials; Rational Expressions

In this chapter we shall review some properties of simple algebraic expressions—polynomials and rational expressions. We consider their behavior under the basic arithmetic operations of addition, subtraction, multiplication, and division, and discuss simplification of the results of these operations. You may wish to review the Preliminary Concepts section, which follows the Preface, before beginning your work in this chapter.

DEFINITIONS; SUMS OF POLYNOMIALS

Any grouping of constants and variables obtained by applying a finite number of the elementary operations—addition, subtraction, multiplication, division, or the extraction of roots—is called an **algebraic expression**. For example,

$$\frac{3x^2 + \sqrt{2x-1}}{3x^3 + 7} \qquad \text{and} \qquad xy + 3x^2z - \sqrt[5]{z}$$

are algebraic expressions. If two expressions have equal values for all values of the variables for which both expressions are defined, then we say that the expressions are **equivalent**. For example, the expressions

$$x + 2 \qquad \text{and} \qquad \frac{1}{2}(2x + 4)$$

are equivalent, whereas the expressions

$$x + 2 \qquad \text{and} \qquad 3x + 6$$

are not equivalent because 0 is in the replacement set of each, but

$$0 + 2 \neq 3(0) + 6.$$

Positive Integer Powers

Recall that an expression of the form a^n is called a **power** of a, where a is the **base** of the power and n is the **exponent** of the power.

Definition 1.1

If n is a positive integer and a is a real number, then

$$a^n = \underbrace{a \cdot a \cdot a \cdots \cdot a}_{n\ factors}.$$

Monomial Expressions

An algebraic expression of the form cx^n, where n is a nonnegative integer, is called a **monomial** in the variable x. The number c is called the **coefficient** of the monomial, and the number n is called the **degree** of the monomial in x. A constant other than 0 is said to be a monomial of degree 0; no degree is assigned to the special monomial 0.

An algebraic expression of the form $cx^m y^n$, where m and n are nonnegative integers, is called a monomial in the variables x and y. The degree of the monomial in x is m, and the degree of the monomial in y is n. The degree of the monomial in both x and y is $m + n$. We define a monomial in any number of variables and its degree in a similar way.

EXAMPLE 1 (a) The degree of the monomial $-4x^2 y^3$ is 2 in x, 3 in y, and 5 in x and y.

(b) The degree of the monomial $2xy^2 z^3$ is 1 in x, 2 in y, 3 in z, 3 in x and y, 4 in x and z, 5 in y and z, and 6 in x, y, and z. ■

Polynomial Expressions

In any algebraic expression of the form $A + B + C + \ldots$, where $A, B, C, \ldots$ are algebraic expressions, $A, B, C, \ldots$ are called **terms** of the expression. For example, in the expression $x + (y + 3)$ the terms are x and $(y + 3)$, but in $x + y + 3$ the terms are x, y, and 3.

A **polynomial** is any algebraic expression that can be written as one which contains only terms that are monomials. The **degree of a polynomial** is the same as the degree of its term of highest degree. Since no degree is assigned to the monomial 0, no degree is assigned to the polynomial 0, either.

A Polynomial is an algebraic expression that can be written as terms that are monomials.

EXAMPLE 2 (a) The degree of the polynomial $3xy^2 + 4xyz^2$ is 1 in x, 2 in y, 3 in x and y, and 4 in x, y, and z.

(b) The degree of the polynomial $x^3 + y^2$ is 3 in x, 2 in y, and 3 in x and y. ■

Because $a - b$ is defined to be $a + (-b)$ (see the Notation section of the Preliminary Concepts), we shall view the signs in any polynomial as the signs of the coefficients and the operation involved as addition. Thus,

$$3x - 5y + 4z = (3x) + (-5y) + (4z),$$

and $3x - 5y + 4z$ is a polynomial with terms $3x$, $-5y$, and $4z$. Again, an expression such as

$$a - (bx + cx^2),$$

in which a set of parentheses is preceded by a negative sign, can be written as

$$a + (-bx) + (-cx^2).$$

Thus, $a - (bx + cx^2)$ is equivalent to a polynomial with terms a, $-bx$, and $-cx^2$. Also, since

$$\frac{a}{b} = a\left(\frac{1}{b}\right),$$

we can view division by a constant b as multiplication by the constant $1/b$ and, for example, write

$$\frac{3x^2}{4} + \frac{x}{2} \quad \text{as} \quad \frac{3}{4}x^2 + \frac{1}{2}x.$$

Thus, a polynomial can be considered to involve only the operations of addition and multiplication. Since the set R of real numbers is closed with respect to these operations (see the Preliminary Concepts), it follows that, for any specific real value of x, a polynomial with real coefficients represents a real number. Therefore, the properties for the real numbers are applicable to the terms in such polynomials and to the polynomials themselves.

By applying the commutative, associative, and distributive laws in various ways, we can frequently rewrite polynomials and sums of polynomials in what might be termed "simpler" forms.

EXAMPLE 3 (a) $(2x^2 + 3x + 5) + (6x^2 + 7) = 2x^2 + 6x^2 + 3x + 5 + 7$
$$= (2 + 6)x^2 + 3x + 5 + 7$$
$$= 8x^2 + 3x + 12$$

(b) $(3x^2 + 2xy - y^2) - (x^2y - xy + x^2) = \left(3x^2 + 2xy - y^2\right) - \left(x^2y + xy - x^2\right)$
$$= (3 - 1)x^2 + (2 + 1)xy - y^2 - x^2y$$
$$= 2x^2 + 3xy - y^2 - x^2y \quad ■$$

A polynomial of degree n in x can be represented—when its terms are rearranged, if need be—by an expression in the **standard form**

$$a_n x^n + a_{n-1} x^{n-1} + a_{n-2} x^{n-2} + \cdots + a_1 x + a_0 \qquad (a_n \neq 0),$$

where it is understood that the a's are the (constant) coefficients of the powers of x in the polynomial.

The term $a_n x^n$ is called the **leading term** and the coefficient a_n is called the **leading coefficient** in the polynomial.

EXAMPLE 4   Write each polynomial in standard form.

(a) $(x^2 - 3x + 1) - (x^2 - 3x + 4)$ (b) $(x^2 - 2x + 4) - (2x^2 - 2)$

Solution   (a) $(x^2 - 3x + 1) - (x^2 - 3x + 4) = x^2 - 3x + 1 - x^2 + 3x - 4$

$$= (x^2 - x^2) + (-3x + 3x) + (1 - 4)$$

$$= -3$$

(b) $(x^2 - 2x + 4) - (2x^2 - 2) = x^2 - 2x + 4 - 2x^2 + 2$

$$= (x^2 - 2x^2) + (-2x) + (4 + 2)$$

$$= -x^2 - 2x + 6 \qquad ■$$

Nested Expressions

In simplifying a polynomial expression that contains nested grouping symbols— that is, one set of grouping symbols inside another set—it is generally best to simplify the expression within the innermost grouping symbols first and work outward.

EXAMPLE 5   Write each polynomial in standard form.

(a) $x - [2x - (3x - 1)]$ (b) $3x - 2[2 - x - 2(x + 5)]$

Solution   In both cases, simplify the innermost expression first.

(a) $x - [2x - (3x - 1)]$ (b) $3x - 2[2 - x - 2(x + 5)]$

$$= x - [2x - 3x + 1] \qquad\qquad = 3x - 2[2 - x - 2x - 10]$$

$$= x - [-x + 1] \qquad\qquad\quad = 3x - 2[-3x - 8]$$

$$= x + (x - 1) \qquad\qquad\quad = 3x + (6x + 16)$$

$$= 2x - 1 \qquad\qquad\qquad\quad = 9x + 16 \qquad ■$$

Symbols for Polynomials

Polynomials are frequently represented by symbols such as

$$P(x), \qquad D(y), \qquad \text{and} \qquad Q(z),$$

where the symbol in parentheses designates the variable. Thus, we might write

$$P(x) = 2x^3 - 3x + 2,$$
$$D(y) = y^6 - 2y^2 + 3y - 2,$$
$$Q(z) = 8z^4 + 3z^3 - 2z^2 + z - 1.$$

EXAMPLE 6 Let $P(x) = x^2 - 2x - 3$ and $Q(x) = 2x^2 - x + 1$, and write each polynomial in standard form.

(a) $P(x) + Q(x)$

(b) $P(x) - Q(x)$

Solution (a) $P(x) + Q(x) = (x^2 - 2x - 3) + (2x^2 - x + 1)$
$$= x^2 - 2x - 3 + 2x^2 - x + 1$$
$$= (1 + 2)x^2 + (-2 - 1)x + (-3 + 1)$$
$$= 3x^2 - 3x - 2$$

(b) $P(x) - Q(x) = (x^2 - 2x - 3) - (2x^2 - x + 1)$
$$= x^2 - 2x - 3 - 2x^2 + x - 1$$
$$= (1 - 2)x^2 + (-2 + 1)x + (-3 - 1)$$
$$= -x^2 - x - 4 \qquad ■$$

The notation $P(x)$ can be used to denote values of the polynomial for specific values of x. Thus, for example, $P(a)$ denotes the value of the polynomial $P(x)$ when x is replaced by a.

EXAMPLE 7 Let $P(x) = x^2 - 5x + 4$. Find

(a) $P(3)$ (b) $P(-2)$

Solution (a) $P(3) = (3)^2 - 5 \cdot 3 + 4$ (b) $P(-2) = (-2)^2 - 5 \cdot (-2) + 4$
$$= -2 \qquad\qquad\qquad\qquad = 18$$

EXAMPLE 8   Let $P(x) = x^2 + x$ and $Q(x) = x - 1$. Find

(a) $P(Q(2))$ 　　　　　　　　　(b) $Q(P(2))$

Solution   (a)   We first compute $Q(2)$:　　(b)   We first compute $P(2)$:

$$Q(2) = 2 - 1 = 1.$$ 　　　　　$$P(2) = 2^2 + 2 = 6.$$

Next we compute $P(Q(2))$: 　　Next we compute $Q(P(2))$:

$$\begin{aligned} P(Q(2)) &= P(1) \\ &= 1^2 + 1 \\ &= 2. \end{aligned}$$ 　　$$\begin{aligned} Q(P(2)) &= Q(6) \\ &= 6 - 1 \\ &= 5. \quad ■ \end{aligned}$$

In some applications, the notation $P(x)|_a^b$ denotes $P(b) - P(a)$.

EXAMPLE 9   Let $P(x) = \dfrac{x^2}{2} - 4x$. Find

(a) $P(x)|_0^1$ 　　　　　　　　(b) $P(x)|_{-2}^3$

Solution   (a) $P(x)|_0^1 = P(1) - P(0)$ 　　(b) $P(x)|_{-2}^3 = P(3) - P(-2)$

$$= \left(\frac{1}{2} - 4\right) - \left(\frac{0}{2} - 0\right)$$ 　　$$= \left(\frac{9}{2} - 12\right) - \left(\frac{4}{2} + 8\right)$$

$$= -\frac{7}{2}$$ 　　　　　　$$= -\frac{35}{2} \quad ■$$

1 - 60 odd

EXERCISE 1.1

A

■   *Simplify each expression and write each polynomial in standard form. See Examples 3 and 4.*

1. $(x^2 + 6x + 4) + (x^2 + 4x - 4)$ 　　　　**2.** $(x^3 + x^2 + x) + (x^3 - x^2 + 2x)$

3. $(x^2 - 7x - 5) - (3x^2 - x + 7)$ 　　　　**4.** $(x^3 - x - 4) - (2x^3 + 2x^2 + 4x + 5)$

5. $(x^3 + x^2 + 3x - 1) - 2x^3$ 　　　　　　**6.** $(x^5 - x^3 + 1) + (x^5 + 2x^3 + 4)$

7. $(4x^4 + 3x^2 + 1) - (3x^3 - x) - (4x^4 + 2x^3)$ 　　**8.** $(3x^2 - 3x + 5) - (4x^2 + 2x - 5) - 10$

9. $(x^2 + 3x + 2) - (x^2 + 2x - 3)$ 　　　　**10.** $(x^3 - 1) - (x^3 - 3x^2 + 3x - 1)$

11. $(x^2 - 2x + 3) - (x^2 + x - 1)$ 　　　　**12.** $(x^3 - x + 4) - (2x^3 - 2x^2 + 2x - 1)$

13. $(2x^2y + xy^2 + y^3) - (x^3 + xy^2 - 3x^2y) - (y^3 - x^3)$ 　**14.** $(2xy + x^2 - y^2) + (xy - x^2) - (y^2 - 3xy)$

15. $(x^2yz + xy^2z) - (2x^3y + 3x^2yz) + (x^3y + xy^2z)$

16. $(x^2y + x^2z) - (xyz + 2x^2z + 3xz^2) - (x^2y - xyz + xz^2)$

17. $(x^2z^2 - 2xyz + x^2y^2z^2) - (2x^2y^2z^2 + xyz + x^2y^2) - (xyz + x^2y^2z^2)$

18. $(2x^2z^3 - 3x^2yz^2 + 4xy) + (3x^2z^3 + 2x^2yz^2 - 2xz) + (3xz - 2xy)$

19. $(xyz - xy + xz - yz) - (2xyz + 3xy - xz - 2yz) + (x^2y^2 + y^2z^2)$

20. $(x^2y^2z^2 + x^3y^3 + x^2z^2 + yz) - (2x^2y^2z^2 - 3x^2z^2) + (x^2y^2z^2 - x^3y^3 - yz)$

■ *Write each polynomial in standard form. See Example 5.*

21. $2x - [3 + x - (4x - 5)]$ **22.** $4x - [(x + 2) - (3x + 1)]$

23. $x - [2x - \{3x - (4x - 1)\}]$ **24.** $x^2 - [1 - (1 - x^2)]$

25. $x^2 - 1 - [x - (2x^2 - 3x + 1)]$ **26.** $2x^2 - x + 5 - [x^2 - x - (2x^2 - 3x + 1)]$

27. $-\{-[(x^2 - 2x) + (x^3 - x^2)] + 3x\} - \{-(x - 2) + (x^2 - 4) - 3\}$

28. $x^2 - \{x - (x^3 - [x^2 - (x + 1) - x^3] + x^2) - x\}$

29. $-\{[(x^2 + x + 1) - (x^3 + x - 1)] - x + [-(x - \{x - (2x + 1)\}) + x]\}$

30. $2x^3 - \{x - [x^2 - (x^3 + 1)] - (x + 1) - [x^4 - (x^4 + x^2 - 1)]\}$

■ *Let* $P(x) = x^2 + 2x - 1$, $Q(x) = 2x^2 - x + 3$, *and* $R(x) = -3x^2 + 4x - 3$. *Write each polynomial in standard form. See Example 6.*

31. $P(x) + Q(x)$ **32.** $R(x) + Q(x)$ **33.** $P(x) + R(x)$

34. $R(x) - Q(x)$ **35.** $P(x) + [Q(x) + R(x)]$ **36.** $P(x) + [Q(x) - R(x)]$

37. $P(x) - [Q(x) + R(x)]$ **38.** $P(x) - [Q(x) - R(x)]$ **39.** $[P(x) - Q(x)] - R(x)$

40. $Q(x) + Q(x)$ **41.** $R(x) + R(x)$ **42.** $[P(x) - P(x)] + [Q(x) - Q(x)]$

■ *For Exercises 43–52, see Examples 7–9.*

43. Given $P(x) = x^2 + x - 1$, find $P(-1)$, $P(0)$, $P(-2)$, $P(x)|_{-1}^{0}$.

44. Given $P(x) = x^2 + 2x - 4$, find $P(-1)$, $P(0)$, $P(-2)$, $P(x)|_{-1}^{0}$.

45. Given $P(x) = x^3 - x^2 + 2$, find $P(-1)$, $P(0)$, $P(-2)$, $P(x)|_{-1}^{0}$.

46. Given $P(x) = x^4 - 2x^3 + x$, find $P(-2)$, $P(1)$, $P(3)$, $P(x)|_{1}^{3}$.

47. Given $P(x) = x^7$, find $P(-1)$, $P(0)$, $P(1)$, $P(x)|_{-1}^{1}$.

48. Given $P(x) = x^{14}$, find $P(-1)$, $P(0)$, $P(1)$, $P(x)|_{-1}^{1}$.

49. Given $P(x) = 2x + 3$ and $Q(x) = x^2$, find $P(Q(2))$ and $Q(P(-1))$.

50. Given $P(x) = 3x - 2$ and $Q(x) = x^2 + 1$, find $P(Q(3))$ and $Q(P(0))$.

51. Given $P(x) = x^2 - 2x + 1$ and $Q(x) = x - 3$, find $P(Q(1))$ and $Q(P(3))$.

52. Given $P(x) = x^2 - 3x$ and $Q(x) = x^2 - 1$, find $P(Q(1))$ and $Q(P(3))$.

B

■ *Given* $P(x) = x^3 - 4x^2 + 1$, $Q(x) = x^2 - 1$, *and* $R(x) = x^2 + 1$, *compute each of the following.*

53. $P(Q(x))|_{0}^{1}$ **54.** $P(R(x))|_{1}^{3}$ **55.** $P(Q(R(1)))$

56. $Q(P(R(-1)))$ **57.** $P(Q(R(x)))|_{0}^{1}$ **58.** $P(R(Q(x)))|_{0}^{1}$

59. If $P(x)$ is of degree n and $Q(x)$ is of degree $n - 2$, what is the degree of $P(x) + Q(x)$? Of $P(x) - Q(x)$? Of $Q(x) + Q(x)$? Of $Q(x) - Q(x)$?

60. If $P(x)$ and $Q(x)$ are each of degree n, what can be said about the degree of $P(x) + Q(x)$?

1.2

PRODUCTS OF POLYNOMIALS

Laws of Exponents for Positive-Integer Exponents

By Definition 1.1 (page 2), we have

$$a^n = \underbrace{a \cdot a \cdot a \cdot \cdots \cdot a}_{n \text{ factors}}.$$

The product $a^m \cdot a^n$, where m and n are positive integers, is then given by

$$a^m \cdot a^n = \underbrace{(a \cdot a \cdot a \cdot \cdots \cdot a)}_{m \text{ factors}} \underbrace{(a \cdot a \cdot a \cdot \cdots \cdot a)}_{n \text{ factors}}$$

$$= \underbrace{a \cdot a \cdot a \cdot \cdots \cdot a}_{(m + n) \text{ factors}} = a^{m+n}.$$

We state the result formally, together with two related properties whose proofs are similar to the one above and are omitted.

Theorem 1.1

If m and n are positive integers and a and b are any real numbers,

$$\text{I.} \quad a^m \cdot a^n = a^{m+n},$$

$$\text{II.} \quad (a^m)^n = a^{mn},$$

$$\text{III.} \quad (ab)^n = a^n b^n.$$

EXAMPLE 1 Rewrite the given expression using Theorem 1.1.

(a) $x^2 x^3$ (b) $(y^3)^4$ (c) $(xy)^4$ (d) $x^3 y^3$

Solution (a) Using Theorem 1.1-I we have   (b) Using Theorem 1.1-II we have

$$x^2 x^3 = x^{2+3} = x^5. \qquad\qquad (y^3)^4 = y^{3 \cdot 4} = y^{12}.$$

(c) Using Theorem 1.1-III we have   (d) Using Theorem 1.1-III we have

$$(xy)^4 = x^4 y^4. \qquad\qquad x^3 y^3 = (xy)^3. \qquad ■$$

Products of Monomials

In rewriting the product of two monomials, we use the commutative and associative laws and Theorem 1.1 to simplify the product.

E X A M P L E 2 Simplify each product. Assume all variable exponents are positive integers.

(a) $(-4xy)\left(\dfrac{1}{2}y^2\right)$

(b) $(3a^n)(ab)^{n+3}$

Solution (a) $(-4xy)\left(\dfrac{1}{2}y^2\right)$

$\qquad = -4 \cdot \dfrac{1}{2} \cdot x \cdot y \cdot y^2$

$\qquad = -2 \cdot x \cdot y^{1+2}$

$\qquad = -2xy^3$

(b) $(3a^n)(ab)^{n+3}$

$\qquad = 3 \cdot a^n \cdot a^{n+3} \cdot b^{n+3}$

$\qquad = 3 \cdot a^{n+n+3} \cdot b^{n+3}$

$\qquad = 3a^{2n+3}b^{n+3}$ ■

Products of Polynomials

The distributive law (see Preliminary Concepts) can be generalized to

$$a(b_1 + b_2 + \cdots + b_n) = ab_1 + ab_2 + \cdots + ab_n$$

and can be applied to simplify the product of a monomial and a polynomial.

E X A M P L E 3 (a) $3xy(x^2 + y + z^2)$

$\qquad = 3xy(x^2) + 3xy(y) + 3xy(z^2)$

$\qquad = 3x^3y + 3xy^2 + 3xyz^2$

(b) $4x(x^2 + 2x + 1)$

$\qquad = 4x(x^2) + 4x(2x) + 4x(1)$

$\qquad = 4x^3 + 8x^2 + 4x$ ■

The distributive law can be applied successively to simplify the products of two or more polynomials.

E X A M P L E 4 (a) $(3x + 2y)(x - y) = 3x(x - y) + 2y(x - y)$

$\qquad = 3x^2 - 3xy + 2xy - 2y^2 = 3x^2 - xy - 2y^2$

(b) $(x + 2)(x - 1)(2x + 3) = [x(x - 1) + 2(x - 1)](2x + 3)$

$\qquad = (x^2 + x - 2)(2x + 3)$

$\qquad = (x^2 + x - 2)(2x) + (x^2 + x - 2)(3)$

$\qquad = 2x^3 + 5x^2 - x - 6$ ■

Alternative Format

An alternative format similar to one used for multiplying multidigit numbers can be used to find the product of two polynomials.

EXAMPLE 5   Write the product $(x + 2)(x^2 + 3x + 4)$ in standard form.

Solution   Write the two factors one above the other as shown and multiply term by term to obtain

$$\begin{array}{l}
x^2 + 3x \;\; + \;\; 4 \\
x \;\; + 2 \\
\hline
x^3 + 3x^2 + \;\; 4x \qquad \leftarrow \text{This is } x^2 + 3x + 4 \text{ times } x. \\
\quad\;\; 2x^2 + \;\; 6x + 8 \leftarrow \text{This is } x^2 + 3x + 4 \text{ times } 2. \\
\hline
x^3 + 5x^2 + 10x + 8 \qquad ■
\end{array}$$

Nested Expressions

If there are several sets of grouping symbols in an expression, we work from the innermost grouping outward, as suggested in Section 1.1, to simplify the expression.

EXAMPLE 6   Write $(x + 2)[(x + 2)^2 - (4x + 4)]$ as a polynomial in standard form.

Solution   We first simplify $(x + 2)^2$. Thus,

$$\begin{aligned}
(x + 2)[(x + 2)^2 - (4x + 4)] &= (x + 2)[(x^2 + 4x + 4) - (4x + 4)] \\
&= (x + 2)(x^2) \\
&= x^3 + 2x^2. \qquad ■
\end{aligned}$$

$P(x)$ Notation

We are now able to simplify expressions of the form $P(Q(x))$ where P and Q are polynomials.

EXAMPLE 7   Given $P(x) = x^2 + 2x + 3$, find the following.

(a) $P(a + 1)$ (b) $P(a + k)$ (c) $P(x)|_a^{a+k}$

Solution   (a) $\begin{aligned}[t] P(a + 1) &= (a + 1)^2 + 2(a + 1) + 3 \\ &= a^2 + 2a + 1 + 2a + 2 + 3 \\ &= a^2 + 4a + 6 \end{aligned}$

(b) $P(a + k) = (a + k)^2 + 2(a + k) + 3$
$$= a^2 + 2ak + k^2 + 2a + 2k + 3$$

(c) $P(x)|_a^{a+k} = P(a + k) - P(a)$
$$= (a^2 + 2ak + k^2 + 2a + 2k + 3) - (a^2 + 2a + 3)$$
$$= 2ak + k^2 + 2k \qquad ■$$

**Standard
Forms**

Some products of polynomials can be simplified mentally. The following binomial products are forms so frequently encountered that you should learn to recognize them on sight.

$$(x + a)(x + b) = x^2 + (a + b)x + ab \tag{1}$$
$$(x + a)^2 = x^2 + 2ax + a^2 \tag{2}$$
$$(x + a)(x - a) = x^2 - a^2 \tag{3}$$
$$(ax + b)(cx + d) = acx^2 + (bc + ad)x + bd \tag{4}$$

EXAMPLE 8 (a) $(x + 2)(x + 3) = x^2 + (2 + 3)x + (2)(3)$ **From (1)**
$$= x^2 + 5x + 6$$

(b) $(x + 3)^2 = x^2 + 2 \cdot 3x + 3^2$ **From (2)**
$$= x^2 + 6x + 9$$

(c) $(x + 4)(x - 4) = x^2 - (4)^2$ **From (3)**
$$= x^2 - 16$$

(d) $(2x - 3)(3x - 2) = 2 \cdot 3x^2 + (-9 - 4)x + (-3)(-2)$ **From (4)**
$$= 6x^2 - 13x + 6 \qquad ■$$

1-60 odd

EXERCISE 1.2

A

■ *Simplify each product. For Exercises 1–12, see Example 2.*

1. $(-xy)(2x^2y^4)$ **2.** $(x^4y^2)(3x^2y)$ **3.** $(2x^2y)(-3x^4y^4)(xy^3)$

4. $(-2x^2)(x^4y^3)(-3xy^2)$ **5.** $(2xy)^3(-xy^2)(3x^2y)$ **6.** $(-3x^2y)^3(x^2y)(x^4y^2)$

7. $x^{n+2}(3x^{n+4})$ **8.** $(2x^2)(4x^{2n})(x^{n+1})$ **9.** $(-x^ny^{n+1})(2x^{2+n}y^n)$

10. $(3x^2y^{n+1})(-3x^{2n}y^{n+2})$ **11.** $(x^2y)^n(-xy^2)^{2n}$ **12.** $(-xy^4)^{n+1}(-x^2y^3)^n$

■ *For Exercises 13–20, see Example 3.*

13. $5(2x + 3)$ **14.** $2(x^2 - 4)$ **15.** $x(2x + 6)$

16. $2x(x^2 - 4)$ **17.** $x^2(x^3 - x^2 - x)$ **18.** $x^2(x^2 - x - 1)$

19. $x^3(x^2 - x + 1)$ **20.** $-x^3(x^4 + x^2 + 2)$

■ *Write each product as a polynomial in standard form. For Exercises 21–38, see Examples 4, 5, and 8.*

21. $(x - 2)(x + 3)$ **22.** $(x + 1)(x + 5)$ **23.** $(3x - 1)(x + 2)$

24. $(x + 3)(2x - 1)$ **25.** $3(2x + 1)(x - 2)$ **26.** $5(x - 1)(3x + 1)$

27. $-(x - 1)(x + 1)$ **28.** $-(x - 1)(x - 2)$ **29.** $-(2x - 1)(2x + 3)$

30. $-(4x + 3)(3x - 1)$ **31.** $(x + 2)(2x^2 + x + 3)$ **32.** $(x + 5)(x^2 + 2x + 4)$

33. $(2x + 1)(x^2 - 2x + 1)$ **34.** $(2x + 1)(2x^2 - 3x + 1)$ **35.** $x(x + 1)(x^2 + x + 1)$

36. $x(x - 2)(x^2 - 2)$ **37.** $(x^2 + x + 1)(x^2 - x + 2)$ **38.** $(x^2 + x)(x^2 - 2x + 3)$

■ *For Exercises 39–48, see Example 6.*

39. $(x + 1)[(x^2 - 1) + (2x + 1)]$ **40.** $(x - 2)[(2x + 1) - (x^2 + x)]$

41. $(x + 2)[(x - 1)^2 + 3]$ **42.** $(x - 1)[(x - 1)^2 - 4]$

43. $(x + 1)[(x + 1)^2 - (2x - 1)]$ **44.** $(x - 3)[(2x + 1)^2 - (x^2 - x - 1)]$

45. $(2x - 3)[(2x + 1) - (x^2 + x)](x^2 - 4)$

46. $(x^2 - x + 4)[(x^2 + x + 4) - (x^2 + x - 4)](x - 1)$

47. $(x^2 + x - 4)[(x^2 - 2x + 1) - (x^2 + x - 1)](x + 2)$

48. $(x^2 - 2x + 2)[(2x^2 - x + 1) - (x^2 - x + 1)](2x - 1)$

■ *For Exercises 49–52, see Example 7.*

49. Given $P(x) = 2x^2 + x + 2$, find $P(c)$, $P(c + h)$, and $P(x)|_c^{c+h}$.

50. Given $P(x) = x^2 + 2x + 3$, find $P(a)$, $P(a - h)$, and $P(x)|_{a-h}^a$.

51. Given $P(x) = x^2 - 1$, find $P(x + 2)$, $P(x^2)$, and $(P(x))^2$.

52. Given $P(x) = x^2 + x - 3$, find $P(x + 2)$, $P(x^2)$, and $(P(x))^2$.

B

■ *For Exercises 53–56, let $P(x) = x^2 - 1$, $Q(x) = x^2 + x - 1$, and $R(x) = x^2 + 2x - 2$.*

53. Find $P(Q(x))$.   **54.** Find $Q(P(x))$.   **55.** Find $P(R(x))$.   **56.** Find $R(P(x))$.

57. If $P(x)$ and $Q(x)$ have degrees m and n, respectively, what is the degree of $P(x) \cdot Q(x)$?

58. If $P(x)$ has degree n, what is the degree of $[P(x)]^2$? Of $[P(x)]^k$, $k \in N$?

59. If $P(x)$ has degree n, what is the degree of $P(x^2)$? Of $P(x^k)$, $k \in N$?

60. If $P(x)$ has degree n and $Q(x)$ has degree k, what is the degree of $P(Q(x))$? Of $Q(P(x))$?

FACTORING POLYNOMIALS

What do we mean when we say that we have *factored* an integer or a polynomial? It is true, for example, that

$$2 = 4\left(\frac{1}{2}\right),$$

but we would not ordinarily say that 4 and $\frac{1}{2}$ are factors of 2 because we usually restrict our attention to integer factors. On the other hand, since

$$10 = (2)(5),$$

we do say that 2 and 5 are integer factors of 10.

Now consider the polynomial

$$2x^2 - 10,$$

which is completely factored as

$$2x^2 - 10 = 2(x^2 - 5)$$

if we are limited to **integral** coefficients—that is, coefficients that are integers. In this book, we are primarily concerned with polynomials whose coefficients are integers.

In factoring polynomials having integer coefficients, we consider as factors only polynomials having integer coefficients with no common integer factor other than 1 or -1. We say that such a polynomial is **prime** if it is not the product of two polynomials of this sort and its leading coefficient is positive. For example, $x + 1$, $2x + 1$, and $x^2 + 1$ are prime, while

$$2x + 2 = 2(x + 1) \qquad \text{and} \qquad x^2 - 1 = (x + 1)(x - 1)$$

are not.

We say that a polynomial other than 0 or 1 is **completely factored** if it is written equivalently as a product of prime polynomials or as such a product times -1.

EXAMPLE 1   (a) $6x^4 - 6 = 6(x^4 - 1)$

$$= 6(x^2 - 1)(x^2 + 1)$$

$$= 6(x - 1)(x + 1)(x^2 + 1)$$

(b) $3x^2 + 18x + 27 = 3(x^2 + 6x + 9)$

$$= 3(x + 3)^2 \quad ■$$

Standard Forms

Several forms are especially useful in factoring certain polynomials of degree 2 and degree 3. We first consider special forms of degree 2. In particular, the following forms are often encountered in this and subsequent chapters.

Memorize

$$(x + a)(x + b) = x^2 + (a + b)x + ab \tag{1}$$

$$(x + a)^2 = x^2 + 2ax + a^2 \tag{2}$$

$$(x - a)(x + a) = x^2 - a^2 \tag{3}$$

$$(ax + b)(cx + d) = acx^2 + (bc + ad)x + bd \tag{4}$$

Recall that when multiplying binomials in Section 1.2, we used forms (1)–(4) viewed from left to right. In this section we are interested in viewing these relationships from right to left. Notice that the left-hand members of each of these forms involve prime polynomial factors.

EXAMPLE 2   Use Form (1) to completely factor:

(a) $x^2 - 3x - 4$ (b) $x^2 - x - 2$

Solution   (a) $x^2 - 3x - 4$ (b) $x^2 - x - 2$
$\qquad\qquad = x^2 + (-4 + 1)x + (-4)(1) \qquad\qquad = x^2 + (-2 + 1)x + (-2)(1)$
$\qquad\qquad = (x - 4)(x + 1) \qquad\qquad\qquad\quad = (x - 2)(x + 1)$

EXAMPLE 3   Use Form (2) to completely factor:

(a) $x^2 + 4x + 4$ (b) $x^4 + 2x^2 + 1$

Solution   (a) $x^2 + 4x + 4 = x^2 + 2(2)x + 2^2$ (b) $x^4 + 2x^2 + 1 = (x^2)^2 + 2x^2 + 1$
$\qquad\qquad\qquad\qquad = (x + 2)^2 \qquad\qquad\qquad\qquad\qquad\quad = (x^2 + 1)^2$

EXAMPLE 4   Use Form (3) to completely factor:

(a) $x^2 - 64$ (b) $4x^2 - 9y^2$

Solution   (a) $x^2 - 64 = x^2 - 8^2$ (b) $4x^2 - 9y^2 = (2x)^2 - (3y)^2$
$\qquad\qquad\qquad\quad = (x - 8)(x + 8) \qquad\qquad\qquad\qquad = (2x - 3y)(2x + 3y)$

EXAMPLE 5   Use Form (4) to completely factor:

(a) $2x^2 - 5x + 2$ (b) $2x^2 + 11x + 14$

Solution   (a) $2x^2 - 5x + 2 = (2)(1)x^2 + [(-1)(1) + (2)(-2)]x + (-1)(-2)$
$$= (2x - 1)(x - 2)$$

(b) $2x^2 + 11x + 14 = (2)(1)x^2 + [(7)(1) + (2)(2)]x + (7)(2)$
$$= (2x + 7)(x + 2)   ■$$

We next consider special forms involving polynomials of degree 3. As in the case of Forms (1)–(4), we view these relationships from right to left when factoring.

$$(x + a)^3 = x^3 + 3ax^2 + 3a^2x + a^3 \tag{5}$$

$$(x - a)(x^2 + ax + a^2) = x^3 - a^3 \tag{6}$$

$$(x + a)(x^2 - ax + a^2) = x^3 + a^3 \tag{7}$$

E X A M P L E   6   Use Form (5) to completely factor:

(a) $x^3 + 9x^2 + 27x + 27$ (b) $x^3 - 6x^2 + 12x - 8$

Solution   (a) $x^3 + 9x^2 + 27x + 27 = x^3 + 3(3)x + 3(3)^2x + 3^3$
$$= (x + 3)^3$$

(b) $x^3 - 6x^2 + 12x - 8 = x^3 + 3(-2)x^2 + 3(-2)^2x + (-2)^3$
$$= (x - 2)^3$$

E X A M P L E   7   Use Form (6) to completely factor:

(a) $27x^3 - 1$ (b) $125 - 8x^3$

Solution   (a) $27x^3 - 1 = (3x)^3 - 1^3$
$$= (3x - 1)[(3x)^2 + (3x)(1) + 1^2]$$
$$= (3x - 1)(9x^2 + 3x + 1)$$

(b) $125 - 8x^3 = 5^3 - (2x)^3$
$$= (5 - 2x)[(5)^2 + (2x)(5) + (2x)^2]$$
$$= (5 - 2x)(25 + 10x + 4x^2)$$

E X A M P L E   8   Use Form (7) to completely factor:

(a) $8a^3 + b^3$ (b) $64x^3 + 27$

Solution   (a) $8a^3 + b^3 = (2a)^3 + b^3$
$$= (2a + b)[(2a)^2 - (2a)(b) + b^2]$$
$$= (2a + b)(4a^2 - 2ab + b^2)$$

(b) $64x^3 + 27 = (4x)^3 + 3^3$
$$= (4x + 3)[(4x)^2 - (4x)(3) + 3^2]$$
$$= (4x + 3)(16x^2 - 12x + 9)$$ ■

Factoring by Grouping

Expressions such as

$$3x^2y + 2y + 3xy^2 + 2x$$

are factorable by grouping terms with a common monomial factor, factoring the common monomials from each grouping, and then factoring the resulting common binomial from each resulting term.

EXAMPLE 9   Factor $3x^2y + 2y + 3xy^2 + 2x$.

Solution   We first write the polynomial in the form

$$(3x^2y + 2x) + (3xy^2 + 2y).$$

Next, we factor the common monomial x from the first group of two terms and y from the second group of two terms, obtaining

$$x(3xy + 2) + y(3xy + 2).$$

Finally, we factor the common binomial $(3xy + 2)$ from each term, obtaining

$$(3xy + 2)(x + y),$$

in which both factors are prime. Thus,

$$3x^2y + 2y + 3xy^2 + 2x = (3xy + 2)(x + y).$$

EXAMPLE 10   Factor $x^2 - y^2 - 2x + 1$.

Solution   We first group terms and rewrite the polynomial as

$$(x^2 - 2x + 1) - y^2.$$

From Form (2) this is equivalent to

$$(x - 1)^2 - y^2,$$

which is the difference of two squares. Thus, from Form (3) we have

$$x^2 - y^2 - 2x + 1 = (x - 1 + y)(x - 1 - y). \qquad ■$$

Strategy for Factoring

It is helpful to approach factoring systematically. The following steps are useful.

Steps for Factoring

1. Factor any common monomial out of each term.
2. Determine whether the resulting polynomial can be written in any of the seven standard forms.
3. Attempt to factor the polynomial by regrouping terms.

It is sometimes necessary to use a combination of the preceding steps, and Steps 2 and 3 are not necessarily done in the order given.

EXAMPLE 11   Factor $2x^4 + 2yx^3 - 2x^2y^2 - 2xy^3$.

Solution   We first factor the common monomial $2x$ from each term to obtain

$$2x(x^3 + yx^2 - xy^2 - y^3).$$

Next we regroup the terms in the expression in parentheses to obtain

$$\begin{aligned} x^3 + yx^2 - xy^2 - y^3 &= (x^3 - xy^2) + (yx^2 - y^3) \\ &= x(x^2 - y^2) + y(x^2 - y^2) \\ &= (x + y)(x^2 - y^2). \end{aligned}$$

Finally, we note from Form (3) that

$$x^2 - y^2 = (x + y)(x - y).$$

Thus, the complete factorization is

$$2x^4 + 2yx^3 - 2x^2y^2 - 2xy^3 = 2x(x + y)(x + y)(x - y). \qquad ■$$

Adding and Subtracting Monomials

In some cases it is helpful to add and subtract an appropriate monomial to change the form of the polynomial to a form that is more easily factored.

EXAMPLE 12   Factor $x^4 + x^2y^2 + 25y^4$.

Solution   We add and subtract $9x^2y^2$ to obtain the difference of two squares. Thus,

$$
\begin{aligned}
x^4 + x^2y^2 + 25y^4 &= x^4 + x^2y^2 + 25y^4 + 9x^2y^2 - 9x^2y^2 \\
&= (x^4 + 10x^2y^2 + 25y^4) - 9x^2y^2 \\
&= (x^2 + 5y^2)^2 - (3xy)^2 \\
&= [(x^2 + 5y^2) - 3xy][(x^2 + 5y^2) + 3xy] \\
&= (x^2 - 3xy + 5y^2)(x^2 + 3xy + 5y^2). \quad \blacksquare
\end{aligned}
$$

1 - 128

EXERCISE 1.3   even

A

■ *Factor completely into products of polynomials with integer coefficients. For Exercises 1–8, see Example 2.*

1. $x^2 + 3x + 2$ **2.** $x^2 + 5x + 6$ **3.** $x^2 - 6x + 8$ **4.** $x^2 - 5x + 4$
5. $x^2 + 4x - 5$ **6.** $x^2 + x - 12$ **7.** $x^2 - 2x - 8$ **8.** $x^2 - x - 6$

■ *For Exercises 9–14, see Example 3.*

9. $x^2 + 2x + 1$ **10.** $x^2 + 14x + 49$ **11.** $x^2 - 4x + 4$ **12.** $x^2 - 6x + 9$
13. $4x^2 + 12x + 9$ **14.** $9x^2 - 6x + 1$

■ *For Exercises 15–20, see Example 4.*

15. $x^2 - 1$ **16.** $x^2 - 4$ **17.** $4x^2 - 9$ **18.** $9x^2 - 4$
19. $25 - 4x^2$ **20.** $16 - 25x^2$

■ *For Exercises 21–28, see Example 5.*

21. $3x^2 + 4x + 1$ **22.** $2x^2 + 7x + 6$ **23.** $3x^2 - 5x + 2$
24. $2x^2 - 7x + 6$ **25.** $6x^2 + 13x + 6$ **26.** $12x^2 + 17x + 6$
27. $6x^2 - x - 12$ **28.** $12x^2 + 29x - 8$

■ *For Exercises 29–34, see Example 6.*

29. $x^3 + 3x^2 + 3x + 1$ **30.** $x^3 + 6x^2 + 12x + 8$ **31.** $x^3 - 3x^2 + 3x - 1$
32. $x^3 - 9x^2 + 27x - 27$ **33.** $8x^3 - 12x^2 + 6x - 1$ **34.** $27x^3 + 27x^2 + 9x + 1$

■ *For Exercises 35–40, see Example 7.*

35. $x^3 - 8$ **36.** $x^3 - 27$ **37.** $8x^3 - 1$ **38.** $64x^3 - 1$
39. $27 - 8x^3$ **40.** $64 - 27x^3$

■ *For Exercises 41–46, see Example 8.*

41. $x^3 + 27$ **42.** $x^3 + 64$ **43.** $27x^3 + 1$ **44.** $8x^3 + 1$
45. $125x^3 + 27$ **46.** $8x^3 + 27$

■ *For Exercises 47–52, see Examples 9–11.*

47. $x^2 - xy - 3x + 3y$ **48.** $y^2 + xy + 2y + 2x$ **49.** $x^2y + 2x^2 - y - 2$
50. $y^2x - y^2 - 4x + 4$ **51.** $x^3 + 3x^2 - 4x - 12$ **52.** $x^3 + 2x^2 - 9x - 18$

Miscellaneous Factoring Exercises

■ *Factor completely into products of polynomials with integer coefficients.*

53. $2x^3 - x^2$ **54.** $3x^4 - 2x^2$ **55.** $8x^6 + 2x^2$
56. $12x^5 + 4x^3$ **57.** $2x^3y^2 - 8x^2y - 32x^2y^2$ **58.** $28y^4z^2 + 14y^3z - 14y^2z^2$
59. $x^2 - 16$ **60.** $36x^2 - 9$ **61.** $3x^2y^4 - 27x^2$
62. $x^2y^2z^2 - 4x^2z^2$ **63.** $2x^2 - 12x + 18$ **64.** $x^2 - 14x + 49$
65. $4y^3x^2 + 8y^3x + 4y^3$ **66.** $x^2y - 16xy + 64y$ **67.** $x^2yz + 10xyz + 25x$
68. $2x^3y^2 - 16x^2y^3 + 32x^2y^2$ **69.** $x^2 + 3x - 18$ **70.** $x^2 - 5x - 14$
71. $x^2 + 8x + 15$ **72.** $2x^2 - 6x + 4$ **73.** $2x^2 + 3x + 1$
74. $3x^2 + 5x + 2$ **75.** $3x^2 + 7x + 2$ **76.** $2x^2 - 3x + 1$
77. $2z^2 - 3z - 2$ **78.** $3z^2 - z - 2$ **79.** $6y^2 - 11y + 3$
80. $8y^2 + 14y + 3$ **81.** $16a^2 + 10a + 1$ **82.** $12a^2 - 17a + 6$
83. $12x^2 + 2x - 2$ **84.** $12x^2 - 8x - 15$ **85.** $3x^2 + 12x - 15$
86. $x^3 - 2x^2 - 8x$ **87.** $ab^2 + 5ab - 14a$ **88.** $a^2b^2 + 5ab^2 - 14b^2$
89. $ab - b + 3a - 3$ **90.** $ab + 3b - a - 3$ **91.** $2xy + 6x - 4y - 12$
92. $xy^2 - 2y^2 + 2xy - 4y$ **93.** $x^3 - 2xy - x^2y^2 + 2y^3$ **94.** $2x^3 - x^2y - 2xy + y^2$
95. $x^2 + 10x + 25 - 4y^2$ **96.** $4x^2 - 4x + 1 - 9z^2$ **97.** $y^2 - 4x^2 + 4x - 1$
98. $z^2 - 9y^2 + 12y - 4$ **99.** $9x^2 - 4y^2 + 6x + 1$ **100.** $16y^2 - 25x^2 - 24y + 9$
101. $8x^3 + 12x^2 + 6x + 1$ **102.** $27x^3 - 27x^2 + 9x - 1$ **103.** $8x^3 - 36x^2 + 54x - 27$
104. $27x^3 + 108x^2 + 144x + 64$ **105.** $125x^3 - 150x^2 + 60x - 8$ **106.** $64x^3 - 144x^2 + 108x - 27$
107. $27 - x^3$ **108.** $x^3y^3 + 8$ **109.** $64x^3 + 8y^3$
110. $64x^3 - y^3$ **111.** $1 + (x - 1)^3$ **112.** $(x - 1)^3 + x^3$
113. $(x^3 + 8)^3 + 1$ **114.** $(y^3 + 1)^3 + 27$ **115.** $x^3 + 3x^2 + yx + 3y$
116. $2xy^2 + 2x^2 + y^2 + x$ **117.** $x^2 + 11x + 30$ **118.** $x^4 + 6x^2 + 8$
119. $x^2y^2 + 3x^2 + y^2 + 3$ **120.** $1 + xy + x + y$ **121.** $x^5 - x$
122. $x - xy^2$

B

■ *Add and subtract an appropriate monomial, and factor. See Example 12.*

123. $x^4 + x^2y^2 + y^4$ **124.** $x^4 - 3x^2y^2 + y^4$ **125.** $x^4 + x^2 + 1$
126. $x^4 + 3x^2 + 4$ **127.** $9x^4 + 2x^2 + 1$ **128.** $4x^4 + 8x^2 + 9$

1.4

QUOTIENTS OF POLYNOMIALS

In this section we shall examine some ways in which we can rewrite quotients of polynomials, even if the resulting expressions are not always themselves polynomials.

Laws of Exponents for Quotients

We begin with the simplest case, that in which the polynomials are monomials and their quotient is a monomial. Consider

$$\frac{a^m}{a^n}$$

for m and n positive integers with $m > n$ and a a nonzero real number. We have

$$\frac{a^m}{a^n} = \frac{\overbrace{a \cdot \cdots \cdot a}^{m \text{ factors}}}{\underbrace{a \cdot \cdots \cdot a}_{n \text{ factors}}} = \frac{\overbrace{(a \cdot \cdots \cdot a)}^{n \text{ factors}}\overbrace{(a \cdot \cdots \cdot a)}^{m-n \text{ factors}}}{\underbrace{(a \cdot \cdots \cdot a)}_{n \text{ factors}}}$$

$$= 1 \cdot \underbrace{(a \cdot \cdots \cdot a)}_{m-n \text{ factors}} = a^{m-n}.$$

Hence,

$$\frac{a^m}{a^n} = a^{m-n}.$$

This establishes the first part of the following result. Proof of the second part is similar to the proof of Theorem 1.1 in Section 1.2 and is omitted.

Theorem 1.2   *If a and b are nonzero real numbers and m and n are positive integers with m > n, then*

$$\text{I.} \quad \frac{a^m}{a^n} = a^{m-n},$$

$$\text{II.} \quad \left(\frac{a}{b}\right)^m = \frac{a^m}{b^m}.$$

EXAMPLE 1   (a) $\dfrac{12a^5b^3}{4a^2b^2} = \dfrac{12}{4}a^{5-2}b^{3-2} = 3a^3b$ $(a, b \neq 0)$

(b) $4\left(\dfrac{x-1}{x}\right)^2 \dfrac{x^3}{x-1} = 4\dfrac{(x-1)^2}{x^2} \cdot \dfrac{x^3}{x-1}$

$$= 4(x-1)^{2-1}x^{3-2}$$

$$= 4(x-1)x \qquad (x \neq 0, 1) \qquad ■$$

Note that in part (a) of Example 1, the variables a and b are not permitted to take the value 0, because if they were, $3a^3b$ would represent a real number, 0, although $(12a^5b^3)/(4a^2b^2)$ would not be defined. In part (b), the variable x can take neither the value 0 nor the value 1, because if x takes either of these values, $4(x-1)x$ would represent the value 0 although $4[(x-1)/x]^2[x^3/(x-1)]$ would not be defined.

Quotients of Polynomials   Properties of the real numbers (see Preliminary Concepts) permit us to rewrite quotients of polynomials whose numerators are not monomials. For example,

$$\frac{2x^3 + 4x^2 + 8x}{2x} = \frac{2x^3}{2x} + \frac{4x^2}{2x} + \frac{8x}{2x},$$

and, by an application of Theorem 1.2-I, the right-hand member can then be denoted by the expression

$$x^2 + 2x + \frac{8x}{2x} \qquad (x \neq 0).$$

Though Theorem 1.2-I is not applicable to the variable factors in expressions such as $(8x)/(2x)$, by the fact that $x/x = 1$ for all $x \neq 0$, we have

$$\frac{8x}{2x} = \frac{8}{2} \cdot \frac{x}{x} = \frac{8}{2} \cdot 1 = 4,$$

for every $x \neq 0$. Thus,

$$\frac{2x^3 + 4x^2 + 8x}{2x} = x^2 + 2x + 4 \qquad (x \neq 0).$$

EXAMPLE 2   Write each quotient as a polynomial.

(a) $\dfrac{10x^4y^2 + 4x^2y}{2x^2y}$

(b) $\dfrac{x^3 + 3x^2 + 4x}{x}$

Solution   (a) $\dfrac{10x^4y^2 + 4x^2y}{2x^2y} = \dfrac{10x^4y^2}{2x^2y} + \dfrac{4x^2y}{2x^2y} = 5x^2y + 2 \qquad (x, y \neq 0)$

(b) $\dfrac{x^3 + 3x^2 + 4x}{x} = \dfrac{x^3}{x} + \dfrac{3x^2}{x} + \dfrac{4x}{x}$

$= x^2 + 3x + 4 \qquad (x \neq 0) \qquad$ ■

As we have noted, quotients of polynomials cannot always be written as polynomials. For example, we have

$$\frac{2x^3 + 4x + 1}{x} = \frac{2x^3}{x} + \frac{4x}{x} + \frac{1}{x}$$

$$= 2x^2 + 4 + \frac{1}{x} \qquad (x \neq 0),$$

where the resulting expression is not a polynomial, but rather an expression of the form

$$Q(x) + \frac{R(x)}{D(x)},$$

where the polynomial $Q(x)$ (equal to $2x^2 + 4$) may be called a "partial quotient" and $R(x)$ (equal to 1) is called a "remainder."

EXAMPLE 3 Write each quotient in the form $Q(x) + \dfrac{R(x)}{D(x)}$, where Q, R, and D are polynomials with the degree of R less than the degree of D.

(a) $\dfrac{x^3 + 3x + 5}{x}$

(b) $\dfrac{x^4 - 3x^2 + x - 5}{6x^2}$

Solution (a) $\dfrac{x^3 + 3x + 5}{x} = \dfrac{x^3}{x} + \dfrac{3x}{x} + \dfrac{5}{x}$

$= x^2 + 3 + \dfrac{5}{x} \qquad (x \neq 0)$

(b) $\dfrac{x^4 - 3x^2 + x - 5}{6x^2} = \dfrac{x^4}{6x^2} - \dfrac{3x^2}{6x^2} + \dfrac{x}{6x^2} - \dfrac{5}{6x^2}$

$= \dfrac{1}{6}x^2 - \dfrac{1}{2} + \dfrac{x - 5}{6x^2} \qquad (x \neq 0) \qquad$ ■

If the divisor of a quotient contains more than one term, the familiar **long-division algorithm** involving successive subtractions can be used to rewrite the quotient. Notice that the algorithm can be continued until the degree of the remainder is less than that of the divisor.

EXAMPLE 4   Write $\dfrac{x^2 + x^4 + 2x - 1}{x - 3}$ in the form $Q(x) + \dfrac{R(x)}{x - 3}$, where $Q(x)$ is a polynomial and $R(x)$ is a constant.

Solution   Using the long-division algorithm, where the dividend (numerator) has been written in decreasing powers and $0x^3$ is used in the dividend for a term involving x^3, even though the dividend contains no such term explicitly, we have

$$
\begin{array}{r}
x^3 + 3x^2 + 10x\ + 32 \\
x - 3 \overline{\smash{\big)}\ x^4 + 0x^3 +\quad x^2 +\ 2x -\ \ 1} \\
\underline{x^4 - 3x^3}\qquad\qquad\qquad\qquad \\
3x^3 +\quad x^2\qquad\qquad\ \\
\underline{3x^3 -\ 9x^2}\qquad\qquad \\
10x^2 +\ 2x\qquad \\
\underline{10x^2 - 30x}\qquad \\
32x -\ \ 1 \\
\underline{32x - 96} \\
95\quad \text{(Remainder).}
\end{array}
$$

Thus, for $x - 3 \neq 0$,

$$
\frac{x^2 + x^4 + 2x - 1}{x - 3} = \frac{x^4 + x^2 + 2x - 1}{x - 3}
$$

$$
= x^3 + 3x^2 + 10x + 32 + \frac{95}{x - 3}. \quad ■
$$

The term $0x^3$ is inserted in Example 4 so that the terms with like powers are automatically aligned correctly when using the long-division algorithm.

1-25

★ EXERCISE 1.4

A

■   *For all of the exercises in this set, assume the denominators do not vanish.
Write each quotient as a polynomial. See Examples 1 and 2.*

1. $\dfrac{18y^4z^3}{3yz}$ **2.** $\dfrac{28x^3y^2}{2xy}$ **3.** $\dfrac{6x^4y^2}{2xy}$ **4.** $\dfrac{3x^5y^6z^7}{xy^4z^5}$

5. $\dfrac{4x^4 + 8x^3}{2x^2}$

6. $\dfrac{x^3 - 5x^2 + 6x}{x}$

7. $\dfrac{3x^2y - 2xy^2 + 3x^2y^2}{xy}$

8. $\dfrac{2x^2yz^3 - xy^2z^2 + xyz^2}{xyz^2}$

■   *Write each fraction as a quotient Q plus a remainder R where the degree of R is less than that of the denominator D. For Exercises 9–32, see Examples 3 and 4.*

9. $\dfrac{x^2 - 2x + 4}{x}$

10. $\dfrac{2x^2 - 3x - 6}{x}$

11. $\dfrac{4x^3 - 2x + 6}{x^2}$

12. $\dfrac{z^3 - z^2 + z - 1}{z^3}$

13. $\dfrac{4x^5 + 2x^3 + x - 4}{2x^2}$

14. $\dfrac{3x^2 + 8x + 4}{3x^3}$

15. $\dfrac{x^2y + xy^2 + y}{xy}$

16. $\dfrac{3x^4y^2 - 2x^2y^3 + xy}{xy^2}$

17. $\dfrac{x^3y^3 + x^2y + x + y}{x^2y}$

18. $\dfrac{6x^4y^2 + 6x^2y^4 - 4x^2 + 4y^2}{x^2y^2}$

19. $\dfrac{3xy^3 + y^2 + y - x}{xy^2}$

20. $\dfrac{x^4y^4 + x^3y^3 + x^2y^2 + x + y}{xy}$

21. $\dfrac{x^2 + 2x + 1}{x + 1}$

22. $\dfrac{2x^2 - 3x + 4}{x + 1}$

23. $\dfrac{4x^3 - 2x + 6}{2x - 3}$

24. $\dfrac{z^3 + z^2 + z + 1}{z - 1}$

25. $\dfrac{x^5 + 2x^3 + x - 4}{x - 4}$

26. $\dfrac{3x^2 + 8x + 4}{x - 2}$

27. $\dfrac{x^5 + 2x^3 + x - 4}{x^3 - x}$

28. $\dfrac{x^5 - 4x^3 + x}{x^2 + x + 1}$

29. $\dfrac{x^5 + 1}{x + 1}$

30. $\dfrac{x^5 - 1}{x - 1}$

31. $\dfrac{4x^4 + 3x^3 + 2x^2 + x}{2x^2 + 1}$

32. $\dfrac{x^5 + x^3 + 1}{x^2 + 1}$

B

33. $\dfrac{x^4 + x^2 - 1}{x^3 + x - 1}$

34. $\dfrac{x^5 + 2x^4 - x + 1}{x^3 + 1}$

35. $\dfrac{x^5 + x^2 + 1}{x^3 + x}$

36. $\dfrac{x^4 - 2x^2 + 2}{x^3 - x + 1}$

37. $\dfrac{x^4}{x^2 + 2x + 1}$

38. $\dfrac{x^5}{x^3 + x^2 + 1}$

39. $\dfrac{3x^3}{3x + 1}$

40. $\dfrac{4x^5}{3x^2 + 1}$

41. $\dfrac{(x^2 + 1)^2 + 4}{x^2 + 1}$

42. $\dfrac{(x^2 - 4)^2 - 25}{x^2 - 4}$

43. $\dfrac{3(x^2 + 9)^3 - x}{(x^2 + 9)^2}$

44. $\dfrac{(x^2 + 15)^4 - 3x}{(x^2 + 15)}$

45. $\dfrac{(x + 1)^2 - (x + 1) - 1}{(x + 1)^2 - 1}$

46. $\dfrac{(2x - 1)^3 - (2x - 1) + 4}{(2x - 1)^2 + 1}$

47. $\dfrac{(x^2 + 1)^2 + (x^2 + 1) - 1}{2(x^2 + 1) - 1}$

48. $\dfrac{(x^2 - 2)^3 + (x^2 - 2)^2 + 2}{3(x^2 - 2) - 4}$

1.5

EQUIVALENT FRACTIONS; SUMS AND DIFFERENCES

Rational Expressions

A fraction is an expression denoting a quotient. If the numerator (dividend) and the denominator (divisor) are polynomials, then the fraction is said to be a **rational expression**. Any polynomial can be considered as being a rational expression, since it is the quotient of itself and 1.

Since for each replacement of the variable(s) for which its denominator is not zero, a rational expression represents a real number, the following relationships for rational expressions follow directly from the properties of real numbers.

> For values of the variables for which the denominators do not vanish,
>
> I. $\dfrac{P(x)}{Q(x)} = \dfrac{R(x)}{S(x)}$ *if and only if* $P(x) \cdot S(x) = Q(x) \cdot R(x)$,
>
> II. $-\dfrac{P(x)}{Q(x)} = \dfrac{-P(x)}{Q(x)} = \dfrac{P(x)}{-Q(x)} = -\dfrac{-P(x)}{-Q(x)}$,
>
> III. $\dfrac{P(x)}{Q(x)} = \dfrac{-P(x)}{-Q(x)} = -\dfrac{-P(x)}{Q(x)} = -\dfrac{P(x)}{-Q(x)}$,
>
> IV. $\dfrac{P(x) \cdot R(x)}{Q(x) \cdot R(x)} = \dfrac{P(x)}{Q(x)}$ (fundamental principle of fractions).

Reducing Fractions

A fraction is said to be in **lowest terms** when the numerator and denominator do not contain certain types of factors in common. The fraction a/b, where a and b are integers and $b \neq 0$, is in lowest terms provided a and b contain no common positive integral factors other than 1. If the numerator and denominator of a fraction are polynomials with integral coefficients, then the fraction is said to be in lowest terms if the numerator and denominator have no common factors other than ± 1.

To express a given fraction in lowest terms (called **reducing** the fraction), we can factor the numerator and denominator and apply the fundamental principle of fractions.

EXAMPLE 1   (a) $\dfrac{y}{y^2} = \dfrac{1 \cdot y}{y \cdot y} = \dfrac{1}{y}$ $(y \neq 0)$

(b) $\dfrac{5x^3 y}{15xy^2} = \dfrac{(5xy)(x^2)}{(5xy)(3y)} = \dfrac{x^2}{3y}$ $(x, y \neq 0)$

(c) $\dfrac{(x-1)^2}{x-1} = \dfrac{(x-1) \cdot (x-1)}{1 \cdot (x-1)} = x-1 \qquad (x \neq 1)$

(d) $\dfrac{a+3}{(a+3)^2} = \dfrac{(a+3) \cdot 1}{(a+3)(a+3)} = \dfrac{1}{a+3} \qquad (a \neq -3)$ ■

Diagonal lines are sometimes used to abbreviate the procedure of reducing a fraction. Thus, in part (a) of Example 1,

$$\dfrac{y}{y^2} = \dfrac{\overset{1}{\cancel{y}}}{\underset{y}{\cancel{y^2}}} = \dfrac{1}{y} \qquad (y \neq 0).$$

To reduce fractions with polynomial numerators and denominators, you should, when possible, write them in factored form. Common factors are then evident by inspection.

E X A M P L E   2   (a) $\dfrac{2x^2 + x - 15}{2x + 6} = \dfrac{(2x-5)(x+3)}{2(x+3)} = \dfrac{2x-5}{2} \qquad (x \neq -3)$

(b) $\dfrac{2x+4}{2x^2 + 8x + 8} = \dfrac{2(x+2)}{2(x+2)(x+2)} = \dfrac{1}{x+2} \qquad (x \neq -2)$

(c) $\dfrac{x^3 - 1}{x^2 - 1} = \dfrac{(x-1)(x^2 + x + 1)}{(x-1)(x+1)} = \dfrac{x^2 + x + 1}{x+1} \qquad (x \neq 1, -1)$

(d) $\dfrac{x^2 + 2xy + y^2}{(x+y)^2} = \dfrac{(x+y)^2}{(x+y)^2} = 1 \qquad (x \neq -y)$ ■

Building Fractions

We can also change fractions to equivalent fractions in higher terms by applying the fundamental principle in the form

$$\dfrac{P(x)}{Q(x)} = \dfrac{P(x) \cdot R(x)}{Q(x) \cdot R(x)}.$$

In general, to change $P(x)/Q(x)$ to an equivalent fraction with $Q(x) \cdot R(x)$ as a denominator, we can determine the factor $R(x)$ by inspection, and then can multiply the numerator and the denominator of the original fraction by this factor.

EXAMPLE 3   Express $\dfrac{x}{x+1}$ as an equivalent fraction with denominator $x^2 + 3x + 2$.

Solution   We observe that the desired denominator

$$x^2 + 3x + 2 = (x+1)(x+2).$$

Thus, we use the fundamental principle of fractions to obtain

$$\frac{x}{x+1} = \frac{x(x+2)}{(x+1)(x+2)} = \frac{x^2 + 2x}{x^2 + 3x + 2}.$$ ■

Sums and Differences

Since rational expressions represent real numbers for each replacement of the variable(s) for which the denominators are not zero, the following rules for finding the sum and difference of two rational expressions are a direct consequence of the properties of the real numbers.

For values of the variables for which $Q(x) \neq 0$,

$$\frac{P(x)}{Q(x)} + \frac{R(x)}{Q(x)} = \frac{P(x) + R(x)}{Q(x)} \quad \text{and} \quad \frac{P(x)}{Q(x)} - \frac{R(x)}{Q(x)} = \frac{P(x) - R(x)}{Q(x)}.$$

For example,

$$\frac{3x^2}{5} + \frac{x}{5} = \frac{3x^2 + x}{5}, \quad \text{and} \quad \frac{3}{y} - \frac{x}{y} = \frac{3-x}{y} \quad (y \neq 0).$$

This principle, of course, extends to any number of fractions. If the fractions in a sum or difference have unlike denominators, we can use the fundamental principle of fractions to replace the given fractions with equivalent fractions having common denominators and then write the sum or difference as a single fraction.

For values for which denominators do not equal zero,

$$\frac{P(x)}{Q(x)} + \frac{R(x)}{S(x)} = \frac{P(x) \cdot S(x)}{Q(x) \cdot S(x)} + \frac{R(x) \cdot Q(x)}{S(x) \cdot Q(x)}$$

$$= \frac{P(x) \cdot S(x) + R(x) \cdot Q(x)}{Q(x) \cdot S(x)},$$

continued

and

$$\frac{P(x)}{Q(x)} - \frac{R(x)}{S(x)} = \frac{P(x) \cdot S(x) - R(x) \cdot Q(x)}{Q(x) \cdot S(x)}.$$

Least Common Multiple

When rewriting fractions in a sum or difference so that they share a common denominator, any such denominator may be used. If the **least common multiple** of the denominators (called the **least common denominator**) is used, however, the resulting fraction will be in simpler form than if any other common denominator is employed.

The least common multiple of two or more positive integers is the least positive integer that is exactly divisible by each of the given numbers (that is, each quotient is a positive integer). We find the least common multiple of two or more positive integers by first completely factoring each integer. Then we create a product by choosing as factors each different factor occurring in any of the given integers, including each factor the greatest number of times it occurs in any of the given integers.

EXAMPLE 4 Find the least common multiple.

(a) 24, 45, 60   (b) 54, 20, 63

Solution (a) Factor   24,   45,   60,

as $2^3 \cdot 3$, $3^2 \cdot 5$, $2^2 \cdot 3 \cdot 5$.

The least common multiple is $2^3 \cdot 3^2 \cdot 5 = 360$.

(b) Factor   54,   20,   63

as $2 \cdot 3^3$, $2^2 \cdot 5$, $3^2 \cdot 7$.

The least common multiple is $2^2 \cdot 3^3 \cdot 5 \cdot 7 = 3780$.   ■

The notion of a least common multiple among several polynomial expressions is, in general, meaningless without further specification of what is desired. We can, however, define the least common multiple of a set of polynomials with integer coefficients to be the polynomial of lowest degree with integer coefficients which has each of the given polynomials as a factor and which, among all such polynomials, has the least possible leading coefficient.

Very often, the least common multiple of a set of positive integers or polynomials can be determined by inspection. When inspection fails us, however, we can find the least common multiple of a set of polynomials with integer coefficients as follows.

Steps for Finding the Least Common Multiple

1. Express each polynomial in completely factored form.
2. Write as factors of a product each *different* factor occurring in any of the polynomials, including each factor the greatest number of times it occurs in any one of the given polynomials.

Notice that this is precisely the same procedure used to find the least common multiple of a set of positive integers.

EXAMPLE 5   Find the least common multiple.

(a) $x^3 - x, x^2 + x, x^2 - 3x + 2$ (b) $x^3, x^2 + x, x^4 + 2x^3 + x^2$

Solution   (a) Factor $x^3 - x,$ $x^3 + x^2,$ $x^2 - 3x + 2$

as $x(x - 1)(x + 1),$ $x^2(x + 1),$ $(x - 1)(x - 2).$

The least common multiple is $x^2(x - 1)(x + 1)(x - 2).$

(b) Factor $x^3,$ $x^2 + x,$ $x^4 + 2x^3 + x^2$

as $x^3,$ $x(x + 1),$ $x^2(x + 1)^2.$

The least common multiple is $x^3(x + 1)^2.$ ■

To simplify sums of fractions having different denominators we proceed as follows.

Steps for Simplifying Sums or Differences of Fractions

1. Find the least common denominator (least common multiple of the denominators).
2. Determine the factor necessary to express each of the fractions as having this common denominator.
3. Write each fraction as an equivalent fraction whose denominator is the least common denominator.
4. Express the sum or difference as a single fraction.

EXAMPLE 6   Write $\dfrac{x+1}{x} - \dfrac{x-1}{x+1}$ as a single fraction.

Solution   The least common denominator of the fractions is $x(x+1)$. Thus,

$$\frac{x+1}{x} - \frac{x-1}{x+1} = \frac{(x+1)(x+1)}{x(x+1)} - \frac{(x-1)x}{(x+1)x}$$

$$= \frac{x^2 + 2x + 1}{x(x+1)} - \frac{x^2 - x}{x(x+1)}$$

$$= \frac{(x^2 + 2x + 1) - (x^2 - x)}{x(x+1)}$$

$$= \frac{3x+1}{x(x+1)} \qquad (x \neq 0, -1).$$

EXAMPLE 7   Write $\dfrac{1}{x} + \dfrac{-2}{x+1} + \dfrac{2x}{(x+1)^2}$ as a single fraction.

Solution   The least common denominator of the fractions is $x(x+1)^2$. Thus,

$$\frac{1}{x} + \frac{-2}{x+1} + \frac{2x}{(x+1)^2} = \frac{1(x+1)^2}{x(x+1)^2} + \frac{-2x(x+1)}{x(x+1)^2} + \frac{2x \cdot x}{x(x+1)^2}$$

$$= \frac{x^2 + 2x + 1}{x(x+1)^2} + \frac{-2x^2 - 2x}{x(x+1)^2} + \frac{2x^2}{x(x+1)^2}$$

$$= \frac{x^2 + 1}{x(x+1)^2} \qquad (x \neq 0, -1). \qquad ■$$

 /-?Y

★ **EXERCISE 1.5**

A

■   *For all exercises in this set assume that no variable in a denominator takes a value for which the denominator is 0. Reduce the fractions to lowest terms. See Examples 1 and 2.*

1. $\dfrac{8x^3y^4}{2x^2y}$ 　　　　 **2.** $\dfrac{3x^{11}y^3}{33y^2}$ 　　　　 **3.** $\dfrac{x^2-1}{x-1}$ 　　　　 **4.** $\dfrac{x^2-x}{x-1}$

5. $\dfrac{4x^2+2x}{2x+1}$ 　　　　 **6.** $\dfrac{-(4x^2-1)}{2x-1}$ 　　　　 **7.** $\dfrac{a^2-b^2}{a+b}$ 　　　　 **8.** $\dfrac{1-d^2}{(1-c)(1-d)}$

9. $\dfrac{ab^3 - a^3b^5}{ab + a^2b^2}$

10. $\dfrac{xy^3 - xy}{xy - x}$

11. $\dfrac{x^2 - 2x + 1}{(x - 1)^2}$

12. $\dfrac{x^2 + 5x + 6}{x^2 + 4x + 4}$

13. $\dfrac{x^2 - 3x - 4}{x^2 - 1}$

14. $\dfrac{x^2 + 3x - 4}{x^2 - 1}$

15. $\dfrac{x + 2}{x^2 + 5x + 6}$

16. $\dfrac{x + 5}{x^2 + 14x + 45}$

17. $\dfrac{x^4 - 1}{2x^2 + 2}$

18. $\dfrac{3x^2 - 6x - 45}{9x^2 + 27x}$

19. $\dfrac{a^4 - x^4}{x^2 - a^2}$

20. $\dfrac{1 - 9x^2}{3x - 1}$

21. $\dfrac{x^4 - 4x^2 - 21}{x^3 + 3x}$

22. $\dfrac{3x^4 + x^2 - 2}{x^3 + x}$

23. $\dfrac{1 - a^4}{(a + 1)^3}$

24. $\dfrac{x^3 - y^3}{(x - y)^2}$

25. $\dfrac{x^6 - 1}{x^2 - 1}$

26. $\dfrac{x^3 - 1}{x^4 - 1}$

27. $\dfrac{-y^3 + 2xy^2 + 6x - 3y}{2x^2 - 3xy + y^2}$

28. $\dfrac{x^4 + 3x^2y^2 + 4y^4}{x^3y - x^2y^2 + 2xy^3}$

29. $\dfrac{x^4 + x^2y^2 + y^4}{x^3 - y^3}$

30. $\dfrac{x^5 - x^2y^3 + x^3y^2 - y^5}{x^4 - y^4}$

■ *Express each fraction as an equivalent fraction with the indicated denominator. See Example 3.* opposite

31. $\dfrac{2}{9}, \dfrac{}{54}$

32. $\dfrac{3}{7}, \dfrac{}{49}$

33. $\dfrac{a^2}{ab^2c}, \dfrac{}{a^2b^2c^2}$

34. $\dfrac{xyz}{xz}, \dfrac{}{x^2yz^2}$

35. $\dfrac{x}{x + 1}, \dfrac{}{x^2 + x}$

36. $\dfrac{x + 1}{x - 1}, \dfrac{}{x^2 - 1}$

37. $\dfrac{3}{x - y}, \dfrac{}{x^2 - y^2}$

38. $\dfrac{1}{x + 1}, \dfrac{}{x^3 + 1}$

■ *Find the least common multiple. See Examples 4 and 5.*

39. 18, 27, 35

40. 15, 20, 45

41. $x^2 - 1, x^2 + x$

42. $x^2 - 1, x^3 - x^2$

43. $x^4 - x^3, x^2 + 3x - 4$

44. $2x^3, 4x + 2, 4x^2 + 8x + 1$

45. $x - 1, x + 1, x^2 + 1$

46. $x - 1, x^2 + x + 1, x^3 + 1$

■ *Write each sum or difference as a single fraction in lowest terms. See Examples 6 and 7.*

47-79 81-90

47. $\dfrac{x - 1}{x^2} + \dfrac{1}{x^2 - x}$

48. $\dfrac{1}{x^2 - 1} + \dfrac{x}{x + 1}$

49. $\dfrac{y - 3}{4y} - \dfrac{1}{2y^2}$

50. $\dfrac{x + 2}{x^2} - \dfrac{1}{x}$

51. $\dfrac{1}{x - 1} + \dfrac{1}{x + 1}$

52. $\dfrac{y - 1}{y + 2} - \dfrac{y + 3}{y + 4}$

53. $\dfrac{4x + 1}{x + 2} - \dfrac{3x + 1}{x - 1}$

54. $\dfrac{x + 1}{x + 3} + \dfrac{x + 2}{x + 6}$

55. $\dfrac{-4x + 2}{2x - 1} + \dfrac{2x + 1}{x + 3}$

56. $\dfrac{-3x + 3}{x + 1} + \dfrac{x}{x - 2}$

57. $\dfrac{1}{2 - x} + \dfrac{x}{x - 2}$

58. $\dfrac{2}{x - x^2} + \dfrac{1}{x^2 - x}$

59. $\dfrac{1}{w^2 - 3w + 2} - \dfrac{w}{w^2 - 1}$

60. $\dfrac{1}{x^2 + 5x + 4} + \dfrac{x}{x^2 - 1}$

61. $\dfrac{x + 1}{x^2 - 5x + 6} + \dfrac{2}{x^2 - 6x + 9}$

62. $\dfrac{1}{u^2 - u - 2} - \dfrac{u}{(u + 1)^2}$

63. $\dfrac{1}{x + 1} - \dfrac{x}{x^2 - 6x + 9}$

64. $\dfrac{1}{x} - \dfrac{x}{x^2 + 2x + 1}$

65. $\dfrac{1}{x+1} - \dfrac{1}{x^2-1} + \dfrac{x}{x-1}$

66. $\dfrac{2}{x} + \dfrac{x+1}{x^2+x} + \dfrac{1}{x+1}$

67. $z - \dfrac{1}{z} - \dfrac{1}{z^2}$

68. $1 + \dfrac{x}{1+x} + \dfrac{x^2}{1+2x+x^2}$

69. $1 - \dfrac{1}{x-1} + \dfrac{x}{x^2-1}$

70. $a + 1 + \dfrac{1}{a} + \dfrac{1-a^3}{a^2+a}$

71. $\dfrac{x+3}{3x^2+7x+4} - \dfrac{x-7}{3x^2+13x+12}$

72. $\dfrac{z+4}{2z^2-5z-3} + \dfrac{2z-1}{2z^2+3z+1}$

73. $\dfrac{x}{x-a} - \dfrac{a}{x+a}$

74. $\dfrac{4ax}{x^2-a^2} + \dfrac{x-a}{x+a}$

B

■ *Reduce each fraction to lowest terms.*

75. $\dfrac{6x^2(3x-4) - 2x(3x-4)^2}{x^4}$

76. $\dfrac{2(x+2)(x^2-1)^3 - 6x(x^2-1)^2(x+2)^2}{(x^2-1)^6}$

77. $\dfrac{(x+1)^3(x-4) - 4(x+1)(x-4)}{4x^2-4x-48}$

78. $\dfrac{(2x-3)^4(x^4-1) - x^3(2x-3)(x^4-1)}{x^4-x^3+x^2-x}$

79. $\dfrac{4x(x^2-1)(x-1)^2 - 2(x-1)(x^2-1)^2}{(x-1)^4}$

80. $\dfrac{(2x+3)(x^2-x-20) - (2x-1)(x^2+3x-4)}{(x^2-x-20)^2}$

■ *Write each sum or difference as a single fraction in lowest terms.*

81. $\dfrac{x+1}{x^3-x^2+x-1} + \dfrac{x-1}{x^4-x^3-x+1}$

82. $\dfrac{2x+1}{2x^3-3x^2-2x+3} - \dfrac{x+2}{2x^2-5x+1}$

83. $\dfrac{x}{x^3-1} - \dfrac{x}{x^4+x^2+1}$

84. $\dfrac{x}{x^4+2x^2+9} - \dfrac{x^2+1}{x^3-2x^2+3x}$

85. $\dfrac{x}{x^2-2x+xy-2y} - \dfrac{y}{y^2-2y+xy-2x}$

86. $\dfrac{x^2}{y^3+xy^2+xy+x^2} - \dfrac{y^2}{x^3+yx^2-xy-y^2}$

87. $\dfrac{1}{x^2-y^2} - \dfrac{x}{x^3+xy^2+x^2y+y^3} + \dfrac{y}{x^4-y^4}$

88. $\dfrac{1}{x^2-y^2} - \dfrac{x}{x^3-x^2y-xy^2+y^3} - \dfrac{y}{x^4-y^4}$

89. $\dfrac{2x-y}{2x^2-3xz-4xy+6yz} - \dfrac{x-2y}{4x^2-6xz-2xy+3yz}$

90. $\dfrac{x+2z}{3x^2-xz+3xy-yz} - \dfrac{3x-z}{x^2+2xz+xy+2yz}$

1.6

PRODUCTS AND QUOTIENTS OF RATIONAL EXPRESSIONS

The following properties also follow from similar properties of real numbers.

For values of the variables for which the denominators do not vanish,

$$\frac{P(x)}{Q(x)} \cdot \frac{R(x)}{S(x)} = \frac{P(x) \cdot R(x)}{Q(x) \cdot S(x)},$$

and

$$\frac{P(x)}{Q(x)} \div \frac{R(x)}{S(x)} = \frac{P(x) \cdot S(x)}{Q(x) \cdot R(x)}.$$

We can factor the numerators and the denominators and use the above relationships to rewrite a product or quotient of fractions as a single fraction in lowest terms.

EXAMPLE 1 (a)

$$\frac{x^2 - 4}{x^2 + x - 12} \cdot \frac{x - 3}{x^2 + 4x + 4} = \frac{(x - 2)(x + 2)}{(x + 4)(x - 3)} \cdot \frac{x - 3}{(x + 2)^2}$$

$$= \frac{(x - 2)(x + 2)(x - 3)}{(x + 4)(x - 3)(x + 2)^2}$$

$$= \frac{(x + 2)(x - 3)}{(x + 2)(x - 3)} \cdot \frac{(x - 2)}{(x + 4)(x + 2)}$$

$$= \frac{x - 2}{(x + 4)(x + 2)} \qquad (x \neq -2, 3, -4)$$

$$= \frac{x - 2}{x^2 + 6x + 8} \qquad (x \neq -2, 3, -4)$$

(b)

$$\frac{x^2 - 5x + 4}{x^2 + 5x + 6} \div \frac{x - 1}{x + 2} = \frac{(x - 4)(x - 1)}{(x + 3)(x + 2)} \cdot \frac{x + 2}{x - 1}$$

$$= \frac{(x - 4)(x - 1)(x + 2)}{(x + 3)(x + 2)(x - 1)}$$

$$= \frac{(x - 1)(x + 2)}{(x - 1)(x + 2)} \cdot \frac{x - 4}{x + 3} = \frac{x - 4}{x + 3}$$

$$(x \neq -3, -2, 1) \qquad ■$$

Since the factors of the numerator and denominator of the product of two fractions are just the factors of the numerators and denominators (respectively) of the fractions, we can divide common factors out of the numerators and denominators before writing the product as a single fraction. Thus, in part (a) of Example 1, we could write

$$\frac{x^2 - 4}{x^2 + x - 12} \cdot \frac{x - 3}{x^2 + 4x + 4} = \frac{(x - 2)\overset{1}{\cancel{(x + 2)}}}{(x + 4)\underset{1}{\cancel{(x - 3)}}} \cdot \frac{\overset{1}{\cancel{x - 3}}}{\underset{1}{\cancel{(x + 2)}}(x + 2)}$$

$$= \frac{x - 2}{(x + 4)(x + 2)} \qquad (x \neq -2, 3, -4).$$

Of course, this procedure can be used to simplify the expression obtained by rewriting a quotient as a product. Hence, in part (b) of Example 1 we could write

$$\frac{x^2 - 5x + 4}{x^2 + 5x + 6} \div \frac{x - 1}{x + 2} = \frac{\overset{1}{\cancel{(x - 1)}}(x - 4)}{\underset{1}{\cancel{(x + 2)}}(x + 3)} \cdot \frac{\overset{1}{\cancel{x + 2}}}{\underset{1}{\cancel{x - 1}}}$$

$$= \frac{x - 4}{x + 3} \qquad (x \neq -2, -3, 1).$$

The product or quotient of a rational expression and the sum or difference of rational expressions can be simplified by first simplifying the sum or difference and then rewriting the resulting product or quotient as a single rational expression in lowest terms. Alternatively, we may do the multiplication first using the distributive law and then simplify the resulting sum or difference.

EXAMPLE 2   (a) $\left(\dfrac{2}{x - 1} + \dfrac{3}{x + 1}\right) \div \dfrac{x + 2}{x^2 - 1} = \left[\dfrac{2(x + 1) + 3(x - 1)}{(x - 1)(x + 1)}\right] \div \dfrac{x + 2}{x^2 - 1}$

$$= \frac{5x - 1}{x^2 - 1} \cdot \frac{x^2 - 1}{x + 2}$$

$$= \frac{5x - 1}{x + 2} \qquad (x \neq -2, -1, 1)$$

(b) $\left(\dfrac{1}{x} - \dfrac{1}{x + 1}\right)x = \dfrac{1}{x} \cdot x - \dfrac{1}{x + 1} \cdot x$

$$= 1 - \frac{x}{x + 1}$$

$$= \frac{x + 1 - x}{x + 1} = \frac{1}{x + 1} \qquad (x \neq -1, 0) \qquad ■$$

Complex Fractions

The quotient $\dfrac{P(x)}{Q(x)} \div \dfrac{R(x)}{S(x)}$ can also be denoted by

$$\frac{\dfrac{P(x)}{Q(x)}}{\dfrac{R(x)}{S(x)}}.$$

This latter form is called a **complex fraction**; that is, it is a fraction containing a fraction in either the numerator or denominator or both.

When the quotient of two fractions is given in the form of a complex fraction, we have a choice of procedures available to us for writing the quotient in the form of a simple (not complex) fraction.

EXAMPLE 3   Write $\dfrac{\dfrac{x+1}{4}}{x - \dfrac{2}{3}}$ as a simple fraction in lowest terms.

Solution 1   We can apply the fundamental principle of fractions to multiply the numerator and denominator by the least common denominator of the simple fractions involved. Then we have

$$\frac{\left(\dfrac{x+1}{4}\right)12}{\left(x - \dfrac{2}{3}\right)12} = \frac{3x+3}{12x-8} \qquad \left(x \neq \frac{2}{3}\right).$$

Solution 2   Alternatively, we can simplify the numerator and denominator separately and then find the quotient. Thus, we write the complex fraction as

$$\frac{\dfrac{x+1}{4}}{x - \dfrac{2}{3}} = \left(\frac{x+1}{4}\right) \div \left(x - \frac{2}{3}\right) = \left(\frac{x+1}{4}\right) \div \left(\frac{3x-2}{3}\right)$$

$$= \frac{x+1}{4} \cdot \frac{3}{3x-2} = \frac{3x+3}{12x-8} \qquad \left(x \neq \frac{2}{3}\right). \qquad ■$$

In the event we have a more complicated expression involving a complex fraction, we can rewrite the expression by simplifying small parts of it at a time.

EXAMPLE 4   Write $\dfrac{1}{x + \dfrac{1}{x + \dfrac{1}{x}}}$ as a simple fraction in lowest terms.

Solution   We can begin by concentrating on the lower right-hand expression, $\dfrac{1}{x + \dfrac{1}{x}}$. We

have

$$\frac{1}{x + \dfrac{1}{x}} = \frac{(1)x}{\left(x + \dfrac{1}{x}\right)x} = \frac{x}{x^2 + 1} \qquad (x \neq 0).$$

Thus,

$$\frac{1}{x + \dfrac{1}{x + \dfrac{1}{x}}} = \frac{1}{x + \dfrac{x}{x^2 + 1}} \qquad (x \neq 0).$$

From this point, we can apply either of the methods shown in Example 3 to the right-hand member above. Using the first method, we have

$$\frac{1}{x + \dfrac{1}{x + \dfrac{1}{x}}} = \frac{1(x^2 + 1)}{\left(x + \dfrac{x}{x^2 + 1}\right)(x^2 + 1)}$$

$$= \frac{x^2 + 1}{x^3 + x + x} = \frac{x^2 + 1}{x^3 + 2x} \qquad (x \neq 0). \qquad ■$$

EXERCISE 1.6

A

■   *Write each product or quotient as a simple fraction in lowest terms. Assume that no denominator vanishes.*

For Exercises 1–22, see Example 1.

1. $\dfrac{x^2 y^3 z^4}{a^2 x^2 b^3} \cdot \dfrac{a^4 y^2 b^4}{x^4 y^3 z^2}$

2. $\dfrac{x^3 y^5 z^3}{a^2 b^2 c^4} \cdot \dfrac{a^2 b^2 c^4}{x y^4 z^2}$

3. $\dfrac{x^2 yz}{ab^2} \cdot \dfrac{a^2 bc^2}{x y^2 z^2}$

4. $\dfrac{x^2}{yz} \cdot \dfrac{2x^2y^2}{z}$

5. $\dfrac{4xyz}{3a} \div \dfrac{6xy^2}{10ab}$

6. $\dfrac{xy^2}{a} \div \dfrac{ay^2}{x}$

7. $\dfrac{x^2y^4a^3}{b^2c^2z^6} \div \dfrac{x^4y^2a}{b^3c^4z^5}$

8. $\dfrac{a^4b^5c^3}{x^4yz^3} \div \dfrac{xy}{bc}$

9. $\dfrac{x-1}{x} \cdot \dfrac{x^2}{x^2-1}$

10. $\dfrac{x+4}{x^3} \cdot \dfrac{x}{x^2-16}$

11. $\dfrac{x^2-2x+1}{x+1} \cdot \dfrac{x^2+2x+1}{x-1}$

12. $\dfrac{x^2-6x+9}{x-4} \cdot \dfrac{x^2-8x+16}{x-3}$

13. $\dfrac{x-2}{2x^2-8} \div \dfrac{x+2}{x}$

14. $\dfrac{2x^2-2}{x-1} \div \dfrac{x}{6x-6}$

15. $\dfrac{x^2-1}{x^2-4} \div \dfrac{x^2-2x+1}{x^2-4x+4}$

16. $\dfrac{x^2+6x+9}{x^2+18x+81} \div \dfrac{x^2-9}{x^2-81}$

17. $\dfrac{a^3b^2}{c^4} \cdot \dfrac{b^3c^2}{a^4} \div \dfrac{ab}{c}$

18. $\dfrac{x^2yz^4}{a^2b^2c} \cdot \dfrac{a^4b^2c}{xy} \div \dfrac{ab^2}{x}$

19. $\dfrac{x^2+y^2}{x^2} \cdot \dfrac{y^2}{x^2+xy+y^2} \div \dfrac{x^2+y^2}{x^3-y^3}$

20. $\dfrac{x^4-y^4}{x^2} \cdot \dfrac{y^2}{x^3-y^3} \div \dfrac{y^2}{x^2+xy+y^2}$

21. $\dfrac{a^2b^2}{a^2} \cdot \dfrac{a^2+ab+b^2}{a^2} \div \dfrac{a^3-b^3}{a^4}$

22. $\dfrac{x^3+8}{x^2-9} \cdot \dfrac{x^3-27}{x^2+5x+6} \div \dfrac{x^2-2x+4}{x^2+6x+9}$

■   *For Exercises 23–34, see Examples 2–4.*

23. $\left(1-\dfrac{1}{z}\right)\left(1+\dfrac{1}{z}\right)$

24. $\left(\dfrac{1}{z}-1\right) \div \left(\dfrac{1}{z}+1\right)$

25. $\left(\dfrac{5}{x+5}-\dfrac{4}{x+4}\right) \cdot \dfrac{x+4}{x}$

26. $\left(\dfrac{x}{x^2+1}-\dfrac{1}{x+1}\right) \cdot \dfrac{x+1}{x-1}$

27. $\left(\dfrac{5}{x^2-9}+1\right) \div \dfrac{x+2}{x-3}$

28. $\left(\dfrac{y}{y+1}+\dfrac{1}{y-1}\right) \div \dfrac{y^2+1}{y+1}$

29. $\left(\dfrac{5}{x^2-9}+\dfrac{x^2-4}{x}\right) \div \dfrac{x-2}{x^2}$

30. $\left(\dfrac{x+2}{x^2-1}+\dfrac{x+2}{x-4}\right) \div \dfrac{x^2-4}{x^2-16}$

31. $\dfrac{\dfrac{a^2}{b}}{\dfrac{b^2}{a}}$

32. $\dfrac{\dfrac{x^4}{y^2}}{\dfrac{x}{y^4}}$

33. $\dfrac{\dfrac{x^2}{y^2}}{\dfrac{y^3}{x^3}}$

34. $\dfrac{\dfrac{x^4}{y^2}}{\dfrac{x^3}{y^4}}$

35. $\dfrac{\dfrac{x^2y^3}{a^2b^2}}{\dfrac{xy^2}{ab}}$

36. $\dfrac{\dfrac{a^2b^2}{x^4y^4}}{\dfrac{a^2b^3}{x^2y}}$

37. $\dfrac{\dfrac{x+1}{(y-3)^2}}{\dfrac{(x+1)^2}{y-3}}$

38. $\dfrac{\dfrac{(x-1)^2}{y}}{\dfrac{y^2}{(x-1)^2}}$

39. $\dfrac{\dfrac{x^2+2x-3}{x^2+2x+1}}{\dfrac{x^2+x-6}{x^2-1}}$

40. $\dfrac{\dfrac{x^2+3x-4}{x^2-2x-3}}{\dfrac{x^2+x-2}{x^2+x-12}}$

41. $\dfrac{1}{1-\dfrac{1}{x}}$

42. $\dfrac{3}{2+\dfrac{3}{x^2}}$

43. $2 - \dfrac{2}{2 + \dfrac{1}{x}}$

44. $3 + \dfrac{1}{1 - \dfrac{2}{x}}$

45. $\dfrac{1 + \dfrac{1}{2x + 1}}{1 + \dfrac{1}{x}}$

46. $\dfrac{1 - \dfrac{2}{x + 1}}{\dfrac{x - 1}{2}}$

47. $1 + \dfrac{2 + \dfrac{1}{x}}{x - \dfrac{1}{x}}$

48. $\dfrac{a + 3 - \dfrac{1}{a + 3}}{a + \dfrac{2}{a + 3}}$

49. $\dfrac{a + 3 + \dfrac{12}{a + 5}}{a + 5 + \dfrac{16}{a + 5}}$

50. $\dfrac{x - \dfrac{5}{x + 1}}{x + 2 - \dfrac{3}{x + 1}}$

51. $\dfrac{x - 7 + \dfrac{36}{x + 6}}{x - 1 + \dfrac{12}{x + 6}}$

52. $\dfrac{x - 2 + \dfrac{4}{x + 3}}{x + 3 - \dfrac{2}{x + 3}}$

53. $\dfrac{1 + \dfrac{1}{1 - \dfrac{x}{y}}}{1 + \dfrac{3}{1 - \dfrac{x}{y}}}$

54. $\dfrac{1 - \dfrac{1}{\dfrac{x}{y} + 2}}{1 + \dfrac{3}{\dfrac{x}{y} + 2}}$

B

55. $\dfrac{\dfrac{1}{x} - \dfrac{1}{y}}{\dfrac{2}{x^2} + \dfrac{1}{y^2}}$

56. $\dfrac{\dfrac{3}{x^2} - \dfrac{3}{y^2}}{\dfrac{6}{y} - \dfrac{6}{x}}$

57. $\dfrac{\dfrac{1}{1 + \dfrac{1}{x}} - \dfrac{1}{1 - \dfrac{1}{x}}}{1 + \dfrac{1}{x^2}}$

58. $\dfrac{\dfrac{3}{2 + \dfrac{3}{x - 1}} - \dfrac{2}{3 - \dfrac{2}{x - 1}}}{2 - \dfrac{3}{(x - 1)^2}}$

59. $\dfrac{\dfrac{1}{1 + \dfrac{2}{x}} + \dfrac{1}{1 - \dfrac{2}{x}}}{\dfrac{1}{2 + \dfrac{1}{x}} - \dfrac{1}{2 - \dfrac{1}{x}}}$

60. $\dfrac{\dfrac{2}{1 + \dfrac{2}{x + 2}} - \dfrac{1}{2 - \dfrac{2}{x + 2}}}{\dfrac{4}{1 + \dfrac{2}{x + 2}} - \dfrac{2}{2 - \dfrac{1}{x + 2}}}$

■   Each of the following expressions is a **partial continued fraction**. Continued fractions can be used to find rational approximations to irrational numbers.

   The value of π (the ratio of the circumference of a circle to its diameter) is 3.141592654 to nine decimal places. Write each of the following as a simple fraction. Then use a calculator to compare the decimal value of the fraction with the preceding approximation of π.

61. $3 + \dfrac{1}{7}$

62. $3 + \dfrac{1}{7 + \dfrac{1}{15}}$

63. $3 + \cfrac{1}{7 + \cfrac{1}{15 + \cfrac{1}{1}}}$

64. $3 + \cfrac{1}{7 + \cfrac{1}{15 + \cfrac{1}{1 + \cfrac{1}{293}}}}$

■ *Write each expression as the quotient of two polynomials.*

65. $1 + \cfrac{1}{2 + \cfrac{1}{2 + \cfrac{1}{2 + \cfrac{1}{x}}}}$

66. $1 + \cfrac{1}{3 + \cfrac{1}{1 + \cfrac{1}{5 + \cfrac{1}{x}}}}$

67. $2 + \cfrac{1}{4 + \cfrac{1}{4 + \cfrac{1}{4 + \cfrac{1}{x}}}}$

68. $1 + \cfrac{1}{1 + \cfrac{1}{2 + \cfrac{1}{2 + \cfrac{1}{x}}}}$

CHAPTER REVIEW

Key Words, Phrases, and Symbols

■ *Define or explain each of the following words, phrases, and symbols.*

1. algebraic expression *Page 1*
2. equivalent expressions
3. a^n for n a positive integer and a a real number
4. monomial *Page 2*
5. coefficient of a monomial
6. degree of a monomial
7. degree of a polynomial
8. standard form of a polynomial in x
9. leading term of a polynomial
10. leading coefficient of a polynomial
11. prime polynomial
12. completely factored polynomial
13. rational expression
14. fraction in lowest terms
15. least common multiple of two numbers
16. least common multiple of two polynomials
17. complex fraction

Review Exercises

A

[1.1] ■ *Simplify.*

1. $(x^2 - 5x + 1) - (2 - 3x - x^2)$

2. $3\{x^2 + 1 - [x^2 - 2(x^2 - 1)] + 1\}$

■   *Given $P(x) = 2x^2 - 3x + 4$, find the following.*

3. $P(3)$ **4.** $P(x)|_{-1}^{2}$

[1.2]   ■   *Simplify.*

5. $2(x + 1)(3x + 2)$ **6.** $(x + 2)(3x^2 + x + 1)$

[1.3]   ■   *Factor completely.*

7. $y^2 - 8y + 15$ **8.** $x^3 + 4x^2 + 4x$

9. $x^4 - 16$ **10.** $z^3 - 64$

[1.4]   ■   *Write each quotient as a polynomial or in the form $Q(x) + \dfrac{R(x)}{D(x)}$.*

11. $\dfrac{30x^2y^3z^4}{5xyz^3}$ **12.** $\dfrac{18y^3 + 21y^2 + 3y}{3y}$

13. $\dfrac{2x^2 + x - 15}{x + 3}$ **14.** $\dfrac{6x^2 + x - 2}{2x - 1}$

15. $\dfrac{x^2 - x + 3}{x - 1}$ **16.** $\dfrac{x^3 - 2x^2 + 1}{x - 2}$

[1.5]   ■   *Reduce to lowest terms. Assume that no denominator vanishes.*

17. $\dfrac{6y^2z + 33yz^2}{3yz}$ **18.** $\dfrac{x^2 + 10x + 21}{x + 3}$

19. $\dfrac{x^2 + 4x - 21}{x^2 - 9}$ **20.** $\dfrac{3x^2 + 12x + 12}{3x^2 - 12}$

■   *Simplify. Assume that no denominator vanishes.*

21. $\dfrac{x}{3} + \dfrac{x + 1}{4}$ **22.** $\dfrac{x + 1}{x - 1} - \dfrac{x - 2}{x + 2}$

23. $\dfrac{x^2 + 3}{x + 1} - 2$ **24.** $\dfrac{y + 2}{y^2 - y - 2} + \dfrac{y - 3}{y + 1}$

[1.6]   ■   *Simplify. Assume that no denominator vanishes.*

25. $\dfrac{x^3y}{z} \cdot \dfrac{xz^2}{y^3}$ **26.** $\dfrac{x^2 + 2x + 1}{x - 3} \cdot \dfrac{x^2 + 4x - 21}{x^2 + 6x + 5}$

27. $\dfrac{x^2 - 1}{x + 2} \div \dfrac{x^2 + 4x + 3}{x^2 + 4x + 4}$ **28.** $\dfrac{x - \dfrac{1}{x + 1}}{x + 4 + \dfrac{11}{x - 3}}$

B

■   *Factor completely into products of polynomials with integer coefficients.*

29. $x^6 - y^6$ **30.** $x^4 - 3x^2 + 2$

31. $x^{16} - 1$ **32.** $x^{27} - 1$

■ *Write each expression as a single fraction reduced to lowest terms.*

33. $\dfrac{x^2}{x^3 + x^2 - x - 1} + \dfrac{x^2 - 1}{x^3 - 4x^2 - x + 4}$

34. $\dfrac{1}{x^8 - 1} - \dfrac{1}{x^3 - 1}$

35. $\dfrac{x^5 - x^3 - x^2 + 1}{x^2 - 4x + 4} \cdot \dfrac{x^4 - x^3 - 8x + 8}{-2x^2 - 2x - 2}$

36. $\dfrac{x^6 - 4x^4 - 16x^2 + 64}{x^3 - 2x^2 + 4x - 8} \div \dfrac{x^3 + 3x^2 + 4x + 12}{x^3 + x^2 - 6x}$

37. Write $5 + \dfrac{1}{3 + \dfrac{1}{2 + \dfrac{1}{3}}}$ as a rational number and compare the decimal equivalent to

$\sqrt{28} \approx 5.2915$.

38. Write $1 + \dfrac{1}{1 + \dfrac{1}{1 + \dfrac{1}{1 + \dfrac{1}{x}}}}$ as the quotient of two polynomials.

2 Rational Exponents— Radicals

In Chapter 1, powers of real numbers were defined for positive integer exponents, and some simple properties of products and quotients of powers were examined. In this chapter, powers with integral and rational exponents are defined in a manner consistent with these properties.

POWERS WITH INTEGRAL EXPONENTS

In Section 1.4 we saw that if a is a nonzero real number and m and n are positive integers, $m > n$, then we have

$$\frac{a^m}{a^n} = a^{m-n}. \tag{1}$$

Reason for Defining a^0 To Be 1

In this section we remove the restriction $m > n$. If Equation (1) is to hold for $m = n$, then we must have

$$\frac{a^n}{a^n} = a^{n-n} = a^0.$$

Since $a^n/a^n = 1$ for $a \neq 0$, we make the following definition.

Definition 2.1   *If a is a nonzero real number, then*

$$a^0 = 1.$$

Note that the only restriction on a in Definition 2.1 is $a \neq 0$. Thus, *any number, except* 0, to the 0th power is defined to be 1, and 0^0 is *undefined*.

EXAMPLE I (a) $2^0 = 1$ (b) $(x - 3)^0 = 1$ (c) $(x - 3)^0$ is undefined
for $x \neq 3$.   for $x = 3$.   ■

Reason for Defining a^{-n} To Be $1/a^n$

In a similar way, if Equation (1) is to hold for $m = 0$, then we must have

$$\frac{a^0}{a^n} = a^{0-n} = a^{-n}.$$

Since $a^0 = 1$, we make the following definition.

Definition 2.2

If a is a nonzero real number and n is a positive integer, then

$$a^{-n} = \frac{1}{a^n}.$$

Note that the only restriction on a in Definition 2.2 is $a \neq 0$. Thus, 0 to a negative power is undefined.

EXAMPLE 2 (a) $3^{-2} = \frac{1}{3^2}$ (b) $(x + 1)^{-1} = \frac{1}{x + 1}$ (c) $(x + 1)^{-1}$ is undefined
for $x \neq -1$.   for $x = -1$.   ■

Laws of Exponents for Integral Exponents

The laws of exponents for positive-integer exponents given by Theorems 1.1 and 1.2 are also valid for negative integers. For example, if m and n are both negative integers, then $-m$ and $-n$ are both positive integers and we can write

$$a^m \cdot a^n = \frac{1}{a^{-m}} \cdot \frac{1}{a^{-n}} \qquad \text{By Definition 2.2.}$$

$$= \frac{1}{a^{-m} \cdot a^{-n}}$$

$$= \frac{1}{a^{(-m)+(-n)}} \qquad \text{By Theorem 1.1-I.}$$

$$= \frac{1}{a^{-(m+n)}} = a^{m+n} \qquad \text{By Definition 2.2.}$$

Thus, $a^m \cdot a^n = a^{m+n}$ for m and n negative integers. A complete discussion of this law and of all the others given by Theorems 1.1 and 1.2 as they are applicable to negative integers requires consideration of several cases. The arguments are similar to the preceding one, and we shall not give them here. We now restate these theorems for integer exponents m, n.

Theorem 2.1   *If a and b are any nonzero real numbers and m and n are integers, then*

$$\text{I.} \quad a^m \cdot a^n = a^{m+n}, \qquad\qquad \text{IV.} \quad (ab)^n = a^n b^n,$$

$$\text{II.} \quad \frac{a^m}{a^n} = a^{m-n}, \qquad\qquad \text{V.} \quad \left(\frac{a}{b}\right)^n = \frac{a^n}{b^n}.$$

$$\text{III.} \quad (a^m)^n = a^{mn},$$

Some applications of Theorem 2.1 follow.

EXAMPLE 3 (a) $(x^2)^3 = x^{2 \cdot 3} = x^6$ By Theorem 2.1-III.

(b) $(xy)^3 = x^3 y^3$ By Theorem 2.1-IV.

(c) $\left(\dfrac{x^2}{y}\right)^{-4} = \dfrac{(x^2)^{-4}}{(y)^{-4}}$ By Theorem 2.1-V.

$= \dfrac{x^{-8}}{y^{-4}}$ By Theorem 2.1-III.

$= \dfrac{1/x^8}{1/y^4}$ By Definition 2.2.

$= \dfrac{y^4}{x^8} \qquad (x, y \neq 0)$ ■

Parts IV and V of Theorem 2.1 can apply to expressions involving more than two factors.

EXAMPLE 4 (a) $\left(\dfrac{a^3 b}{3c^2}\right)^{-2} = \dfrac{a^{-6} b^{-2}}{3^{-2} c^{-4}}$

$= \dfrac{3^2 c^4}{a^6 b^2}$

$= \dfrac{9c^4}{a^6 b^2}$

$(a, b, c \neq 0)$

(b) $\left(\dfrac{x^2 y^{-1}}{3y^2 x^{-1}}\right)^{-2} = \dfrac{x^{-4} y^2}{3^{-2} y^{-4} x^2}$

$= \dfrac{3^2 y^2 y^4}{x^2 x^4}$

$= \dfrac{9y^6}{x^6}$

$(x, y \neq 0)$ ■

Multiplication, Factoring

Theorem 2.1 along with the distributive law can be used to multiply and/or factor expressions involving sums of powers with integer exponents.

E X A M P L E 5   (a) $x^{-2}(x^3y + xy^2)$

$$= x^{-2+3}y + x^{-2+1}y^2$$

$$= xy + x^{-1}y^2$$

(b) $x^{-3}y^{-1}(x^3y^2 + xy^3)$

$$= (x^{-3+3}y^{-1+2} + x^{-3+1}y^{-1+3})$$

$$= y + x^{-2}y^2$$

E X A M P L E 6   Factor each expression as indicated.

(a) $x^{-2}y + x^2y^{-3} = x^{-2}y^{-3}(?)$

(b) $x^2y^{-2} + y^2x^{-2} = x^{-3}y^{-2}(?)$

Solution   (a) $x^{-2}y + x^2y^{-3}$

$$= x^{-2}y^{-3}y^4 + x^{-2}y^{-3}x^4$$

$$= x^{-2}y^{-3}(y^4 + x^4)$$

(b) $x^2y^{-2} + y^2x^{-2}$

$$= x^{-3}y^{-2}x^5 + x^{-3}y^{-2}xy^4$$

$$= x^{-3}y^{-2}(x^5 + xy^4)$$ ■

Complex Fractions

Negative exponents on powers in expressions involving sums and differences result in complex fractions. To simplify such expressions, we first write all powers in equivalent forms with positive exponents and simplify the resulting complex fraction.

E X A M P L E 7   (a) $\dfrac{x^{-1} - 1}{x^{-2}y} = \dfrac{\dfrac{1}{x} - 1}{\dfrac{y}{x^2}}$

$$= \left(\frac{1}{x} - 1\right) \cdot \frac{x^2}{y}$$

$$= \frac{1 - x}{x} \cdot \frac{x^2}{y}$$

$$= \frac{(1 - x) \cdot x}{y}$$

$$(x, y \neq 0)$$

(b) $\dfrac{x^{-3}y^2}{x^{-1} + y^{-1}} = \dfrac{\dfrac{y^2}{x^3}}{\dfrac{1}{x} + \dfrac{1}{y}}$

$$= \frac{\dfrac{y^2}{x^3}}{\dfrac{y + x}{xy}}$$

$$= \frac{y^2}{x^3} \cdot \frac{xy}{y + x}$$

$$= \frac{y^3}{x^2(y + x)}$$

$$(x, y \neq 0, x \neq -y)$$ ■

Scientific Notation

In many applications of mathematics, we deal with numbers that are very close to zero or that are very large. These numbers are usually expressed as a product of a number between 1 and 10 (excluding 10) and a power of 10. When a number is expressed in this way, we say it is expressed in **scientific notation**.

Steps for Converting to Scientific Notation

A number can be expressed in scientific notation as follows:

1. Move the decimal point to a position where the resulting number is between 1 and 10 (not including 10).
2. Multiply the result of Step 1 by a power of 10, using as the exponent the number of positions the decimal point was moved. The exponent is positive if the decimal point was moved to the left and negative if it was moved to the right.

EXAMPLE 8 (a) $7{,}450{,}000{,}000 = 7{,}450{,}000{,}000.$

$$= 7.45 \times 10^9$$

(b) $0.00000612 = 0.00000612$

$$= 6.12 \times 10^{-6} \qquad ■$$

Many calculators display results of operations in the form of a number between 1 and 10 (excluding 10) and an integer which represents a power of 10, that is, essentially in scientific notation. Thus, for example, a calculator display of

$$\boxed{2.1459 \qquad -15}$$

should be interpreted as

$$2.1459 \times 10^{-15}.$$

To translate a number from scientific notation to standard decimal notation, we simply reverse the procedure just described. In other words, we simply move the decimal point the number of places indicated by the exponent on 10. We move it to the right if the exponent is positive and to the left if the exponent is negative.

EXAMPLE 9 Write the number represented by the following calculator displays in standard decimal notation.

(a) $\boxed{6.7345 \qquad 2}$ (b) $\boxed{4.3571 \qquad -5}$

Solution   In each case we first interpret the display in scientific notation and then convert to standard decimal notation.

(a) $6.7345 \times 10^2 = 6.7345$

$= 673.45$

(b) $4.3571 \times 10^{-5} = 00004.3571$

$= 0.000043571$ ■

Scientific notation, along with the laws of exponents, can sometimes be used to simplify the operations of multiplication and division, as shown in the following example.

EXAMPLE 10   (a) $(0.000051)(2{,}000{,}000)$

$= (5.1 \times 10^{-5})(2 \times 10^6)$

$= (5.1)(2) \times 10^{-5} \times 10^6$

$= 10.2 \times 10^1$

$= 1.02 \times 10^2$

(b) $\dfrac{25{,}000{,}000}{0.0002}$

$= \dfrac{2.5 \times 10^7}{2 \times 10^{-4}}$

$= \dfrac{2.5}{2} \times \dfrac{10^7}{10^{-4}}$

$= 1.25 \times 10^{11}$ ■

EXERCISE 2.1

A

■   *In order to avoid the necessity of constantly noting exceptions, we shall assume that in the exercises in this section the variables are restricted so that no denominator vanishes.*

■   *For the given expression, write a number with exponent 1. See Examples 1 and 2.*

1. $(-13)^0$

2. 144^0

3. 3^{-1}

4. 4^{-2}

5. $(-4)^{-3}$

6. $(-5)^{-4}$

7. $\dfrac{1}{6^{-2}}$ < 36

8. $\dfrac{1}{4^{-1}}$

9. $\left(\dfrac{4}{3}\right)^{-1}$ ≈ 3/4

10. $\left(\dfrac{5}{2}\right)^{-2}$

11. $\left(\dfrac{2}{3}\right)^{-3}$ 8

12. $\left(\dfrac{3}{8}\right)^{-2}$ = 64/9

■   *Write the expression as a product or quotient in which each variable occurs at most once in the expression and all exponents are positive. See Examples 3 and 4.*

13. $x^5 x^{-8}$

14. $y^{-2} y^4$

15. $\dfrac{y^{-2}}{x^4}$

16. $\dfrac{x^3}{y^{-3}}$

17. $(x^2 y)^3$

18. $(xy^{-2})^2$

19. $\left(\dfrac{x^2}{2y}\right)^5$

20. $\left(\dfrac{x^{-1} y^2}{2z^2}\right)^{-3}$

21. $\left(\dfrac{2x}{y^2}\right)^2 \cdot \dfrac{y^2}{x}$

22. $\left(\dfrac{x^0 y^2}{z^2}\right)^{-2} \cdot \dfrac{x^2}{y^{-2}}$

23. $\left(\dfrac{x^2 y}{z^3}\right)^{-2} \cdot \left(\dfrac{xy^0}{z}\right)^{-1}$

24. $\left(\dfrac{2x^{-1}}{y^2}\right)^{-1} \cdot \dfrac{x}{y^2}$

25. $\left(\dfrac{x}{y^2}\right)^{-2} \div \left(\dfrac{y}{x}\right)^2$

26. $\left(\dfrac{x}{y}\right)^{-2} \div \left(\dfrac{x^2}{y^2}\right)^4$

27. $\left(\dfrac{xz^2}{y^{-3}}\right)^{-3} \div \left(\dfrac{x^3}{y^2 z}\right)^{-4}$

28. $\left(\dfrac{x^2 y}{z^0}\right)^{-2} \div \left(\dfrac{xy^4 z^3}{q^8}\right)^0$

29. $\dfrac{(x^2 y^3)(xy^{-2})(x^{-2}y^2)}{(x^{-1}y^2)(y^{-2}x^{-3})}$

30. $\dfrac{(x^2 y^{-3})(x^{-3}y^2)(xy^{-1})}{(x^2 y^{-4})(x^{-3}y^{-6})}$

31. $\dfrac{(xy^{-1})^{-1}(x^2 y^2)^{-2}(x^{-1}y)^0}{(x^{-3}y^2)^{-1}(x^{-1}y^{-3})^{-1}}$

32. $\dfrac{(x^2 y^3)^{-2}(xy^{-3})^0 (x^3 y^2)^{-1}}{(x^5 y^{-2})^{-2}(x^4 y)^2}$

33. $\dfrac{(xyz^0)^{-1}(x^2 y^{-1}z^{-2})^{-3}(x^2 y^3 z)^3}{(x^2 y^2 z^4)^{-1}(x^3 y^0 z^2)^2}$

34. $\dfrac{(x^{-4}yz)^0 (x^{-3}y^2 z)^{-1}(x^2 yz)^2}{(x^5 y^{-3}z^2)^{-1}(x^4 y^2 z^3)}$

■ *Write each product as a sum of powers. See Example 5.*

35. $x^{-1}(x - xy)$

36. $y^{-2}(x + xy)$

37. $x^{-1}y^{-2}(x^2 y - xy)$

38. $x^{-2}y^{-3}(x^3 - y^3)$

39. $x^{-2}y^2(x^2 y^{-2} + xy)$

40. $x^2 y^{-3}(x^3 - y^3)$

41. $(x^2 + y^2)(x^{-2} + y^{-2})$

42. $(x^3 - y^3)(x^{-3} + y^3)$

■ *Factor each expression as indicated. See Example 6.*

43. $x^{-3}y^2 - 3xy^{-1} = x^{-3}y^{-1}(?)$

44. $xy^{-1} + x^{-1}y^{-2} = x^{-1}y^{-2}(?)$

45. $2x^{-3}y^{-4} - 6x^{-4}y^{-3} = 2x^{-4}y^{-4}(?)$

46. $4y^{-1} - 2x^2 y^{-3} = 2y^{-3}(?)$

47. $xy^2 z^3 + x^2 y^{-1}z^{-3} = xy^{-1}z^{-3}(?)$

48. $x^{-1}z^{-3} - 2y^{-2}z = x^{-1}y^{-2}z^{-3}(?)$

■ *Write the expression as a product or quotient in which each variable occurs at most once in the expression and all exponents are positive. See Example 7.*

49. $\dfrac{x^{-1} - y}{2x^{-1}y}$

50. $\dfrac{2x^{-2} - y^{-2}}{xy}$

51. $\dfrac{2x^{-2}y^{-2}}{x^{-2} - y^{-2}}$

52. $\dfrac{2x^3 y^{-3}}{6x^2 - 2y^{-2}}$

53. $\dfrac{x^{-1} - y^{-1}}{x^{-2} - y^{-2}}$

54. $\dfrac{x^3 + y^{-1}}{x^{-1} + y^2}$

■ *Write each number in scientific notation. See Example 8.*

55. 123,456

56. 525

57. −1945

58. −32,768

59. 0.04321

60. 0.0000004

61. −0.010010001

62. −0.9

■ *Write each number in decimal notation. See Example 9.*

63. 4.28×10^8

64. 6.75×10^3

65. -5.642×10^6

66. -6.371×10^2

67. 5.63×10^{-4}

68. 1.67×10^{-6}

69. -4.555×10^{-8}

70. -6.501×10^{-3}

■ *Write the product or quotient as a single number in scientific notation. See Example 10.*

71. $(0.00043)(2000)$

72. $(21,000,000)(8,000,000)$

73. $\dfrac{201,000,000}{0.0003}$

74. $\dfrac{0.000000046}{0.00002}$

75. $\dfrac{201,000,000}{(0.0003)(67,000)}$

76. $\dfrac{(0.0006)(32,000,000)}{0.0000008}$

■ *Use a calculator to perform the indicated calculation. Write the result in scientific notation. Round off the multiplier between 1 and 10 to two decimal places. See Examples 9 and 10.*

77. $(1.41 \times 10^8)(1.41 \times 10^{-8})$

78. $(2.65 \times 10^4)(3.14 \times 10^{-3})$

79. $\dfrac{1.71 \times 10^{15}}{2.35 \times 10^{12}}$

80. $\dfrac{3.75 \times 10^{30}}{1.25 \times 10^{-10}}$

81. $\dfrac{(1.23 \times 10^9)(1.11 \times 10^4)}{2.54 \times 10^{13}}$

82. $\dfrac{(2.73 \times 10^4)(3.63 \times 10^{-9})}{3.00 \times 10^5}$

■ *Write your answer in scientific notation. Round off the multiplier between 1 and 10 to the nearest hundredth.*

83. The speed of light in a vacuum is approximately 186,000 miles per second. Find the distance light travels in a vacuum in each of the following time periods.

(a) One hour   (b) One day   (c) One week   (d) One year

84. The star Alpha Centauri is the closest star (except for the sun) to our solar system. It is approximately 4.3 light-years from our system. Determine its distance from our system in miles. A **light-year** is the distance light travels in one year.

85. Luna City and Selena are two colonies on the moon that are 120 miles from each other. A message is transmitted at the speed of light from Luna City. How long does it take for the message to reach Selena?

86. A scientist in Selena transmits data to a colleague in Newton (a third lunar colony), which is 25 miles from Selena. The data is being transmitted at one-half the speed of light. How long does it take for the data to reach Newton?

87. All the water on earth would fill a cubical tank about 700 miles on a side. The density of water is approximately 62.5 lb/ft³. What is the total weight of all the water on earth? (There are 5280 feet in 1 mile.)

88. A star whose mass is between 1.4 and 2 times the mass of our sun will eventually go supernova and collapse to a neutron star. Theoretical physicists have constructed models of neutron stars which suggest that they are spherical with a kind of iron crust on the surface which is extremely smooth and about a yard thick. Most of the interior of the neutron star consists of subnuclear particles. The density of the matter in a neutron star is approximately 3×10^8 kilograms per cubic centimeter. A typical neutron star has a radius of about 6 kilometers. What is the mass of such a star? [*Hint:* The volume of a sphere is approximately $4.19r^3$ where r is the radius and

$$\text{mass} = (\text{density}) \cdot (\text{volume}).]$$

■ *The maximum rate of transfer of information to or from a memory device for a computer is called the **bandwidth.***

89. If a computer has a central-memory bandwidth of one word per 12.5 nanosecond cycle, how many words can be transferred to or from central memory in 1 second? (One second is 10^9 nanoseconds.)

90. If a computer can transfer 800 million words per second to or from memory and it has a cycle time of 20 nanoseconds, how many words can it transfer per cycle?

B

■ *Write the expression as an equivalent product in which each variable occurs only once. Assume all exponents are integers.*

91. $x^n x^{n+3}$

92. $x^n x^{n-1}$

93. $y^{n+1} y^{-n-1}$

94. $\dfrac{x^{n+1} x^{n-2}}{x^{n+3}}$

95. $\dfrac{y^{3n} y^{2n-1}}{y^2}$

96. $\dfrac{x^{-2}}{x^{n+2}}$

97. $\left(\dfrac{x^n}{x^{-1}}\right)^{2+n}$

98. $\dfrac{(x^2)^{n+1}}{(x^n)^{-1}}$

99. $\dfrac{x^n y^{n+2}}{(xy)^{-2}}$

100. $\dfrac{x^{2n+1} y^{2n+1}}{x^n y^{n-1}}$

101. $\left(\dfrac{xy}{x^n}\right)^{-1}$

102. $\left(\dfrac{x^{n+1} y}{xy^{n+1}}\right)^{-1}$

$$\frac{xy^{n+1}}{x^{n+1}y} = x^{1-(n+1)}y^{n+1-1}$$

$$\boxed{x^{-n}y^n}$$

POWERS WITH RATIONAL EXPONENTS

In Section 2.1 we defined powers of real numbers with 0 and negative-integer exponents, so that the laws of exponents are valid for all integer exponents. In this section we define powers of real numbers with rational numbers as exponents. As before, we shall want the definition to be consistent with all the laws of exponents for integers.

We observed in Section 1.2 that for any real number a and positive integers m and n,

$$(a^m)^n = a^{mn}. \tag{1}$$

If (1) is to hold for $m = 1/n$, and $a^{1/n}$ is a real number, then we must have

$$(a^{1/n})^n = a^{(1/n)(n)} = a^{n/n} = a^1 = a, \tag{2}$$

so that the nth power of $a^{1/n}$ must be a. Thus, we make the following definition.

Definition 2.3   *If a is a real number and n is a positive integer, then a real **nth root** of a is any real number b so that*

$$b^n = a.$$

*For n equal to 2 or 3, respectively, an nth root is called a **square root** or a **cube root**.*

Number of nth Roots

The number of real nth roots of a real number a depends on whether n is even or odd and whether a is negative, zero, or positive. We shall look at four cases that cover all possibilities.

Case I: n odd, a any real number
In this case, a has exactly one real nth root. For example,

$$(-2)^3 = -8,$$

and -2 is the *only* real number whose cube is -8. Thus, -2 is the only real cube root of -8.

Case II: n even, $a < 0$
In this case, a has no real nth roots since *any nonzero real number raised to an even power is positive.*

Case III: n an even integer, $a = 0$
In this case, a has exactly one nth root, namely 0.

Case IV: n even, $a > 0$
In this case, a has two real nth roots. For example,

$$(-2)^4 = 16 \qquad \text{and} \qquad 2^4 = 16,$$

so that -2 and 2 are both fourth roots of 16.

These four cases are summarized in the following table. The entries in the table are the number of real nth roots of a.

	$a < 0$	$a = 0$	$a > 0$	
n odd	1	1	1	(Case I)
n even	0	1	2	(Cases II, III, IV)

Definition of $a^{1/n}$

We now use this information and the fact that $a^{1/n}$ should satisfy Equation (2) to make the following definition, so that $a^{1/n}$ represents one and only one real number.

Definition 2.4

If a is a real number and n is a positive integer, then we define the principal nth root of a as $a^{1/n}$ where

$a^{1/n}$ is the nth root of a in Cases I and III,

$a^{1/n}$ is the *positive* nth root of a in Case IV, and

$a^{1/n}$ is undefined in Case II.

EXAMPLE 1   (a) $25^{1/2} = 5$ (b) $-25^{1/2} = -5$

(c) $(-25)^{1/2}$ is undefined.   (d) $27^{1/3} = 3$

(e) $-27^{1/3} = -3$ (f) $(-27)^{1/3} = -3$ ■

Notice in parts (b) and (e) of Example 1 that $-25^{1/2}$ and $-27^{1/3}$ denote $-(25^{1/2})$ and $-(27^{1/3})$, respectively.

Often the *principal* nth root of a is referred to simply as *the* nth root of a. Thus, 3 is *a* square root of 9 and -3 is a square root of 9, whereas the principal square root or *the* square root of 9 is 3.

Definition of $a^{m/n}$

Let a be a real number, n a positive integer, and m any integer. We want to define $a^{m/n}$ so that Theorem 2.1-III holds true. Hence, we must have

$$a^{m/n} = a^{m \cdot (1/n)} = (a^{1/n})^m.$$

Thus, we make the following definition.

Definition 2.5

If $a^{1/n}$ is a real number and m and n are integers with $n > 0$, then

$$a^{m/n} = (a^{1/n})^m.$$

Requiring that n be positive does not alter the fact that m/n can represent every rational number—negative, zero, or positive—since with $n > 0$ the sign of m/n is determined by the sign of m and m can be negative, zero, or positive.

Definition 2.5 requires that $a^{1/n}$ be a real number. If it is not defined, then $a^{m/n}$ is undefined.

EXAMPLE 2   (a) $16^{3/4} = [(16)^{1/4}]^3$ (b) $(16)^{2/4} = [(16)^{1/4}]^2$
                 $\quad\quad\ = (2)^3 = 8$ $\quad\quad\quad\ = (2)^2 = 4$

(c) $(-3)^{3/2}$ is undefined because   (d) $(-8)^{2/3} = [(-8)^{1/3}]^2$
      $(-3)^{1/2}$ is undefined. $\quad\quad\quad\ = (-2)^2 = 4$

(e) $(-16)^{8/4}$ is undefined because   (f) $(-27)^{6/3} = [(-27)^{1/3}]^6$
      $(-16)^{1/4}$ is undefined. $\quad\quad\quad\quad = (-3)^6 = 729$ ■

Observe that in part (e) of Example 2, $(-16)^{8/4}$ is undefined even though $8/4 = 2$ and $(-16)^2 = 256$. This example illustrates that care must be taken when working with fractional exponents and negative bases. For this reason most calculators will return an error message when trying to calculate a fractional power of a negative base even in the case where the quantity is well defined.

Notice that if n is an even natural number and a is any real number, then a^n is positive. Thus, by Definition 2.4, $(a^n)^{1/n}$ is nonnegative *regardless of the sign of a*, and we have

$$(a^n)^{1/n} = |a|, \qquad n \text{ an even natural number.}$$

EXAMPLE 3　**(a)**　$(a^2)^{1/2} = |a|^{2/2} = |a|$　　**(b)**　$(a^6)^{1/6} = |a|^{6/6} = |a|$　■

Observe that in Example 3 it is necessary to use absolute values because the results have to be positive. For example, if $a = -2$, then

$$[(-2)^2]^{1/2} = |-2|^{2/2} = 2^1 = 2 = |-2|$$

and

$$[(-2)^6]^{1/6} = |-2|^{6/6} = 2^1 = 2 = |-2|.$$

Laws of Exponents

Powers with rational exponents have the same properties as those with integral exponents, *as long as the base is positive*. Theorem 2.1 can be used to rewrite powers with rational exponents whenever $a, b > 0$. If the base is negative, however, care must be taken in dealing with any exponents m/n, where n is even.

EXAMPLE 4　**(a)**　$x^{2/3}y^{1/2} \cdot x^{1/3}y^{1/4}$
$\qquad = x^{(2/3)+(1/3)}y^{(1/2)+(1/4)}$
$\qquad = xy^{3/4}$

(b)　$\dfrac{x^{3/5}}{x^{2/5}} = x^{(3/5)-(2/5)}$
$\qquad = x^{1/5}$

(c)　$(x^6)^{1/3} = x^{6(1/3)}$
$\qquad = x^2$

(d)　$(x^2y^3)^{1/3} = (x^2)^{1/3}(y^3)^{1/3}$
$\qquad = x^{2/3}y$

(e)　$\left(\dfrac{x^2}{y^6}\right)^{2/3} = \dfrac{(x^2)^{2/3}}{(y^6)^{2/3}}$
$\qquad = \dfrac{x^{4/3}}{y^4}$

(f)　$(x^6y^2)^{1/2} = |x|^{6/2}|y|^{2/2}$
$\qquad = |x|^3|y|$　■

Multiplication, Factoring

We use the distributive law along with Theorem 2.1 for rational exponents to multiply and/or factor algebraic expressions involving powers with rational exponents.

EXAMPLE 5 Apply the distributive law to write the product as a sum or difference.

(a) $x^{1/4}(x^{3/4} - x)$

(b) $(x^{1/2} - x)(x^{1/2} + x)$

Solution
(a) $x^{1/4}(x^{3/4} - x)$
$$= x^{1/4}x^{3/4} - x^{1/4}x$$
$$= x - x^{5/4}$$

(b) $(x^{1/2} - x)(x^{1/2} + x)$
$$= x^{1/2}x^{1/2} - x^2$$
$$= x - x^2$$

EXAMPLE 6 Factor each expression as indicated.

(a) $y^{-1/2} + y^{1/2} = y^{-1/2}(?)$

(b) $x^{3/2} - x^{-1/2} = x^{-1/2}(?)$

Solution
(a) $y^{-1/2} + y^{1/2}$
$$= (y^{-1/2} + y \cdot y^{-1/2})$$
$$= y^{-1/2}(1 + y)$$

(b) $x^{3/2} - x^{-1/2}$
$$= (x^2 \cdot x^{-1/2} - x^{-1/2})$$
$$= x^{-1/2}(x^2 - 1) \qquad ■$$

Use of a Calculator

In many cases, rational powers of real numbers are themselves irrational. For example, $3^{1/2}$, $2^{1/3}$, and $1.5^{1/2}$ are all irrational.

Calculators that have a y^x key can be used to approximate powers with a rational exponent *and a positive base* by first approximating any base or exponent with a decimal. The accuracy of the approximation to the power depends on the accuracy of the approximation of the base and the exponent. Note that if a negative base is used, many calculators will return an error message even if an integer exponent is used.

For convenience of notation we shall use "$=$" instead of "approximately equals ($\approx$)" throughout this book.

EXAMPLE 7 Approximate each of the following using a calculator. Round off each exponent and the resulting power to the nearest hundredth.

(a) $2^{1/2}$

(b) $3^{2/3}$

(c) $4^{-1/3}$

Solution
(a) $2^{1/2} = 2^{0.50}$
$$= 1.41$$

(b) $3^{2/3} = 3^{0.67}$
$$= 2.09$$

(c) $4^{-1/3} = 4^{-0.33}$
$$= 0.63 \qquad ■$$

(handwritten: Ex. $27^{1/3} = 3$ 1-50   59-72)

EXERCISE 2.2

(handwritten: $x^{1/2} = \sqrt{x}$ $x^{1/n}$ - The Nth Root of X)

(A)

■ Write the expression as a number not involving exponents. See Examples 1 and 2.

1. $16^{1/2}$ **2.** $16^{-1/2}$ **3.** $9^{-3/2}$ **4.** $9^{4/2}$

5. $-27^{2/3}$ **6.** $-27^{-2/3}$ **7.** $(-8)^{2/3}$ **8.** $(-8)^{-2/3}$

9. $(-32)^{-2/5}$ **10.** $(-32)^{2/5}$ **11.** $\left(\dfrac{9}{4}\right)^{3/2}$ **12.** $\left(\dfrac{9}{4}\right)^{-3/2}$

■ Write the given expression as a product or quotient in which each variable occurs only once and all exponents are positive or the sum (or difference) of such expressions. Assume all variable bases are positive. See Example 4.

13. $x^{3/2} \cdot x^{1/3}$ **14.** $x^{-1/2} \cdot x^{1/3}$ **15.** $y^3 \cdot y^{1/3}$ **16.** $y^{-1/3} \cdot y^2$

17. $\dfrac{y^{1/3}}{y^{1/9}}$ **18.** $\dfrac{y^{-3/5}}{y^{-2/5}}$ **19.** $\dfrac{(4x)^{3/2}}{(16x)^{-3/4}}$ **20.** $\dfrac{(8x)^{2/3}}{(8x)^{-1/3}}$

21. $\left(\dfrac{x^3 y^{1/2}}{x^{1/2}}\right)^4$ **22.** $\left(\dfrac{x^{1/2} y^{1/3}}{x^{1/4}}\right)^4$ **23.** $\left(\dfrac{125 x^3 y^4}{27 x^{-6} y}\right)^{2/3}$ **24.** $\left(\dfrac{16 x^3 y^2}{x^{1/2}}\right)^{1/2}$

25. $\dfrac{x^{1/2}}{y^2} \cdot \dfrac{y^{-1/3}}{x^3} \div \dfrac{y^3}{x^{-2}}$ **26.** $\dfrac{y^{-2/3}}{x^{1/3}} \cdot \dfrac{x^{-2/3}}{y^{-1/3}} \div \dfrac{x^{-1}}{y^{-1}}$

27. $\dfrac{(x^{1/2} y^{1/4})^{2/3} \cdot (x^{-1/2} y^{2/3})^{-6}}{x^{2/3} y}$ **28.** $\dfrac{(x^{-1/3} y^{1/5})^4 (x^{-2/3} y^{2/5})^{-2}}{x^{-3} y^2}$

■ For Exercises 29–42, see Example 5.

29. $(x^{1/2} + y^{1/2})^2 (x^{-1} y^{-1})$ **30.** $(x^{2/3} + y^{2/3})^2 (x^{-2/3} y^{-2/3})$

31. $(x^{1/2} - y^{1/2})(x^{-1/2} + y^{-1/2})$ **32.** $(x^{-1/3} + y^{-1/3})(x^{1/3} - y^{1/3})$

33. $x(x^{1/2} - x)$ **34.** $x^{1/2}(1 - x)$

35. $x^{3/2}(x^{1/2} + 1)$ **36.** $x^{2/3}(x^{1/3} - 2)$

37. $y^{-2/3}(y^{1/3} - y^{2/3})$ **38.** $y^{-1/2}(y^2 + y^{1/2})$

39. $(x^{1/2} - y^{1/2})(x^{1/2} + y^{1/2})$ **40.** $(x^{-1/2} + y^{1/2})(x^{-1/2} - y^{1/2})$

41. $(x + y)^{1/2}[(x + y)^{1/2} - (x + y)]$ **42.** $(x - y)^{2/3}[(x - y)^{-1/3} + (x - y)]$

■ Factor each expression as indicated. See Example 6.

43. $x^{7/3} + x^2 = x(?)$ **44.** $y^2 - y^{10/3} = y^2(?)$ **45.** $x^{3/2} - x = x(?)$

46. $y + y^{2/3} = y(?)$ **47.** $x^{-3/2} - x^{1/2} = x^{-3/2}(?)$ **48.** $y^{3/4} + y^{1/4} = y^{1/4}(?)$

49. $(x + 1)^{1/2} + (x + 1)^{-1/2} = (x + 1)^{1/2}(?)$ **50.** $(y + 2)^{1/5} + (y + 2)^{-4/5} = (y + 2)^{1/5}(?)$

■ *Exercises 51–58 require the use of a calculator with a y^x key. Approximate each power. Round off each exponent and the resulting power to the nearest hundredth. See Example 7.*

51. $4^{1/3}$　　　　**52.** $10^{1/6}$　　　　**53.** $3^{-2/3}$　　　　**54.** $9^{-5/6}$

55. $\left(\dfrac{1}{4}\right)^{2/3}$　　**56.** $\left(\dfrac{3}{5}\right)^{1/4}$　　**57.** $\left(\dfrac{4}{3}\right)^{-2/3}$　　**58.** $\left(\dfrac{9}{8}\right)^{-2/3}$

B

■ *Write the given expression as a product or quotient in which each variable occurs only once and all exponents are positive. Assume all variable bases are positive and all variables in exponents are nonzero rational numbers.*

59. $(x^{1/n}y^n)^{1/n}(z^2)^{n/2}$　　　　**60.** $(x^{3n})^2(y^4)^{n/2}$　　　　**61.** $(x^{n/2}y^{n+1})^3$

62. $\left(\dfrac{a^n}{b^{2n}}\right)^{1/2}$　　　　**63.** $\dfrac{(a^n)^3}{a^{3-2n}}$　　　　**64.** $(x^{2n}y^{-n})^{-1/n}$

■ *In the preceding exercises, the variables were restricted to positive numbers. In the remaining exercises, consider each variable base to be any real number. Simplify.*

65. $[(-3)^6]^{1/3}$　　**66.** $[(-2)^8]^{1/4}$　　**67.** $[(-2)^6]^{1/2}$　　**68.** $[(-3)^{12}]^{1/4}$

69. $(9x^2)^{1/2}$　　**70.** $[x^2(x^2+1)]^{1/2}$　　**71.** $[x^4(1-x)]^{-1/2}$　　**72.** $[x^{-2}(x+1)]^{-1/2}$

■ *Use a calculator to perform the indicated calculation and show the result in scientific notation. Round off the factor between 1 and 10 to the nearest hundredth.*

73. $(2.00 \times 10^8)^{1/2}$　　　**74.** $(3.14 \times 10^{10})^{1/2}$　　　**75.** $(2.75 \times 10^9)^{-2/3}$

76. $(3.74 \times 10^{24})^{-4/3}$　　**77.** $(-3.24 \times 10^8)^{2/3}$　　**78.** $(-8.37 \times 10^{11})^{-3/5}$

79. $(-2.78 \times 10^{-10})^{-1/3}$　　**80.** $(-7.58 \times 10^{-7})^{-4/3}$

2.3

RADICAL EXPRESSIONS

The radical sign, $\sqrt{}$, is often used in denoting powers with rational exponents.

Definition 2.6　*If n is a positive integer and $a^{1/n}$ is a real number, then*

$$\sqrt[n]{a} = a^{1/n}.$$

The symbol $\sqrt[n]{a}$ is not defined if $a^{1/n}$ is undefined. In the symbol $\sqrt[n]{a}$, a is called the **radicand** and n the **index** of the radical, and the expression is called a radical expression of order *n*. If no index is shown with a radical expression, then the index is assumed to be 2.

E X A M P L E   1   (a) $\sqrt[4]{16} = 16^{1/4}$ (b) $\sqrt[3]{-27} = (-27)^{1/3}$ (c) $\sqrt{5} = 5^{1/2}$

$\qquad\qquad\qquad\qquad = 2$ $\qquad\qquad\qquad\qquad\qquad\quad = -3$ ■

We now have two different symbols for the principal *n*th root of a—$a^{1/n}$: the exponential form defined in Section 2.2; and $\sqrt[n]{a}$, the radical form. Either of these forms can be used in working with principal *n*th roots.

E X A M P L E   2   Write each expression in exponential form.

(a) $\sqrt{y^5}$ $\qquad\qquad\qquad$ (b) $\sqrt[3]{3x^2}$ $\qquad\qquad\qquad$ (c) $\sqrt[5]{x^2 + y^2}$

Solution   (a) $\sqrt{y^5} = (y^5)^{1/2}$ $\qquad$ (b) $\sqrt[3]{3x^2} = (3x^2)^{1/3}$ $\qquad$ (c) $\sqrt[5]{x^2 + y^2}$

$\qquad\qquad\qquad = y^{5/2}$ $\qquad\qquad\qquad\qquad = 3^{1/3}x^{2/3}$ $\qquad\qquad\qquad\qquad = (x^2 + y^2)^{1/5}$

E X A M P L E   3   Write each expression in radical form.

(a) $4^{1/3}$ $\qquad\qquad\qquad$ (b) $x^{1/2}y$ $\qquad\qquad\qquad$ (c) $(x^2 + y^2)^{-1/2}$

Solution   (a) $4^{1/3} = \sqrt[3]{4}$ $\qquad$ (b) $x^{1/2}y = \sqrt{x}\,y$ $\qquad$ (c) $(x^2 + y^2)^{-1/2}$

$\qquad\qquad\qquad\qquad\qquad\qquad\qquad = y\sqrt{x}$ $\qquad\qquad\qquad\qquad\quad = \dfrac{1}{\sqrt{x^2 + y^2}}$ ■

Properties of Radicals

The properties of radicals in Theorem 2.2 follow from Definition 2.6 and the corresponding properties of exponents stated in Theorem 2.1.

Theorem 2.2   *For integers m, n, and c, with n > 0, c > 0, and for real values of a and b for which all the radical expressions denote real numbers,*

I_A. $\sqrt[n]{a^n} = a$ (*n an odd positive integer*), $\qquad$ III. $\sqrt[n]{a} \cdot \sqrt[n]{b} = \sqrt[n]{ab}$,

I_B. $\sqrt[n]{a^n} = |a|$ (*n an even positive integer*),

II. $\sqrt[n]{a^m} = (\sqrt[n]{a})^m$, $\qquad$ IV. $\dfrac{\sqrt[n]{a}}{\sqrt[n]{b}} = \sqrt[n]{\dfrac{a}{b}}$,

$\qquad\qquad\qquad\qquad\qquad\qquad\qquad\qquad\qquad$ V. $\sqrt[cn]{a^{cm}} = \sqrt[n]{a^m}$.

Note in Parts III and IV that if $a, b < 0$ and *n* is even, then the radicals in the left-hand members are not defined, even though the radical in the right-hand member is. Thus, Parts III and IV do not apply in this case. For example,

$$\sqrt{-4}\sqrt{-4} \neq \sqrt{16}$$

because the expressions in the left-hand member are not defined.

The parts of Theorem 2.2 can be used to rewrite radical expressions in what is called "standard form."

A radical expression is said to be in **standard form** if:

1. There are no polynomial factors in the radicand raised to a power higher than or equal to the index of the radical.
2. There are no fractions under the radical sign.
3. No denominator contains a radical.
4. The index of the radical is as small as possible.

EXAMPLE 4   (a) $\sqrt[3]{x^4} = \sqrt[3]{x^3}\sqrt[3]{x} = x\sqrt[3]{x}$ By Parts III and IA of Theorem 2.2.

(b) $\sqrt[3]{\dfrac{1}{x^3}} = \dfrac{\sqrt[3]{1}}{\sqrt[3]{x^3}} = \dfrac{1}{x}$ By Parts IV and IA of Theorem 2.2.

(c) $\dfrac{1}{\sqrt[3]{x}} = \dfrac{(1)\sqrt[3]{x}\sqrt[3]{x}}{\sqrt[3]{x}\sqrt[3]{x}\sqrt[3]{x}} = \dfrac{(\sqrt[3]{x})^2}{x}$ By Part IA of Theorem 2.2.

(d) $\sqrt[6]{a^3} = \sqrt[3\cdot2]{a^3} = \sqrt{a}$ By Part V of Theorem 2.2.

EXAMPLE 5   Write each expression in standard form.

(a) $\sqrt{8x^5}$ (b) $\sqrt{3xy}\sqrt{6xy^3}$

(c) $\sqrt[12]{81}$ (d) $\sqrt[4]{x^2y^2}$

Solution   (a) $\sqrt{8x^5} = \sqrt{4x^4}\sqrt{2x}$ (b) $\sqrt{3xy}\sqrt{6xy^3} = \sqrt{18x^2y^4}$
$\qquad\qquad = 2x^2\sqrt{2x}$ $= \sqrt{9x^2y^4}\sqrt{2}$
$\qquad\qquad\qquad\qquad\qquad\qquad\qquad\qquad = 3xy^2\sqrt{2}$

(c) $\sqrt[12]{81} = \sqrt[3\cdot4]{3^{1\cdot4}}$ (d) $\sqrt[4]{x^2y^2} = \sqrt[2\cdot2]{x^{1\cdot2}y^{1\cdot2}}$
$\qquad\quad = \sqrt[3]{3}$ $= \sqrt{xy}$ ■

The process used to simplify the expressions in parts (**b**) and (**c**) of Example 5 is called "rationalizing the denominator." The result is a fraction with the denominator free of radicals.

E X A M P L E 6   Rationalize the denominator of each expression.

(**a**) $\dfrac{1}{\sqrt{3}}$ (**b**) $\dfrac{x}{\sqrt{y}}$ (**c**) $\dfrac{1}{\sqrt[3]{x}}$

Solution   (**a**) $\dfrac{1}{\sqrt{3}} = \dfrac{1\sqrt{3}}{\sqrt{3}\sqrt{3}}$ (**b**) $\dfrac{x}{\sqrt{y}} = \dfrac{x\sqrt{y}}{\sqrt{y}\sqrt{y}}$ (**c**) $\dfrac{1}{\sqrt[3]{x}} = \dfrac{1\sqrt[3]{x^2}}{\sqrt[3]{x}\sqrt[3]{x^2}}$

$\qquad\qquad = \dfrac{\sqrt{3}}{3}$ $= \dfrac{x\sqrt{y}}{y}$ $= \dfrac{\sqrt[3]{x^2}}{\sqrt[3]{x^3}}$

$\qquad\qquad\qquad\qquad\qquad\qquad\qquad\qquad\qquad\qquad\qquad\qquad = \dfrac{\sqrt[3]{x^2}}{x}$ ■

Note that it is not *always* preferable, in working with fractions, to have their denominators rationalized. Sometimes, in fact, it is desirable to rationalize the numerator. Thus, for example,

$$\frac{\sqrt[3]{2}}{\sqrt{3}} = \frac{\sqrt[3]{2}\,\sqrt[3]{4}}{\sqrt{3}\,\sqrt[3]{4}}$$

$$= \frac{2}{\sqrt{3}\,\sqrt[3]{4}},$$

where the numerator of the fraction has been rationalized.

E X A M P L E 7   Rationalize the numerator of each expression.

(**a**) $\dfrac{\sqrt{x}}{2}$ (**b**) $\dfrac{\sqrt[3]{y}}{y}$

Solution   (**a**) $\dfrac{\sqrt{x}}{2} = \dfrac{\sqrt{x}\sqrt{x}}{2\sqrt{x}}$ (**b**) $\dfrac{\sqrt[3]{y}}{y} = \dfrac{\sqrt[3]{y}\,\sqrt[3]{y^2}}{y\sqrt[3]{y^2}}$

$\qquad\qquad = \dfrac{x}{2\sqrt{x}}$ $= \dfrac{1}{\sqrt[3]{y^2}}$ ■

1-60

EXERCISE 2.3

A

■   (In Exercises 1–64, assume all variables represent positive real numbers and all radicands are positive.)

Write each expression in exponential form. See Examples 1 and 2.

1. $\sqrt[3]{27}$ **2.** $\sqrt{25}$ **3.** $\sqrt[3]{8^2}$ **4.** $\sqrt[4]{(16)^3}$

5. $\sqrt{(-2)^4}$ **6.** $\sqrt[4]{(-1)^2}$ **7.** $\sqrt[3]{8x^4}$ **8.** $8\sqrt[3]{x^4}$

9. $\sqrt[4]{x^2y^3}$ **10.** $\sqrt{4y^4}$ **11.** $\sqrt{(x+y)^3}$ **12.** $\sqrt[5]{x^6y^4}$

13. $\sqrt[3]{x^2+1}$ **14.** $\sqrt{1-x^2}$ **15.** $\sqrt[3]{x^3+y^3}$ **16.** $\sqrt[4]{x^4+16}$

17. $\sqrt{(x^2+1)^3}$ **18.** $\sqrt[4]{(4-x^2)^3}$ **19.** $\sqrt[4]{-(x^2-1)^5}$ **20.** $\sqrt[5]{-(25-x^4)^3}$

■   Write each expression in radical form. See Example 3.

21. $5^{1/2}$ **22.** $2^{1/4}$ **23.** $x^{3/5}$ **24.** $x^{5/4}$

25. $2x^{1/3}$ **26.** $9y^{1/2}$ **27.** $(9y)^{1/2}$ **28.** $(2x)^{1/3}$

29. $xy^{1/3}$ **30.** $x^{1/3}y$ $y\sqrt[3]{x}$ **31.** $(xy)^{1/3}$ **32.** $(xy)^{4/3}$

33. $(x-y)^{1/4}$ **34.** $(x+y)^{1/3}$ **35.** $(x^2+y^2)^{1/2}$ **36.** $(1+4x^2)^{1/2}$

37. $(x^3+y^3)^{4/3}$ **38.** $(16x^2+25)^{5/2}$ **39.** $(x^2-y^2)^{-3/4}$ **40.** $(x^3+y^3)^{-2/3}$

■   Write each expression in standard form. See Examples 4 and 5.

41. $\sqrt{x^7}$ **42.** $\sqrt{9x^5}$ **43.** $\sqrt[3]{-27x^4}$ **44.** $\sqrt[3]{2x^3y^4}$

45. $\sqrt{x}\sqrt{xy}$ **46.** $\sqrt[3]{xy^2}\sqrt[3]{x^2y}$ **47.** $\sqrt[3]{4x^2}\sqrt[3]{16x^{10}}$ **48.** $\sqrt{5xy^3}\sqrt{20x}$

49. $\dfrac{\sqrt{xy}\sqrt{x}}{\sqrt{y}}$ **50.** $\dfrac{\sqrt[3]{xy}\sqrt[3]{x^4y}}{\sqrt[3]{x^2y^5}}$ **51.** $\dfrac{\sqrt[3]{2x^4}\sqrt[3]{16y}}{\sqrt[3]{4x}}$ **52.** $\dfrac{\sqrt{3y}\sqrt{6x}}{\sqrt{2x^3y^3}}$

53. $\sqrt[6]{81}$ **54.** $\sqrt[10]{32}$ **55.** $\sqrt[4]{16x^2}$ **56.** $\sqrt[9]{8a^3}$

57. $\sqrt[6]{8x^3}$ **58.** $\sqrt[6]{125z^3}$ **59.** $\sqrt{(x-1)^4}$ **60.** $\sqrt[12]{(x-2y)^4}$

■   Rationalize the denominator of each expression. See Example 6.

61. $\dfrac{1}{\sqrt{2}}$ **62.** $\dfrac{6}{\sqrt{3}}$ **63.** $\dfrac{4}{\sqrt{2x}}$ **64.** $\dfrac{\sqrt{x}}{\sqrt{2y}}$ **65.** $\dfrac{1}{\sqrt[3]{4}}$

66. $\dfrac{3}{\sqrt[3]{9}}$ **67.** $\dfrac{1}{\sqrt[3]{y^2}}$ **68.** $\dfrac{1}{\sqrt[3]{xy^2}}$ **69.** $\dfrac{1}{\sqrt{x^2+y^2}}$ **70.** $\dfrac{1}{\sqrt{2x^2+1}}$

■   Rationalize the numerator of each expression. See Example 7.

71. $\dfrac{\sqrt{2}}{2}$ **72.** $\dfrac{\sqrt[3]{2}}{2}$ **73.** $\dfrac{\sqrt[3]{x}}{y}$ **74.** $\dfrac{\sqrt{xy}}{x}$

75. $\dfrac{\sqrt[3]{xy}}{x^2}$ **76.** $\dfrac{\sqrt[3]{xy^2}}{y}$ **77.** $\sqrt{x^2+y^2}$ **78.** $-\sqrt{x^2-y^2}$

■   *A conical pendulum is a mass m suspended from a fixed point O by a string of fixed length l, rotating along a circle of radius r with uniform angular speed ω radians/sec. The frequency of revolution is given by*

$$f = \frac{30}{\pi} \sqrt{\frac{g}{l^2 - r^2}}$$

where g is the acceleration due to gravity, g ≈ 32.174 ft/sec².

79.   What is the frequency of a conical pendulum with $l = 10$ ft and $r = 1$ ft to the nearest hundredth of a cycle per second?

80.   What is the frequency of a conical pendulum with $l = 8$ ft and $r = \frac{1}{2}$ ft to the nearest hundredth of a cycle per second?

81.   What is the frequency of a conical pendulum with $l = 6$ ft and $r = 5$ ft to the nearest hundredth of a cycle per second?

82.   What is the frequency of a conical pendulum with $l = 6$ ft and $r = 5\frac{1}{2}$ ft to the nearest hundredth of a cycle per second?

■   *The escape velocity of a planet is the minimum speed at which a body must be shot from the planet to avoid recapture by the gravitational field of the planet. The escape velocity is given by*

$$V_e = \sqrt{\frac{2\mu_p}{R}}$$

where μ_p is the gravitational parameter of the planet and R is the distance of the launch site from the center of the planet. The gravitational parameter of the earth is $\mu_{earth} = 1.41 \times 10^{16} ft^3/sec^3$.

83.   The radius of the earth at the equator is approximately 2.09×10^7 ft. What is the escape velocity (to the nearest tenth of a foot per sec) of a probe launched from a point on the equator?

84.   If the probe in the preceding problem is launched from a parking orbit 150 nautical miles above the equator, what is the escape velocity? (1 nautical mile = 6076.10 feet.)

■   *The time of flight for travel from one planet to another using the Hohman transfer between the orbits of the planets (which are assumed to be circular and in the same plane) is given by*

$$T = \pi \sqrt{\frac{(r_1 + r_2)^3}{8\mu_{sun}}}$$

where r_1 is the radius of the departure planet's orbit about the sun, r_2 is the radius of the arrival planet's orbit about the sun, and $\mu_{sun} = 1.327 \times 10^{11} km^3/sec^2$ is the gravitational parameter of the sun.

85.   The radius of Earth's orbit around the sun is 1.495×10^8 km while that of Mars is 2.278×10^8 km. What is the time of flight (to the nearest day) using a Hohman transfer from Earth to Mars?

86. The radius of Venus' orbit around the sun is 1.081×10^8 km. What is the time of flight (to the nearest day) using a Hohman transfer from Earth to Venus?

87. The radius of Neptune's orbit around the sun is 4.494×10^9 km. What is the time of flight (to the nearest year) using a Hohman transfer from Earth to Neptune?

88. What is the time of flight (to the nearest year) using a Hohman transfer from Venus to Neptune?

B

■ *Write in standard form.*

89. $\sqrt[4]{\dfrac{(x^2 - 1)^2}{(x + 1)^3}}$ **90.** $\sqrt{\dfrac{(x^4 - 1)^6}{x^2}}$ **91.** $\dfrac{\sqrt[3]{x^2(y + 1)^2}}{\sqrt{x^4}}$ **92.** $\dfrac{\sqrt[5]{x^2(x + 1)^2}}{\sqrt[3]{(x - 1)^2}}$

93. Is $\sqrt{a^2 + b^2} = a + b$ for all a and b with $a \geq 0$, $b \geq 0$? Why or why not?

94. Is there a natural number $n \geq 2$ for which $\sqrt[n]{a^n + b^n} = a + b$ for all $a \geq 0$ and $b \geq 0$? If so, what is it? If not, why not?

2.4

OPERATIONS ON RADICAL EXPRESSIONS

We have defined radical expressions so that they represent real numbers, and therefore the properties of the real numbers can be applied. For example, the distributive law permits us to express certain sums as products.

EXAMPLE 1 (a) $2\sqrt{5} + 4\sqrt{5}$ (b) $5\sqrt{2} - 9\sqrt{2}$ (c) $5\sqrt{2} + \sqrt{75}$

$\quad\quad\quad\quad = (2 + 4)\sqrt{5} \quad\quad\quad\quad = (5 - 9)\sqrt{2} \quad\quad\quad\quad = 5\sqrt{2} + 5\sqrt{3}$

$\quad\quad\quad\quad = 6\sqrt{5} \quad\quad\quad\quad\quad\quad = -4\sqrt{2} \quad\quad\quad\quad\quad\; = 5(\sqrt{2} + \sqrt{3})$ ■

The distributive law in the form $c(a + b) = ca + cb$ permits us to write certain products as sums or differences.

EXAMPLE 2 (a) $\sqrt{2x}(\sqrt{x} + \sqrt{2x})$ (b) $(\sqrt{x} + 2)(\sqrt{x} - 1)$

$\quad\quad\quad\quad = \sqrt{2x \cdot x} + \sqrt{2x \cdot 2x} \quad\quad\quad = \sqrt{x}(\sqrt{x} - 1) + 2(\sqrt{x} - 1)$

$\quad\quad\quad\quad = x\sqrt{2} + 2x \quad\quad\quad\quad\quad\quad\quad = x - \sqrt{x} + 2\sqrt{x} - 2$

$\quad\quad\quad\quad\quad\quad\quad\quad\quad\quad\quad\quad\quad\quad\quad\quad\quad = x + \sqrt{x} - 2$ ■

Binomial Denominators

The distributive law also provides us with a means of rationalizing denominators of fractions in which radicals occur in one or both of two terms. To accomplish this, we first observe that

$$(b - \sqrt{c})(b + \sqrt{c}) = b^2 - (\sqrt{c})^2 = b^2 - c,$$

where the expression in the right-hand member contains no radical term. Each of the two factors of a product exhibiting this property is said to be the **conjugate** of the other. Now consider a fraction of the form

$$\frac{a}{b + \sqrt{c}},$$

where c is positive and $b \neq -\sqrt{c}$. If we multiply the numerator and denominator of this fraction by the conjugate of the denominator, then the denominator of the resulting fraction will contain no term involving $\sqrt{c}$ and hence will be free of radicals. That is,

$$\frac{a}{b + \sqrt{c}} = \frac{a(b - \sqrt{c})}{(b + \sqrt{c})(b - \sqrt{c})} = \frac{ab - a\sqrt{c}}{b^2 - c},$$

where the denominator has been rationalized. This process is equally applicable to radical fractions of the form

$$\frac{a}{\sqrt{b} + \sqrt{c}},$$

since

$$\frac{a}{\sqrt{b} + \sqrt{c}} = \frac{a(\sqrt{b} - \sqrt{c})}{(\sqrt{b} + \sqrt{c})(\sqrt{b} - \sqrt{c})} = \frac{a\sqrt{b} - a\sqrt{c}}{b - c}.$$

EXAMPLE 3 (a) $\dfrac{1}{\sqrt{2} + 1} = \dfrac{1(\sqrt{2} - 1)}{(\sqrt{2} + 1)(\sqrt{2} - 1)}$

$$= \frac{\sqrt{2} - 1}{2 - 1}$$

$$= \sqrt{2} - 1$$

(b) $\dfrac{\sqrt{x}}{\sqrt{x} + x} = \dfrac{\sqrt{x}(\sqrt{x} - x)}{(\sqrt{x} + x)(\sqrt{x} - x)}$

$$= \frac{x - x\sqrt{x}}{x - x^2} \quad ■$$

Binomial Numerators

As we noted in Section 2.3, it is sometimes preferable, in working with fractions, to have their numerators rationalized. Thus, for example,

$$\frac{\sqrt{b} + \sqrt{c}}{a} = \frac{(\sqrt{b} + \sqrt{c})(\sqrt{b} - \sqrt{c})}{a(\sqrt{b} - \sqrt{c})} = \frac{b - c}{a(\sqrt{b} - \sqrt{c})}.$$

EXAMPLE 4   (a) $\dfrac{1 - \sqrt{2}}{2}$ (b) $\dfrac{1 - \sqrt{x + 1}}{\sqrt{x + 1}}$

$$= \frac{(1 - \sqrt{2})(1 + \sqrt{2})}{2(1 + \sqrt{2})} \qquad\qquad = \frac{(1 - \sqrt{x + 1})(1 + \sqrt{x + 1})}{\sqrt{x + 1}(1 + \sqrt{x + 1})}$$

$$= \frac{1 - 2}{2 + 2\sqrt{2}} \qquad\qquad\qquad = \frac{1 - (x + 1)}{\sqrt{x + 1} + x + 1}$$

$$= \frac{-1}{2 + 2\sqrt{2}} \qquad\qquad\qquad = \frac{-x}{\sqrt{x + 1} + x + 1} \qquad ■$$

1–49

EXERCISE 2.4

A

🖐 *Write each sum as a product. See Example 1.*

1. $2\sqrt{5} + 6\sqrt{5}$ **2.** $4\sqrt{6} - 2\sqrt{6}$ **3.** $4\sqrt{3} - 6\sqrt{3} + 3\sqrt{3}$

4. $4\sqrt{2} + 2\sqrt{2} + \sqrt{2}$ **5.** $4\sqrt{3} - 3\sqrt{12}$ **6.** $\sqrt{18} - 5\sqrt{2}$

7. $2\sqrt{98} + \sqrt{8} - \sqrt{2}$ **8.** $4\sqrt{75} - 3\sqrt{27} + \sqrt{12}$ **9.** $3\sqrt[3]{2} - 3\sqrt[3]{16}$

10. $2\sqrt[3]{54} + 4\sqrt[3]{250}$ **11.** $\sqrt[4]{32} - \sqrt[4]{2}$ **12.** $6\sqrt[4]{3} - 6\sqrt[4]{48}$

■ *Multiply factors and write all radicals in the result in simplest form. Assume all variables represent positive real numbers and all radicands are positive. See Example 2.*

13. $3(2 - \sqrt{6})$ **14.** $\sqrt{5}(3 - 3\sqrt{5})$ **15.** $(\sqrt{5} - 1)(\sqrt{5} + 1)$

16. $(\sqrt{2} - \sqrt{3})(2\sqrt{2} + \sqrt{3})$ **17.** $(\sqrt{7} + 2)(1 - \sqrt{7})$ **18.** $(3 - \sqrt{2})(3 + 2\sqrt{2})$

19. $\sqrt{x}(\sqrt{2x} - \sqrt{8x})$ **20.** $\sqrt{x}(\sqrt{3x} + \sqrt{6x})$ **21.** $(\sqrt{x} - 1)(2\sqrt{x} - 5)$

22. $(\sqrt{2} + \sqrt{x})(1 + 2\sqrt{x})$ **23.** $(\sqrt{5} - \sqrt{x})(\sqrt{5} + \sqrt{x})$ **24.** $(2\sqrt{3} - \sqrt{x})(\sqrt{x} + \sqrt{3})$

■ *Rationalize the denominators. See Example 3.*

25. $\dfrac{-1}{1 + \sqrt{7}}$ **26.** $\dfrac{3}{\sqrt{3} + 3}$ **27.** $\dfrac{x}{4 + \sqrt{x}}$ **28.** $\dfrac{2x}{\sqrt{x} + 1}$

29. $\dfrac{\sqrt{2x}}{\sqrt{x+2}}$

30. $\dfrac{\sqrt{2x}}{\sqrt{2x+3}}$

31. $\dfrac{x+1}{\sqrt{x+1}}$

32. $\dfrac{x-1}{\sqrt{x-1}}$

33. $\dfrac{2x-1}{\sqrt{x-1}}$

34. $\dfrac{2x+1}{\sqrt{x+1}}$

35. $\dfrac{x-2}{2-\frac{1}{2}\sqrt{x}}$

36. $\dfrac{x+3}{3-\frac{1}{4}\sqrt{x}}$

■ *Rationalize the numerators. See Example 4.*

37. $\dfrac{\sqrt{5}-1}{2}$

38. $\dfrac{\sqrt{5}+2}{5}$

39. $\dfrac{1+2\sqrt{x}}{x^2}$

40. $\dfrac{3+\sqrt{xy}}{xy}$

41. $\dfrac{2-\sqrt{2x+1}}{\sqrt{2x+1}}$

42. $\dfrac{\sqrt{x-4}+1}{\sqrt{x}}$

43. $\dfrac{\sqrt{x+1}}{x+1}$

44. $\dfrac{\sqrt{x-1}}{x-2}$

45. $\dfrac{\sqrt{2x}-1}{3x-1}$

46. $\dfrac{\sqrt{2x}+1}{2x+1}$

47. $\dfrac{1-\frac{1}{3}\sqrt{x}}{3x+1}$

48. $\dfrac{3-\frac{1}{2}\sqrt{x}}{3x-1}$

B

■ *Rationalize the denominators.*

(49) $\dfrac{\sqrt{x}-\sqrt{y}}{\sqrt{x}+\sqrt{y}}$

50. $\dfrac{4\sqrt{x}+\sqrt{3x+1}}{\sqrt{x}-\sqrt{3x+1}}$

51. $\dfrac{\sqrt{x+1}-\sqrt{x}}{\sqrt{x+1}+\sqrt{x}}$

52. $\dfrac{\sqrt{2x+1}}{\sqrt{x}-\sqrt{2x+1}}$

53. $\dfrac{x}{\sqrt{x}+\sqrt{y}-1}$

54. $\dfrac{y}{\sqrt{x}-\sqrt{y}-1}$

■ *Rationalize the numerators.*

55. $\dfrac{\sqrt{x-1}-1}{\sqrt{x-1}}$

56. $\dfrac{\sqrt{x+1}}{\sqrt{x+1}}$

57. $\dfrac{\sqrt{x}+\sqrt{y}}{\sqrt{x}-\sqrt{y}}$

58. $\dfrac{\sqrt{x+1}-\sqrt{x}}{\sqrt{x+1}+\sqrt{x}}$

59. $\dfrac{\sqrt{x}+\sqrt{y}-1}{x}$

60. $\dfrac{\sqrt{x}-\sqrt{y}+1}{y}$

2.5

COMPLEX NUMBERS

In Section 2.3 we defined a real number $\sqrt{b}$, where b is a real number with $b \geq 0$, as the positive real number whose square is b. Note that there is no real number whose square is negative. For this reason, the expression $\sqrt{-b}$, for $b > 0$, is undefined in the set of real numbers. In this section, we wish to consider a set of numbers that satisfies the axioms and properties of the real numbers (see Preliminary Concepts), that contains the set of real numbers, and that contains members whose squares are negative real numbers. We shall call this new set of numbers the set C of complex numbers.

Definition of $\sqrt{-b}$

We begin by using the symbol i to represent a number whose square is equal to -1. Thus,

$$i^2 = -1,$$

or, using radical notation,

$$i = \sqrt{-1}.$$

We use this symbol to make the following definition.

Definition 2.7

For any positive real number b

$$\sqrt{-b} = (\sqrt{b})i.$$

Note that, from the axioms of real numbers (which we assume are true) and the definition of i, we have

$$(\sqrt{-b})^2 = [(\sqrt{b})i]^2 = (\sqrt{b})^2 i^2$$
$$= (b)(-1) = -b.$$

Similarly, we have

$$(-\sqrt{-b})^2 = -b.$$

EXAMPLE 1   (a) $\sqrt{-4} = (\sqrt{4})i = 2i$ (b) $-\sqrt{-3} = -(\sqrt{3})i = -i\sqrt{3}$ ■

Definition of the Complex Numbers

A number of the form $\sqrt{-b}$ or $-\sqrt{-b}$, where b is a positive real number, is called a **pure imaginary number**.

An expression formed by adding a real number and a pure imaginary number is called a **complex number**. Thus, we have the following definition.

Definition 2.8

An expression of the form $a + bi$, *where* a *and* b *are real numbers, is a* ***complex number***. *The set* $C = \{a + bi \,|\, a, b$ *are real numbers*$\}$ *is called the* ***set of complex numbers***.

Any expression of the form $a + \sqrt{-b}$, $b < 0$, can be expressed as the complex number $a + bi$.

EXAMPLE 2   (a) $3\sqrt{-18} = 3i\sqrt{18}$ (b) $2 - 3\sqrt{-16} = 2 - 3i\sqrt{16}$
$$= 3i(3)\sqrt{2} \qquad\qquad\qquad = 2 - 3i(4)$$
$$= 9i\sqrt{2} \qquad\qquad\qquad\qquad = 2 - 12i \quad ■$$

If $b = 0$, then $a + bi = a$, and it is evident that the set R of real numbers is contained in the set C of complex numbers. If $b \neq 0$, then $a + bi$ is called an **imaginary number,** where a is the **real part** of the number and b is the **imaginary part.** For example, the numbers -7, $3 + 2i$, and $4i$ are all complex numbers. However, -7 is also a real number, and $3 + 2i$ and $4i$ are also imaginary numbers. Furthermore, $4i$ is a pure imaginary number. Figure 2.1 shows the relationships among these sets of numbers.

FIGURE 2.1

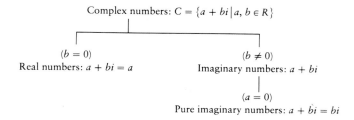

We say two complex numbers are equal if their real parts are equal and their imaginary parts are equal.

Definition 2.9   *The complex numbers $a + bi$ and $c + di$ are **equal** if and only if*

$$a = c \quad and \quad b = d.$$

Sums and Differences

To add or subtract complex numbers, we simply add or subtract their real parts and their imaginary parts. In general:

$$(a + bi) + (c + di) = (a + c) + (b + d)i$$

and

$$(a + bi) - (c + di) = (a - c) + (b - d)i.$$

EXAMPLE 3   (a) $(2 + 3i) + (5 - 4i)$

$\qquad\qquad = (2 + 5) + (3 - 4)i$

$\qquad\qquad = 7 - i$

(b) $(2 + 3i) - (5 - 4i)$

$\qquad\qquad = (2 - 5) + [3 - (-4)]i$

$\qquad\qquad = -3 + 7i$ ■

Products

To multiply complex numbers, we treat them as though they were binomials and replace i^2 with -1.

EXAMPLE 4 (a) $(2 - i)(1 + 3i) = 2 + 6i - i - 3i^2$

$\qquad = 2 + 6i - i - 3(-1)$

$\qquad = 2 + 6i - i + 3$

$\qquad = 5 + 5i$

(b) $(3 - i)^2 = (3 - i)(3 - i)$

$\qquad = 9 - 3i - 3i + i^2$

$\qquad = 9 - 6i + (-1)$

$\qquad = 8 - 6i$ ■

More formally, we express $(a + bi)(c + di)$ as follows:

$$(a + bi)(c + di) = (ac - bd) + (ad + bc)i.$$

Ordinarily, we compute such products by multiplying, as we did in Example 4, rather than from this formula.

Conjugates   Recall from Section 2.4 that, for $b > 0$, the *conjugate* of $a + \sqrt{b}$ is $a - \sqrt{b}$. Similarly, the conjugate of $a + \sqrt{-b}$ is $a - \sqrt{-b}$. The conjugate of the complex number z is denoted by $\bar{z}$. The conjugate of z is used when computing quotients.

EXAMPLE 5 (a) If $z = 2 + 3i$, then $\bar{z} = 2 - 3i$ (b) If $z = -3 - i$, then $\bar{z} = -3 + i$.

(c) If $z = 2i$, then $\bar{z} = -2i$.   (d) If $z = -4 + i$, then $\bar{z} = -4 - i$.

■

The conjugate operation has a number of interesting and useful properties. For example, if z is a real number, then $z = a + 0i$, where a is a real number. Thus,

$$\bar{z} = a - 0i = a + 0i = z.$$

We have shown that the following statement is true.

If z is a real number then $\bar{z} = z$.

Quotients   The quotient of two complex numbers can be found by using the following property, which is analogous to the *fundamental principle of fractions* in the set of real numbers. The quotient

$$\frac{a + bi}{c + di}$$

of two complex numbers can be simplified by multiplying the numerator and the denominator by $c - di$, the conjugate of the denominator. That is,

$$\frac{a + bi}{c + di} = \frac{(a + bi)(c - di)}{(c + di)(c - di)}.$$

EXAMPLE 6   (a) $\dfrac{4 + i}{2 + 3i} = \dfrac{(4 + i)(2 - 3i)}{(2 + 3i)(2 - 3i)}$

$$= \frac{8 - 10i - 3i^2}{4 - 9i^2}$$

$$= \frac{8 - 10i + 3}{4 + 9}$$

$$= \frac{11}{13} - \frac{10}{13}i$$

(b) $\dfrac{3}{1 + \sqrt{-2}} = \dfrac{3}{1 + i\sqrt{2}}$

$$= \frac{3(1 - i\sqrt{2})}{(1 + i\sqrt{2})(1 - i\sqrt{2})}$$

$$= \frac{3(1 - i\sqrt{2})}{1 - i^2(2)}$$

$$= \frac{3(1 - i\sqrt{2})}{3} = 1 - i\sqrt{2} \qquad ■$$

More formally, we express the quotient as follows:

$$\frac{a + bi}{c + di} = \frac{ac + bd}{c^2 + d^2} + \frac{bc - ad}{c^2 + d^2}i.$$

In practice, we usually compute a quotient by multiplying the numerator and denominator by the conjugate of the denominator as in Example 6.

Radical Notation

The symbol $\sqrt{-b}$ $(b > 0)$ should be used with care, since certain relationships involving the square root symbol that are valid for real numbers are not valid when the symbol does not represent a real number. For instance,

$$\sqrt{-2}\sqrt{-3} = (i\sqrt{2})(i\sqrt{3}) = i^2\sqrt{6} = -\sqrt{6};$$

however,

$$\sqrt{-2}\sqrt{-3} \neq \sqrt{(-2)(-3)} = \sqrt{6}.$$

To avoid difficulty with this point:

Rewrite all expressions of the form $\sqrt{-b}$ $(b > 0)$ in the form $i\sqrt{b}$ before performing any computations.

EXAMPLE 7   (a) $\sqrt{-2}\,(3 - \sqrt{-5}\,)$

$$= i\sqrt{2}\,(3 - i\sqrt{5}\,)$$
$$= 3i\sqrt{2} - i^2\sqrt{10}$$
$$= 3i\sqrt{2} - (-1)\sqrt{10}$$
$$= \sqrt{10} + 3i\sqrt{2}$$

(b) $(2 + \sqrt{-3}\,)(2 - \sqrt{-3}\,)$

$$= (2 + i\sqrt{3}\,)(2 - i\sqrt{3}\,)$$
$$= 4 - 3i^2$$
$$= 4 - 3(-1)$$
$$= 7$$

EXERCISE 2.5

A

■ *Write each expression in the form* $a + bi$ *or* $a + ib$. *For Exercises 1–14, see Examples 1 and 2.*

1. $\sqrt{-16}$

2. $\sqrt{-25}$

3. $\sqrt{-27}$

4. $\sqrt{-72}$

5. $3\sqrt{-32}$

6. $4\sqrt{-20}$

7. $3\sqrt{-28}$

8. $2\sqrt{-18}$

9. $3 - 2\sqrt{-1}$

10. $5 + 2\sqrt{-1}$

11. $5\sqrt{-50} + 1$

12. $-5\sqrt{-12} + 6$

13. $\sqrt{9} + \sqrt{-9}$

14. $\sqrt{24} - \sqrt{-24}$

■ *For Exercises 15–20, see Example 3.*

15. $(3 + 2i) + (2 - i)$

16. $(1 - i) + (5 - 6i)$

17. $(3 - 4i) - (2 - 3i)$

18. $(5 - i) - (4 + 2i)$

19. $8 - (5 + 2i)$

20. $(9 - 6i) - 3$

■ *For Exercises 21–54, see Examples 4–6.*

21. $(2 - 3i)(4 - 2i)$

22. $(1 - 3i)(3 - 5i)$

23. $(3 + 4i)(5 + 2i)$

24. $(3 + i)(2 - 3i)$

25. $(2 - 3i)(4 - 3i)$

26. $(7 + 3i)(2 - 3i)$

27. $(2 + 5i)^2$

28. $(4 - 3i)^2$

29. $(4 - i)(4 + i)$

30. $(5 - 3i)(5 + 3i)$

31. $i(2 - i)(2 + 3i)$

32. $-i(1 + i)(3 - i)$

33. $3i(2 + i)(1 + 2i)$

34. $-2i(2 - 3i)(-1 + i)$

35. $(1 + i)(1 + 2i)(1 - 2i)$

36. $(1 + 2i)(3 - 2i)(1 + i)$

37. $\dfrac{1}{8i}$

38. $\dfrac{-2}{3i}$

39. $\dfrac{3 + i}{4i}$

40. $\dfrac{3 + 2i}{4i}$

41. $\dfrac{3}{1 + i}$

42. $\dfrac{-5}{2 - i}$

43. $\dfrac{1 - 3i}{2 - i}$

44. $\dfrac{1 - i}{3 + i}$

45. $\dfrac{3 + 2i}{2 + 3i}$

46. $\dfrac{2 + 5i}{6 - i}$

47. $\dfrac{5 + 3i}{3 - 2i}$

48. $\dfrac{2-7i}{-4+3i}$

49. $\dfrac{1+i}{1-i} \cdot \dfrac{1}{2+i}$

50. $\dfrac{1+2i}{3+2i} \cdot \dfrac{1}{1-2i}$

51. $\dfrac{3+2i}{2-i} \cdot \dfrac{3-i}{1+i}$

52. $\dfrac{2+3i}{2-i} \cdot \dfrac{2+i}{1-3i}$

53. $\dfrac{2-3i}{5+i} \cdot \dfrac{1-5i}{1+2i}$

54. $\dfrac{3+i}{3-2i} \cdot \dfrac{1+4i}{2+3i}$

■ *Write the given product or quotient in the form $a + bi$ or $a + ib$. See Example 7.*

55. $\sqrt{-4}(1 - \sqrt{-4})$

56. $\sqrt{-9}(3 + \sqrt{-16})$

57. $(2 + \sqrt{-9})(3 - \sqrt{-9})$

58. $(4 - \sqrt{-2})(3 + \sqrt{-2})$

59. $\dfrac{3}{\sqrt{-4}}$

60. $\dfrac{-1}{\sqrt{-25}}$

61. $\dfrac{2 - \sqrt{-1}}{2 + \sqrt{-1}}$

62. $\dfrac{1 + \sqrt{-2}}{3 - \sqrt{-2}}$

63. $\dfrac{\sqrt{-1}\sqrt{-6}}{\sqrt{6}}$

64. $\dfrac{\sqrt{-4}\sqrt{-2}}{-\sqrt{8}}$

65. $\dfrac{\sqrt{-12}\sqrt{-5}}{\sqrt{-3}}$

66. $\dfrac{-\sqrt{-35}\sqrt{-12}}{\sqrt{-20}}$

B

67. Simplify.

    (a) i^6 (b) i^{12} (c) i^{15} (d) i^{102}

68. Express with a positive exponent and simplify.

    (a) i^{-1} (b) i^{-2} (c) i^{-3} (d) i^{-6}

69. Evaluate $x^2 + 2x + 3$ for $x = 1 + i$.

70. Evaluate $2y^2 - y + 2$ for $y = 2 - i$.

■ *Prove each of the following statements.*

71. If z_1, z_2 are complex numbers, then $\overline{z_1 \pm z_2} = \overline{z_1} \pm \overline{z_2}$.

72. If z_1, z_2 are complex numbers, then $\overline{z_1 \cdot z_2} = \overline{z_1} \cdot \overline{z_2}$.

73. If z is a complex number, then $\overline{z^n} = \overline{z}^n$.   **74.** If z is a complex number, then $\overline{\overline{z}} = z$.

75. If $z = a + bi$, then $z + \overline{z} = 2a$.   **76.** If $z = a + bi$, then $z - \overline{z} = 2bi$.

CHAPTER REVIEW

exponents
Radicals

Key Words, Phrases, and Symbols

■ *Define or explain each of the following words, phrases, and symbols.*

1. a^0, $a \neq 0$ **3.** complex fraction

2. a^{-n}, $a \neq 0$ **4.** scientific notation

5. nth root
6. $a^{1/n}$
7. $a^{m/n}$
8. $\sqrt[n]{a}$
9. standard form of radical expressions

10. rationalizing the denominator (numerator)
11. i (imaginary number)
12. pure imaginary number
13. complex number
14. conjugate of a complex number

Review Exercises

A

[2.1]   ■   *Write the expression as a product or quotient in which each variable occurs at most once in the expression and all exponents are positive.*

1. $(x^3y^2)^4$ **2.** $\dfrac{x^{-1}y^2}{x^2y^{-1}}$ **3.** $\dfrac{(xy)^{-2}}{x^0y^{-3}}$ **4.** $\dfrac{(x^{-1}y^0)^{-1}}{x^{-2}}$

■   *Represent the expression as a single fraction involving positive exponents only.*

5. $x^{-1} + x^{-3}$ **6.** $\dfrac{x^{-1} + y^{-2}}{(xy)^{-2}}$

■   *Write the given number in scientific notation.*

7. 35,100   **8.** 0.000181

■   *Write the given number in decimal form.*

9. 3.14×10^8 **10.** 6.75×10^{-6}

[2.2]   ■   *Write the expression as a product or quotient of powers in which each variable occurs only once and all exponents are positive. Assume that all variable bases are positive.*

11. $x^{1/2} \cdot x^{2/3}$ **12.** $\left(\dfrac{x^3}{y^6}\right)^{-1/6}$ **13.** $(x^8y^{16})^{1/2}$ **14.** $\left(\dfrac{y^8}{x^4y^{12}}\right)^{1/4}$

■   *Apply the distributive law to write the product as a sum.*

15. $y^{1/3}(y + y^{2/3})$ **16.** $(2y + y^{1/2})(2y - y^{1/2})$

■   *Factor as indicated.*

17. $x^{-1/4} + x^{1/2} = x^{-1/4}(?)$ **18.** $x^{2/3} - x^{-2/3} = x^{-2/3}(?)$

[2.3]   ■   *Write in simplest form.*

19. $\sqrt{4x^5y^3}$ **20.** $\sqrt{3x} \cdot \sqrt{12xy}$ **21.** $\dfrac{\sqrt{2x}\sqrt{3xy}}{\sqrt{2y}}$

22. $\dfrac{\sqrt[3]{6x}\sqrt[3]{9x^2y}}{\sqrt[3]{2y}}$ **23.** $\sqrt[4]{25}$ **24.** $\sqrt[6]{27x^3y^3}$

[2.4]   ■   *Rationalize the denominator.*

25. $\dfrac{1}{3 - \sqrt{x}}$ **26.** $\dfrac{\sqrt{x} + 1}{2\sqrt{x} - 1}$ **27.** $\dfrac{1}{\sqrt[3]{x}}$ **28.** $\dfrac{2}{\sqrt[3]{(x - 1)^2}}$

■   *Rationalize the numerator.*

29. $\dfrac{\sqrt{x} + 1}{2}$ **30.** $\dfrac{\sqrt{x} - 1}{\sqrt{x} + 2}$ **31.** $\dfrac{\sqrt[3]{x}}{x + 1}$ **32.** $\dfrac{\sqrt[3]{(x + 1)^2}}{x}$

[2.5]   ■   *Write each expression in the form $a + bi$ or $a + ib$.*

33. $\sqrt{9} - \sqrt{-9}$ **34.** $-3\sqrt{-18}$ **35.** $(2 + 3i) + (3 - 2i)$

36. $(2 - i) - (4 + 2i)$ **37.** $(2 - i)(2 + i)$ **38.** $\dfrac{2i}{2 + i}$

39. $(1 + \sqrt{-4})\sqrt{-9}$ **40.** $\dfrac{2 - \sqrt{-9}}{\sqrt{-16}}$

B

41. Write $(8{,}000{,}000{,}000{,}000)^{2/3}$ in scientific notation.

42. Write $(0.00000000000004)^{-1/2}$ in scientific notation.

43. Rationalize the denominator of $\dfrac{1}{\sqrt{x + y} + 1}$.

44. Rationalize the denominator of $\dfrac{1}{1 - \sqrt{x + y}}$.

45. Rationalize the numerator of $\dfrac{\sqrt{x + h} - \sqrt{x}}{h}$.

46. Rationalize the numerator of $\dfrac{\sqrt{x - h} - \sqrt{x}}{h}$.

47. Evaluate $(x^2 + x + 6)^2$ for $x = -\dfrac{1}{2} + \dfrac{\sqrt{7}}{2}i$.

48. Evaluate $(x^2 - 2x + 2)^3$ for $x = 1 - \sqrt{3}\,i$.

49. Simplify i^{25}.

50. Simplify i^{-10}.

99-101 Homework for Oct. 10th

3

Equations and Inequalities in One Variable

In this chapter we consider real-number solutions of first- and second-degree equations, and of equations involving radicals and absolute values. We also consider inequalities involving rational expressions and absolute values. Interval notation is introduced as a form for expressing the solution sets of inequalities.

EQUIVALENT EQUATIONS; SOLUTIONS OF FIRST-DEGREE EQUATIONS

In Section 1.1 we introduced symbols such as $P(x)$, $Q(x)$, and $R(x)$ to represent algebraic expressions in the variable x. If we replace x in $P(x) = Q(x)$ with a real number for which both members of the equation are defined, and if the resulting statement is true, then the number is a **solution** of the equation. The set of all solutions of an equation is the **solution set** of the equation.

Equivalent Equations

Equations that have the same solution set are called **equivalent equations**. For example, the equations

$$x + 3 = -3 \quad \text{and} \quad x = -6$$

are equivalent, because the solution set of each is $\{-6\}$.

Elementary Transformations

To solve an equation, we first try to determine the solution by inspection. Failing that, we generate a sequence of equivalent equations until we arrive at one with an obvious solution. The properties of equality (see Preliminary Concepts) guarantee that the following operations do not alter the solution set of an equation. They are called **elementary transformations** and are used to generate equivalent equations.

I. Any expression may be added to both sides of an equation.
II. Both sides of an equation may be multiplied by any expression that does not represent zero.

EXAMPLE 1   Solve $\dfrac{x+8}{3} - 5 = \dfrac{1}{2}$.

Solution   We first multiply by the least common denominator of all the fractions. By Transformation II, the given equation is equivalent to

$$(6)\left(\frac{x+8}{3} - 5\right) = (6)\left(\frac{1}{2}\right),$$

from which

$$2x - 14 = 3. \tag{1}$$

Next, we add 14 to both members of Equation (1). By Transformation I, Equation (1) is equivalent to

$$2x - 14 + 14 = 3 + 14$$

or

$$2x = 17. \tag{2}$$

Finally, we multiply both members of Equation (2) by $\frac{1}{2}$. By Transformation II, Equation (2) is equivalent to

$$x = \frac{17}{2}.$$

Thus, the equation $x = \frac{17}{2}$ is equivalent to the original equation. Therefore, the solution set is $\{\frac{17}{2}\}$.   ■

An elementary transformation always produces an equivalent equation. Care must be taken in the application of Transformation II, however, for we have specifically excluded multiplication by zero.

EXAMPLE 2   Solve $\dfrac{x}{x-3} = \dfrac{3}{x-3} + 2$. $\tag{3}$

Solution   We might first multiply each member by $(x-3)$ to find an equation that is free of fractions. We have

$$(x-3)\frac{x}{x-3} = (x-3)\frac{3}{x-3} + (x-3)2$$
$$x = 3 + 2x - 6, \tag{4}$$

from which

$$x = 3.$$

Thus, 3 is a solution of Equation (4). But substituting 3 for x in Equation (3), we have

$$\frac{3}{3-3} = \frac{3}{3-3} + 2 \qquad \text{or} \qquad \frac{3}{0} = \frac{3}{0} + 2,$$

and neither member is defined. Thus, 3 is not a solution of Equation (3). In obtaining Equation (4), each member of Equation (3) was multiplied by $(x - 3)$; but if x is 3, then $(x - 3)$ is zero, and Transformation II does not apply. Equation (4) is not equivalent to Equation (3), and in fact Equation (3) has no solution. ■

Example 2 shows that if we multiply or divide both members of an equation by an expression that is zero for some value or values of the variable, we should always check to see if the result obtained is a solution of the equation by substituting it in the original equation and determining whether or not the resulting statement is true.

EXAMPLE 3   Solve $\dfrac{4}{2x+3} + \dfrac{4x}{4x^2-9} = \dfrac{12}{2x+3}$.   (5)

Solution   We first multiply by the least common denominator of all the fractions, $(2x+3)(2x-3)$, to obtain

$$\left(\frac{4}{2x+3} + \frac{4x}{4x^2-9}\right)(2x+3)(2x-3) = \frac{12}{2x+3}(2x+3)(2x-3)$$

Simplifying both members yields

$$4(2x-3) + 4x = 12(2x-3)$$
$$12x - 12 = 24x - 36.$$

We now add 12 and $-24x$ to both members to obtain

$$12x - 12 + 12 - 24x = 24x - 36 + 12 - 24x$$

or

$$-12x = -24.$$

Finally, we multiply both members by $-\frac{1}{12}$ to obtain

$$x = 2. \tag{6}$$

The solution set of (6) is $\{2\}$. However, to obtain (6) we multiplied both members of (5) by an expression in x that is zero for some values of x. Thus, we must check our solution.

Check: $\dfrac{4}{2(2) + 3} + \dfrac{4(2)}{4(2)^2 - 9} = \dfrac{12}{2(2) + 3}.$

The solution set is $\{2\}$.   ■

Equations with Other Variables

An equation containing more than one variable, or containing symbols such as a, b, and c, representing constants, can often be solved for one of the symbols in terms of the remaining symbols. This is accomplished by using Transformations I and II to isolate the desired variable in one member of the equation.

E X A M P L E 4   Solve $ay = b + y$, for y.

Solution   We generate the following sequence of equivalent equations.

$$ay - y = b \qquad \text{(Transformation I)}$$
$$y(a - 1) = b$$
$$y = \frac{b}{a - 1} \quad (a \neq 1) \qquad \text{(Transformation II)}$$

E X A M P L E 5   Solve $x'(x - 3) + 4 = 5(x + x') - 9$, for x'.

Solution   We generate the following sequence of equivalent equations.

$$x'x - 3x' + 4 = 5x + 5x' - 9$$
$$x'x - 3x' - 5x' = 5x - 9 - 4 \qquad \text{(Transformation I)}$$
$$x'(x - 8) = 5x - 13$$
$$x' = \frac{5x - 13}{x - 8} \quad (x \neq 8) \qquad \text{(Transformation II)}\qquad ■$$

Applications

Linear equations provide mathematical models for both physical and social areas of science.

E X A M P L E   6   The estimated demand D for a certain radio receiver is related to its price P (in dollars) by

$$D + 10P = 100{,}000. \tag{7}$$

At what price should the receiver be sold to produce a demand of 1000 receivers?

Solution   Substituting 1000 for D in Equation (7) we obtain the linear equation

$$1000 + 10P = 100{,}000.$$

We then solve for P, obtaining

$$10P = 99{,}000 \qquad P = 9900.$$

Thus, the price for the receiver should be $9900.   ■

Solution of a Linear Equation

The equation

$$ax + b = 0, \tag{8}$$

where $a \neq 0$, is a **first-degree**, or **linear**, equation. We can show that any equation equivalent to a linear equation has one and only one solution. By Transformation I on page 77 [adding $-b$ to both members of Equation (8)],

$$ax = -b \tag{9}$$

is equivalent to Equation (8). Further, by Transformation II [multiplying both members of Equation (9) by $1/a$],

$$x = \frac{-b}{a} \tag{10}$$

is equivalent to Equation (9). Equation (10) has the unique solution $-b/a$, and Equations (8), (9), and (10) are equivalent. Thus, Equation (8) has the unique solution $-b/a$.

E X E R C I S E   3.1

A

■ *Solve. See Examples 1–3.*

1. $2x = 3$

2. $-3x = 4$

3. $2x - 4 = -3$

4. $-3x + 5 = -4$

5. $-2x + 3 = -x + 4$

6. $3x - 2 = -5x + 6$

7. $3x - 2(8 + x) = 3(x + 2)$

8. $5(x - 1) = 4(x - 2) + 3$

9. $-3[x - (x - 3)] = x - 4$

10. $-[2x - (3 - x)] = -4x + 11$

11. $x^2 + 3 = 2 + (x - 1)(x + 2)$

12. $-2 + x - x^2 = 3 - (x + 1)^2$

13. $2[3x + (4x - 1)] = 10x + 3$

14. $5x - (6 + 4x) = x + [2x - (3x - 4)]$

15. $\dfrac{x + 5}{4} = 1 - \dfrac{x + 9}{12}$

16. $\dfrac{2x - 3}{6} = \dfrac{x - 1}{9}$

17. $\dfrac{x}{6} + 1 = \dfrac{2x - 6}{3}$

18. $\dfrac{x}{3} - \dfrac{2x - 1}{2} = 3$

19. $\dfrac{2 + x}{x} = -5$

20. $\dfrac{x + 2}{x + 4} = \dfrac{2}{3}$

21. $\dfrac{2}{x + 1} = \dfrac{x}{x + 1} + 1$

22. $\dfrac{3}{x - 2} = \dfrac{1}{2} + \dfrac{2x - 7}{2x - 4}$

23. $\dfrac{2x + 1}{x - 4} - \dfrac{4x - 3}{2(x - 4)} = 0$

24. $\dfrac{3x}{x + 3} + \dfrac{-9x - 1}{3(x + 3)} = \dfrac{4}{x + 3}$

25. $\dfrac{1}{x - 1} + \dfrac{2}{x + 1} = \dfrac{3x - 1}{x^2 - 1}$

26. $\dfrac{4}{x + 2} - \dfrac{1}{x} = \dfrac{2x - 1}{x^2 + 2x}$

27. $\dfrac{3}{x + 1} + \dfrac{x - 4}{x^2 - x - 2} = \dfrac{-10}{x - 2}$

28. $\dfrac{1}{x} + \dfrac{1}{x^2 + x} = \dfrac{3}{x + 1}$

■   *Solve for the indicated variable. Leave the results in the form of an equation equivalent to the given equation. Indicate any restrictions on the variables. See Examples 4 and 5.*

29. $S = 2\pi rh$,   for r

30. $S = 2\pi rh$,   for h

31. $v = k + gt$,   for k

32. $v = k + gt$,   for t

33. $A = \dfrac{h}{2}(b + c)$,   for c

34. $S = \dfrac{a}{1 - r}$,   for r

35. $l = a + (n - 1)d$,   for n

36. $\dfrac{1}{r} = \dfrac{1}{r_1} + \dfrac{1}{r_2}$,   for r

37. $V = \pi R^2 h - \pi r^2 h$,   for r^2

38. $V = \dfrac{4}{3}\pi R^3 - \dfrac{4}{3}\pi r^3$,   for R^3

39. $2y'y + 2x = 0$,   for y'

40. $2x + 2y + 2xy' + 2yy' = 0$,   for y'

41. $\dfrac{2xy - 2x^2 yy'}{y^4} = 0$,   for y'

42. $\dfrac{(2x + 2yy')2y - 2y(x^2 + y^2)}{4y^2} = 1$,   for y'

43. $x^2 y' - 3x - 2y^3 y' = 1$,   for y'

44. $2xy' - 3y' + x^2 = 0$,   for y'

45. $x_1 x_2 - 2x_1 x_3 = x_4$,   for x_1

46. $3x_1 x_3 + x_1 x_2 = x_4$,   for x_1

■   *Solve.*

Three units are normally used as a measure of temperature, namely Celsius (C), Fahrenheit (F), and Kelvin (K). These units are related by the formulas

$$C = \frac{5F - 160}{9} \qquad and \qquad K = C + 273.$$

47. A normal temperature for a human being is 98.6°F. What is the normal temperature on the Celsius scale?

48. The boiling point of water at sea level is 212°F. What is the boiling point of water at sea level on the Celsius scale?

49. What Fahrenheit temperature corresponds to a temperature of 290° Kelvin?

50. Find the Kelvin temperature that corresponds to a temperature of 98.6° Fahrenheit.

■ *A projectile launched upward from a position s_0 feet from ground level at a velocity of v_0 feet per second is located at a height s above the ground at time t, as given by*

$$s = -16t^2 + v_0 t + s_0.$$

51. With what velocity must an object be launched from the ground to attain a height of 200 feet in 8 seconds?

52. With what velocity must an object be launched from 100 feet above the ground to attain a height of 300 feet in 5 seconds?

53. From what initial height (s_0) must a projectile be launched with an initial velocity of 50 feet per second to attain a height of 500 feet in 2 seconds?

54. From what initial position (s_0) must a projectile be launched with an initial velocity of 100 feet per second to attain a height of 500 feet in 5 seconds?

■ *The cutting speed S (in feet per minute) of a circular saw blade is given by*

$$S = \frac{C\omega D}{12d},$$

where C (in feet) is the circumference of the saw blade, ω is the angular velocity in revolutions per minute of the driving motor, D is the diameter of the driving pulley, and d is the diameter of the driven pulley.

55. What angular velocity must the driving motor attain to achieve a cutting speed of 7000 feet per minute if $C = 3$ feet, $D = 4$ inches, and $d = 3$ inches?

56. What angular velocity must the driving motor attain to achieve a cutting speed of 150 feet per minute if $C = 1$ foot, $D = 2$ inches, and $d = 1$ inch?

57. What must the diameter of the driven pulley be to achieve a cutting speed of 800 feet per minute if $\omega = 1000$ revolutions per minute, $D = 1$ foot, and $C = 2$ feet?

58. What must the diameter of the driven pulley be to achieve a cutting speed of 8000 feet per minute if $\omega = 2000$ revolutions per minute, $D = \frac{1}{2}$ foot, and $C = 4$ feet?

■ *A rigid bar rotating about a fixed point is called a lever. The fixed point is called the fulcrum of the lever. If a force F_1 is applied to one side of a lever at a distance d_1 from the fulcrum and a force F_2 is applied on the other side of the lever at a distance d_2 from the fulcrum (as shown in the figure), then the lever will be in equilibrium if and only if $F_1 d_1 = F_2 d_2$.*

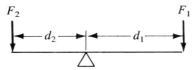

This relationship is called the law of the lever and has many practical applications.

59. A 24-gram weight is placed 2 centimeters further from the fulcrum on one side of a balanced lever than a 32-gram weight on the other side. How far is each weight from the fulcrum?

60. A 48-gram weight and a 36-gram weight are placed on opposite ends of a beam, each weight at a distance of 8 centimeters from the fulcrum. Where should a 24-gram weight be placed to balanced the beam? [*Hint:* $F_1 d_1 = F_2 d_2 + F_3 d_3$.]

61. Where should the fulcrum of a 6-foot crowbar be placed so that a 200-pound man can just balance a 2200-pound weight?

62. Where should the fulcrum of a 9-foot crowbar be placed so that a 180-pound man can just balance a 900-pound weight?

63. The estimated demand D for a television set is related to its price P (in dollars) by

$$D = -60P + 12,000, \qquad 80 < P < 140.$$

At what price should the set be sold to produce a demand for 4800 sets?

64. A business estimates that its total sales S of units is related to the amount A spent on advertising and the number C of competitors in its marketing area by

$$S = \frac{15A}{C}.$$

How much should the business allocate to advertising if it has three competitors and wishes to have total sales of 5000 units?

65. Consumers are usually charged for their use of electricity in units of kilowatt-hours (kWh), which are calculated using the formula

$$kWh = \frac{E \cdot I \cdot t}{1000},$$

where E is the voltage, I is the current in amperes, and t is the time in hours. Find the amperes, I, if a motor operates for 2 hours on a 120-volt line and uses 2.88 kilowatt-hours of electricity.

66. The horsepower (hp) generated by water flowing over a dam of height s, in feet, is given by

$$hp = \frac{62.4N \cdot s}{33,000},$$

where N is the number of cubic feet of water flowing over the dam per minute. Find the number of cubic feet of water that flowed over a 165-foot dam if 468 horsepower were generated.

67. The net resistance R_n of an electrical circuit is given by

$$\frac{1}{R_n} = \frac{1}{R_1} + \frac{1}{R_2},$$

where R_1 and R_2 are individual resistors in parallel. If one of the individual resistors is 40 ohms, what must the other be if the net resistance is 30 ohms?

68. The net resistance R_n of an electrical circuit containing resistors R_1, R_2, and R_3 in parallel is given by

$$\frac{1}{R_n} = \frac{1}{R_1} + \frac{1}{R_2} + \frac{1}{R_3}.$$

If two of the individual resistors are 20 ohms and 40 ohms. what must the third resistor be if the net resistance is 10 ohms?

69. The monthly benefit B paid on a certain retirement plan is determined by

$$B = \frac{2}{5}w\left(1 + \frac{2n}{100}\right),$$

where w is the average monthly wage of the worker and n is the number of years worked. How many years must a person work to receive a monthly benefit of $512 if his average monthly wage is $800?

70. A manufacturer plans to depreciate a large lathe that cost $12,000. The value V in t years is given by the formula

$$V = C\left(1 - \frac{t}{15}\right),$$

where C is the original cost. In how many years will the value be $5000? In how many years will the lathe be completely depreciated?

71. The force F necessary to raise a weight w using a special pulley system is given by

$$F = \frac{w}{2}\left(\frac{D_1 - D_2}{D_1}\right).$$

Find the weight that can be lifted using a force of 45 kilograms, where $D_1 = 25$ and $D_2 = 22$.

72. Using the special pulley system of Exercise 71, find the weight that can be lifted using a force of 100 kilograms if $D_2/D_1 = 0.75$.

B

■ *Find the value of k that makes the two equations equivalent.*

73. $kx = 3,\ x = 4$

74. $-kx = 2,\ x = 3$

75. $kx + 1 = 2,\ x = 2$

76. $-kx + 3 = 6,\ x = 9$

77. $2x + k = 1,\ x = k$

78. $-3x - 2k = 2,\ x = 2k$

79. $2x - k = 3,\ 2x + k = 1$

80. $3x - 2k = 1,\ 2x + k = 4$

3.2

CONSTRUCTION OF MATHEMATICAL MODELS; LINEAR EQUATIONS

In Section 3.1 we considered methods for solving linear equations. The exercises included several physical and economic situations that are modeled by linear equations. In each of these exercises the mathematical model was given in the problem. Often we must construct a model from a verbal or written description of a problem. The following suggestions are frequently helpful for finding an equation that provides a model for a given description.

Steps for Constructing Models

1. Represent the unknown quantities by using word phrases.
2. Represent each unknown quantity in terms of symbols.
3. Where applicable, draw a sketch and label all known quantities in terms of symbols.
4. Find a quantity that can be represented in two different ways, and write these representations as an equation (mathematical model). Summarizing given information in a table is sometimes helpful.

In the following examples, we illustrate the use of these steps to construct models. The resulting equations are then solved by the methods discussed in Section 3.1.

EXAMPLE 1   A small business calculator sells for $28 less than a scientific model. If the cost to buy both is $158, how much does each cost?

Solution   We first express the *two* quantities asked for in *two* simple phrases and represent these quantities using one variable.

$$\text{Cost of scientific calculator:} \quad c$$

$$\text{Cost of business calculator:} \quad c - 28$$

Next, we write an equation relating the unknown costs of the two types of calculators and the total cost of the two calculators.

$$\begin{bmatrix} \text{Cost of scientific} \\ \text{calculator} \end{bmatrix} + \begin{bmatrix} \text{Cost of business} \\ \text{calculator} \end{bmatrix} = [\text{Total cost}]$$

$$c \qquad + \qquad (c - 28) \qquad = \qquad 158$$

We then solve the resulting equation.

$$c + c - 28 = 158$$
$$2c - 28 = 158$$
$$2c = 186$$
$$c = 93$$
$$c - 28 = 65$$

The scientific calculator costs $93 and the business model costs $65.

EXAMPLE 2   A man has an annual income of $12,000 from two investments. He has $10,000 more invested at 8% than he has invested at 12%. How much does he have invested at each rate?

Solution   We first express the two quantities asked for in two simple phrases and represent these quantities using symbols.

$$\text{Amount invested at } 12\%: \quad A$$
$$\text{Amount invested at } \ 8\%: \quad A + 10{,}000$$

Second, we set up a table showing the interest earned on each account.

Investment	Rate	Amount	Interest
12%	0.12	A	$0.12A$
8%	0.08	$A + 10{,}000$	$0.08(A + 10{,}000)$

Next, we write an equation relating the interest from each investment and the total interest received. The total interest received is given to be $12,000.

$$\begin{bmatrix} \text{Interest from} \\ 12\% \text{ investment} \end{bmatrix} + \begin{bmatrix} \text{Interest from} \\ 8\% \text{ investment} \end{bmatrix} = [\text{Total interest}]$$

$$0.12A \qquad + \quad 0.08(A + 10{,}000) = \qquad 12{,}000$$

Finally, we solve for A. We first multiply each member by 100 to clear the equation of decimals.

$$12A + 8(A + 10,000) = 1,200,000$$
$$12A + 8A + 80,000 = 1,200,000$$
$$20A = 1,120,000$$
$$A = 56,000$$
$$A + 10,000 = 66,000$$

Therefore, $56,000 is invested at 12% and $66,000 is invested at 8%.

EXAMPLE 3   The vertex angle of an isosceles triangle measures 30° more than the sum of the measures of its other two angles. What is the measure of each angle in the triangle?

Solution   We first express the two quantities asked for in two simple phrases and represent these quantities using symbols. Next, we draw an isosceles triangle and label the angles. Use the fact that in an isosceles triangle the two angles opposite the sides of equal length have equal measure.

Measure of each base angle: x

Measure of the vertex angle: $2x + 30$

Next, we write an equation using the fact that the sum of the measures of the angles in any triangle is 180°.

$$(x) + (x) + (2x + 30) = 180$$

Finally, we solve the equation.

$$4x + 30 = 180$$
$$4x = 150$$
$$x = 37\tfrac{1}{2}$$
$$2x + 30 = 105$$

Therefore, the angles in the triangle measure $37\tfrac{1}{2}°$, $37\tfrac{1}{2}°$, and $105°$.

EXAMPLE 4   How many liters of a 10% solution of acid should be added to 20 liters of a 60% solution of acid to obtain a 50% solution?

Solution   First, we express the quantity asked for in a simple phrase and represent the quantity using a symbol.

Number of liters of 10% solution: n

Next, we set up a table to show the amount of acid in each component solution of the mixture and in the final solution resulting from the mixture.

Mixture	Part of Acid in Mixture	Amount of Mixture	Amount of Acid
10%	0.10	n	$0.10n$
60%	0.60	20	$0.60(20)$
50%	0.50	$n + 20$	$0.50(n + 20)$

We then write an equation relating the amount of *pure* acid before and after combining the solutions.

$$\begin{bmatrix} \text{Pure acid in} \\ 10\% \text{ solution} \end{bmatrix} + \begin{bmatrix} \text{Pure acid in} \\ 60\% \text{ solution} \end{bmatrix} = \begin{bmatrix} \text{Pure acid in} \\ 50\% \text{ solution} \end{bmatrix}$$

$$0.10n + 0.60(20) = 0.50(n + 20)$$

Finally, we solve the equation. We first multiply each member by 100.

$$10n + 60(20) = 50(n + 20)$$

$$10n + 1200 = 50n + 1000$$

$$-40n = -200$$

$$n = 5$$

Therefore, 5 liters of the 10% solution are needed.

EXAMPLE 5   One pipe can empty a tank in 6 hours, and a second pipe can empty the same tank in 9 hours. How long will it take both pipes to empty the tank?

Solution   We first express the quantity asked for in a simple phrase and represent the quantity using a symbol.

Number of hours for both pipes to empty the tank: t

We then construct a table showing the portion of the tank emptied by each pipe.

	Part of Tank Emptied in 1 Hour	Hours	Part of Tank Emptied
Pipe 1	$\dfrac{1}{6}$	t	$\left(\dfrac{1}{6}\right)t$
Pipe 2	$\dfrac{1}{9}$	t	$\left(\dfrac{1}{9}\right)t$

Next, we use the information in the table to write an equation expressing the conditions of the problem.

$$\begin{bmatrix} \text{Part of tank} \\ \text{emptied by pipe 1} \end{bmatrix} + \begin{bmatrix} \text{Part of tank} \\ \text{emptied by pipe 2} \end{bmatrix} = \begin{bmatrix} \text{Entire tank} \\ \text{emptied by both} \end{bmatrix}$$

$$\left(\frac{1}{6}\right)t \quad + \quad \left(\frac{1}{9}\right)t \quad = \quad 1$$

Finally, we solve the equation.

$$\left(\frac{1}{6} + \frac{1}{9}\right)t = 1$$

$$\left(\frac{5}{18}\right)t = 1$$

$$t = \frac{18}{5}$$

Thus, it takes $3\frac{3}{5}$ hours for both pipes to empty the tank.   ■

E X E R C I S E   3.2

A

■ (*a*) *Write an equation that models the given conditions.*
  (*b*) *Solve for the specified quantity.*

For Exercises 1–16, see Examples 1 and 2.

1. A company sells its used telephones. The rotary dial models are $15 cheaper than the Touch-Tone models. If it costs $75 to buy one of each, how much does each cost?

2. A dealer is selling a standard edition car and a luxury edition. If the luxury edition costs $3500 more than the standard edition and it costs $31,500 to buy one of each edition, how much does each cost?

3. A manager spends $1510 purchasing calculators for her store's employees. She buys 10 business models and 25 scientific models. If the business model is $24 cheaper than the scientific model, how much does each cost?

4. A pilot spends $7050 to obtain his license. He puts in 60 hours of flying time and 15 hours of ground school. If flying time costs $30 per hour more than ground school, how much per hour is each?

5. In a student body election, 584 students voted for one or the other of two candidates for president. If the winner received 122 more votes than the loser, how many votes were cast for each candidate?

6. The profit in selling an item is $15 less than the cost of the item. If the item sells for $85, what is its cost, and how much profit results from the sale?

7. One year the Dean of Admissions at City College selected $\frac{3}{4}$ of all applicants for the freshman class. How many students applied for entrance if the college selected 600 students?

8. A secretary takes home $\frac{4}{5}$ of her total salary. What is her total salary if she takes home $120 per week?

9. On an airplane flight for which first-class fare is $80 and tourist fare is $64 there were 42 passengers. If receipts for the flight totaled $2880, how many first-class and how many tourist passengers were on the flight?

10. A hardware retailer bought 50 machine screws — 1-inch screws at 18¢ each and $\frac{3}{4}$-inch screws at 15¢ each. If the total cost was $8.10, how many screws of each length did the retailer purchase?

11. A church member donates 10% of her weekly net earnings to her church. If her payroll deductions are 28% of her gross earnings and if she gives $45 per week to her church, how much is her gross income per week?

12. A taxable estate after credits totals $500,000. The estate is left to a single heir. If the federal tax rate on such an estate is 37%, at what interest rate (to the nearest quarter of a percent) does the heir have to invest the inheritance to receive an annual income of $45,000 in interest?

13. A woman has invested $8000, part in a bank at 10% and part in a savings and loan association at 12%. If her annual return is $844, how much has she invested at each rate?

14. A sum of $2400 is split between an investment in a mutual fund paying 14% and one in corporate bonds paying 11%. If the return on the 14% investment exceeds that on the 11% investment by $111 per year, how much is invested at each rate?

15. If $3000 is invested in bonds at 8%, how much additional money must be invested in stocks paying 13% to make the earnings on the total investment equal 10%?

16. If $6000 is invested in bonds at 9%, how much additional money must be invested in stocks at 12% to earn a return of 11% on the total investment?

■ *For Exercises 17–24, see Example 3.*

17. One angle of a triangle measures two times another angle, and the third angle measures 12° more than the sum of the measures of the other two. Find the measure of each angle.

18. One angle of a triangle measures two times the sum of the other two. The other two angles are equal. Find the measure of each angle.

19. The measure of the largest angle of a triangle is 70° greater than the measure of another angle and 50° greater than the measure of the third angle. Find the measure of each angle.

20. The measure of the smallest angle of a triangle is 25° less than the measure of another angle and 50° less than the measure of the third angle. Find the measure of each angle.

21. Each of the equal sides of an isosceles triangle is 3 centimeters longer than two times the length of the base. Find the length of each side of the triangle if its perimeter is 86 centimeters.

22. Each of the equal sides of an isosceles triangle is 5 cm longer than $\frac{1}{3}$ of the length of the third side. If the perimeter is 100 cm, find the length of each side of the triangle.

23. When the length of each side of a square is decreased by 2 centimeters, the area is decreased by 20 square centimeters. Find the length of a side of the original square.

24. When the length of each side of a square is increased by 5 centimeters, the area is increased by 85 square centimeters. Find the length of a side of the original square.

■ *For Exercises 25–32, see Example 4.*

25. How much water should be added to 50 gallons of a solution that is 50% acid to obtain a 12% solution?

26. How much water should be added to 6 gallons of pure acid to obtain a 15% solution?

27. A tank contains 20 gallons of a liquid fertilizer with a 16% nitrogen content. How many gallons of this solution should be drained from the tank and replaced with water so that the fertilizer will have a 12% nitrogen content?

28. An automobile radiator contains 8 quarts of a 40% antifreeze solution. How many quarts of this solution should be drained from the radiator and replaced with water if the resulting solution is to be 20% antifreeze?

29. How many pounds of an alloy containing 32% silver must be melted with 25 pounds of an alloy containing 48% silver to obtain an alloy containing 42% silver?

30. How many pounds of an alloy containing 80% silver must be melted with 5 pounds of an alloy containing 55% silver to obtain an alloy containing 75% silver?

■ *A karat is a measure comprising 24 units used to specify the proportion of gold in an alloy, e.g., 12K (12 karat) gold is 12/24 or 50% gold.*

31. How much 22K gold must be melted with 1 ounce of 14K gold to produce 18K gold?

32. How much 12K gold must be melted with 2 ounces of 20K gold to produce 18K gold?

■ *For Exercises 33–36, see Example 5.*

33. It takes one pipe 30 hours to fill a tank, while a second pipe can fill the same tank in 45 hours. How long will it take both pipes running together to fill the tank?

34. One pipe can fill a tank in 4 hours and another can empty it in 6 hours. If both pipes are open, how long will it take to fill the empty tank?

35. A new billing machine can process a firm's monthly billings in 10 hours. By using an older machine together with the new machine, the billings can be completed in 6 hours. How long would it take the older machine to do the job alone?

36. A tractor plows $\frac{5}{9}$ of a field in 10 hours. By adding another tractor the job is finished in another 3 hours. How long would it take the second tractor to do the job alone?

B

$$d = rt$$

37. Georgia runs 10-kilometer races at a rate of 7 kilometers per hour (kph). Jack runs 10-kilometer races at a rate of 6 kph. In their next 10-kilometer race, how far ahead of Jack will Georgia finish?

38. Suppose that due to starting positions in their next 10-kilometer race Jack has a 100-meter head start on Georgia. If they run at the rates given in Exercise 37, who will win the race and by how much?

39. An airplane travels 1260 miles in the same time that an automobile travels 420 miles. If the rate of the airplane is 120 miles per hour greater than the rate of the automobile, find the rate of each.

40. Two planes leave an airport at the same time and travel in opposite directions. If one plane averages 440 miles per hour over the ground and the other 560 miles per hour, in how long will they be 2500 miles apart?

41. A ship traveling at 20 knots is 5 nautical miles out from a harbor when another ship leaves the harbor at 30 knots sailing on the same course. How long does it take the second ship to overtake the first?

42. A boat sails due west from a harbor at 36 knots. An hour later, another boat leaves the harbor at 45 knots sailing on the same course. How far out at sea will the second boat overtake the first?

3.3

SECOND-DEGREE EQUATIONS

The equation

$$ax^2 + bx + c = 0 \qquad (a \neq 0)$$

is a **second-degree**, or **quadratic**, equation. Any equation that can be reduced to this form by elementary transformations is equivalent to a quadratic equation. The form shown above is called the **standard form** for such equations.

Solution by Factoring

The following theorem, which is a consequence of the properties of real numbers (see Preliminary Concepts), will be useful in solving quadratic equations.

Theorem 3.1   *If $P(x)$ and $Q(x)$ are expressions in x, then $P(x)Q(x) = 0$ if and only if*

$$P(x) = 0 \qquad or \qquad Q(x) = 0 \qquad or\ both$$

and both expressions are defined.

EXAMPLE I   Solve $x^2 + 2x - 15 = 0$.

Solution   We first factor the left-hand member of the equation, obtaining

$$x^2 + 2x - 15 = (x + 5)(x - 3).$$

Thus, the equation

$$x^2 + 2x - 15 = 0$$

is equivalent to

$$(x + 5)(x - 3) = 0.$$

By Theorem 3.1, $(x + 5)(x - 3) = 0$ if and only if

$$x + 5 = 0 \quad \text{or} \quad x - 3 = 0.$$

The solutions of these two equations are -5 and 3, respectively. Thus, the solution set of the original equation is $\{-5, 3\}$.   ■

Number of Solutions

The solution set of a quadratic equation can contain one or two members. The equation in Example 1 has two solutions.

By contrast, the equation

$$x^2 - 2x + 1 = 0 \tag{1}$$

is equivalent to

$$(x - 1)^2 = 0, \tag{2}$$

and since the only value of x for which Equation (2) holds is 1, this is the only member of the solution set of Equation (1). It is customary to say that such a solution is of **multiplicity 2**.

Extraction of Roots

If we cannot easily factor a quadratic equation, we must use other methods of solution. Quadratic equations of the form

$$(x - p)^2 = q$$

can be solved by observing that if x is a solution of the equation, then $x - p$ must be one of the square roots of q. That is, either

$$x - p = \sqrt{q} \quad \text{or} \quad x - p = -\sqrt{q}.$$

Thus, the solution set of $(x - p)^2 = q$ is $\{p + \sqrt{q}, p - \sqrt{q}\}$.

The preceding method of solving a quadratic equation is sometimes called **extraction of roots**.

E X A M P L E 2   Solve $3x^2 - 7 = 0$.

Solution   We first rewrite the equation in the form

$$x^2 = \frac{7}{3}.$$

By extraction of roots, we have

$$x = \sqrt{\frac{7}{3}} \quad \text{or} \quad x = -\sqrt{\frac{7}{3}},$$

and the solution set is $\left\{ \sqrt{\frac{7}{3}}, -\sqrt{\frac{7}{3}} \right\}$.

EXAMPLE 3   Solve $(x - 3)^2 = -4$.

Solution   We first use extraction of roots, obtaining

$$x - 3 = \sqrt{-4} \quad \text{or} \quad x - 3 = -\sqrt{-4},$$

from which

$$x - 3 = 2i \quad \text{or} \quad x - 3 = -2i.$$

Then, $x = 3 + 2i$ or $x = 3 - 2i$, and the solution set is $\{3 + 2i, 3 - 2i\}$.   ■

Completing the Square

By using a process called **completing the square**, we can write any quadratic equation in the equivalent form

$$(x - p)^2 = q,$$

which we can solve by using extraction of roots. The procedure for completing the square on a quadratic equation of the form $ax^2 + bx + c = 0$ is as follows.

Steps for Completing the Square

1. Multiply both sides of the equation by the reciprocal of the leading coefficient.
2. Add the negative of the constant term to both sides of the equation.
3. Add the square of one-half times the coefficient of x to both sides of the equation.
4. Write the left-hand number of the equation as the square of a binominal.

Note that each of these steps is an elementary transformation so that the resulting equation is equivalent to the original.

EXAMPLE 4   Solve $2x^2 + 3x - 2 = 0$.

Solution   We first complete the square as follows.

$$2x^2 + 3x - 2 = 0$$

$$x^2 + \frac{3}{2}x - 1 = 0 \qquad \text{(Step 1)}$$

$$x^2 + \frac{3}{2}x = 1 \qquad \text{(Step 2)}$$

$$x^2 + \frac{3}{2}x + \left(\frac{1}{2} \cdot \frac{3}{2}\right)^2 = 1 + \left(\frac{1}{2} \cdot \frac{3}{2}\right)^2 \qquad \text{(Step 3)}$$

$$\left(x + \frac{3}{4}\right)^2 = \frac{25}{16} \qquad \text{(Step 4)}$$

We now use extraction of roots to solve the last equation.

$$x + \frac{3}{4} = \frac{5}{4}$$

or

$$x + \frac{3}{4} = -\frac{5}{4}$$

The solution set is $\{\frac{1}{2}, -2\}$.   ■

Equations with Variable Coefficients   We occasionally must solve quadratic equations with coefficients that are themselves variables. In these cases we can use either of the methods, factoring or extraction of roots.

EXAMPLE 5   Solve $x^2 - 4yx = 0$ for x in terms of y.

Solution   We first factor the left-hand member of the equation, obtaining

$$x(x - 4y) = 0.$$

Thus, either

$$x = 0 \qquad \text{or} \qquad x = 4y.$$ ■

1-49 Homework

EXERCISE 3.3

A

■ *Solve by factoring. See Example 1.*

1. $x^2 - x = 2$

2. $x^2 = 6 + x$

3. $x^2 + 8x = 0$

4. $x^2 - 5x - 3 = 11$

5. $x^2 - 3x = -1 - x$

6. $-x(1 - 2x) = 1$

7. $6x^2 + 5x + 9 = 8$

8. $3x^2 - 16x + 6 = 1$

9. $2x^2 - x - 3 = 0$

10. $12x^2 + 5x - 2 = 0$

11. $6x^2 + 11x + 4 = 0$

12. $6x^2 - 19x + 3 = 0$

13. $2x^2 = 6 - 11x$

14. $3x^2 - 5 = 14x$

15. $x(2x + 9) = -10$

16. $(x - 5)(x + 1) = -5$

17. $-1 = \dfrac{6}{x^2} + \dfrac{7}{x}$

18. $3 = \dfrac{-10}{x - 2} + \dfrac{10}{x - 5}$

■ *Solve by extraction of roots. See Examples 2 and 3.*

19. $x^2 = 16$

20. $x^2 = 25$

21. $2x^2 = 18$

22. $5x^2 = 20$

23. $x^2 = -9$

24. $x^2 = -8$

25. $(x - 2)^2 = 9$

26. $(x - 4)^2 = 36$

27. $(2x + 1)^2 = 4$

28. $(x - 3)^2 = 11$

29. $(x - 2)^2 = -1$

30. $(x + 3)^2 = -7$

31. $\left(x - \dfrac{1}{2}\right)^2 = \dfrac{3}{4}$

32. $\left(x + \dfrac{1}{3}\right)^2 = \dfrac{2}{9}$

33. $(2x - 1)^2 = \dfrac{4}{25}$

34. $(3x - 2)^2 = \dfrac{16}{9}$

35. $(x - 2)^2 = -\dfrac{1}{4}$

36. $(x + 4)^2 = -\dfrac{25}{9}$

■ *Solve by completing the square. See Example 4.*

37. $x^2 + x - 2 = 0$

38. $x^2 + 4x + 4 = 0$

39. $x^2 + 7x + 12 = 0$

40. $x^2 + 6x + 8 = 0$

41. $3x^2 - 5x - 2 = 0$

42. $5x^2 + 8x = 4$

43. $x^2 - x - 4 = 0$

44. $x^2 - 2x - 2 = 0$

45. $x^2 - 3x + 3 = 0$

46. $3x^2 + x + 7 = 0$

47. $x^2 + 6x + 13 = 0$

48. $9x^2 - 12x + 5 = 0$

■ *Solve each equation for x in terms of y. See Example 5.*

49. $y^2x - x^2 = 0$

50. $y^2x^2 - x = 0$

51. $-y^2x^2 + y = 0$

52. $-x^2y^2 + 2y = 0$

53. $(y - x)^2 = 1$

54. $(y + x)^2 = 4$

55. $(2y + 3x)^2 = 9$

56. $(3y - 2x)^2 = 4$

B

■ *Complete the square for each equation.*

57. $x^2 + 2x + c = 0$

58. $x^2 - 4x + c = 0$

59. $x^2 + bx + 1 = 0$

60. $x^2 + bx - 4 = 0$

61. $ax^2 + 2x - 4 = 0$

62. $ax^2 - 4x + 9 = 0$

■   *Complete the square in both x and y.*

63. $x^2 + y^2 + 2x + 2y = 0$

64. $x^2 + y^2 + 4x - 4y = 0$

65. $x^2 + 3y^2 + 4x + 6y + 1 = 0$

66. $4x^2 + y^2 - 8x + 6y - 9 = 0$

3.4

THE QUADRATIC FORMULA

The process of completing the square can be used with arbitrary coefficients to find a formula for the solutions of a quadratic equation in standard form. This formula, called the **quadratic formula**, is given in the following theorem. The proof is left as an exercise.

Theorem 3.2   *If a, b, and c are real numbers (a ≠ 0), then*

$$x = \frac{-b \pm \sqrt{b^2 - 4ac}}{2a}$$

are the solutions to $ax^2 + bx + c = 0$.

The symbol $\pm$ in the quadratic formula is used to condense the two equations

$$x = \frac{-b + \sqrt{b^2 - 4ac}}{2a} \quad \text{and} \quad x = \frac{-b - \sqrt{b^2 - 4ac}}{2a}$$

into a single equation. We need only substitute the coefficients a, b, and c of a given quadratic equation in the formula to find the solution set for the equation.

EXAMPLE 1   Solve $x^2 + 3x - 2 = 0$ by using the quadratic formula.

Solution   Substituting 1 for a, 3 for b, and -2 for c in the quadratic formula and simplifying, we have

$$x = \frac{-3 \pm \sqrt{9 - 4 \cdot 1 \cdot (-2)}}{2 \cdot 1} = \frac{-3 \pm \sqrt{17}}{2}.$$

The solution set is $\left\{ \dfrac{-3 + \sqrt{17}}{2}, \dfrac{-3 - \sqrt{17}}{2} \right\}$.

EXAMPLE 2 Solve $x^2 = 2x - 2$ by using the quadratic formula.

Solution We first write the equation in the standard form

$$x^2 - 2x + 2 = 0.$$

Substituting 1 for a, -2 for b, and 2 for c in the quadratic formula and simplifying, we obtain

$$x = \frac{-(-2) \pm \sqrt{4 - 4 \cdot 1 \cdot 2}}{2 \cdot 1} = \frac{2 \pm \sqrt{-4}}{2}$$

$$= \frac{2 \pm 2i}{2} = 1 \pm i.$$

The solution set is $\{1 + i, 1 - i\}$. ■

Determination of Number of Real Solutions

An examination of the quadratic formula,

$$x = \frac{-b \pm \sqrt{b^2 - 4ac}}{2a},$$

shows that if $ax^2 + bx + c = 0$, where a, b, and c are real numbers, is to have a nonempty solution set in the set of real numbers, then $\sqrt{b^2 - 4ac}$ must be real. This, in turn, implies that only those quadratic equations for which $b^2 - 4ac \geq 0$ have real solutions. The number represented by $b^2 - 4ac$ is called the **discriminant** of the quadratic equation $ax^2 + bx + c = 0$. It yields the following information about the nature of the solution set of the equation.

1. If $b^2 - 4ac = 0$, then there is precisely one real solution (multiplicity two).
2. If $b^2 - 4ac > 0$, then there are two real solutions.
3. If $b^2 - 4ac < 0$, then there are two imaginary solutions.

EXAMPLE 3 Determine the number of distinct real solutions of each equation.

(a) $x^2 - 4x + 4 = 0$ (b) $x^2 - 2x - 3 = 0$ (c) $x^2 + x + 1 = 0$

Solution In each case, we compute the value of the discriminant and compare it to 0.

(a) $b^2 - 4ac = (-4)^2 - 4(1)(4) = 0$
Since $b^2 - 4ac = 0$, the equation has one real solution of multiplicity two.

(b) $b^2 - 4ac = (-2)^2 - 4(1)(-3) = 16$
Since $b^2 - 4ac > 0$, the equation has two distinct real solutions.

(c) $b^2 - 4ac = (1)^2 - 4(1)(1) = -3$

Since $b^2 - 4ac < 0$, the equation has two imaginary solutions but no real solutions. ■

Equations with Variable Coefficients

We can use the quadratic formula directly to solve quadratic equations with coefficients that are variables.

E X A M P L E 4   Solve $y^2 - yx + x^2 = 0$ for y in terms of x, where x represents a real number.

Solution   We shall use the quadratic formula. Substituting 1 for a, $-x$ for b, and x^2 for c, we have

$$y = \frac{-(-x) \pm \sqrt{x^2 - 4 \cdot 1 \cdot x^2}}{2 \cdot 1}$$

$$= \frac{x \pm \sqrt{-3x^2}}{2}$$

$$= \frac{x \pm |x|\sqrt{3}\,i}{2}.   \qquad ■$$

Applications

Quadratic equations arise in the application of mathematics to many areas, such as physics, business, and economics. Since these equations often have two solutions, we must take care to choose the solution that makes sense in the context of the stated problem. In the exercises for this section, applications are considered for which the model has been provided. In Section 3.5 we will discuss applications for which we must construct the model.

E X A M P L E 5   A stone thrown down from the edge of a cliff travels a distance (in feet) of $h = 16t^2 - 2t$ in t seconds. How long will it take the stone to travel 20 feet?

Solution   Substituting 20 for h, we have the quadratic equation

$$20 = 16t^2 - 2t$$

or, in the standard form,

$$16t^2 - 2t - 20 = 0.$$

We use the quadratic formula. Substituting 16 for a, -2 for b, and -20 for c, we obtain

$$t = \frac{-(-2) \pm \sqrt{4 - 4 \cdot 16 \cdot (-20)}}{2 \cdot 16}$$

$$= \frac{2 \pm \sqrt{1284}}{32}.$$

Thus, $t \approx 1.18$ or $t \approx -1.06$. Since t is measured in seconds after the stone is thrown, we must have $t > 0$. Therefore, it takes the stone approximately 1.18 seconds to travel 20 feet. ■

1-35 odd Homework

EXERCISE 3.4

A

must Put in Standard form.

■ *Solve for x by using the quadratic formula. See Examples 1 and 2.*

1. $x^2 + 7x + 12 = 0$ **2.** $x^2 - x - 20 = 0$ **3.** $4x^2 - 11x + 6 = 0$

4. $6x^2 - 7x - 3 = 0$ **5.** $6x^2 + 5x - 1 = 0$ **6.** $3x^2 + 4x - 4 = 0$

7. $x^2 + 2x - 6 = 0$ **8.** $x^2 - 4x - 2 = 0$ **9.** $2x^2 + 6x + 2 = 0$

10. $x^2 + x - 1 = 0$ **11.** $x^2 - 8x + 18 = 0$ **12.** $x^2 - 6x + 12 = 0$

13. $5x^2 - 6x + 2 = 0$ **14.** $3x^2 + 10x + 15 = 0$ **15.** $x^2 + 3x + 6 = 0$

16. $x^2 + x + 1 = 0$ **17.** $-2x^2 + 3x = 3$ **18.** $x^2 = 10x - 30$

19. $\dfrac{2x}{x^2 + 2x - 2} = -2$ **20.** $\dfrac{3x - 1}{x^2 + x + 5} = -1$ **21.** $2x^2 - (x + 1)^2 = 1$

22. $2x^2 - (2x + 1)^2 = 4$ **㉓** $(x - 4)^2 + (x + 2)^2 = -1$

24. $(2x - 3)^2 + (x + 4)^2 = -2$

25. $\dfrac{2x}{x + 1} + \dfrac{x - 1}{2} = 0$

26. $\dfrac{3}{2x - 3} + \dfrac{2x + 3}{x} = 0$

27. $\dfrac{2x}{x - 1} - \dfrac{4}{x^2 - x} = \dfrac{x - 1}{x}$

28. $\dfrac{3x + 1}{3x - 1} + \dfrac{3x}{2x + 1} = 1$

■ *Use the discriminant to determine the nature of the solutions of each equation. See Example 3.*

29. $x^2 + 2x - 8 = 0$ **30.** $2x^2 + 3x - 2 = 0$ **31.** $x^2 + 2 = -2x$

32. $x^2 = x + \dfrac{5}{4}$ **33.** $9x^2 + 1 = -6x$ **34.** $4x^2 = -12x - 9$

■ *Solve for y in terms of x. See Example 4.*

㉟ $xy^2 + y = 1$ **36.** $2x^2y^2 - y = -1$ **37.** $-x^2y^2 + 2y = 2$

38. $-xy^2 + 2y = 1$ **39.** $y^2 + 2yx + x^2 = 0$ **40.** $2y^2 + xy - 3x^2 = 0$

41. $x^2 - yx - y^2 = 0$ **42.** $2x^2 - yx - 2y^2 = 0$

■   *For Exercises 43–54, see Example 5.*

43. A ball thrown vertically upward reaches a height h in feet given by $h = 56t - 16t^2$, where t is the time in seconds after the throw. How long will it take the ball to reach a height of 24 feet on its way up? How long after the throw will the ball return to the height from which it was thrown?

44. A stone thrown down from the edge of a cliff travels a distance (in feet) of $h = 16t^2 - 8t$ in t seconds. If the cliff is 64 feet high, how long will it take the stone to fall halfway to the ground below? How long to fall all the way to the ground below?

■   *The distance s (from its starting point) of a body falling in a vacuum is given by*

$$s = v_0 t + \frac{1}{2}gt^2,$$

where s is measured in feet, t is measured in seconds, v_0 is the initial velocity in feet per second, and g is the constant of acceleration due to gravity (approximately 32 feet per second per second on earth).

45. On earth, how long will it take a body to fall 150 feet if v_0 is 20 feet per second? How long will it take if the body starts from rest?

46. On earth, how long would it take a body to return to its starting point if $v_0 = -10$ feet per second (the negative initial velocity indicates the velocity is directed upward)?

47. On the moon, the value of g is approximately 5.6 feet per second per second. How long will it take a body to fall 150 feet if v_0 is 20 feet per second on the moon? How long will it take if the body starts from rest?

48. On the moon, how long would it take a body to return to its starting point if $v_0 = -10$ feet per second?

49. The number of diagonals, D, of a polygon of n sides is given by

$$D = \frac{n}{2}(n - 3).$$

How many sides does a polygon with 90 diagonals have?

50. The formula

$$s = \frac{n}{2}(n + 1)$$

gives the sum of the first n natural numbers 1, 2, 3, . . . , n. How many consecutive natural numbers starting with 1 will give the sum of 406?

B

■ *If two resistors with resistances R_1 ohms and R_2 ohms, respectively, are connected in series, the equivalent resistance is $R = R_1 + R_2$. If the same two resistors are connected in parallel, the equivalent resistance R satisfies the equation $1/R = 1/R_1 + 1/R_2$.*

51. A resistor with resistance R is connected in series with a 10-ohm resistor, and the combination is connected in parallel with another resistor of resistance R. If the equivalent resistance is 15/4 ohms, find the value of R.

52. A resistor with resistance R is connected in series with a 5-ohm resistor, and the combination is connected in parallel with another resistor of resistance R. If the equivalent resistance is 100/9 ohms, find the value of R.

53. A resistor with resistance R is connected in parallel with a resistor of resistance $R + 5$. If the equivalent resistance is 5/3 ohms, find the value of R.

54. A resistor with resistance $R + 5$ is connected in parallel with a resistor of resistance $R + 10$. If the equivalent resistance is 60/7 ohms, find the value of R.

55. Show that if a, b, c are real numbers, then the equation $ax^2 + bx + c = 0 \ (a \neq 0)$ can be written equivalently as

$$x = \frac{-b \pm \sqrt{b^2 - 4ac}}{2a}.$$

56. Show that if r_1 and r_2 are roots of the quadratic equation $ax^2 + bx + c = 0$, then

$$r_1 + r_2 = -b/a \qquad \text{and} \qquad r_1 r_2 = c/a.$$

■ *Find the value(s) of k so that each equation has one solution of multiplicity two.*

57. $x^2 + kx + 1 = 0$ **58.** $x^2 - kx + 9 = 0$

59. $kx^2 + x + 4 = 0$ **60.** $kx^2 - 3x + 8 = 0$

61. $2x^2 + kx + k = 0$ **62.** $3x^2 - 2kx + 2k = 0$

63. $x^2 + (k + 1)x + k = 0$ **64.** $x^2 - (k + 1)x + 2k - 2 = 0$

3.5

CONSTRUCTING MODELS INVOLVING QUADRATIC EQUATIONS

In Section 3.2 we listed suggestions for constructing models. We then applied those suggestions to conditions that produced linear equations. In this section we shall use the same methods to construct models in which the resulting equation is quadratic. In these situations the resulting equation may have two solutions. We must take care to choose the solution that makes sense in the context of the stated problem.

EXAMPLE 1   The product of two consecutive positive integers is 132. Find the integers.

Solution   We first express the quantities asked for in simple phrases and represent the quantities with symbols.

$$\text{Smaller integer:} \quad n$$

$$\text{Larger integer:} \quad n + 1$$

Next, we use the relationship given in the problem statement to write an equation in the unknown n.

$$n(n + 1) = 132$$

Finally, we solve the equation for the desired variable.

$$n^2 + n - 132 = 0$$

$$(n + 12)(n - 11) = 0$$

The solution set is $\{-12, 11\}$. Since we are seeking two consecutive *positive* integers, we know $n \neq -12$. Thus, $n = 11$ and the desired integers are 11 and 12.

EXAMPLE 2   The length of a rectangle is 4 centimeters greater than the width, and the area is 77 square centimeters. Find the dimensions of the rectangle.

Solution   First, we express the quantities asked for in simple phrases and represent the quantities with symbols.

$$\text{Width:} \quad x$$

$$\text{Length:} \quad x + 4$$

Next, we sketch and label a figure. We then write an equation expressing the relationship between the quantities x and $x + 4$. Since the area (77 square centimeters) of a rectangle is the product of its width and length, the equation is

$$x(x + 4) = 77.$$

Finally, we solve the equation to obtain

$$x^2 + 4x - 77 = 0$$
$$(x + 11)(x - 7) = 0.$$

The solution set is $\{-11, 7\}$. Since a dimension cannot be negative, the only acceptable value for the width x is 7. If $x = 7$, then $x + 4 = 11$, and the dimensions are 7 centimeters and 11 centimeters.

EXAMPLE 3   A landscaper is to use 800 square feet of a rectangular field for a flower garden. The garden is to have a sidewalk of width 5 feet enclosing it. The length of the field is 20 feet longer than the width. What are the dimensions of the field?

Solution   First, we express the quantities asked for in simple phrases and represent the quantities with symbols.

Width of field: x

Length of field: $x + 20$

Next, we sketch and label a figure. We then write an equation expressing the relationship between the area used for the garden and the desired variable x. To accomplish this, we first express the length and width of the garden in terms of the variable x.

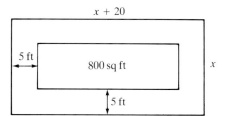

$$\begin{bmatrix} \text{Length of} \\ \text{garden} \end{bmatrix} = \begin{bmatrix} \text{Length of} \\ \text{field} \end{bmatrix} - 2\begin{bmatrix} \text{Width of} \\ \text{sidewalk} \end{bmatrix}$$
$$= x + 20 - 2(5)$$
$$= x + 10$$
$$\begin{bmatrix} \text{Width of} \\ \text{garden} \end{bmatrix} = \begin{bmatrix} \text{Width of} \\ \text{field} \end{bmatrix} - 2\begin{bmatrix} \text{Width of} \\ \text{sidewalk} \end{bmatrix}$$
$$= x - 2(5)$$
$$= x - 10.$$

Thus, we have

$$\begin{bmatrix} \text{Length of} \\ \text{garden} \end{bmatrix} \cdot \begin{bmatrix} \text{Width of} \\ \text{garden} \end{bmatrix} = \begin{bmatrix} \text{Area of} \\ \text{garden} \end{bmatrix}$$

or

$$(x + 10) \cdot (x - 10) = 800$$

Finally, we solve this equation for the desired variable.

$$x^2 - 100 = 800$$

$$x^2 = 900$$

The solution set is $\{-30, 30\}$. Since x is a measure of length, we must have $x > 0$. Thus, the desired solution is $x = 30$, and the dimensions of the field are 30 feet and 50 feet.

EXAMPLE 4 A shipment of computer chips was purchased by a company for $10,000. Inspectors determined that two of the chips were damaged in shipment and could not be resold. The remaining chips were sold at a profit of $50 each, and a total profit of $2000 was realized when all of the chips were sold. How many chips were purchased?

Solution First, we express the quantity asked for in a simple phrase and represent the quantity with a symbol.

$$\text{Number of chips purchased:} \quad x$$

We shall use the relationship between profit, and cost and selling price to obtain an equation relating these variables. This relationship is

$$\begin{bmatrix} \text{Profit} \\ \text{per chip} \end{bmatrix} = \begin{bmatrix} \text{Selling price} \\ \text{per chip} \end{bmatrix} - \begin{bmatrix} \text{Cost} \\ \text{per chip} \end{bmatrix} \tag{1}$$

Next, we express these quantities in symbols.

$$\text{Cost of each chip} = \frac{\text{Total cost}}{\text{Number purchased}} = \frac{10,000}{x}$$

$$\text{Selling price of each chip} = \frac{\text{Total revenue}}{\text{Number sold}}$$

Since the total profit from the transaction is

$$\text{Total profit} = \text{Total revenue} - \text{Total cost},$$

we have

$$\text{Total revenue} = \text{Total cost} + \text{Total profit}$$

$$= 10,000 + 2,000.$$

Further, since two chips cannot be sold,

$$\text{Selling price of each chip} = \frac{10,000 + 2,000}{x - 2} = \frac{12,000}{x - 2}.$$

We now write an equation relating these quantities using the fact that the undamaged chips were sold at a profit of $50 per chip. From Equation (1) we have

$$50 = \frac{12,000}{x - 2} - \frac{10,000}{x}.$$

Finally, we solve the equation for the desired variable.

$$50(x - 2)x = 12,000x - 10,000(x - 2)$$

$$x^2 - 2x = 240x - 200x + 400$$

$$x^2 - 42x - 400 = 0$$

$$(x + 8)(x - 50) = 0$$

The solution set is $\{-8, 50\}$. Since the company could not purchase a negative number of chips, the only acceptable value for the number of chips purchased is 50. Hence, it purchased 50 chips.

EXAMPLE 5   A rope 9 feet long is to be cut into two pieces of unequal length, and each piece is to be shaped to enclose a square. The area of the smaller square is to be exactly one-half the area of the larger square. Determine the lengths of the two pieces of rope.

Solution   We first express the quantities asked for in simple phrases and represent the quantities with symbols.

$$\text{Length of short piece:} \quad x$$

$$\text{Length of long piece:} \quad 9 - x$$

Next, we draw and label a figure showing the two squares and their dimensions. Recall that for a square

$$\text{Side length} = \frac{1}{4}[\text{perimeter}].$$

Thus, we have

Larger area: A_1

Smaller area: A_2

Next, we express the relationships between these quantities in symbols.

$$A_2 = \frac{1}{2}A_1, \qquad A_1 = \left(\frac{9 - x}{4}\right)^2, \qquad A_2 = \left(\frac{x}{4}\right)^2$$

We then write a single equation in the variable x.

$$\left(\frac{x}{4}\right)^2 = \frac{1}{2}\left(\frac{9-x}{4}\right)^2$$

Finally, we solve the equation.

$$\frac{x^2}{16} = \frac{81 - 18x + x^2}{32}$$

$$x^2 + 18x - 81 = 0$$

$$x = \frac{-18 \pm \sqrt{324 + 324}}{2}$$

$$= \frac{-18 \pm 18\sqrt{2}}{2}$$

$$= -9 \pm 9\sqrt{2}$$

If the $-$ sign is chosen, it results in a negative value for x, which is not an admissible solution since $x > 0$. Thus, the length of the short piece is $-9 + 9\sqrt{2}$ feet and the length of the long piece is $9 - (-9 + 9\sqrt{2}) = 18 - 9\sqrt{2}$.   ■

E X E R C I S E   3.5

A

■ (a) *Write an equation that models the given conditions.*

   (b) *Solve for the specified quantity.*

For Exercises 1–8, see Example 1.

1. Find two consecutive positive even integers whose product is 168.

2. Find two consecutive positive even integers whose product is 224.

3. Find two consecutive positive odd integers whose product is 323.

4. Find two consecutive positive odd integers whose product is 255.

5. Find two consecutive positive integers with the property that the sum of the reciprocals is $\frac{17}{72}$.

6. Find two consecutive positive integers with the property that the sum of the reciprocals is $\frac{29}{210}$.

7. Find two consecutive positive integers with the property that the reciprocal of the smaller minus the reciprocal of the larger is $\frac{1}{20}$.

8. Find two consecutive positive integers with the property that the reciprocal of the smaller minus the reciprocal of the larger is $\frac{1}{110}$.

■   *For Exercises 9–14, see Example 2.*

9. A rectangular lawn is 2 meters longer than it is wide. If its area is 63 square meters, what are the dimensions of the lawn?

10. The length of a rectangular steel plate is 2 centimeters greater than two times its width. If the area of the plate is 40 square centimeters, find its dimensions.

11. The length of a rectangle is 1 meter less than four times the width. Its area is 33 square meters. Find the dimensions.

12. The length of a rectangle is 3 meters less than three times the width. Its area is 168 square meters. Find the dimensions.

13. The area of a triangle is 75 square centimeters. Find the lengths of the base and altitude if the altitude is 25 centimeters longer than the base.

14. The area of a triangle is 70 square centimeters. Find the lengths of the base and the altitude if the base is 13 centimeters longer than the altitude.

■   *For Exercises 15–28, see Example 3.*

15. A book designer decides to use 48 square inches of type on a page and to make the height of the page twice the width. If the margin around the type is to be 2 inches uniformly, what are the dimensions of the page?

16. A book designer decides to use 88 square inches of type on a page and to make the height of the page two-thirds of the width. If the top and bottom margins are to be 1 inch and the side margins 2 inches, what are the dimensions of the book?

17. A typist is to leave a uniform margin on all four sides when typing a report. The page is 11 inches by 14 inches and his typing is to cover 108 square inches. What is the size of the margin?

18. Assume in Exercise 17 that the top and bottom margins are to be twice as wide as the side margins and that the bottom of the page is 11 inches wide. How wide should the side margins be?

19. A garden measuring 12 meters by 18 meters has its area increased by 216 square meters by adding a border of uniform width on all sides. Find the width of the border.

20. A garden measuring 30 feet by 50 feet has its area decreased by 444 square feet by paving a border of uniform width on all sides. Find the width of the border.

21. A sprinkler covers a circular area of 113.1 square yards. To increase the area covered by the sprinkler to 201.1 square yards would require an increase of how many yards in the radius?

22. The area of a circular patio is to be reduced by adding flower beds around the outer edge. If the original patio had a radius of 10 feet, how wide should the flower bed be to keep a usable patio area of 200 square feet?

23. A lawn measuring 25 meters by 50 meters has its area increased by 318 square meters by a border of uniform width along both shorter sides and one longer side. Find the width of the border.

24. A lawn measuring 25 meters by 60 meters has its area increased by 453 square meters by a border of uniform width along both longer sides and one shorter side. Find the width of the border.

25. A rectangular field with one edge on a river is to be fenced on the three sides not on the river. If 1000 feet of fencing is available and the longer side is to be along the river, what must the dimensions of the field be to use all of the fence and enclose 120,000 square feet?

26. A rectangular field with one edge on a river is to be fenced on the three sides not on the river. If 600 feet of fencing is available and the longer side is to be along the river, what must the dimensions of the field be to use all of the fence and enclose 40,000 square feet?

27. A table is to be constructed by attaching semicircles at each end of a rectangle. If the length of the rectangular portion of the table is 4 feet, how wide (to the nearest foot) should the rectangle be if the table is to have an area of 19 square feet?

28. A tray is formed from a rectangular piece of metal whose length is 2 centimeters greater than its width by cutting a square with sides 2 centimeters in length from each corner, and then bending up the sides. Find the dimensions of the tray if the volume is 160 cubic centimeters.

■ *For Exercises 29–32, see Example 4.*

29. An oil storage company purchases a shipment of crude oil for $50,000. Due to negligence, 5 barrels are spilled at delivery time. The remaining barrels are resold at a profit of $10 per barrel. A total profit of $9700 is realized when all the oil is sold. How many barrels were originally purchased?

30. A parcel of land is purchased for $95,000 by a real estate agency. The agency decides to use two acres of this land as the location for the corporate headquarters, and hence this land cannot be resold. If the rest of the land is resold at a profit of $200 per acre and a total profit of $17,700 is realized, how many acres did the original parcel of land contain?

31. A produce supplier purchases a truckload of melons for $1500. As a gesture of goodwill the farmer from whom the supplier purchased the melons ships 50 pounds more than the weight contracted for. If the produce supplier resells all of the fruit at a profit of $0.10 per pound and a total profit of $285 is realized, how many pounds of melons were originally contracted for? Assume the capacity of the truck is 400 pounds.

32. An office supply store purchases a case of calculators for $360. To maintain good relations, the salesperson decides to give the office supply store two extra calculators for use in the office. The management has already provided the personnel in the office supply store with calculators, so they resell the two extra calculators. If there is a profit of $4 on each calculator sold and a total profit of $116 is realized, how many calculators were in the case originally purchased? Assume the case contains an even number of calculators.

B

■ *For Exercises 33–36, see Example 5.*

33. A wire 1 foot long is to be cut into two pieces. The shorter piece will be bent into a square and the longer piece will be bent into a rectangle with length twice its width. If the areas of the two figures are to be equal, how long should the shorter piece be (to the nearest hundredth of a foot)?

34. A rope of length 16 feet is to be cut into two pieces. The shorter piece is to be shaped into an equilateral triangle; the longer piece is to be shaped into a square with area twice that of the triangle. How long is the shorter piece?

35. A contractor uses two skiploaders for 1 hour to screen a pile of gravel. If skiploader A can do the job alone in $1\frac{1}{2}$ hours less than skiploader B, how long would it take skiploader B to do the job alone?

36. A man and his son working together can paint their house in 4 days. The man can do the job alone in 6 days less than the son can do it. How long would it take each of them to paint the house alone? [*Hint:* What part of the job could each of them do in 1 day?]

37. A riverboat that travels 18 miles per hour in still water can go 30 miles up a river in 1 hour less time than it can go 63 miles down the river. What is the speed of the current in the river?

38. If a boat travels 20 miles per hour in still water and takes 3 hours longer to go 90 miles up a river than it does to go 75 miles down the river, what is the speed of the current in the river?

39. A commuter takes a train 10 miles to her job in a city. The train returns her home at a rate 10 miles per hour greater than the rate of the train that takes her to work. If she spends a total of 50 minutes a day commuting, what is the rate of each train?

40. On a 50-mile trip, a woman traveled 10 miles in heavy traffic and then 40 miles in less congested traffic. If her average rate in heavy traffic was 20 miles per hour less than her average rate in light traffic, what was each rate if the trip took 1 hour and 30 minutes?

3.6

EQUATIONS INVOLVING OTHER ALGEBRAIC EXPRESSIONS

In this section we discuss some methods of solving equations that contain expressions other than first- and second-degree polynomials.

Equality of Like Powers

In order to find solution sets for equations containing radical expressions, we need the following result.

Theorem 3.3

If $P(x)$ and $Q(x)$ are expressions in x, then the solution set of

$$P(x) = Q(x)$$

is a subset of the solution set of

$$[P(x)]^n = [Q(x)]^n,$$

for each positive integer n.

This theorem permits us to raise both members of an equation to the same natural-number power *without losing any solutions* of the original equation. On the other hand, it does *not* assert that the resulting equation will be equivalent to the original equation. The equation

$$[P(x)]^n = [Q(x)]^n$$

may have additional solutions (called **extraneous solutions**) that are not solutions of $P(x) = Q(x)$. For example, the solution set of $x = 10$ is $\{10\}$, whereas the solution set of $x^2 = 10^2$ or $x^2 = 100$ is $\{10, -10\}$. The equations $x = 10$ and $x^2 = 10^2$ are not equivalent.

EXAMPLE I   Find the solution set of $\sqrt[3]{x - 1} = -1$.

Solution   We raise each member of $\sqrt[3]{x - 1} = -1$ to the third power, obtaining

$$(\sqrt[3]{x - 1})^3 = (-1)^3$$
$$x - 1 = -1,$$

which is equivalent to

$$x = 0.$$

Since $\sqrt[3]{0 - 1} = -1$, 0 is a solution of the original equation. By Theorem 3.3, every solution of $\sqrt[3]{x - 1} = -1$ is a solution of $x = 0$. Thus, the solution set of $\sqrt[3]{x - 1} = -1$ is $\{0\}$.   ■

Necessity of Checking Solutions   Since raising both members of an equation to a natural-number power does not always produce an equivalent equation, each solution obtained using Theorem 3.3 *must* be substituted for the variable in the original equation to check its validity.

EXAMPLE 2   Find the solution of $\sqrt{x + 2} + 4 = x$.

Solution   We first rewrite the equation so the radical expression is by itself as one member of an equation,

$$\sqrt{x + 2} = x - 4,$$

and then apply Theorem 3.3. We obtain

$$(\sqrt{x + 2})^2 = (x - 4)^2$$
$$x + 2 = x^2 - 8x + 16.$$

This is equivalent to

$$x^2 - 9x + 14 = 0$$

or

$$(x - 2)(x - 7) = 0,$$

which clearly has solutions 2 and 7. Checking 2 in the original equation, however, we obtain

$$\sqrt{2 + 2} + 4 = 2,$$

which is false. Hence, 2 is not a solution of the original equation; it is an extraneous root. However, 7 does satisfy the original equation, so the solution set is $\{7\}$.   ■

It is sometimes necessary to apply Theorem 3.3 more than once in solving certain equations. This occurs most frequently when dealing with equations having more than one radical expression.

EXAMPLE 3   Find the solution set of $\sqrt{x + 4} + \sqrt{9 - x} = 5$.

Solution   We first rewrite this equation so that each member of the equivalent equation contains only one of the radical expressions. Thus, by adding $-\sqrt{9 - x}$ to each member, we obtain

$$\sqrt{x + 4} = 5 - \sqrt{9 - x}.$$

An application of Theorem 3.3 leads to

$$(\sqrt{x + 4})^2 = (5 - \sqrt{9 - x})^2$$
$$x + 4 = 25 - 10\sqrt{9 - x} + 9 - x$$
$$2x - 30 = -10\sqrt{9 - x}$$
$$x - 15 = -5\sqrt{9 - x}.$$

Applying Theorem 3.3 again, we have

$$(x - 15)^2 = (-5\sqrt{9 - x})^2$$
$$x^2 - 30x + 225 = 25(9 - x)$$
$$x^2 - 5x = 0$$
$$x(x - 5) = 0.$$

This equation has 0 and 5 as solutions. Both of these satisfy the original equation. Hence, the solution set is $\{0, 5\}$.   ■

Equations with Several Variables

In some cases we will need to solve an equation involving radical expressions for one variable in terms of one or more other variables. To do this, we employ the same methods discussed in Examples 1–3.

EXAMPLE 4   Solve $\sqrt{\dfrac{A}{\pi} + r^2} = R$, for $r > 0$.

Solution   Squaring both members we have

$$\frac{A}{\pi} + r^2 = R^2 \qquad \text{or} \qquad r^2 = R^2 - \frac{A}{\pi}.$$

Thus,

$$r = \pm\sqrt{R^2 - \frac{A}{\pi}}.$$

However, we assumed $r > 0$. Thus, we have

$$r = \sqrt{R^2 - \frac{A}{\pi}}. \qquad \blacksquare$$

Substitution Method

Some equations that are not polynomial equations can be solved by means of related polynomial equations. For example, though $y + 2\sqrt{y} - 8 = 0$ is not a polynomial equation, if the variable p is substituted for the radical expression $\sqrt{y}$ and p^2 is substituted for y, we then have $p^2 + 2p - 8 = 0$, which is a polynomial equation in p.

EXAMPLE 5   Find the solution set of $y + 2\sqrt{y} - 8 = 0$, $y \geq 0$.

Solution   If we set $p = \sqrt{y}$, then $p^2 = y$ and substituting in the given equation yields

$$p^2 + 2p - 8 = 0$$
$$(p + 4)(p - 2) = 0,$$

which has 2 and -4 as solutions. The value $p = 2$ leads to $\sqrt{y} = 2$ or $y = 4$. The value $p = -4$ leads to $\sqrt{y} = -4$, which has no solution since $\sqrt{y}$ is always nonnegative. The solution set is therefore $\{4\}$.   ■

The technique of substituting a variable for an expression is called the **method of substitution**. It is not limited to cases involving radicals. For example, in the equation

$$\left(x + \frac{1}{x}\right)^{-2} + 6\left(x + \frac{1}{x}\right)^{-1} + 8 = 0,$$

we would set

$$p = \left(x + \frac{1}{x}\right)^{-1}$$

to obtain

$$p^2 + 6p + 8 = 0.$$

Similarly, in the equation

$$(y + 3)^{1/2} - 4(y + 3)^{1/4} + 4 = 0,$$

we would set

$$p = (y + 3)^{1/4}$$

to obtain

$$p^2 - 4p + 4 = 0.$$

For some equations of the form $P(x) = 0$, the technique of substitution is simply a step that makes factoring $P(x)$ easier.

E X A M P L E 6   Solve $x^4 - 3x^2 - 18 = 0$.

Solution I   If we set $p = x^2$, then $p^2 = x^4$ and substituting in the given equation yields

$$p^2 - 3p - 18 = 0$$
$$(p - 6)(p + 3) = 0,$$

which has 6 and -3 as solutions. The value $p = 6$ leads to $x^2 = 6$, or

$$x = \sqrt{6}$$

or

$$x = -\sqrt{6}.$$

The value $p = -3$ leads to $x^2 = -3$, or

$$x = i\sqrt{3} \qquad \text{or} \qquad x = -i\sqrt{3}.$$

Thus, the solution set is $\{\sqrt{6}, -\sqrt{6}, i\sqrt{3}, -i\sqrt{3}\}$.

Solution 2   Factoring the left-hand member of the given equation, we have

$$(x^2 - 6)(x^2 + 3) = 0,$$

from which we also obtain the solution set $\{\sqrt{6}, -\sqrt{6}, i\sqrt{3}, -i\sqrt{3}\}$.   ■

Applications   Equations involving radical expressions are sometimes used in mathematical models. When solving such equations, care must be taken not to include extraneous solutions or solutions that have no meaning in the context of the given problem.

EXAMPLE 7   A man has a cable 102 feet long that he wants to attach to his house, to a point on his neighbor's house at the same height as the point on his own house, and to the ground at a point 25 feet from his house. If the two houses are 100 feet apart and the ground between the houses is level, how high above the ground should the man attach the cable to the two houses?

Solution   We first express the quantity asked for in a simple phrase and represent the quantity with a symbol.

$$\text{Height above ground to attach the cable:} \quad x$$

We next draw a figure and label it with the information given in the problem statement.

Length of cable from first house to ground: y

Length of cable from second house to ground: z

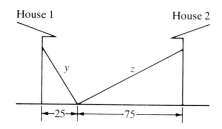

Next, we use the Pythagorean theorem and the fact that the length of the cable is 102 feet to express the relationships between these quantities.

$$y = \sqrt{x^2 + (25)^2}, \qquad z = \sqrt{x^2 + (75)^2}, \qquad y + z = 102$$

We then write a single equation in the variable x.

$$\sqrt{x^2 + (25)^2} + \sqrt{x^2 + (75)^2} = 102$$

Finally, we solve the equation.

$$\sqrt{x^2 + 625} = 102 - \sqrt{x^2 + 5625}$$

$$x^2 + 625 = 10{,}404 - 204\sqrt{x^2 + 5625} + x^2 + 5625$$

$$204\sqrt{x^2 + 5625} = 15{,}404$$

$$\sqrt{x^2 + 5625} \approx 75.5$$

$$x^2 + 5625 \approx 5700.25$$

$$x \approx \pm 8.7$$

Since $x > 0$ we find $x \approx 8.7$. The man should attach the cable about 8.7 feet above ground level.   ■

1–77 odd

E X E R C I S E   3.6

A

■   *Solve by using Theorem 3.3 and check. If there is no solution, so state. See Examples 1–3.*

1. $\sqrt{x} = 5$

2. $4\sqrt{x} = 6$

3. $\sqrt{y - 5} = 1$

4. $\sqrt{y - 2} = 2$

5. $x - 1 = \sqrt{2x - 3}$

6. $2y - 3 = \sqrt{-2y + 9}$

7. $\sqrt[3]{2 + y} = 2$

8. $\sqrt[5]{7 + x} = 3$

9. $\sqrt{x + 3}\sqrt{x + 9} = 4$

10. $\sqrt{x}\sqrt{x - 4} = 1$

11. $\sqrt{y^2 + y} = \sqrt{y^2 - 3y + 4}$

12. $\sqrt{2x^2 - 3} = \sqrt{2x^2 + 3x + 2}$

13. $\sqrt{y - 1} = \sqrt{y + 2} + 2$

14. $\sqrt{x} - \sqrt{2} = \sqrt{x - 2}$

15. $\sqrt{x - 4} + \sqrt{x + 4} = 4$

16. $\sqrt{x + 1} - \sqrt{x - 2} = 3$

17. $(5 + x)^{1/2} - x^{1/2} = 2$

18. $(y + 7)^{1/2} - (y - 4)^{1/2} = 1$

■   *Solve for the indicated variable. Leave the results in the form of an equation. Assume that denominators are not zero. See Example 4.*

19. $r = \sqrt{\dfrac{A}{\pi}}$, for A

20. $t = \sqrt{\dfrac{2v}{g}}$, for g

21. $x\sqrt{xy} = 1$, for y

22. $P = \pi \sqrt{\dfrac{l}{g}}$,   for g **23.** $x = \sqrt{a^2 - y^2}$,   for $y \geq 0$ **24.** $y = \dfrac{1}{\sqrt{1 - x}}$,   for x

25. $R = \sqrt{\dfrac{V}{\pi h} + r^2}$,   for V **26.** $R = \sqrt[3]{\dfrac{3V}{4\pi} + r^3}$,   for V

■   *Solve by the method of substitution. See Examples 5 and 6.*

27. $x + 6\sqrt{x} - 7 = 0$ **28.** $y - 12y^{1/2} + 35 = 0$ **29.** $z^{2/3} - 3z^{1/3} - 18 = 0$

30. $x^{2/3} - x^{1/3} - 6 = 0$ **31.** $\dfrac{1}{y^2} - \dfrac{7}{y} - 18 = 0$ **32.** $z^{-2} - 5z^{-1} - 14 = 0$

33. $\dfrac{x^2}{(x + 1)^2} + \dfrac{x}{x + 1} = 30$ **34.** $\left(1 + \dfrac{1}{y}\right)^2 + 3\left(1 + \dfrac{1}{y}\right) = 40$ **35.** $\sqrt{x} - 6\sqrt[4]{x} + 8 = 0$

36. $\sqrt{x} - 3\sqrt[4]{x} - 4 = 0$ **37.** $\sqrt{x - 6} + 3\sqrt[4]{x - 6} - 18 = 0$ **38.** $\sqrt[3]{x^2} - 12\sqrt[3]{x} + 20 = 0$

■   *For Exercises 39–46, see Example 7.*

39. The base of an isosceles triangle is one-half the length of the altitude. If the perimeter is $1 + \sqrt{17}$ centimeters, find the length of the base.

40. The longer leg of a right triangle is 1 centimeter shorter than the hypotenuse. Find the hypotenuse if the perimeter is 30 centimeters.

41. A plane flying due north at a rate of 200 mph passes a plane flying due east at a rate of 100 mph. How long will it take for the two planes to be 200 miles apart?

42. If a plane flying due north at the rate of 200 mph passes over city A at noon and a plane flying due east at the rate of 150 mph passes over city A at 1 PM, at what time will the two planes be 300 miles apart?

43. A woman on one side of a river wants to swim across and then walk to the next town, which is 2 miles downstream from the point directly opposite her position. If the river is $\frac{1}{2}$ mile wide and she wants her total distance traveled to be $2\frac{1}{4}$ miles, how far from the town should she come out of the water?

44. In Exercise 43, if the town were 4 miles from the point opposite the woman and she wanted to travel a total of $4\frac{1}{3}$ miles, how far from the town would she leave the river?

45. A wire of length 54 feet is to be attached to the top of an 8-foot vertical pole, then to the ground, and along the ground to a house. If the pole is 50 feet from the house, how far from the house should the wire be attached to the ground?

46. A wire of length 103 feet is to be attached to the ground between a 10-foot vertical pole and a house, to the top of the pole, and to the roof of the house. If the roof level of the house is 10 feet and the house is 100 feet from the pole, how far from the house should the wire be attached to the ground?

B

■   *Find all the real solutions of the given equation. If there are none, so state. Use a calculator to check the solutions in Exercises 47–60.*

47. $\sqrt{5 + \sqrt{x}} = \sqrt{x} - 1$ **48.** $\sqrt{13 + \sqrt{x}} = \sqrt{x} + 1$

49. $\sqrt{x + \sqrt{x}} = 1 - \sqrt{x}$ **50.** $\sqrt{\sqrt{x} - x} = \sqrt{x} + 3$

51. $x - x^{2/3} = x^{1/3} - 1$ **52.** $x^{4/3} + x^{2/3} = -x^2 - 1$

53. $x^{1/2} + x^{1/3} = 4x^{1/6} + 4$ **54.** $x^{3/4} + 9 = x^{1/2} + 9x^{1/4}$

55. $\sqrt{x^{1/2} + (x + 1)^{1/2}} = \sqrt[4]{x}$ **56.** $\sqrt{x^{1/2} - (x - 1)^{1/2}} = \sqrt[4]{x}$

57. $\sqrt{(x + 1)^{1/2} - (x - 1)^{1/2}} = \sqrt[4]{x}$ **58.** $\sqrt{(x + 1)^{1/2} + (x - 1)^{1/2}} = \sqrt[4]{x}$

59. $\sqrt{(x + 2)^{1/2} - (x - 2)^{1/2}} = \sqrt[4]{x}$ **60.** $\sqrt{(x + 2)^{1/2} + (x - 2)^{1/2}} = \sqrt[4]{x}$

3.7

SOLUTION OF LINEAR INEQUALITIES

A real number is called a **solution** of an inequality if, when the variable is replaced by that number, the resulting statement is true. The set of all solutions of an inequality is called the **solution set** of the inequality. Two inequalities are called **equivalent** if they have the same solution set.

Interval Notation

The solution set of an inequality is often expressed using a notation called **interval notation**. Sets such as $\{x \mid -2 < x \le 5\}$ are called **intervals** of real numbers. Interval notation involves the use of a parenthesis to denote an endpoint that is not in the set and a bracket to denote an endpoint that is in the set. Thus,

$$(-2, 5] = \{x \mid -2 < x \le 5\}.$$

For an infinite interval such as $\{x \mid x < 6\}$, we would write $(-\infty, 6)$, where the symbol $-\infty$ denotes the inclusion of all real numbers less than 6 in the interval. Similarly, $\{x \mid x > 4\}$ can be written as $(4, +\infty)$.

Intervals of real numbers can be graphed on a number line as shown in Table 3.1 on page 120. Note that an open circle on the graph indicates that the number corresponding to the indicated point is *not* in the set; a closed dot indicates the corresponding number *is* in the set.

EXAMPLE 1   Represent the solution set of $-1 \le x < 2$ in interval notation and as a graph on a number line.

Solution   The solution set of $-1 \le x < 2$ is the set of numbers between -1 and 2 including -1 but not including 2. In interval notation we write

$$[-1, 2).$$

The graph of the solution set is shown on the number line.

TABLE 3.1

Interval Notation	Graph
1. (a, b)	1.
2. $[a, b)$	2.
3. $(a, b]$	3.
4. $[a, b]$	4.
5. $(-\infty, a)$	5.
6. $(-\infty, a]$	6.
7. (a, ∞)	7.
8. $[a, \infty)$	8.

Elementary Transformations

As in the case with equations, we solve a given inequality by generating a series of equivalent inequalities until the solution set is obvious. To do this, we shall use the operations that follow, which from the order properties of the real numbers (see Preliminary Concepts) produce equivalent inequalities.

> **I.** Any expression may be added to each member of an inequality.
>
> **II.** Each member of an inequality may be multiplied by the same expression representing a positive number.
>
> **III.** Each member of an inequality may be multiplied by the same expression representing a negative number, if the direction of the inequality is reversed.

EXAMPLE 2 Find and graph the solution set of $\dfrac{-x + 3}{4} > -\dfrac{2}{3}$.

Solution We first multiply each member by the least common denominator of the fractions. By Transformation II, the given inequality is equivalent to

$$3(-x + 3) > -8$$
$$-3x + 9 > -8. \tag{1}$$

Next, we add -9 to each member of Inequality (1). By Transformation I, Inequality (1) is equivalent to

$$-3x > -17. \tag{2}$$

Finally, we multiply each member of Inequality (2) by $-\frac{1}{3}$ and reverse the direction of the inequality. By Transformation III, Inequality (2) is equivalent to

$$x < \frac{17}{3}.$$

The solution set is written

$$S = \left\{x \mid x < \frac{17}{3}\right\} = \left(-\infty, \frac{17}{3}\right).$$

The graph of the solution set is shown in the figure. The open dot on the right-hand endpoint indicates that $\frac{17}{3}$ is not a member of the solution set.

Inequalities sometimes appear in a form such as

$$-6 < 3x \le 15, \tag{3}$$

where an expression is between two inequality symbols. In Inequality (3), each expression may be multiplied by $\frac{1}{3}$ to obtain

$$-2 < x \le 5.$$

The solution set,

$$S = \{x \mid -2 < x \le 5\} = (-2, 5],$$

is shown on a line graph in Figure 3.1.

FIGURE 3.1

EXAMPLE 3 Find and graph the solution set of

$$-1 \le \frac{-2x + 5}{3} \le 1.$$

Solution We first multiply each member of the inequalities by 3 to obtain the equivalent inequalities

$$-3 \leq -2x + 5 \leq 3.$$

Next, we add -5 to each member of the inequalities to obtain

$$-8 \leq -2x \leq -2.$$

Finally, we multiply each member of the inequalities by $-\frac{1}{2}$ *and reverse the sense of the inequalities* to obtain the equivalent inequalities

$$(-8)\left(-\frac{1}{2}\right) \geq x \geq (-2)\left(-\frac{1}{2}\right)$$

or, equivalently,

$$1 \leq x \leq 4.$$

The solution set is then written

$$\{x \mid 1 \leq x \leq 4\} = [1, 4].$$

The graph of the solution set is shown in the figure.

Applications Inequalities can often be used to create mathematical models in cases where a range of values is sought.

E X A M P L E 4 A student has grades of 80%, 92%, 78%, and 86% on the first four tests. What grade must the student make on the fifth test if her average grade is to be greater than or equal to 80% but less than 86%?

Solution We first express the quantities used to obtain a solution in simple phrases and represent them using symbols.

Grade on fifth test: t_5

Average test grade: $\bar{t}$

We next use the fact that the average of five tests is the sum of the five test scores divided by 5 to write

$$\bar{t} = \frac{80 + 92 + 78 + 86 + t_5}{5}$$

$$= \frac{336 + t_5}{5}.$$

Next, we use the conditions given in the problem statement to write an inequality.

$$80 \le \frac{336 + t_5}{5} < 86$$

Finally, we solve the inequality.

$$80 \le \frac{336 + t_5}{5} < 86$$

$$400 \le 336 + t_5 < 430$$

$$64 \le t_5 < 94$$

Thus, the student's fifth test score must lie in the interval [64, 94).   ■

Inequalities with Variable Coefficients

In certain cases we want to solve inequalities that involve variable coefficients. In such cases we generally have some constraints on the variables that represent the coefficients, and we use the same methods employed in the preceding examples.

EXAMPLE 5   Assume ε (epsilon) is a positive real number and solve $-\varepsilon < 2x + 3 < \varepsilon$.

Solution   We first add -3 to each member of the inequalities to obtain

$$-3 - \varepsilon < 2x < -3 + \varepsilon.$$

We next multiply each member by $\frac{1}{2}$ to obtain

$$\frac{1}{2}(-3 - \varepsilon) < x < \frac{1}{2}(-3 + \varepsilon).$$

The solution set in interval notation is

$$\left(\frac{1}{2}(-3 - \varepsilon), \frac{1}{2}(-3 + \varepsilon) \right).$$ ■

11-33 odd   49-59 odd

EXERCISE 3.7

A

■ *Represent the solution set in interval notation. See Example 1.*

I. $0 \le x \le 1$ **2.** $-1 \le x \le 4$ **3.** $-2 < x < 5$ **4.** $3 < x < 6$

5. $2 \le x < 7$ **6.** $-2 < x \le 4$ **7.** $-5 \le x$ **8.** $5 < x$

9. $x \le -1$ **10.** $x < 2$

■ *Solve each inequality. Write the solution set in interval notation and represent it on a line graph. See Examples 2 and 3.*

I I. $x + 11 > 5$ **12.** $x - 3 < -2$ **13.** $3x + 1 \le 10$

14. $5x - 4 \le 11$ **15.** $2x + 7 \ge -5$ **16.** $3 - x \le -1$

17. $x + 6 \le 3x + 1$ **18.** $2x + 7 \le x + 4$ **19.** $1 - 4x > x + 6$

20. $4x \ge -5x - 1$ **21.** $-(x + 3) \le -1$ **22.** $-(2x + 4) \ge 3$

23. $-2 \le \dfrac{3x + 1}{-4}$ **24.** $1 \le \dfrac{-2x - 1}{-3}$ **25.** $\dfrac{3x + 4}{2} < 6$

26. $\dfrac{-3x + 2}{-4} < 5$ **27.** $\dfrac{2x - 3}{3} \le \dfrac{3x}{2}$ **28.** $\dfrac{3x - 4}{-2} \le \dfrac{-2x}{-5}$

29. $-1 \le x + 5 \le 6$ **30.** $0 \le 2x - 3 < 1$ **31.** $-4 < \dfrac{3x + 2}{5} \le -2$

32. $4 \le \dfrac{3x - 1}{2} \le 10$ **(33)** $5 \le \dfrac{x + 1}{-3} \le 8$ **34.** $-2 \le \dfrac{x + 5}{2} \le 0$

■ *For Exercises 35–42, see Example 4.*

35. The Alpha car company rents a compact car for $24 per day. The Beta car company rents a similar car for $18 per day plus an initial fee of $90. For what rental period would it be cheaper to rent from Beta?

36. Employment agency A charges a commission of $50 plus 15% of the first month's salary for its services. Agency B charges a commission of $80 plus 12% of the first month's salary. When is it cheaper to use agency B?

37. A woman wishes to invest $10,000, part at 9% and part at 12%. What is the least amount she can invest at 12%, if she wishes an annual return of at least $1008?

38. A man wants to invest $5000, with equal parts at 11% and 12% and the remainder at 10%. What is the least amount he can invest at 12%, if he wishes an annual return of at least $550?

■ *For Exercises 39 and 40, assume that a student must have an average of 80%–90% inclusive on five tests to receive a grade of B.*

39. What grade on the fifth test would qualify a student for a B if his grades on the first four tests were 78%, 64%, 88%, and 76%?

40. What grade on the fifth test would qualify a student for a B if her first three test scores were 78%, 96%, and 65%, and she missed the fourth test but the instructor agreed to substitute the grade she received on the fifth test for her fourth test score?

41. The Fahrenheit and Celsius temperatures are related by $C = \frac{5}{9}(F - 32)$. Within what range must the temperature be in Fahrenheit degrees for the temperature in Celsius degrees to be between $-10°$ and $20°$?

42. Within what range must the temperature be in Fahrenheit degrees for the temperature in Celsius degrees to be between $0°$ and $100°$?

B

■ *Assume that ε (epsilon) is a positive real number and refer to Example 5. Solve each inequality. Write the solution set in interval notation.*

43. $-\varepsilon < 4x - 5$ **44.** $4x - 5 < \varepsilon$ **45.** $-\varepsilon < 3x + 2 < \varepsilon$

46. $-\varepsilon < 2x + 4 < \varepsilon$ **47.** $-\varepsilon < -3x + 7 < \varepsilon$ **48.** $-\varepsilon < -2x + 1 < \varepsilon$

■ *Determine the values of k so that each equation has real solutions. Write the solution in interval notation.*

49. $x^2 + 2x + k = 0$ **50.** $x^2 - 3x - k = 0$ **51.** $2x^2 - x + k = 0$

52. $2x^2 + 3x - k = 0$ **53.** $3x^2 - 2x + 2k = 4$ **54.** $3x^2 + 4x - 2k = 2$

■ *Determine the values of k so that each equation has two imaginary solutions. Write the solution in interval notation.*

55. $x^2 + 3x + k = 0$ **56.** $x^2 - 2x + k = 0$ **57.** $2x^2 + 3x - k = 2$

58. $3x^2 - 2x + k = 1$ **59.** $-3x^2 + 2x - 2k = 1$ **60.** $-x^2 + 4x - 2k = 3$

3.8

SOLUTION OF NONLINEAR INEQUALITIES

There are several methods for solving nonlinear inequalities. We shall consider only one of the methods here.

Critical Numbers

We shall first discuss the concept of a critical number. To illustrate this concept, consider the inequality

$$x^2 + 4x - 5 < 0.$$

The left-hand member of this inequality factors to give us

$$(x + 5)(x - 1) < 0.$$

Notice that the algebraic sign of the product on the left is determined by the algebraic signs of the individual factors; that is, the product can be less than zero, if and only if the factors $x + 5$ and $x - 1$ are opposite in sign. We can illustrate the set of real numbers that satisfies this condition with a sign array (Figure 3.2).

FIGURE 3.2

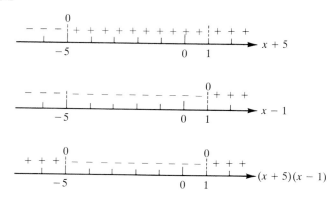

Observe from Figure 3.2 that the numbers -5 and 1, which can be obtained by solving the equation $(x + 5)(x - 1) = 0$, separate the set of real numbers into the three intervals

$$(-\infty, -5), \quad (-5, 1), \quad (1, +\infty).$$

Further, each of these intervals either is or is not part of the solution set of the inequality $x^2 + 4x - 5 < 0$.

Notice that the sign of the product in Figure 3.2 changes at points where the product is zero. This happens because it is at these points where one of the factors in the product changes sign. In general, a product $P(x)$ can only change sign at points where it (and hence one of its factors) is either zero or undefined. Thus, we call the values of x for which $P(x) = 0$ or $P(x)$ is undefined **critical numbers** of the inequality

$$P(x) > 0 \quad [\text{or} \quad P(x) < 0, \quad \text{or} \quad P(x) \geq 0, \quad \text{or} \quad P(x) \leq 0].$$

Notice that an expression does not *have* to change sign at a critical point. For example, the inequality

$$x^2 - 2x + 1 > 0$$

has 1 as a critical number, but the left-hand member $x^2 - 2x + 1 = (x - 1)^2$ is positive except at $x = 1$.

EXAMPLE I   Find the critical numbers for each inequality.

(a) $2x^2 - x - 1 > 0$ (b) $\dfrac{x}{x^2 - 4} < 0$

Solution   (a)   Factoring the left-hand member of

$$2x^2 - x - 1 = 0$$

we obtain

$$(2x + 1)(x - 1) = 0,$$

from which

$$x = -\frac{1}{2} \quad \text{or} \quad x = 1.$$

Thus, the critical numbers are $-\frac{1}{2}$ and 1.

(b)   Solving the equation

$$\frac{x}{x^2 - 4} = 0,$$

we obtain $x = 0$. The expression $\dfrac{x}{x^2 - 4}$ is undefined, when $x^2 - 4 = 0$ or when $x = \pm 2$. Thus, the critical numbers are 0, 2, and -2.   ■

The discussion before Example 1 suggests that the critical numbers of an inequality $P(x) > 0$ separate the set of real numbers into nonoverlapping intervals, and each of these intervals either is or is not part of the solution set (since the sign of $P(x)$ cannot change between critical values). This provides us with a method for solving nonlinear inequalities.

EXAMPLE 2   Solve $x^2 - 3x - 4 \geq 0$.

Solution   Factoring the left-hand member yields

$$(x + 1)(x - 4) \geq 0.$$

We observe that -1 and 4 are critical numbers because $(x + 1)(x - 4) = 0$ for these values. Hence, we wish to check the intervals shown on the following number line.

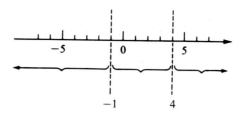

Since each of these intervals either is or is not part of the solution set, it suffices to check a single value from each interval to see whether or not the corresponding interval is part of the solution set.

We now select any value from each of these intervals, say -2, 0, and 5. We then check each of these values in the original inequality.

$$(-2)^2 - 3(-2) - 4 \overset{?}{\geq} 0 \qquad (0)^2 - 3(0) - 4 \overset{?}{\geq} 0 \qquad (5)^2 - 3(5) - 4 \overset{?}{\geq} 0$$

$$\text{Yes.} \qquad\qquad\qquad \text{No.} \qquad\qquad\qquad \text{Yes.}$$

A value satisfies the inequality if and only if the corresponding interval is part of the solution set. Since the inequality is satisfied by the critical values -1 and 4 (because $0 \geq 0$), these values are included in the solution set. Hence, the solution set is

$$\{x \mid x \leq -1\} \cup \{x \mid x \geq 4\} = (-\infty, -1] \cup [4, \infty).$$

The graph is shown in the figure.

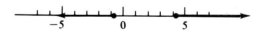

E X A M P L E 3   Solve $x^2 - 2x - 1 \leq 0$.

Solution   We first find the critical numbers by using the quadratic formula to solve $x^2 - 2x - 1 = 0$. We have

$$x = \frac{2 \pm \sqrt{4 + 4}}{2} = 1 \pm \sqrt{2}.$$

Hence, we wish to check the intervals shown on the following number line.

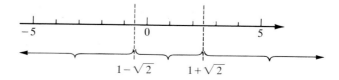

We select any value from each of these intervals, say -4, 1, and 4. We then check each of these values in the original inequality.

$$(-4)^2 - 2(-4) - 1 \overset{?}{\leq} 0 \qquad 1^2 - 2(1) - 1 \overset{?}{\leq} 0 \qquad 4^2 - 2(4) - 1 \overset{?}{\leq} 0$$

No.   Yes.   No.

Finally, we test the critical numbers themselves; since they make the expression 0, they are solutions. The solution set is therefore $[1 - \sqrt{2}, 1 + \sqrt{2}]$. The solution set is shown in the figure.

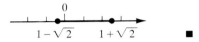

The notion of critical values can be used to solve inequalities involving fractions.

EXAMPLE 4   Solve the inequality $\dfrac{x}{x - 2} \geq 5$.

Solution   We first write the given inequality equivalently as

$$\frac{x}{x - 2} - 5 \geq 0,$$

from which

$$\frac{x - 5(x - 2)}{x - 2} \geq 0$$

$$\frac{-4x + 10}{x - 2} \geq 0.$$

In this case, the critical numbers are $\dfrac{5}{2}$ and 2, because $\dfrac{-4x + 10}{x - 2}$ equals zero for $x = \dfrac{5}{2}$, and $\dfrac{-4x + 10}{x - 2}$ is undefined for $x = 2$. Thus, we want to check the intervals shown on the number line that follows.

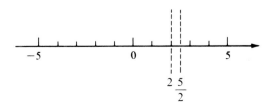

Substituting values from each of the three intervals, say 0, $\frac{9}{4}$, and 3, for the variable in the original inequality and checking the critical numbers 2 and $\frac{5}{2}$, we find the solution set,

$$\left\{x \mid 2 < x \le \frac{5}{2}\right\} = \left(2, \frac{5}{2}\right].$$

The graph is shown in the following figure. Note that the left-hand endpoint is an open dot (2 is not a member of the solution set) because the left-hand member of the original inequality is undefined for $x = 2$.

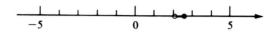

EXAMPLE 5   Solve $\dfrac{(x - 1)(x + 2)}{(x - 3)} \le 0.$

Solution   We first find the critical numbers. The left-hand member of the inequality is 0 when the numerator is 0, that is, when $(x - 1)(x + 2) = 0$. Thus, 1 and -2 are critical numbers. Further, the left-hand member is undefined when the denominator is 0, that is, when $x - 3 = 0$. Thus, 3 is also a critical number.

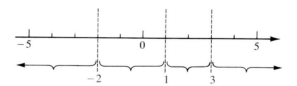

We next check the intervals shown on the number line by substituting a value from each of the intervals, say -3, 0, 2, and 4.

$$\frac{(-3 - 1)(-3 + 2)}{(-3 - 3)} \overset{?}{\le} 0 \qquad\qquad \frac{(0 - 1)(0 + 2)}{(0 - 3)} \overset{?}{\le} 0$$

Yes.   No.

$$\frac{(2 - 1)(2 + 2)}{(2 - 3)} \overset{?}{\le} 0 \qquad\qquad \frac{(4 - 1)(4 + 2)}{(4 - 3)} \overset{?}{\le} 0$$

Yes.   No.

Finally, we test each of the critical values; 1 and -2 are solutions but 3 is not. Thus, the solution set is $(-\infty, -2] \cup [1, 3)$ and is shown in the figure.

1-41 odd

EXERCISE 3.8

A

■ Solve each inequality. Write the solution set in interval notation. For Exercises 1–20, see Examples 2 and 3.

1. $(x + 2)(x - 3) > 0$ **2.** $(x + 4)(x + 7) < 0$ **3.** $(2x - 3)(3x + 8) \le 0$

4. $(3x + 2)(3x - 4) \ge 0$ **5.** $x^2 - x - 12 > 0$ **6.** $x^2 + 3x - 28 \le 0$

7. $x^2 - 2x - 3 > 0$ **8.** $x^2 + x - 20 \ge 0$ **9.** $x^2 + 5x > 4x + 2$

10. $4x + 5 \ge -x^2 - 2x$ **11.** $2x^2 + 8x \le -x^2 + 3$ **12.** $4x^2 + x < -2x^2 + 2$

13. $4x^2 + 10x \ge -4x^2 + 3$ **14.** $6x^2 - 10 > x + 2$ **15.** $x^2 + 4 \le 0$

16. $x^2 + 9 \ge 0$ **17.** $x^2 - 4x + 1 \ge 0$ **18.** $x^2 + 4x + 2 \le 0$

19. $x^2 - x + 4 < 0$ **20.** $-x^2 + x + 1 > 0$

■ For Exercises 21–48, see Examples 4 and 5.

21. $\dfrac{1}{x + 2} \ge 0$ **22.** $\dfrac{3}{x - 4} \ge 0$ **23.** $\dfrac{-x}{2x + 1} < 0$ **24.** $\dfrac{-x}{3x - 1} < 0$

25. $\dfrac{2x}{2x + 3} \ge 0$ **26.** $\dfrac{3x}{2x + 5} \ge 0$ **27.** $\dfrac{x + 1}{x} \le 4$ **28.** $\dfrac{x}{x - 2} \ge 4$

29. $\dfrac{x - 1}{x + 1} < 2$ **30.** $\dfrac{x + 1}{x - 1} \le 2$ **31.** $\dfrac{2x}{2 - x} + 3 > 2$ **32.** $\dfrac{2x}{3 + x} - 3 \le -6$

33. $\dfrac{(x + 1)}{(x - 1)(x - 3)} > 0$ **34.** $\dfrac{(x - 1)(x + 1)}{(x - 5)} < 0$

35. $\dfrac{(x - 2)(x + 1)}{x} < 0$ **36.** $\dfrac{x}{(x + 2)(x - 4)} \ge 0$

37. $(x - 1)(x + 1)(x + 3) \ge 0$ **38.** $(x + 1)(x - 4)x < 0$

39. $\dfrac{(x + 2)(x - 2)}{x(x + 4)} > 0$ **40.** $x(x + 2)(x - 2)(x - 4) \ge 0$

B

41. $\dfrac{2}{(x - 1)(x + 1)} < 1$ **42.** $\dfrac{3}{(x - 2)(x + 2)} < 2$

43. $\dfrac{-2}{x^2 - 3x + 2} \ge 5$ **44.** $\dfrac{-3}{x^2 + 2x - 8} \ge 1$

■ Solve each inequality. Write the solution in interval notation. Hint: First rewrite the inequality in the form $P(x) < 0$ or $P(x) \le 0$.

45. $\dfrac{1}{x + 1} - \dfrac{1}{x^2 - 1} \le \dfrac{1}{x - 1}$ **46.** $\dfrac{1}{x - 2} + \dfrac{1}{x^2 - 4} \le \dfrac{1}{x + 2}$

47. $\dfrac{3}{x+1} + \dfrac{2}{x-2} > \dfrac{x}{x^2-x-2}$

48. $\dfrac{3}{x-3} - \dfrac{1}{x-1} > \dfrac{4x}{x^2-4x+3}$

49. A ball thrown vertically reaches a height h in feet given by $h = 56t - 16t^2$, where t is time measured in seconds. During what period(s) of time is the ball between 40 feet and 48 feet high?

50. A projectile fired from level ground is at a height of $320t - 16t^2$ feet after t seconds. During what period of time is it higher than 1024 feet?

51. The volume of a right circular cylinder is $V = \pi r^2 h$, where r is the radius of the base and h is the height of the cylinder. If the height is two times the diameter, in what range must the radius be if the volume is to be between 500π cubic feet and 2916π cubic feet?

52. The volume of a can in the shape of a right circular cylinder is to be between 21.2 cubic inches and 21.6 cubic inches. If the diameter of the base is equal to the height, in what range (to the nearest hundredth of an inch) must the radius of the base fall?

53. The volume of a sphere is given by the formula $V = 4\pi r^3/3$, where r is the radius and $\pi \approx 3.1416$. In what range (to the nearest hundredth of a foot) must the radius be if the volume is to be between 4 cubic feet and 4.1 cubic feet?

54. The volume of a spherical tank is to be between 539,200 cubic feet and 539,500 cubic feet. In what range (to the nearest hundredth of a foot) must the radius fall to accomplish this?

■　*Determine the values of k so that each equation has two real solutions. Write the set of values for k in interval notation.*

55. $x^2 + kx + 1 = 0$

56. $x^2 - 2kx + 5 = 0$

57. $kx^2 + 2kx + 1 = 0$

58. $kx^2 - 3kx + 6 = 0$

59. $kx^2 + 3kx - k = 6$

60. $kx^2 - 2kx + 2k = 4$

3.9

EQUATIONS AND INEQUALITIES INVOLVING ABSOLUTE VALUE

The absolute value of a real number x is defined (see Preliminary Concepts) by

$$|x| = \begin{cases} x & \text{if} \quad x \geq 0, \\ -x & \text{if} \quad x < 0. \end{cases}$$

For example, $|-5| = 5$. Since it is also true that $|5| = 5$, the equation $|x| = 5$ has two solutions, 5 and -5.

In general, simple equations involving absolute values are contractions of two equations without absolute values. The same is true of inequalities. Summarizing, for $a \geq 0$ we have the following statements.

Definitions

> **I.** $|x| = a$ is equivalent to $x = a$ or $x = -a$;
>
> **II.** $|x| < a$ is equivalent to $-a < x$ and $x < a$; i.e., $-a < x < a$;

III. $|x| > a$ is equivalent to $x < -a$ or $x > a$.

IV. $|x| \leq a$ is equivalent to $-a \leq x$ and $x \leq a$; i.e., $-a \leq x \leq a$.

V. $|x| \geq a$ is equivalent to $x \leq -a$ or $x \geq a$.

The relationships for $a > 0$ are indicated in Figure 3.3.

FIGURE 3.3

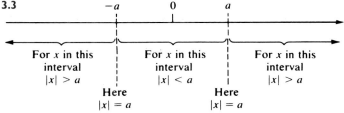

To solve equations involving absolute value, we first use Statement I to write two equivalent equations without absolute value. We then solve each of the two equations using methods studied in the earlier sections of this chapter.

E X A M P L E I   Solve $|x - 3| = 5$.

Solution   By Statement I, this equation is equivalent to

$$x - 3 = 5 \qquad \text{or} \qquad x - 3 = -5.$$

The solution to the first equality is 8; the solution to the second, -2. The solution set is $\{-2, 8\}$.

E X A M P L E 2   (a) $|3x - 4| = 2$ (b) $|x^2 - 1| = 3$

Solution   (a)   From Statement I, we have

$$3x - 4 = -2 \qquad \text{or} \qquad 3x - 4 = 2$$
$$x = \frac{2}{3} \qquad\qquad\qquad x = 2$$

Thus, the solution set is $\{2, \frac{2}{3}\}$.

(b)   From Statement I, we have

$$x^2 - 1 = 3 \qquad \text{or} \qquad x^2 - 1 = -3$$
$$x = \pm 2 \qquad\qquad \text{No real solutions.}$$

Thus, the solution set is $\{-2, 2\}$.   ■

To solve inequalities involving absolute values, we use Statements II–V to write the given inequality in terms of an equivalent pair of inequalities which we then solve using the methods discussed in Sections 3.7 and 3.8.

E X A M P L E   3   Solve $|x - 3| < 5$.

Solution   By Statement II, this inequality is equivalent to

$$-5 < x - 3 \qquad \text{and} \qquad x - 3 < 5,$$

or to

$$-5 < x - 3 < 5,$$
$$-2 < x < 8.$$

The solution set of the given inequality is the interval $(-2, 8)$.

E X A M P L E   4   Solve $|x - 3| > 4$.

Solution   By Statement III, this inequality is equivalent to

$$x - 3 < -4 \qquad \text{or} \qquad x - 3 > 4.$$

The first inequality reduces to $x < -1$, so its solution set is the interval $(-\infty, -1)$. The second reduces to $x > 7$ and has solution set $(7, \infty)$. The solution of the given inequality is therefore $(-\infty, -1) \cup (7, \infty)$.   ■

Inequalities involving absolute values and $\leq$ or $\geq$ are treated similarly.

E X A M P L E   5   (a) $|x - 2| \leq 5$ (b) $|x + 3| \geq 5$

Solution   (a)   From Statement IV, we have

$$-5 \leq x - 2 \leq 5$$

from which we obtain

$$-3 \leq x \leq 7.$$

Thus, the solution set is $[-3, 7]$.

(b)   From Statement V, we have

$$x + 3 \geq 5 \quad \text{or} \quad x + 3 \leq -5$$
$$x \geq 2 \qquad\qquad\qquad x \leq -8$$

Thus, the solution set is $(-\infty, -8] \cup [2, \infty)$.   ■

Alternative Method of Solution

We can also solve absolute-value inequalities by using a method similar to the method of critical numbers used in Section 3.8.

E X A M P L E   6   Solve $|x^2 - 1| < 8$ in the set of real numbers.

Solution   The critical numbers are obtained by solving the equation

$$|x^2 - 1| - 8 = 0,$$

or, equivalently, the two equations

$$x^2 - 1 = 8 \quad \text{or} \quad x^2 - 1 = -8$$
$$x^2 = 9 \qquad\qquad\qquad x^2 = -7$$
$$x = \pm 3 \qquad\qquad \text{No real solutions.}$$

Thus, the critical numbers are the real numbers -3 and 3. These values separate the set of real numbers into three intervals, as shown in the figure.

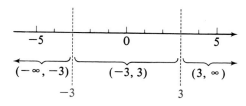

Substituting arbitrary values from each of the three intervals, say, $-4, 0,$ and 4, for the variable in the original inequality, we have

$$|(-4)^2 - 1| < 8 \qquad |(0)^2 - 1| < 8 \qquad |(4)^2 - 1| < 8$$
$$\text{No.} \qquad\qquad \text{Yes.} \qquad\qquad \text{No.}$$

Hence, the solution set is $\{x \mid -3 < x < 3\} = (-3, 3)$.   ■

Solution Set to Inequality

It is sometimes useful to represent an interval as the solution set of an inequality involving absolute values. From Statements II and IV, the intervals $(-a, a)$ and $[-a, a]$ are the solution sets of $|x| < a$ and $|x| \leq a$, respectively. For intervals that are not centered at 0, we add the value to each member, which results in an inequality that has the form

$$-a < x + b < a \quad \text{or} \quad -a \leq x + b \leq a.$$

E X A M P L E   7   Write an inequality involving absolute values that has $-1 \leq x \leq 3$ as its solution set.

Solution   Notice that the length of the interval $[-1, 3]$ is 4 and its center is at 1. Thus, if we subtract 1 from each member of the given inequality we have

$$-2 \leq x - 1 \leq 2,$$

which is equivalent to $|x - 1| \leq 2$.   ■

E X E R C I S E   3.9

A

■ *Solve the equation in the set of real numbers. See Examples 1 and 2.*

1. $|x - 2| = 4$ **2.** $|3 + x| = 4$ **3.** $|6 + x| = 5$

4. $|x - 4| = 2$ **5.** $|2x - 5| = 3$ **6.** $|3x + 2| = 3$

7. $|x^2| = 16$ **8.** $|(x - 2)^2| = 4$ **9.** $|x^2 + 5x + 6| = 6$

10. $|x^2 - 2x - 8| = 8$ **11.** $|x^3 + 6| = 2$ **12.** $|x^3 - 1| = 7$

■ *Solve the inequality. Write the solution set in interval notation. See Examples 3–6.*

13. $|x| \leq 1$ **14.** $|x| \geq 2$ **15.** $|x + 2| < 2$ **16.** $|x - 3| < 4$

17. $\left|x + \dfrac{1}{2}\right| < \dfrac{1}{3}$ **18.** $\left|x - \dfrac{1}{4}\right| < \dfrac{1}{2}$ **19.** $|x + 4| > 4$ **20.** $|x - 2| > 3$

21. $|2 - 3x| > 1$ **22.** $|2x - 1| > 1$ **23.** $|x + 4| \leq 5$ **24.** $|x - 2| \leq 6$

25. $|2x - 3| \leq 3$ **26.** $|3x + 4| \leq \dfrac{1}{2}$ **27.** $\left|x - \dfrac{1}{2}\right| \geq 2$ **28.** $|x + 1| \geq \dfrac{1}{2}$

29. $|3x - 1| \geq 3$ **30.** $\left|\dfrac{1}{2}x - \dfrac{1}{4}\right| \geq \dfrac{1}{4}$

■ *Write an inequality involving absolute values that has the given interval as its solution set. For Exercises 31–38 see Example 7.*

31. $-3 \le x \le 3$ **32.** $-4 \le x \le 4$ **33.** $-2 < x < 6$ **34.** $0 \le x \le 8$

35. $-8 \le x \le -2$ **36.** $-10 \le x \le 0$ **37.** $-5 < x < 2$ **38.** $-7 < x < -2$

B

■ *Solve. If there are no solutions, so state. [Hint: Divide both members by the right-hand member.]*

39. $|2x + 3| = |x + 1|$ **40.** $|3x - 4| = |x - 2|$

41. $|x^2 + x| = |x + 1|$ **42.** $|x^2| = |x - 1|$

■ *Solve. Write the solution set in interval notation.*

43. $|4x^2 - 2| < 2$ **44.** $|2x^2 - 6| < 4$ **45.** $|x^2 - 4x| \ge 4$

46. $|x^2 + 5x| \ge 6$ **47.** $\left|\dfrac{x}{2x + 1}\right| \le 3$ **48.** $\left|\dfrac{2x}{x - 1}\right| \le 4$

■ *Express each of the following statements using an inequality involving absolute values.*

49. x is within 5 units of 4.   **50.** y is within 2 units of -2.

51. $x^2 - 4$ is no more than 0.1 units from 12.   **52.** $y^3 - 2$ is no more than 10^{-4} units from -29.

53. $x - 2$ is within 10^{-4} units of L.   **54.** $x^2 + 4$ is within 10^{-5} units of L.

55. $x^2 + 1$ is within ε units of L.   **56.** $x^4 - 2x$ is within ε units of L.

57. x is more than 4 units from y.   **58.** x is more than 2 units from x^2.

59. $x - 2$ is at least 10^{-3} units from 2.   **60.** $x + 4$ is at least 10^{-5} units from 5.

61. $P(x)$ is within ε units of L.   **62.** $P(x)$ is at least ε units from L.

CHAPTER REVIEW

Key Words and Phrases

■ *Define or explain each of the following words and phrases.*

1. solution of an equation (inequality)

2. solution set of an equation (inequality)

3. equivalent equations (inequalities)

4. elementary transformation of an equation (inequality)

5. linear equation

6. quadratic equation

7. multiplicity

8. completing the square

9. quadratic formula

10. discriminant

11. extraneous solutions

12. interval

13. interval notation

14. critical numbers of an inequality

Review Exercises

A

[3.1] ■ *Solve the equation.*

1. $3 + \dfrac{x}{5} = \dfrac{7}{10}$

2. $\dfrac{3x}{4} - \dfrac{5x - 1}{8} = \dfrac{1}{2}$

3. $\dfrac{x}{x + 1} + \dfrac{1}{3} = 2$

4. $1 - \dfrac{y + 1}{y - 1} = \dfrac{3}{y}$

5. Solve $\dfrac{x + y}{5} = \dfrac{x - y}{3}$ for y in terms of x.

6. Solve $\dfrac{x + y}{5} = \dfrac{x - y}{3}$ for x in terms of y.

[3.2] **7.** A woman swims $\frac{1}{2}$ mile in 45 minutes. At the same rate, how far can she swim in 1 hour and 15 minutes?

8. How much water should be added to 25 gallons of a solution that is 20% acid to obtain a solution that is 5% acid?

[3.3] ■ *Solve by factoring.*

9. $(x - 2)(x + 1) = 4$

10. $x(x + 4) = 21$

■ *Solve by extraction of roots.*

11. $6x^2 = 30$

12. $(x - 5)^2 = 2$

■ *Write the equation in the form $(x - p)^2 = q$ by completing the square.*

13. $x^2 - 4x + 1 = 0$

14. $4x^2 + 12x + 8 = 0$

[3.4] ■ *Solve for x by using the quadratic formula.*

15. $x^2 + 5x - 2 = 0$

16. $2x^2 + x - 1 = 0$

17. $8x^2 - 4x - 7 = 0$

18. $x^2 = 4 - 2x$

■ *Solve for x in terms of k by using the quadratic formula.*

19. $kx^2 - 3x + 1 = 0$

20. $x^2 + kx - 4 = 0$

[3.5] **21.** A rectangular lawn is twice as long as it is wide. If its area is 288 square yards, what are its dimensions?

22. A steel plate is to be in the shape of a square with an equilateral triangle at one edge. If the plate is to have total area of $4 + \sqrt{3}$ square inches, what is the side length of the square?

[3.6] ■ *Solve the equation by using Theorem 3.3.*

23. $x - 5\sqrt{x} + 6 = 0$ **24.** $\sqrt{x + 1} + \sqrt{x + 8} = 7$

■ *Solve the equation by the method of substitution.*

25. $y - 3y^{1/2} + 2 = 0$ **26.** $y^{-2} - y^{-1} - 42 = 0$

[3.7–3.8] ■ *Solve the inequality.*

27. $\dfrac{x - 3}{5} \geq 7$ **28.** $2(x + 1) < \dfrac{1}{3}x$

29. $x^2 + 3x - 10 < 0$ **30.** $\dfrac{2}{1 - x} > 3$

[3.9] ■ *Solve the equation.*

31. $|3x + 1| = 5$ **32.** $\left|x - \dfrac{2}{3}\right| = \dfrac{5}{3}$

33. $|(x - 2)^2| = 4$ **34.** $|x^2 + x - 2| = 2$

■ *Solve the inequality and write the solution set in interval notation.*

35. $|2x + 1| \leq 1$ **36.** $|-2x + 3| \geq 4$

37. $|x^2 - 2| < 1$ **38.** $\left|\dfrac{2}{x + 1}\right| < 3$

■ *Write the given inequality as a single inequality involving an absolute-value symbol.*

39. $-5 \leq x \leq 3$ **40.** $-6 < x < 2$

B

41. Solve $\dfrac{(1 + y')y^2 - 2yy'(x + y)}{y^4} = 0$ for y' in terms of x and y.

42. Solve $\dfrac{(2x - 2yy')y^3 - 3y^2y'(x^2 - y^2)}{y^6} = 0$ for y' in terms of x and y.

43. For what real numbers k does the quadratic equation $x^2 + kx + k^2 = 0$ have a real solution?

44. For what real numbers k does the quadratic equation $x^2 + kx - k^2 = 0$ have two distinct real solutions?

■ *Find the real solutions of each equation.*

45. $\sqrt{x + \sqrt{x}} = \sqrt{x} + 1$ **46.** $\sqrt{\sqrt{x} - 2} = \sqrt{x} - 3$

47. $x^{1/2} + x^{1/3} = x^{1/6} + 1$ **48.** $x^2 + x^{4/3} = x^{2/3} + 1$

■ *Write the solution set of each inequality in interval notation.*

49. $\dfrac{x^2 + x}{x + 2} > \dfrac{-x + 3}{x + 2}$ **50.** $\dfrac{x^2 - 5x}{x + 1} \geq \dfrac{-x^2 + 4x - 4}{x + 1}$

51. $|x^2 + 2x| \leq 3$ **52.** $\left| \dfrac{2x}{x + 2} \right| \leq 4$

Stop

Part 199

4 Relations and Functions

In Chapter 3 the models for applied problems with which we dealt involved only one variable. In Chapter 4 we begin to concern ourselves with problems that require models involving two variables.

PAIRINGS OF REAL NUMBERS

Ordered Pairs

The replacement sets of expressions involving one variable are sets of numbers. The members of replacement sets for expressions involving two variables are pairs of numbers, one number to replace each variable.

When the numbers of a number pair are to be considered in a specified order, the pair is called an **ordered pair**, and the pair is denoted by a symbol such as $(3, 2)$, $(2, 3)$, $(-1, 5)$, or $(0, 3)$. Each of the two numbers in an ordered pair is called a **component** of the ordered pair; in the ordered pair (x, y), x is called the **first component** and y is called the **second component**. In this chapter these components will be restricted to real numbers. The set of all ordered pairs of real numbers is called the **Cartesian product** of R and R, and is denoted by $R \times R$ or R^2 (read "R cross R" and "R squared," respectively).

Relations

In mathematics and applications of mathematics, we are often concerned with relationships that involve the pairing of numbers. Such relationships as Fahrenheit and Celsius temperatures, cost of production and prices of materials, and standard height and weight tables are common examples. The following definition gives us a basis for considering such relationships mathematically.

Definition 4.1

A *relation* is a set of ordered pairs.

Relations Determined by Sets of Ordered Pairs

It follows from Definition 4.1 that any subset of $R \times R$ is a relation. The set of all first components of the pairs in a relation is called the **domain** of the relation, and the set of all second components is called the **range** of the relation. For example,

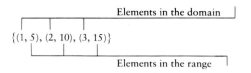

is a relation with domain $\{1, 2, 3\}$ and range $\{5, 10, 15\}$.

EXAMPLE 1   Find the domain of the following relations.

(a) $\{(2, 1), (3, 1), (3, 2)\}$ (b) $\{(1, 2), (2, 4), (3, 6)\}$

Solution   The domain is the set of first components. Thus, we have

(a) $\{2, 3\}$ (b) $\{1, 2, 3\}$ ■

Relations Determined by Rules

Relations are often defined by equations in two variables. Such an equation has ordered pairs as solutions and can be thought of as a "rule" for finding the value or values in the range of a relation paired with a given value in the domain. For brevity, we shall simply refer to equations such as

$$y = \frac{1}{x - 2} \quad \text{or} \quad y^2 = x - 2$$

as relations, where replacement values for x are in the domain and replacement values for y are elements in the range. We can find ordered pairs in such a relation by assigning values to one of the variables and solving for the other variable.

EXAMPLE 2   Find ordered pairs in the relation $y = \sqrt{x - 2}$ by replacing x with the given value.

(a)   3   (b)   6   (c)   9

Solution   (a) $y = \sqrt{3 - 2}$ (b) $y = \sqrt{6 - 2}$ (c) $y = \sqrt{9 - 2}$
                       $= \sqrt{1} = 1$ $= \sqrt{4} = 2$ $= \sqrt{7}$
                       $(3, 1)$ $(6, 2)$ $(9, \sqrt{7})$ ■

If a relation is defined by an equation with variables x and y and the domain (set of values for x) is not specified, we shall understand that *the domain is the set of all real numbers for which a real number exists in the range.*

EXAMPLE 3   Find the domain of each of the following relations.

(a) $y = \dfrac{1}{x - 2}$ (b) $y = \sqrt{x - 2}$

Solution   (a)   Since $1/(x - 2)$ is a real number for every value x except 2, the domain of the relation is $\{x \mid x \neq 2\}$.

(b)   Since $\sqrt{x - 2}$ is a real number only when $x - 2 \geq 0$, the domain of the relation is $\{x \mid x \geq 2\}$.   ■

Functions

A relation may be a pairing in which one or more elements in the domain are paired with *more than one element in the range*, as shown in Figure 4.1. A relation may also be a pairing in which each element in the domain is paired with *only one element in the range*, as illustrated in Figures 4.2a and 4.2b. Such a relation is given a special name.

FIGURE 4.1

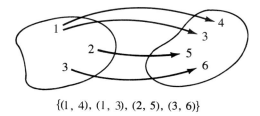

$\{(1, 4), (1, 3), (2, 5), (3, 6)\}$

FIGURE 4.2

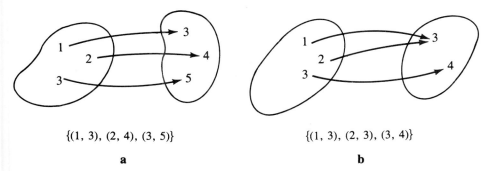

$\{(1, 3), (2, 4), (3, 5)\}$ $\{(1, 3), (2, 3), (3, 4)\}$

a   b

Definition 4.2   A *function* is a relation in which each element of the domain is paired with one and only one element of the range.

It follows from Definition 4.2 that no two ordered pairs belonging to a function can have the same first components and different second components. Thus, each of the sets of ordered pairs in Figures 4.1 and 4.2 is a relation; however, only the sets in Figure 4.2 are functions. The set in Figure 4.1 is not a function because two elements, 3 and 4, in the range are paired with one element, 1, in the domain.

E X A M P L E 4   Is the relation $\{(3, 5), (4, 8), (4, 9), (5, 10)\}$ a function?

Solution   The relation is not a function because two ordered pairs, $(4, 8)$ and $(4, 9)$, have the same first component.   ■

It also follows from Definition 4.2 that some relations defined by equations are functions and some are not.

E X A M P L E 5   Which of the following equations define functions?

(a) $y = x^2 + 4$ (b) $y^2 = x$

Solution   (a)   Since there is only one value of y paired with each value of x, the equation $y = x^2 + 4$ defines a function.

(b)   Since for each value of $x > 0$ there are two values of y ($\sqrt{x}$ and $-\sqrt{x}$) paired with the given value of x, the equation $y^2 = x$ does not define a function.   ■

Function Notation   In general, functions are denoted by single symbols; for example, f, g, h, and F might designate functions. Furthermore, a symbol such as $f(x)$, read "f of x" or "the value of f at x," represents the value in the range of a function f associated with the value x in the domain. Notation such as $f(x)$, $g(x)$, and so on, is called **function notation**. This is the same use we made of $P(x)$ in Section 1.1 when we discussed algebraic expressions and, in particular, polynomial expressions.

E X A M P L E 6   If $f(x) = x^2 + 3$, find the indicated elements in the range.

(a) $f(2)$ (b) $f(3)$ (c) $f(-3)$

Solution   Substituting the given elements in the domain for x, we have

(a) $f(2) = (2)^2 + 3$ (b) $f(3) = (3)^2 + 3$ (c) $f(-3) = (-3)^2 + 3$
$\qquad\quad = 7$ $\qquad\quad = 12$ $\qquad\qquad = 12$ ■

For a given function f it is often useful to know what change in the range value is produced by a given change h in the domain value. As the domain value changes from a to $a + h$, the range value changes from $f(a)$ to $f(a + h)$ and the change in the range value is

$$f(a + h) - f(a).$$

E X A M P L E 7   Find $f(a + h) - f(a)$ for $f(x) = 2x + 1$.

Solution   We first compute

$$f(a + h) = 2(a + h) + 1$$
$$= 2a + 2h + 1.$$

We then subtract $f(a) = 2a + 1$ to obtain

$$f(a + h) - f(a) = (2a + 2h + 1) - (2a + 1)$$
$$= 2h. \quad ■$$

In some applications, for specific domain values a and b, it is useful to know the difference $f(b) - f(a)$ between two specific range values. We introduce the notation $f(x)\Big|_a^b$ for that difference. Thus,

$$f(x)\Big|_a^b = f(b) - f(a).$$

E X A M P L E 8   For the given function f and values a and b, find $f(x)\Big|_a^b$.

(a) $f(x) = x^2; \quad a = 1, b = 2$ (b) $f(x) = 2x - 3; \quad a = 0, b = 1$

Solution   (a) $x^2\Big|_1^2 = (2)^2 - (1)^2$ (b) $(2x - 3)\Big|_0^1 = (2(1) - 3) - (2(0) - 3)$

$$= 4 - 1 \qquad\qquad\qquad\qquad = -1 - (-3)$$
$$= 3 \qquad\qquad\qquad\qquad\qquad = 2 \quad ■$$

Note that $f(x)$ represents an element in the range of a function and hence plays the same role as y in equations involving the variables x and y. For example,

$$y = x + 3 \qquad \text{and} \qquad f(x) = x + 3$$

define the same function. Since the value of y or $f(x)$ depends on the value of x, we sometimes call y or $f(x)$ the **dependent variable** and x the **independent variable**.

Composition

It is sometimes useful to form a new function from two given functions. For example, suppose $f(x) = x^2 + 3$ and $g(x) = x + 2$; then

$$f(g(x)) = f(x + 2)$$
$$= (x + 2)^2 + 3$$
$$= x^2 + 4x + 7.$$

The value $f(g(x))$ is called the **composition** of f and g evaluated at x. A special symbol is used to denote the composition of two functions.

Definition 4.3

If f and g are functions, the **composition** of f and g is

$$[f \circ g](x) = f(g(x)).$$

E X A M P L E 9   If $f(x) = x^2 + 1$ and $g(x) = 2/x$, find each of the indicated values.

(a) $[f \circ g](1)$ (b) $[g \circ f](x)$

Solution   (a) Since $g(1) = \dfrac{2}{1} = 2$,   (b) Since $f(x) = x^2 + 1$,

$$[f \circ g](1) = f(g(1))$$ $$[g \circ f](x) = g(f(x))$$
$$= f(2)$$ $$= g(x^2 + 1)$$
$$= (2)^2 + 1 = 5$$ $$= \frac{2}{x^2 + 1}$$ ■

Notice that $[f \circ g](x)$ is defined only for values of x so that

1. $g(x)$ is defined, and

2. $g(x)$ is in the domain of f.

For example, if $f(x) = 1/x$ and $g(x) = 1/x$,

$$[f \circ g](x) = \frac{1}{1/x} = x,$$

and thus we might expect $[f \circ g](0) = 0$. But in fact, $[f \circ g](0)$ is undefined since $g(0)$ is undefined, and thus Condition 1 above is not satisfied.

EXERCISE 4.1

A

■ *Specify the domain of each relation and state whether or not each relation is a function. See Examples 1 and 4.*

1. $\{(2, 3), (5, 7), (7, 8)\}$

2. $\{(-1, 6), (0, 2), (3, 3)\}$

3. $\{(2, -1), (3, 4), (3, 6)\}$

4. $\{(-4, 7), (-4, 8), (3, 2)\}$

5. $\{(5, 5), (6, 6), (7, 7)\}$

6. $\{(0, 0), (2, 4), (4, 2)\}$

7. $\{(1, -1), (-1, 1), (1, 2), (-1, -2)\}$

8. $\{(0, 1), (1, 0), (0, -1), (-1, 0)\}$

■ *Supply the missing components so that the ordered pairs are members of the given relations. See Example 2.*

a. $(0, \ )$ b. $(1, \ )$ c. $(2, \ )$ d. $(-3, \ )$ e. $(\frac{2}{3}, \ )$

9. $2x - 3y = 4$ **10.** $y = 9 + 2x^2$ **11.** $y = \sqrt{3x + 15}$ **12.** $y = |3x - 2|$

■ *Specify the domain of the relation. See Example 3.*

13. $y = 2x - 6$ **14.** $y = 2x + 9$ **15.** $y = x^2 + 1$

16. $y = x^2 + 9x$ **17.** $y = \dfrac{1}{x - 2}$ **18.** $y = \dfrac{1}{2x - 1}$

19. $y = \sqrt{x + 4}$ **20.** $y = \sqrt{3 - 2x}$ **21.** $y = \sqrt{9 - x^2}$

22. $y = \sqrt{x^2 - 4}$ **23.** $y = \dfrac{4x}{(\dot{x} + 1)(x - 1)}$ **24.** $y = \dfrac{x}{(x + 1)(x - 2)}$

■ *State whether or not the given equation defines y as a function of x. See Example 5.*

25. $x + y = 4$ **26.** $2x + 3y = 6$ **27.** $2x^2 - y = 3$

28. $y = -x^2 + 4x$ **29.** $y = \sqrt{x^2 + 25}$ **30.** $y = \sqrt{16 - 2x^2}$

31. $x^2 - y^2 = -25$ **32.** $y = \pm\sqrt{x^6}$ **33.** $y^2 = -3x^5$

34. $y^2 = x^4$ **35.** $x^2 + y^2 = 0$ **36.** $x^2 - y^2 = 0$

37. $xy = 1$ **38.** $x^2y^2 = 1$ **39.** $x = 2$

40. $x = -4$ **41.** $y = 3$ **42.** $y = -4$

■ *Find the indicated range value for the given function. See Example 6.*

43. $f(x) = x - 1, f(-2)$ **44.** $f(x) = x + 5, f(5)$

45. $f(x) = x + 3, f(2)$ **46.** $f(x) = x - 2, f(-1)$

47. $f(x) = -2x + 4, f(a + 1)$ **48.** $f(x) = -\dfrac{3}{2}x + 1, f(a - 2)$

49. $f(x) = x^2 - 3x, f(-1)$ **50.** $f(x) = -x^3 + 3x^2 + 1, f(-2)$

51. $g(x) = x^2 + 2x, g(-2)$

52. $g(x) = -3x^2 + x + 1, g(1)$

53. $g(x) = x^2 + 1, g(1/a)$

54. $g(x) = -2x^2 + x - 4, g(a + 1)$

55. $s(x) = x^3 - 2x^2 + x - 1, s(1)$

56. $s(x) = 3x^3 - 2x^2 - 2x + 4, s(-2)$

57. $h(t) = t^4 - 2t^2 + 1, h(a - 2)$

58. $h(t) = -3t^4 + 2t^3 + t, h(1/a)$

59. $h(x) = \dfrac{1}{x^2 - 4}, h(-1)$

60. $h(x) = \dfrac{2x}{x - 3}, h(2)$

61. $h(x) = \sqrt{4 - x^2}, h(-1)$

62. $h(x) = \dfrac{1}{\sqrt{x^2 - 1}}, h(2)$

63. $f(x) = |x - 3|, f(-2)$

64. $f(x) = \dfrac{1}{|x - 1|}, f(-1)$

■ Find $f(a + h) - f(a)$ for the given function. See Example 7.

65. $f(x) = 2x - 1$

66. $f(x) = 4x - 2$

67. $f(x) = -3x + 2$

68. $f(x) = -5x + 4$

69. $f(x) = 2 - x^2$

70. $f(x) = x^2 + 2x$

71. $f(x) = x^2 - 3x + 1$

72. $f(x) = -x^2 + 4x + 2$

■ Compute $f(x)\Big|_a^b$ for the given function and values of a and b. See Example 8.

73. $f(x) = x^2 - 4; \quad a = -1, b = 2$

74. $f(x) = x^4; \quad a = -1, b = 1$

75. $f(x) = \dfrac{x^3}{3} + \dfrac{x^2}{2} + x; \quad a = 0, b = 1$

76. $f(x) = \dfrac{2x^3}{3} - 2x^2 + 3x; \quad a = -1, b = 1$

77. $f(x) = \sqrt{x}; \quad a = 4, b = 9$

78. $f(x) = \sqrt{x + 1}; \quad a = 15, b = 24$

■ For each pair of functions, find the indicated values. For Exercises 79–84, see Example 9.

79. $f(x) = x + 4, g(x) = x^2$

    a. $[f \circ g](3)$ b. $[g \circ f](3)$

80. $f(x) = x^2 + 1, g(x) = \dfrac{1}{x}$

    a. $[f \circ g](1)$ b. $[g \circ f](1)$

81. $f(x) = x^2 + 4, g(x) = \sqrt{x - 4}$

    a. $[f \circ g](5)$ b. $[g \circ f](3)$

82. $f(x) = x^3 + x, g(x) = \sqrt[3]{x}$

    a. $[f \circ g](1)$ b. $[g \circ f](-1)$

83. $f(x) = x, g(x) = \dfrac{1}{x}$

    a. $[f \circ g](a)$ b. $[g \circ f](a)$

84. $f(x) = \dfrac{1}{x}, g(x) = \dfrac{1}{x}$

    a. $[f \circ g](a + 4)$ b. $[g \circ f](a + 4)$

B

85. If $f(x) = x^2 - x + 1$, find and simplify $\dfrac{f(a + h) - f(a)}{h}, h \neq 0$.

86. If $f(x) = 1/x$, find and simplify $\dfrac{f(a + h) - f(a)}{h}, h \neq 0$.

87. If $f(x) = \sqrt{x}$, find and simplify $\dfrac{f(a + h) - f(a)}{h}, h \neq 0$.

88. If $f(x) = \dfrac{1}{\sqrt{x}}$, find and simplify $\dfrac{f(a + h) - f(a)}{h}$, $h \neq 0$.

89. If $f(x) = |x^2|$, find and simplify $\dfrac{f(0 + h) - f(0)}{h}$, $h \neq 0$.

90. If $f(x) = |x|$, find and simplify $\dfrac{f(0 + h) - f(0)}{h}$, $h \neq 0$.

4.2

LINEAR EQUATIONS; GRAPHS

In this section we begin our study of functions and their graphs by reviewing the basis for all plane graphing and then proceeding to a discussion of lines in the plane.

R^2 and the Geometric Plane

The facts that each member of R^2 corresponds to a point in the geometric plane and that the coordinates of each point in the geometric plane are the components of a member of R^2 are the basis for all plane graphing. As you probably recall from your earlier study of algebra, the correspondence between points in the plane and ordered pairs of real numbers is usually established through a **Cartesian** (or **rectangular**) **coordinate system,** as shown in Figure 4.3. For this correspondence, the first component of an ordered pair is called the **abscissa** and the second component, the **ordinate**.

FIGURE 4.3

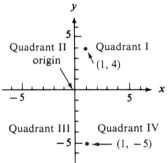

Graphs of First-Degree Equations

A **first-degree equation** in x and y is an equation that can be written equivalently in the form

$$Ax + By + C = 0 \qquad (A \text{ and } B \text{ not both } 0). \qquad (1)$$

We call (1) the **standard form** for a first-degree equation. The graph of any such equation is a straight line, although we do not prove this here. Hence, such equations are called **linear equations,** and functions defined by such equations are called **linear functions.**

Since two distinct points determine a straight line, we need find only two solutions of such an equation to determine its graph. In practice, the two solutions easiest to find are usually those with first and second components, respectively, equal to zero—that is, the solutions $(0, y_1)$ and $(x_1, 0)$. The numbers x_1 and y_1 are called the **x- and y-intercepts** of the graph.

EXAMPLE 1   Find the x- and y-intercepts of the graph of $2x + 6y = 4$.

Solution   Substituting 0 for y we have

$$2x + 6(0) = 4.$$

Solving for x we obtain $x = 2$. Thus, the x-intercept is 2. Substituting 0 for x in the original equation we have

$$2(0) + 6y = 4.$$

Solving for y we obtain $y = \frac{2}{3}$. Thus, the y-intercept is $\frac{2}{3}$.

EXAMPLE 2   Graph the function defined by the equation $3x + 4y = 12$.

Solution   If

$$y = 0,$$

we have $x = 4$, and the x-intercept is 4.
If

$$x = 0,$$

then $y = 3$, and the y-intercept is 3. Drawing a line through the intercepts, we obtain the figure shown.   ■

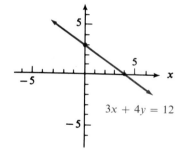

If a graph intersects both axes at the origin, the intercepts do not represent two separate points. It is then necessary to plot at least one other point at a distance far enough removed from the origin to establish the line with pictorial accuracy.

EXAMPLE 3   Graph $8x + 8y = 0$.

Solution   If $y = 0$, we have $x = 0$, and the x-intercept is 0. Thus, the graph intersects both axes at the origin. Hence, it is necessary to plot another point on the line. Letting $x = 2$ and solving for y, we obtain $y = -2$. Graphing the points corresponding to $(0, 0)$ and $(2, -2)$ and drawing the line through them, we obtain the figure shown.   ■

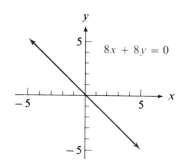

Equations of Horizontal Lines

Two special types of linear equations are worth noting. First, an equation such as

$$y = 4$$

may be considered an equation in two variables:

$$0x + y = 4.$$

For each x, this equation assigns $y = 4$. That is, any ordered pair of the form $(x, 4)$ is a solution of the equation. For instance,

$$(1, 4), \quad (2, 4), \quad (3, 4), \quad \text{and} \quad (4, 4)$$

are all solutions of the equation. If we graph these points and connect them with a straight line, we have the graph shown in Figure 4.4a.

FIGURE 4.4

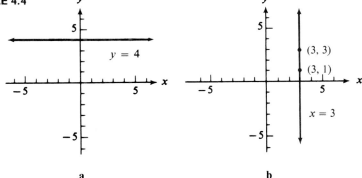

a

b

Since the equation

$$y = 4$$

assigns to each x the same value for y, the function defined by this equation is called a **constant function**. The graph of such a function is a horizontal straight line.

Equations of Vertical Lines

The other special case of the linear equation is of the type

$$x = 3,$$

which may be looked upon as

$$x + 0y = 3.$$

Here, only one value is permissible for x, namely 3, whereas any value may be assigned to y. That is, any ordered pair of the form $(3, y)$ is a solution of this equation. If we choose two solutions, say, (3, 1) and (3, 3), we can draw the graph shown in Figure 4.4b. Notice that the graph of such an equation is a vertical line.

Distance Between Two Points

We next consider two useful properties of the line segment joining two points in the plane. The first of these is the distance between the two points.

FIGURE 4.5

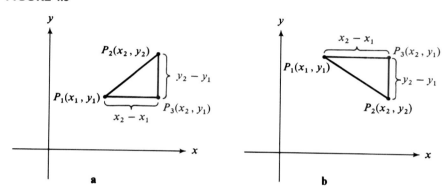

The **distance** between any two points $P_1(x_1, y_1)$ and $P_2(x_2, y_2)$ in a Cartesian plane (see Figure 4.5) is the *length* of the line segment joining those points. This distance is denoted by P_1P_2 (or simply d) and can be obtained by using the Pythagorean theorem (see Preliminary Concepts).

Theorem 4.1   *The distance P_1P_2 between the points $P_1(x_1, y_1)$ and $P_2(x_2, y_2)$ is given by*

$$P_1P_2 = \sqrt{(x_2 - x_1)^2 + (y_2 - y_1)^2}.$$

Note from Theorem 4.1 that a distance (as differentiated from a directed distance) is always positive—or 0 if the points coincide.

EXAMPLE 4   Find the distance between the given points.

(a) $P_1(0, 0)$ and $P_2(1, 1)$ (b) $P_1(2, -3)$ and $P_2(-4, 1)$

Solution   (a) $P_1P_2 = \sqrt{(1 - 0)^2 + (1 - 0)^2}$ (b) $P_1P_2 = \sqrt{(-4 - 2)^2 + [1 - (-3)]^2}$

$= \sqrt{2}$ $= \sqrt{52}$ ■

If the points P_1 and P_2 lie on the same horizontal line (Figure 4.6a), then the distance P_1, P_2 between them is

$$P_1P_2 = \sqrt{(x_2 - x_1)^2 + (y_1 - y_1)^2}$$
$$= \sqrt{(x_2 - x_1)^2} = |x_2 - x_1|.$$

If they lie on the same vertical line (Figure 4.6b), then the distance is

$$P_1P_2 = \sqrt{(x_1 - x_1)^2 + (y_2 - y_1)^2}$$
$$= \sqrt{(y_2 - y_1)^2} = |y_2 - y_1|.$$

FIGURE 4.6

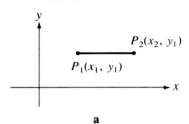

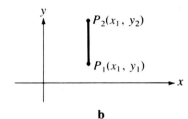

a   b

E X A M P L E   5   Find the distance between the given points.

(a) $P_1(2, 3)$ and $P_2(2, -4)$ (b) $P_1(3, 1)$ and $P_2(-1, 1)$

Solution   (a) $P_1P_2 = \sqrt{(2 - 2)^2 + (-4 - 3)^2} = \sqrt{(-7)^2}$
$$= |-7| = 7$$

(b) $P_1P_2 = \sqrt{(-1 - 3)^2 + (1 - 1)^2} = \sqrt{(-4)^2}$
$$= |-4| = 4 \qquad ■$$

Directed Distances

We next consider the idea of the directed distance between two points on the same vertical or horizontal line.

Let P_1, with coordinates (x_1, y_1), and P_2, with coordinates (x_2, y_2), be endpoints of a line segment. If we construct through P_2 a line parallel to the y-axis and through P_1 a line parallel to the x-axis, the lines will meet at a point P_3, as shown in Figure 4.5. The x-coordinate of P_3 is evidently the same as the x-coordinate of P_2, and the y-coordinate of P_3 is the same as that of P_1. Hence, the coordinates of P_3 are (x_2, y_1).

In general, since $y_2 - y_1$ is positive or negative as P_2 is above or below P_3, respectively, and $x_2 - x_1$ is positive or negative as P_3 is to the right or left of P_1, respectively, it is also convenient to consider the distances from P_1 to P_3 and from P_3 to P_2 as positive or negative as $x_2 - x_1$ and $y_2 - y_1$ are positive or negative, respectively. The values $x_2 - x_1$ and $y_2 - y_1$ are called the **directed distances** from P_1 to P_3 and from P_3 to P_2, respectively.

Slope of a Line Segment

The second useful property of a line segment is its inclination with respect to the *x*-axis. This inclination is a measure of the rate at which the dependent variable (y) changes with respect to the independent variable (x). The inclination of a line can be measured by comparing the *rise* (the directed distance from P_3 to P_2) of the segment joining any two points with the corresponding *run* (the directed distance from P_3 to P_1), as shown in Figure 4.7.

FIGURE 4.7

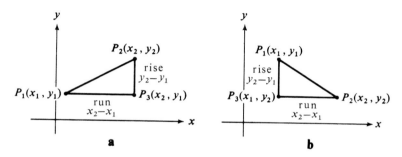

The ratio of rise to run is called the **slope** of the line segment and is designated by the letter m. Since the rise is simply $y_2 - y_1$ and the run is $x_2 - x_1$, the slope of the line segment joining P_1 and P_2 is given by the slope formula, which follows.

$$m = \frac{y_2 - y_1}{x_2 - x_1} \qquad (x_2 \neq x_1) \tag{2}$$

The slope is the quantity that we use to measure the inclination of a line segment. Notice that if the magnitude (absolute value) of m is small, then the line is nearly horizontal (i.e., is not inclined very much to the *x*-axis). But if the magnitude of m is large, then the line is close to vertical (i.e., is steeply inclined to the *x*-axis). This is illustrated in Figure 4.8.

If P_2 is to the right of P_1, then $x_2 - x_1$ is positive, and the slope is positive or negative as $y_2 - y_1$ is positive or negative. Thus, a positive slope indicates that a line rises to the right (Figure 4.8a); a negative slope indicates that it falls to the right (Figure 4.8b).

Since

$$\frac{y_2 - y_1}{x_2 - x_1} = \frac{-(y_1 - y_2)}{-(x_1 - x_2)} = \frac{y_1 - y_2}{x_1 - x_2},$$

FIGURE 4.8

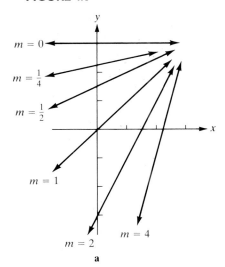

a

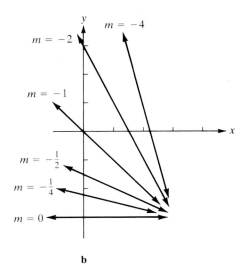

b

the restriction that P_2 be to the right of P_1 is not necessary, and the order in which the points are considered is immaterial in determining slope. Furthermore, it can be shown that the slope of a line does not depend on the points chosen to compute it.

EXAMPLE 6 Find the slope of the line segment joining the given points.

(a) $P(2, -1)$ and $Q(3, 2)$ (b) $P(-1, 1)$ and $Q(2, 2)$

Solution (a) $m = \dfrac{2 - (-1)}{3 - 2}$ (b) $m = \dfrac{2 - 1}{2 - (-1)}$

$\qquad\qquad = \dfrac{3}{1} = 3$ $\qquad\qquad = \dfrac{1}{3}$ ■

If a line segment is parallel to the x-axis, then $y_2 - y_1 = 0$, and the line has slope 0 (Figure 4.9a on page 158). If it is parallel to the y-axis, then $x_2 - x_1 = 0$, and its slope is not defined (Figure 4.9b).

Function Notation

If $B \neq 0$ in Equation (1) on page 151, then the equation can be rewritten equivalently as

$$y = -\frac{A}{B}x - \frac{C}{B},$$

FIGURE 4.9

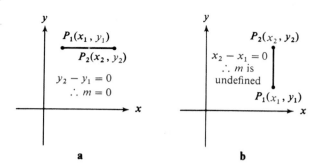

or, using function notation, as

$$f(x) = -\frac{A}{B}x - \frac{C}{B}.$$

When we graph such functions, we sometimes use $f(x)$ instead of y to label the vertical axis. Further, $f(a)$ is the ordinate of the point on the graph with abscissa a.

E X A M P L E   7   Graph $f(x) = 2x - 1$.

Solution   The intercepts of

$$f(x) = y = 2x - 1$$

are $\frac{1}{2}$ and -1. We sketch the line through these points as shown.   ■

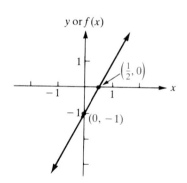

E X E R C I S E   4.2

A

■ *Find the x- and y-intercepts of the graph of each equation. See Example 1.*

1. $3x + 2y = 6$

2. $4x + y = 4$

3. $-2x + y = 1$

4. $-3x - 2y = 4$

5. $2x - 3y = 0$

6. $-3x + 6y = 0$

7. $\frac{3}{2}x + \frac{1}{4}y = \frac{3}{4}$

8. $\frac{4}{3}x - \frac{5}{6}y = \frac{1}{3}$

9. $\frac{1}{3}x - \frac{1}{4}y = \frac{1}{5}$

10. $\frac{2}{5}x - \frac{1}{3}y = \frac{3}{2}$

11. $\sqrt{2x} - \sqrt{5y} = \frac{1}{2}$

12. $\sqrt{6x} + \sqrt{8y} = \frac{2}{3}$

■ *Graph. See Examples 2, 3, and 7.*

13. $y = -2x$

14. $3x = 2y$

15. $2x + 3y = 6$

16. $3x - 2y = 8$

17. $2x + 5y = 10$

18. $2x + 3y = 9$

19. $4x - 4y = 1$

20. $8y - 16x = -3$

21. $x = -2$

22. $x = 3$

23. $y = 4$

24. $y = -3$

25. $f(x) = x - 1$

26. $f(x) = x + 3$

27. $f(x) = 3x + 1$

28. $f(x) = 2x - 5$

29. $f(x) = \frac{1}{2}(x - 3)$

30. $f(x) = \frac{2}{3}(x + 2)$

■ *Find the distance between each of the given pairs of points, and find the slope of the line segment joining them. See Examples 4–6.*

31. $(1, 1)$; $(4, 5)$

32. $(-1, 1)$; $(5, 9)$

33. $(-3, 2)$; $(2, 14)$

34. $(-4, -3)$; $(1, 9)$

35. $(2, 1)$; $(1, 0)$

36. $(-3, 2)$; $(0, 0)$

37. $(5, 4)$; $(-1, 1)$

38. $(2, -3)$; $(-2, -1)$

39. $(3, 5)$; $(-2, 5)$

40. $(2, 0)$; $(-2, 0)$

41. $(0, 5)$; $(0, -5)$

42. $(-2, -5)$; $(-2, 3)$

43. $(2, 3)$; $(-2, 3)$

44. $(-3, 1)$; $(3, 1)$

45. $(1, 2)$; $(1, -2)$

46. $(-3, 2)$; $(-3, -2)$

■ *Determine algebraically whether or not the points lie on the same line.*

47. $(2, 7), (-2, -5), (0, 1)$

48. $(3, 3), (-3, -1), (0, 1)$

49. $(0, 1), (-2, 7), (2, -5)$

50. $(-1, 5), (1, 1), (3, -3)$

51. $(1, 3), (3, 5), (5, 11)$

52. $(0, 4), (2, 2), (-1, 5)$

■ *For each figure, list the line with the largest slope and the one with the smallest slope.*

53.

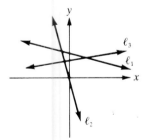

54.

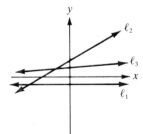

55.

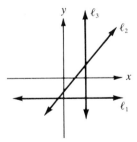

56.

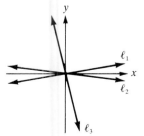

57.

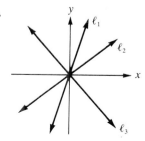

58.

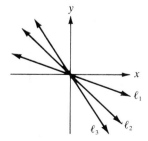

B

59. Let f be the function defined by $f(x) = 2x + 1$ on the interval $[-1, 1]$. Graph $y = f(x)$.

60. Let f be the function defined by $f(x) = -\frac{1}{2}x + 2$ on the interval $[-2, 1]$. Graph $y = f(x)$.

61. Graph $x + 2y = 3$ and $x - y = 0$ on the same set of coordinate axes. Estimate the coordinates of the point of intersection. What can you say about the coordinates of this point in relation to the two linear equations?

62. Graph $x + y = 6$ and $5x - y = 0$ on the same set of axes. Estimate the coordinates of the point of intersection. What can you say about the coordinates of this point in relation to the two linear equations?

■   *Compute and simplify* $\dfrac{f(a + h) - f(a)}{h}$, $h \neq 0$ *for the given function. Interpret the result geometrically.*

63. $f(x) = 3x$ **64.** $f(x) = -2x$ **65.** $f(x) = 2x + 4$ **66.** $f(x) = -3x + 4$

67. $f(x) = \frac{2}{3}x + 2$ **68.** $f(x) = -\frac{1}{3}x + 2$ **69.** $f(x) = -3$ **70.** $f(x) = 2$

4.3

FORMS OF LINEAR EQUATIONS

In Section 4.2 we called $Ax + By + C = 0$ the standard form of a linear equation in two variables. It is sometimes convenient to use a linear equation in a form other than the standard form.

Point-Slope Form

Consider a line in the plane with given slope m that passes through a given point (x_1, y_1), as shown in Figure 4.10. If we choose any other point on the line, say with coordinates (x, y), then the slope of the line is given by

$$\frac{y - y_1}{x - x_1} = m,$$

from which

$$y - y_1 = m(x - x_1). \qquad (1)$$

FIGURE 4.10

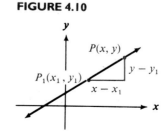

Note that (1) is satisfied also by $(x, y) = (x_1, y_1)$. Since now x and y are the coordinates of *any* point on the line, (1) is an equation of the line passing through

(x_1, y_1) with slope m. This is called the <u>**point-slope form**</u> for a linear equation. Any linear equation in point-slope form can be written equivalently in standard form.

EXAMPLE 1   Find an equation in standard form of the line with slope $\frac{3}{4}$ passing through the point $(-4, 1)$.

Solution   Substituting $\frac{3}{4}$, -4, and 1 for m, x_1, and y_1, respectively, in Equation (1), we have

$$y - 1 = \frac{3}{4}[x - (-4)],$$

$$y - 1 = \frac{3}{4}x + 3,$$

from which

$$3x - 4y + 16 = 0. \qquad ■$$

Two-Point Form   Consider the line through the points (x_1, y_1) and (x_2, y_2), as shown in Figure 4.11. If $x_2 \neq x_1$, then the slope of the line is

$$\frac{y_2 - y_1}{x_2 - x_1}.$$

FIGURE 4.11

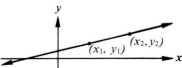

Substituting this value for m into the point-slope form for a linear equation [Equation (1)], we obtain

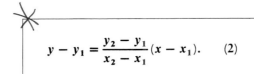

$$y - y_1 = \frac{y_2 - y_1}{x_2 - x_1}(x - x_1). \qquad (2)$$

This is called the <u>**two-point form**</u> of the linear equation.

EXAMPLE 2   Find an equation in standard form of the line passing through the points $(2, 1)$ and $(3, -2)$.

Solution   Substituting 2, 3, 1, and -2 for x_1, x_2, y_1, and y_2, respectively, in Equation (2), we have

$$y - 1 = \frac{-2 - 1}{3 - 2}(x - 2),$$

$$y - 1 = -3x + 6,$$

from which

$$3x + y - 7 = 0. \qquad ■$$

Slope-Intercept Form   Now consider the line with slope m passing through a point on the y-axis having coordinates $(0, b)$, as shown in Figure 4.12. Substituting the components of $(0, b)$ in the point-slope form of a linear equation,

$$y - y_1 = m(x - x_1),$$

we obtain

$$y - b = m(x - 0),$$

from which

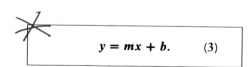

$$y = mx + b. \qquad (3)$$

FIGURE 4.12

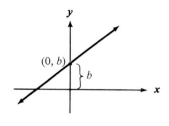

Equation (3) is called the **slope-intercept form** for a linear equation. A linear equation in standard form with $B \neq 0$ can be written in the slope-intercept form by solving for y in terms of x.

The slope of the graph of a linear equation that has a slope can be obtained directly from its slope-intercept form.

EXAMPLE 3   Find the slope and y-intercept of the line with equation $2x + 3y - 6 = 0$.

Solution   We can write $2x + 3y - 6 = 0$ equivalently as

$$y = -\frac{2}{3}x + 2.$$

We can now read the slope of the line, $-\frac{2}{3}$, and the y-intercept, 2, directly from the last equation.   ■

Intercept Form

If the x- and y-intercepts of the graph of

$$y = mx + b \qquad (4)$$

are a and b $(a, b \neq 0)$, respectively, as shown in Figure 4.13, then the slope m is the rise over the run, which is equal to $-b/a$. Replacing m in Equation (4) with $-b/a$, we have

$$y = -\frac{b}{a}x + b,$$

$$ay = -bx + ab,$$

$$bx + ay = ab,$$

and multiplying each member by $1/ab$ produces

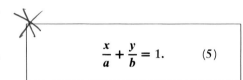

$$\frac{x}{a} + \frac{y}{b} = 1. \qquad (5)$$

FIGURE 4.13

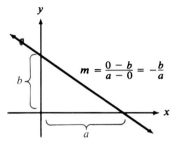

$$m = \frac{0-b}{a-0} = -\frac{b}{a}$$

Equation (5) is called the **intercept form** for a linear equation.

EXAMPLE 4   Find an equation in standard form of the line with x-intercept 3 and y-intercept -2.

Solution   We substitute 3 and -2 for a and b, respectively, in Equation (5) to obtain

$$\frac{x}{3} + \frac{y}{-2} = 1,$$

from which

$$-2x + 3y + 6 = 0, \qquad \text{or} \qquad 2x - 3y - 6 = 0. \qquad ■$$

Any of the forms of linear equations discussed in this section may be used in working with linear functions and their graphs. We summarize the different forms in Table 4.1 on page 164.

Parallel Lines

Similar triangles can be used to show that lines not perpendicular to the x-axis are parallel if and only if they have equal slopes. The proof of the following theorem depends on familiar geometric properties and is left as an exercise.

TABLE 4.1

Form	Data Required	Equation
Point-slope	Slope, m Point, (x_1, y_1)	$y - y_1 = m(x - x_1)$
Two-point	Two points, $\quad (x_1, y_1)$ and (x_2, y_2)	$y - y_1 = \dfrac{y_2 - y_1}{x_2 - x_1}(x - x_1)$
Slope-intercept	Slope, m y-intercept, b	$y = mx + b$
Intercept	x-intercept, a y-intercept, b	$\dfrac{x}{a} + \dfrac{y}{b} = 1$

Theorem 4.2   *Two nonvertical lines with slopes m_1 and m_2 are parallel if and only if $m_1 = m_2$. Two lines perpendicular to the x-axis are parallel to each other.*

E X A M P L E 5   Find the equation in standard form of the line through $(-1, 2)$ that is parallel to $3x - 2y = 6$.

Solution   We first graph the equation $3x - 2y = 6$. The given equation can be written in slope-intercept form as

$$y = \frac{3}{2}x - 3.$$

By inspection, we see that the slope of its graph is $\frac{3}{2}$. From Theorem 4.2 the desired line has slope $\frac{3}{2}$. Using the point-slope form for a linear equation, with $m = \frac{3}{2}$, $x_1 = -1$, and $y_1 = 2$, we have

$$(y - 2) = \frac{3}{2}(x + 1),$$

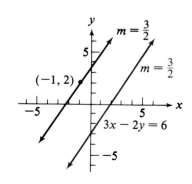

which can be written in the standard form

$$3x - 2y + 7 = 0. \quad \blacksquare$$

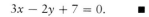

Perpendicular Lines

The relationship between the slopes of two perpendicular lines is given in the following theorem. The proof of this theorem also follows from familiar geometric properties and is left as an exercise.

Theorem 4.3

Two nonvertical lines with slopes m_1 and m_2 are perpendicular if and only if $m_1 \cdot m_2 = -1$. Two lines that are parallel to the x- and y-axes, respectively, are also perpendicular to each other.

E X A M P L E   6

Find the equation in standard form of the line passing through $(3, -2)$ that is perpendicular to the graph of $2x + 5y = 10$.

Solution

We first graph the equation $2x + 5y = 10$. The given equation can be written in slope-intercept form as

$$y = -\frac{2}{5}x + 2.$$

By inspection, the slope m_1 of its graph is $-\frac{2}{5}$. Hence, from Theorem 4.3, the slope of any line perpendicular to this graph is given by

$$m_2 = -\frac{1}{m_1} = -\frac{1}{-\frac{2}{5}} = \frac{5}{2}.$$

Using the point-slope form for a linear equation with $m = \frac{5}{2}$, $x_1 = 3$, and $y_1 = -2$, we have

$$y - (-2) = \frac{5}{2}(x - 3).$$

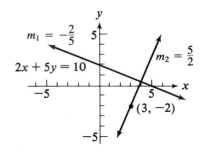

This can be written in the standard form

$$5x - 2y - 19 = 0. \quad ■$$

Midpoint Formula

Similar triangles can be used to prove the following theorem, which gives the coordinates of the midpoint of the line segment joining two points. The proof is left as an exercise.

Theorem 4.4

The midpoint of the line segment joining the points $A(x_1, y_1)$ and $B(x_2, y_2)$ is $M(\bar{x}, \bar{y})$, where

$$\bar{x} = \frac{x_1 + x_2}{2} \qquad and \qquad \bar{y} = \frac{y_1 + y_2}{2}.$$

EXAMPLE 7   Find an equation of the line that is the perpendicular bisector of the segment whose endpoints are $(7, -4)$ and $(-5, -9)$.

Solution   We first find the slope of the line through the given points, obtaining

$$m = \frac{y_2 - y_1}{x_2 - x_1} = \frac{-9 - (-4)}{-5 - 7} = \frac{-5}{-12} = \frac{5}{12}.$$

From Theorem 4.3, we conclude that the slope of the desired bisector is $-\frac{12}{5}$. We use Theorem 4.4 to obtain the coordinates of the midpoint of the given segment. These are

$$\bar{x} = \frac{x_1 + x_2}{2} = \frac{7 + (-5)}{2} = \frac{2}{2} = 1 \quad \text{and} \quad \bar{y} = \frac{y_1 + y_2}{2} = \frac{-4 + (-9)}{2} = -\frac{13}{2}.$$

We then use the point-slope form for a linear equation with $m = -\frac{12}{5}$ and given point $(1, -\frac{13}{2})$, to obtain the equation

$$y - \left(-\frac{13}{2}\right) = -\frac{12}{5}(x - 1).$$

This simplifies to the standard form

$$24x + 10y + 41 = 0. \qquad ■$$

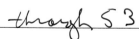 through 53

EXERCISE 4.3

A

■ Find the equation, in standard form, of the line passing through each of the given points and having the given slope. See Example 1.

1. $(2, 1)$, $m = 4$

2. $(-2, 3)$, $m = 5$

3. $(-3, -2)$, $m = \frac{1}{2}$

4. $(0, -1)$, $m = -\frac{1}{2}$

5. $(2, -3)$, $m = 0$

6. $(-4, 2)$, $m = 0$

7. $(-1, -2)$,   parallel to y-axis

8. $(3, 4)$,   parallel to y-axis

9. $(1, 3)$, $m = \sqrt{2}$

10. $(-1, 0)$, $m = -\sqrt{3}$

11. $(\sqrt{3}, 2)$, $m = 2$

12. $(\sqrt{2}, \sqrt{2})$, $m = -1$

■ Find an equation, in standard form, of the line passing through the given points. See Example 2.

13. $(-2, 1)$; $(4, -2)$

14. $(-1, -1)$; $(2, 1)$

15. $(-2, 1)$; $(1, 2)$

16. $(3, 4)$; $(4, 3)$

17. $(2, 1)$; $(3, 1)$

18. $(2, -1)$; $(-5, -1)$

■   *Write each equation in slope-intercept form; specify the slope of the line and the y-intercept. See Example 3.*

19. $x + y = 3$ **20.** $2x + y = -1$ **21.** $3x + 2y = 1$ **22.** $3x - y = 7$

23. $x - 3y = 2$ **24.** $2x - 3y = 0$ **25.** $4x - 3y = 7$ **26.** $-2x + 6y = 8$

27. $y - 2 = 4$ **28.** $3y - 4 = 6$ **29.** $x = 5$ **30.** $2x - 5 = 0$

■   *Find the equation, in standard form, of the line with the given intercepts. The x-intercept is given first. See Example 4.*

31. $2;\ 3$ **32.** $4;\ -1$ **33.** $-2;\ -5$

34. $-1;\ 7$ **35.** $-\dfrac{1}{2};\ \dfrac{3}{2}$ **36.** $\dfrac{2}{3};\ -\dfrac{3}{4}$

■   *Find the equations in standard form of (**a**) the line that is parallel and (**b**) the line that is perpendicular to the graph of the given equation, both of which pass through the given point. See Examples 5 and 6.*

37. $3x + y = 6;\quad (5, 1)$ **38.** $2x - y = -3;\quad (-2, 1)$ **39.** $3x - 2y = 5;\quad (0, 0)$

40. $4x - 2y = -5;\quad (0, 0)$ **41.** $4x + 6y = 3;\quad (-2, -1)$ **42.** $5x - 3y = 2;\quad (-4, -1)$

43. $\sqrt{2}x + \sqrt{8}y = 1;\quad (1, 2)$ **44.** $\sqrt{24}x - \sqrt{6}y = 2;\quad (3, -3)$

■   *Find an equation of the line that is the perpendicular bisector of the segment whose endpoints are given. For Exercises 45–52, see Example 7.*

45. $(1, -2)$ and $(-2, 1)$ **46.** $(3, -1)$ and $(1, -3)$ **47.** $(5, -5)$ and $(-9, 1)$

48. $(9, 4)$ and $(-3, 4)$ **49.** $(a, 0)$ and $(0, b)$ **50.** (a, b) and $(a + k_1, b + k_2)$

51. (a, b) and (b, a) **52.** $(a, -b)$ and $(b, -a)$

B

53. Use slopes to show that the triangle with vertices at $A(0, 8)$, $B(6, 2)$, and $C(-4, 4)$ is a right triangle.

54. Use slopes to show that the triangle with vertices $D(2, 5)$, $E(5, 2)$, and $F(10, 7)$ is a right triangle.

■   *For Exercises 55–62, assume that the variables are approximately related by a linear equation.*

55. A ball is dropped from the edge of a building at time $t = 0$. At time $t = 1$ second, the velocity of the ball is -32 feet per second. Find a linear equation relating the velocity (v) and the time (t).

56. A ball is thrown vertically with an initial velocity (at time $t = 0$) of 5 feet per second. At time $t = 3$ seconds, the velocity is -91 feet per second. Find a linear equation relating the velocity (v) and the time (t).

57. The cost of producing 100 computers is \$190,000, and the cost of producing 200 computers is \$280,000. Find a linear equation relating the cost (c) and the number (n) of computers produced.

58. The cost of producing 20 large TV monitors is $58,000, and the cost of producing 50 of the same monitors is $70,000. Find a linear equation relating the cost (c) and the number (n) of monitors produced.

59. The boiling point of water at sea level on the Fahrenheit scale is 212° and on the Celsius scale it is 100°. The freezing point of water at sea level on the Fahrenheit scale is 32° and on the Celsius scale it is 0°. Find a linear equation relating temperature in degrees Celsius (C) and in degrees Fahrenheit (F).

60. The boiling point of water at sea level on the Fahrenheit scale is 212° and on the Kelvin scale it is 373°. The freezing point of water at sea level on the Fahrenheit scale is 32° and on the Kelvin scale it is 273°. Find a linear equation relating temperature in degrees Fahrenheit (F) and in degrees Kelvin (K).

61. There is a demand for 7625 TV sets when the price is set at $225 per set and a demand for 2675 sets when the price is $315 per set. Find a linear equation relating the demand (D) and the price (p).

62. There is a demand for 500 hi-tech radio receivers when the price is set at $100,000 per receiver. When the price is set at $150,000, the demand drops to 250 units. Find a linear equation relating the demand (D) and the price (p).

63. Recall from geometry that if two parallel lines are cut by a transversal, then corresponding angles are congruent. Use this fact to show, in the figure below, where L_1 is parallel to L_2, that $\triangle P_1 Q_1 P_3$ is similar to $\triangle P_2 Q_2 P_4$ and, hence, that the slopes m_1 and m_2 of the lines are equal.

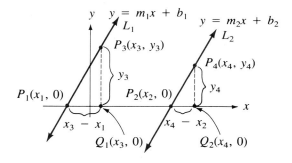

64. In the figure for Exercise 63, assume that $m_1 = m_2$. Deduce that $\triangle P_1 Q_1 P_3$ is similar to $\triangle P_2 Q_2 P_4$ and, hence, that L_1 is parallel to L_2.

65. In the adjoining figure, where L_1 is perpendicular to L_2, use the fact that $\triangle P_1 Q P_2$ is similar to $\triangle P_3 Q P_1$, so that

$$\frac{y_1 - y_2}{x_2 - x_1} = \frac{x_2 - x_1}{y_3 - y_1},$$

to deduce that $m_1 m_2 = -1$.

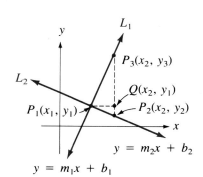

66. Use the figure in Exercise 65 to argue that, because

$$(m_1 - m_2)^2 = m_1^2 - 2m_1 m_2 + m_2^2,$$

and because the converse of the Pythagorean theorem is true, the relation $m_1 m_2 = -1$ implies that L_1 is perpendicular to L_2.

67. In the adjoining figure, $m(\bar{x}, \bar{y})$ is the midpoint of the line segment AB. Show, using similar triangles, that

$$AC = 2AE \qquad \text{and} \qquad BC = 2BD.$$

Assume $\triangle ACB$, $\triangle MDB$, and $\triangle AEM$ are right angles.

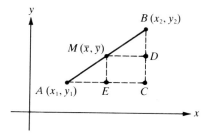

68. Use the result of Exercise 67 to prove Theorem 4.4.

4.4

QUADRATIC FUNCTIONS

Graph of a Quadratic Function

Consider the quadratic equation in two variables,

$$y = x^2 - 4. \tag{1}$$

Solutions of this equation are ordered pairs (x, y). Such ordered pairs can be found by arbitrarily assigning values to x and computing values for y. For instance, assigning the value -3 to x in Equation (1), we obtain

$$y = (-3)^2 - 4,$$

$$y = 5,$$

and so $(-3, 5)$ is a solution. Similarly, we find that

$$(-2, 0), \quad (-1, -3), \quad (0, -4), \quad (1, -3), \quad (2, 0), \quad \text{and} \quad (3, 5)$$

are also solutions of (1). Locating the corresponding points on the plane, we have the graph in Figure 4.14a on page 170.

Clearly, these points do not lie on a straight line. We might ask what pattern the graph of

$$y = x^2 - 4$$

forms on the plane. By graphing more solutions of Equation (1)—with x-components between those already found—we can obtain a clearer picture.

FIGURE 4.14

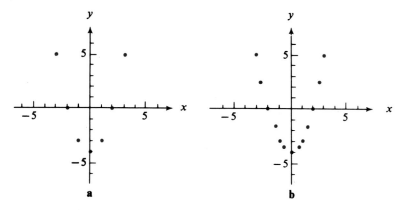

a   b

Accordingly, we find the solutions

$$\left(-\frac{5}{2},\frac{9}{4}\right),\left(-\frac{3}{2},-\frac{7}{4}\right),\left(-\frac{1}{2},-\frac{15}{4}\right),$$

$$\left(\frac{1}{2},-\frac{15}{4}\right),\left(\frac{3}{2},-\frac{7}{4}\right),\left(\frac{5}{2},\frac{9}{4}\right).$$

FIGURE 4.15

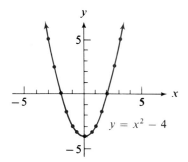

By graphing these points along with those found earlier, we have the graph in Figure 4.14b. We connect these points with a smooth curve, as shown in Figure 4.15. It is reasonable to assume that this curve is a good approximation to the graph of Equation (1). This curve is an example of a **parabola**.

More generally, the graph of any second-degree equation of the form

$$y = ax^2 + bx + c, \tag{2}$$

where a, b, and c are real numbers and $a \neq 0$, is a parabola.

Since for each x, an equation of the form (2) determines only one y, such an equation defines a function, called a **quadratic function**.

Axis of Symmetry

If $a > 0$, the graph of an equation of the form (2) has a lowest point and opens upward. It has a highest point and opens downward if $a < 0$. The lowest (or highest) point on a parabola is called the **vertex** of the parabola. The line through the vertex and parallel to the y-axis is called the **axis of symmetry**, or simply the **axis**, of the parabola, since the parabola is symmetric with respect to this line (Figure 4.16).

FIGURE 4.16

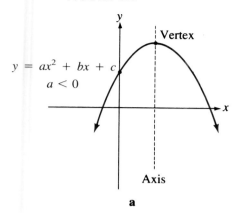

$y = ax^2 + bx + c$
$a < 0$

$y = ax^2 + bx + c$
$a > 0$

a

b

Equation of Axis of Symmetry

If we observe that the graphs of

$$y = ax^2 + bx + c \tag{3}$$

and

$$y = ax^2 + bx \tag{4}$$

have the same axis (Figure 4.17), we can find an equation for the axis of (3) by inspecting Equation (4). Factoring the right-hand member of $y = ax^2 + bx$ yields

$$y = x(ax + b).$$

Thus, we can see that 0 and $-b/a$ are the x-intercepts of the graph. Since the axis of symmetry bisects the segment with these endpoints, an equation for the axis of symmetry is

$$x = -\frac{b}{2a}.$$

FIGURE 4.17

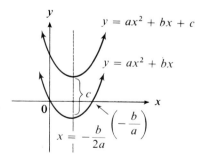

The y-coordinate of the vertex can be obtained by substituting the value $-b/2a$ for x in the equation for the parabola.

EXAMPLE 1 Find an equation for the axis of symmetry, and also find the vertex, of the graph of $y = 2x^2 - 5x + 7$.

Solution   By comparing the given equation with $y = ax^2 + bx + c$, we see that $a = 2$ and $b = -5$. Hence, we have

$$x = -\frac{b}{2a} = -\frac{-5}{2(2)} = \frac{5}{4},$$

and $x = \frac{5}{4}$ is an equation for the axis. The x-coordinate of the vertex is $\frac{5}{4}$. To obtain the y-coordinate, we substitute $\frac{5}{4}$ for x in the given equation to obtain

$$y = 2\left(\frac{5}{4}\right)^2 - 5\left(\frac{5}{4}\right) + 7 = \frac{31}{8}.$$

Thus, the vertex is $(\frac{5}{4}, \frac{31}{8})$.

EXAMPLE 2   Find (a) the x-intercepts (if they exist); (b) the y-intercept; (c) the equation of the axis of symmetry; and (d) the coordinates of the vertex of the graph of $y = x^2 - 2x - 3$.

Solution   (a)   Setting $y = 0$ and solving $x^2 - 2x - 3 = 0$, we have $x = 3$ and $x = -1$. The x-intercepts are 3 and -1.

(b)   Setting $x = 0$, we obtain $y = -3$. The y-intercept is -3.

(c)   Substituting -2 and 1 for b and a, respectively, in the equation $x = -b/2a$, we obtain $x = 1$, the equation for the axis of symmetry.

(d)   Substituting 1 [from part (c)] for x in the equation $y = x^2 - 2x - 3$, we obtain $y = -4$. Since $a > 0$, the graph opens upward and the point $(1, -4)$ is the lowest point.   ■

Graphing a Quadratic Function

When graphing a quadratic function, we begin by sketching the parts that give us the most information about the curve. Thus, to graph $y = ax^2 + bx + c$ proceed as follows.

Steps for Graphing $y = ax^2 + bx + c$

1.   Find and graph the intercepts if they exist.
2.   Find the graph the axis of symmetry $(x = -b/2a)$ and the vertex.
3.   Note whether the parabola opens up $(a > 0)$ or opens down $(a < 0)$.
4.   If there are too few points obtained in Steps 1 and 2 to get an accurate picture of the graph, graph a few more selected points that will be helpful.
5.   Sketch the curve through the points obtained in the preceding steps.

E X A M P L E 3   Graph $y = x^2 - 3x - 4$.

Solution   Setting $x = 0$, we find that the y-intercept is -4. Setting $y = 0$, we have

$$0 = x^2 - 3x - 4 = (x - 4)(x + 1),$$

and the x-intercepts are 4 and -1. The equation of the axis of symmetry is

$$x = -\frac{b}{2a}$$

$$= -\frac{-3}{2(1)} = \frac{3}{2},$$

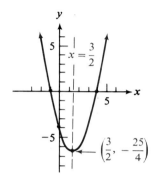

and thus the x-coordinate of the vertex is $\frac{3}{2}$. By substituting $\frac{3}{2}$ for x in $y = x^2 - 3x - 4$, we obtain

$$y = \left(\frac{3}{2}\right)^2 - 3\left(\frac{3}{2}\right) - 4$$

$$= \frac{9}{4} - \frac{9}{2} - 4 = -\frac{25}{4}.$$

We graph the intercepts and the coordinates of the vertex and note that since $a = 1$ is positive, the parabola opens up. Sketching the curve produces the graph shown. ■

In Example 3 we plotted four distinct points in Steps 1 and 2, and thus Step 4 was unnecessary.

E X A M P L E 4   Graph $y = x^2 + x + 3$.

Solution   Setting $x = 0$, we find that the y-intercept is 3. By using the quadratic formula, we find that $x^2 + x + 3 = 0$ has no real solutions; thus, there are no x-intercepts.
The equation of the axis of symmetry is

$$x = -\frac{b}{2a}$$

$$= -\frac{1}{2(1)} = -\frac{1}{2}.$$

Thus, the x-coordinate of the vertex is $-\frac{1}{2}$. Substitute $-\frac{1}{2}$ for x in $y = x^2 + x + 3$

to obtain

$$y = \left(-\frac{1}{2}\right)^2 + \left(-\frac{1}{2}\right) + 3$$

$$= \frac{11}{4}.$$

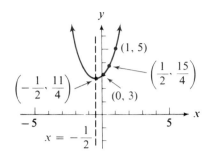

Thus, the vertex is $(-\frac{1}{2}, \frac{11}{4})$.

To obtain a more accurate picture of the graph, we can graph additional points. It is useful to graph several points on the same side of the axis of symmetry and then use the fact that a parabola is symmetric about its axis to sketch the curve. So we let $x = \frac{1}{2}$ and find $y = \frac{15}{4}$, and let $x = 1$ and find $y = 5$. We graph these points and note that since the coefficient of x^2 is positive, the curve opens up. Now we sketch the curve to obtain the graph shown.   ■

Solutions and x-Intercepts

Notice that the x-intercepts of a quadratic function

$$f(x) = ax^2 + bx + c \tag{5}$$

are precisely the *real* solutions of the quadratic equation

$$ax^2 + bx + c = 0. \tag{6}$$

We noted in Chapter 3 that such an equation may have no real solutions, one real solution, or two real solutions. If Equation (6) has no real solutions, then the graph of the corresponding quadratic function (5) has no x-intercepts (Figure 4.18a). If Equation (6) has one real solution, then the graph of (5) is tangent to the x- axis and has exactly one x-intercept (Figure 4.18b). Finally, if Equation (6) has two real solutions, then the graph of (5) has two x-intercepts (Figure 4.18c).

FIGURE 4.18

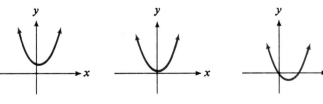

No intercept	One intercept	Two intercepts
No real solutions	One real solution	Two unequal real solutions
a	**b**	**c**

EXERCISE 4.4

A

■ *For the graph of each quadratic function find:*

(a) *The x-intercepts, if they exist.*

(b) *The y-intercept.*

(c) *The equation of the axis of symmetry.*

(d) *The coordinates of the highest (or lowest) point.*

■ *For Exercises 1–12, see Example 2.*

1. $y = x^2 - 3x + 2$ **2.** $y = x^2 + x - 12$ **3.** $y = -2x^2 - 5x - 2$

4. $y = -6x^2 - x + 1$ **5.** $y = x^2 - 2x + 1$ **6.** $y = x^2 - 6x + 9$

7. $f(x) = -4x^2 + 4x - 1$ **8.** $f(x) = -9x^2 + 12x - 4$ **9.** $f(x) = x^2 + 1$

10. $f(x) = x^2 + 4$ **11.** $f(x) = -2x^2 - 9$ **12.** $f(x) = -3x^2 - 4$

■ *Graph. (Obtain the intercepts and the vertex algebraically.) For Exercises 17–32, see Examples 3 and 4.*

13. $y = x^2 - 5x + 4$ **14.** $y = x^2 - 5x - 6$ **15.** $y = x^2 - 6x - 7$

16. $y = x^2 - 3x + 2$ **17.** $y = 2x^2 + 5x - 3$ **18.** $y = 2x^2 + x - 6$

19. $y = -x^2 + 5x - 6$ **20.** $y = -x^2 + 6x - 8$ **21.** $f(x) = -x^2 + 5x - 4$

22. $f(x) = -x^2 - 8x + 9$ **23.** $g(x) = \frac{1}{2}x^2 + 2$ **24.** $g(x) = -\frac{1}{2}x^2 - 2$

25. $f(x) = x^2 + 2$ **26.** $f(x) = -x^2 - 2$ **27.** $f(x) = -x^2$

28. $f(x) = x^2$ **29.** $g(x) = -x^2 - 2x - 1$ **30.** $g(x) = x^2 - 4x + 4$

31. $g(x) = 4x^2 - 12x + 9$ **32.** $g(x) = -9x^2 + 12x - 4$

33. Graph $f(x) = x^2 + 1$. Draw the line segment joining the points $(4, f(4))$ and $(0, f(0))$. Compute $\dfrac{f(4) - f(0)}{4 - 0}$ and interpret this quantity geometrically.

34. Graph $g(x) = x^2 - 1$. Draw the line segment joining the points $(2, g(2))$ and $(-3, g(-3))$. Compute $\dfrac{g(2) - g(-3)}{2 - (-3)}$ and interpret this quantity geometrically.

B

35. On a single set of axes, sketch the family of four curves that are the graphs of

$$y = x^2 + k \qquad (k = -2, 0, 2, 4).$$

What effect does varying k have on the graph?

36. On a single set of axes, sketch the family of six curves that are the graphs of

$$y = kx^2 \qquad \left(k = \frac{1}{2}, 1, 2, -\frac{1}{2}, -1, -2\right).$$

What effect does varying k have on the graph?

37. On a single set of axes, sketch the family of three curves that are the graphs of

$$y = x^2 + kx \qquad (k = -1, 0, 1).$$

What effect does varying k have on the graph?

38. On a single set of axes, sketch the family of three curves that are the graphs of

$$y = -x^2 + kx \qquad (k = -1, 0, 1).$$

What effect does varying k have on the graph?

39. For $f(x) = 2x^2 + x + 1$, compute and simplify $\dfrac{f(a + h) - f(a)}{h}$, $h \neq 0$, and interpret this quantity geometrically for $a = 2$.

40. For $f(x) = -x^2 + 3x + 2$, compute and simplify $\dfrac{f(a + h) - f(a)}{h}$, $h \neq 0$, and interpret this quantity geometrically for $a = -1$.

41. For $f(x) = 3x^2 + 4x - 1$, compute and simplify $\dfrac{f(a + h) - f(a)}{h}$, $h \neq 0$, and interpret this quantity geometrically for $a = 0$.

42. For $f(x) = -2x^2 + 3x - 3$, compute and simplify $\dfrac{f(a + h) - f(a)}{h}$, $h \neq 0$, and interpret this quantity geometrically for $a = 3$.

43. Show that if a parabola opens upward, then the y-coordinate of the vertex is the least of all the y-coordinates of points on the curve. [*Hint:* Complete the square in the right-hand member of $y = ax^2 + bx + c$.]

44. Show that if a parabola opens downward, then the y-coordinate of the vertex is the greatest of all the y-coordinates of points on the curve.

45. A ball is thrown straight up in the air. The height above the ground of the ball (measured in feet) t seconds after being thrown is given by $s(t) = -16t^2 + 12t$. How long does it take the ball to reach its maximum height? What is its maximum height?

46. A ball is thrown straight up from a hot air balloon. The height above the ground of the ball (measured in feet) t seconds after being thrown is given by $s(t) = -16t^2 + 8t + 3$. How long does it take the ball to reach its maximum height? What is its maximum height?

47. The distance in miles between two ships is given by $d(t) = \sqrt{(65 - 10t)^2 + 225t^2}$ where t is time measured in hours. Find the minimum distance between the two ships and the time t at which the ships are closest together. [*Hint:* Find the minimum $[d(t)]^2$.]

48. The distance in miles between two cars is given by $d(t) = \sqrt{(150 - 75t)^2 + 5625t^2}$ where t is measured in hours. Find the minimum distance between the two cars and the time t at which the cars are closest together.

49. What is the minimum value of $f(x) = (ax^2 + bx + c)^2$ if $b^2 - 4ac \geq 0$? Explain your answer.

50. What is the minimum value of $f(x) = (ax^2 + bx + c)^2$ if $b^2 - 4ac < 0$? Explain your answer.

4.5

SPECIAL FUNCTIONS

Functions Defined Piecewise

It is possible to define a function using different rules for different parts of the domain. Such a function is said to be **piecewise defined**. To find the range value paired with a given domain value, we use that part of the function in which the given domain value lies.

EXAMPLE 1 Given that $f(x) = \begin{cases} x^2 & \text{if} \quad x < 0 \\ -x + 1 & \text{if} \quad x \geq 0 \end{cases}$, find the range value.

(a) $f(-1)$

(b) $f(1)$

Solution (a) Since $-1 < 0$, use $f(x) = x^2$.

(b) Since $1 \geq 0$; use $f(x) = -x + 1$.

$$f(-1) = (-1)^2 = 1$$

$$f(1) = -1 + 1 = 0 \qquad ■$$

The graph of a function that is defined piecewise can be obtained by graphing it separately over each subset of the domain determined by the definition of the function.

EXAMPLE 2 Graph $f(x) = \begin{cases} x & \text{if} \quad x \leq 1 \\ -x + 2 & \text{if} \quad x > 1 \end{cases}$.

Solution We first graph

$$y = f(x) = x \qquad \text{for} \qquad x \leq 1$$

to obtain Figure a on page 178. Next, we graph

$$y = f(x) = -x + 2 \qquad \text{for} \qquad x > 1$$

to obtain Figure b. We then sketch both graphs on the same set of coordinate axes to obtain Figure c, which shows the completed graph of the function.

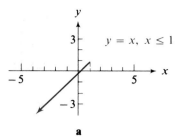

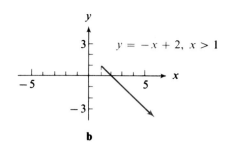

a   **b**

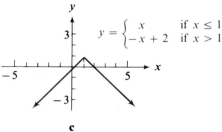

c   ■

Absolute-Value Functions

Functions involving the absolute value of the independent variable are piecewise-defined functions. Consider the function defined by

$$y = |x|. \qquad (1)$$

From the definition of $|x|$, we have

$$y = \begin{cases} x & \text{for} \quad x \ge 0 \\ -x & \text{for} \quad x < 0 \end{cases}. \qquad \begin{matrix} (2) \\ (3) \end{matrix}$$

If we graph (2) and (3) on the same set of axes, we have the graph of $y = |x|$ shown in Figure 4.19.

We can graph any equation involving $|x|$ or $|f(x)|$ by first writing the equation in piecewise-defined form and then using the method described above.

FIGURE 4.19

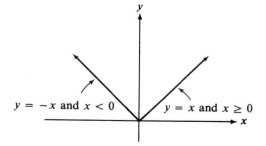

$y = -x$ and $x < 0$ $y = x$ and $x \ge 0$

E X A M P L E 3 Graph $y = |x| + 1$.

Solution   The definition of absolute value implies that $y = |x| + 1$ is equivalent to

$$y = \begin{cases} x + 1 & \text{if} \quad x \geq 0 \\ -x + 1 & \text{if} \quad x < 0 \end{cases}.$$

Thus, we graph $y = x + 1$, $x \geq 0$ and $y = -x + 1$, $x \leq 0$ on the same set of coordinate axes to obtain the figure shown.   ■

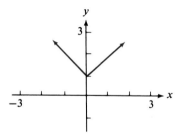

Reflections   If $y < 0$, then the x-axis is the perpendicular bisector of the line segment joining the points (x, y) and $(x, |y|)$ (see Figure 4.20a). Hence, the points (x, y) and $(x, |y|)$ are reflections of one another in the x-axis. Thus, we can obtain the graph of $y = |f(x)|$ by graphing $y = f(x)$ and then reflecting the portion of the graph with $y < 0$ in the x-axis (see Figure 4.20b).

FIGURE 4.20

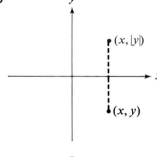

a

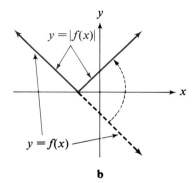

b

EXAMPLE 4   Graph $y = |2x + 1|$.

Solution   We first sketch $y = 2x + 1$, as shown in Figure a. Next, we reflect in the x-axis that portion of the graph below the x-axis to obtain the graph of $y = |2x + 1|$, which is shown in Figure b.

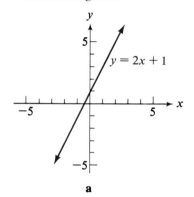

a

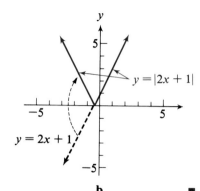

b   ■

Vertical Shifts   Equations of the form

$$y = |f(x)| + c \qquad (4)$$

can be graphed by observing that for each x, $|f(x)| + c$ differs from $|f(x)|$ by c units. Thus, the graph of Equation (4) is simply the graph of $y = |f(x)|$ shifted by c units. The graph is shifted upward if $c > 0$ and downward if $c < 0$.

E X A M P L E 5   Graph $y = |x| - 2$.

Solution   We first sketch $y = |x|$, as shown in Figure a. We then shift the graph 2 units down, as shown in Figure b, to obtain the graph of $y = |x| - 2$.

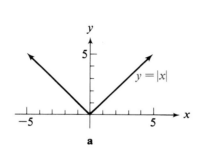

a

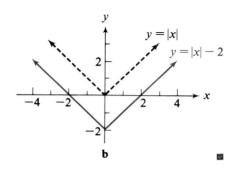

b

Bracket Function   Another interesting piecewise function (sometimes called the **greatest-integer function**) is defined as follows:

For each x, $f(x)$ is the greatest integer less than or equal to x.

The greatest-integer function can be written as a piecewise-defined function as

$$f(x) = j \quad \text{if} \quad j \le x < j + 1, \quad \text{where } j \text{ is an integer.} \qquad (5)$$

The greatest-integer function is sometimes denoted by

$$f(x) = [x],$$

and hence this function is also called the **bracket function**. For example,

$$[2] = 2, \qquad \left[\frac{7}{4}\right] = 1, \qquad [-2] = -2, \qquad \left[\frac{-3}{2}\right] = -2, \qquad \text{and} \qquad \left[-\frac{5}{2}\right] = -3.$$

To graph the bracket function, we consider unit intervals along the x-axis, as shown in Table 4.2. The graph is shown in Figure 4.21. The heavy dot on the left-hand endpoint of each line segment indicates that the endpoint is a part of the graph. The bracket function is also called a "step function," for an obvious reason.

TABLE 4.2

x-interval	[x]
[−2, −1)	−2
[−1, 0)	−1
[0, 1)	0
[1, 2)	1
[2, 3)	2

FIGURE 4.21

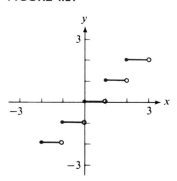

EXAMPLE 6   Graph $f(x) = [3x]$ on the interval $-1 \le x \le 1$.

Solution   We first determine the values in the domain at which the expression $3x$ is an integer value. These are -1, $-\frac{2}{3}$, $-\frac{1}{3}$, 0, $\frac{1}{3}$, $\frac{2}{3}$, and 1. It is at these points that the graph makes jumps. Next, we make a table showing the function values in the intervals between these points. Finally, we use the table to draw the graph as shown.

x-interval	[3x]
$\left[-1, -\frac{2}{3}\right)$	−3
$\left[-\frac{2}{3}, -\frac{1}{3}\right)$	−2
$\left[-\frac{1}{3}, 0\right)$	−1
$\left[0, \frac{1}{3}\right)$	0
$\left[\frac{1}{3}, \frac{2}{3}\right)$	1
$\left[\frac{2}{3}, 1\right)$	2
1	3

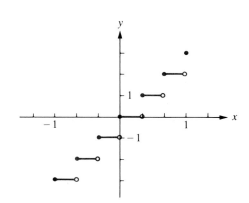

■

E X E R C I S E   4.5

A

■ *Find the indicated range value for the given function. See Example 1.*

1. $f(3)$ for $f(x) = \begin{cases} x - 2 & \text{if } x \geq 2 \\ 2 - x & \text{if } x < 2 \end{cases}$ **2.** $f(1)$ for $f(x) = \begin{cases} -3x - 1 & \text{if } x < -1 \\ 2x + 1 & \text{if } x \geq -1 \end{cases}$

3. $f(2)$ for $f(x) = \begin{cases} x^2 - 1 & \text{if } x \leq 2 \\ x^2 + 1 & \text{if } x > 2 \end{cases}$ **4.** $f(-2)$ for $f(x) = \begin{cases} x^2 - x & \text{if } x < -2 \\ x^2 + x & \text{if } x \geq -2 \end{cases}$

5. $f(2)$ for $f(x) = \begin{cases} x & \text{if } x < 0 \\ 2x + 1 & \text{if } 0 \leq x < 2 \\ 3x + 2 & \text{if } x \geq 2 \end{cases}$ **6.** $f(-1)$ for $f(x) = \begin{cases} -2x & \text{if } x \leq -1 \\ x & \text{if } -1 < x < 1 \\ 2x & \text{if } x \geq 1 \end{cases}$

7. $f(2)$ for $f(x) = \begin{cases} \dfrac{x^2 - 4}{x - 2} & \text{if } x \neq 2 \\ 4 & \text{if } x = 2 \end{cases}$ **8.** $f(-1)$ for $f(x) = \begin{cases} \dfrac{x^2 - 1}{x + 1} & \text{if } x \neq -1 \\ -2 & \text{if } x = -1 \end{cases}$

■ *Graph the function. For Exercises 9–14, see Example 2.*

9. $f(x) = \begin{cases} -x & \text{if } x \leq -1 \\ 3x + 4 & \text{if } x > -1 \end{cases}$ **10.** $f(x) = \begin{cases} -2x + 1 & \text{if } x \leq 2 \\ -3 & \text{if } x > 2 \end{cases}$

11. $f(x) = \begin{cases} x^2 - 2 & \text{if } x \leq 1 \\ -x^2 & \text{if } x > 1 \end{cases}$ **12.** $f(x) = \begin{cases} -x^2 + 1 & \text{if } x \leq 0 \\ x^2 & \text{if } x > 0 \end{cases}$

13. $f(x) = \begin{cases} x^2 & \text{if } x < 0 \\ x^2 + 1 & \text{if } x \geq 0 \end{cases}$ **14.** $f(x) = \begin{cases} -x^2 & \text{if } x < -1 \\ x^2 + 1 & \text{if } x \geq -1 \end{cases}$

■ *For Exercises 15–22, see Examples 3–5.*

15. $y = |x| + 2$ **16.** $y = -|x| + 3$ **17.** $f(x) = |x + 1|$ **18.** $f(x) = |x - 2|$

19. $y = -|2x - 1|$ **20.** $y = -|3x + 2|$ **21.** $g(x) = |2x| - 3$ **22.** $y = |3x| + 2$

■ *Graph the function over the interval* $-5 \leq x \leq 5$. *See Example 6.*

23. $y = 2[x]$ **24.** $y = [2x]$ **25.** $y = [x + 1]$ **26.** $f(x) = [x] + 1$

B

■ *Graph the function. For Exercises 27–44, graph the function over the interval* $-5 \leq x \leq 5$.

27. $f(x) = \begin{cases} -1 & \text{if } x < -1 \\ 1 & \text{if } -1 \leq x \leq 1 \\ \dfrac{1}{2} & \text{if } 1 < x \end{cases}$ **28.** $f(x) = \begin{cases} 0 & \text{if } x \leq 0 \\ 1 & \text{if } 0 < x < 1 \\ 0 & \text{if } 1 \leq x \end{cases}$

29. $f(x) = \begin{cases} 1 - x & \text{if } x \le 0 \\ 1 + x & \text{if } 0 < x \le 2 \\ -3x + 9 & \text{if } x > 2 \end{cases}$

30. $f(x) = \begin{cases} x + 1 & \text{if } x \le 0 \\ 1 - x & \text{if } 0 < x \le 1 \\ x - 1 & \text{if } 1 < x \end{cases}$

31. $f(x) = \begin{cases} -x^2 & \text{if } x < -1 \\ -1 & \text{if } -1 \le x < 1 \\ x^2 - 2 & \text{if } 1 \le x \end{cases}$

32. $f(x) = \begin{cases} x & \text{if } x < 0 \\ x^2 & \text{if } 0 \le x < 2 \\ -x + 6 & \text{if } x \ge 2 \end{cases}$

33. $f(x) = |2x| + |x|$

34. $f(x) = |3x| - |x|$

35. $y = -2|x| + x$

36. $y = 3|x| - x$

37. $y = |x|^2$

38. $y = |x^2|$

39. $g(x) = |x + 1| - x$

40. $g(x) = x + 1 - |x|$

41. $y = [x] + x$

42. $y = \left[\frac{1}{2}x\right] + x$

43. $y = [x] - x$

44. $y = |[x]|$

■ *You-Buy-We-Fly Pizza charges to deliver its pizza according to the function*

$$P(x) = \begin{cases} 0.50[x] & \text{if } 0 \le x \le 5 \\ 3 + 0.50[x - 5] & \text{if } 5 < x \le 15 \\ 10 + 0.50[x - 15] & \text{if } 15 < x \end{cases}$$

where x is the distance from the pizza parlor to the delivery point.

45. If you live $5\frac{1}{3}$ miles from the pizza parlor, how much will you pay for delivery?

46. If it costs 30¢ to drive your car $\frac{1}{2}$ mile, how much do you save by having the pizza delivered to a friend who lives $\frac{1}{2}$ mile from you (and $4\frac{5}{6}$ miles from the pizza parlor) and driving to pick it up and taking it home?

47. If you live 1 block from the pizza parlor, how much do you save by picking up your own pizza?

48. You can buy a pizza from We-Serve pizza parlor, which delivers free, for $4 more than the same pizza from You-Buy-We-Fly. If you live 6.9 miles from You-Buy-We-Fly, which is the cheaper pizza, including delivery cost, and by how much?

■ *A charter airline charges for charters from Los Angeles to London according to the function*

$$P(x) = \begin{cases} 2000 & \text{if } 0 \le x < 25 \\ 1500 - 10[x - 25] & \text{if } 25 \le x < 75 \\ 900 - 10[x - 75] & \text{if } 75 \le x \le 125 \end{cases}$$

where P(x) is the price in dollars charged each passenger and x is the number of passengers. The revenue the charter company generates is $R(x) = xP(x)$ where x is the number of passengers.

49. How much revenue does the company generate if 5 people charter a flight?

50. How much revenue does the company generate if 15 people charter a flight?

51. If children pay half fare and are counted as half of a passenger, how much revenue does the company generate when 11 families each with 2 adults and 3 children charter a flight?

52. How much revenue does a group of 40 children and 10 adults generate? See Exercise 51 for children's rate.

■ *The cost to the charter company of providing a flight from Los Angeles to London is*

$$C(x) = \begin{cases} 20{,}000 + 10x & \text{if } 0 \le x < 25 \\ 25{,}000 + 25x & \text{if } 25 \le x < 75 \\ 35{,}000 + 50x & \text{if } 75 \le x \le 125 \end{cases}$$

where x is the number of passengers on the flight. The profit the company makes is the revenue (use the price function from Exercises 49–52) minus the cost.

53. What profit does the company make if 9 people charter a flight?

54. How many people must charter a flight in order for the company to make a profit and what is their profit?

55. What profit does the company make if 125 people charter a flight?

56. How many people must charter the flight in order for the company to make the largest possible profit and what is that profit?

4.6

INVERSE FUNCTIONS

In Section 4.1 we saw that a function can be considered as a rule for determining a range value *y* corresponding to a given domain value *x*. In some applications, it is useful to reverse this process for a given function *f*, that is, for a given range value *y*, to find a domain value *x* for which $y = f(x)$. We first note that for some functions there is more than one *domain value* that corresponds to a given range value.

E X A M P L E 1   If $y = f(x) = x^2$ and $y = 4$, then there are two domain values, $x = 2$ and $x = -2$, that correspond to the range value 4.   ■

One-to-One Functions   Functions that have the property that for every range value there is *only one* corresponding domain value are called **one-to-one functions.** Thus, $f(x) = x^2$ in Example 1 is not a one-to-one function.

Horizontal Line Test   Since a one-to-one function associates only one element of its domain with any given element of its range, no two points on the graph of a one-to-one function lie on the same horizontal line.

FIGURE 4.22

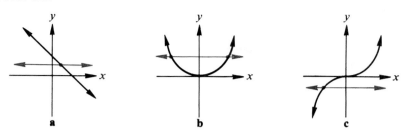

a   b   c

Imagine a horizontal line moving across each graph in Figure 4.22 from top to bottom. If the line at any position cuts the graph in more than one point, then the curve is not the graph of a one-to-one function. Thus, Figure 4.22a and 4.22c are graphs of one-to-one functions, but the curve in Figure 4.22b is the graph of a function that is not one-to-one.

E X A M P L E 2   Is the graph in the figure the graph of a one-to-one function?

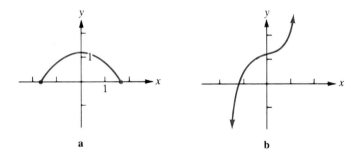

a   b

Solution   (a)   The horizontal line $y = 1$ intersects the graph two times. Thus, the graph is not that of a one-to-one function.

(b)   No horizontal line crosses the graph more than one time. Thus, the graph is the graph of a one-to-one function.   ■

Inverse Functions   Notice that if f is a one-to-one function, then each element of the range of f is paired with only one element of the domain of f. Thus, no two ordered pairs of f have the same second component. Hence, the relation obtained by interchanging the first and second components of every pair in f is itself a function.

Definition 4.4   *If a function is a one-to-one function, then the set of ordered pairs obtained by interchanging the first and second components of each ordered pair in f is called the **inverse function of f** and is denoted f^{-1}.*

EXAMPLE 3   If $f = \{(2, 1), (3, 2), (4, 3)\}$, find f^{-1}.

Solution   We first note that f is a one-to-one function. We then interchange the components of each ordered pair of f:

$$f^{-1} = \{(1, 2), (2, 3), (3, 4)\}. \quad ■$$

It is evident from Definition 4.4 and Example 3 that the domain and range of f^{-1} are just the range and domain, respectively, of f.

Finding f^{-1} from f   Recall that we write $y = f^{-1}(x)$ when (x, y) is a member of f^{-1} and $x = f(y)$ when (y, x) is a member of f. Since from Definition 4.4, (x, y) is a pair in f^{-1} if and only if (y, x) is a pair in f, we have

$$y = f^{-1}(x) \quad \textit{if and only if} \quad x = f(y). \tag{1}$$

Thus, $y = f^{-1}(x)$ and $x = f(y)$ are equivalent equations that define the inverse of the function $y = f(x)$. If f is a one-to-one function defined by $y = f(x)$, we find f^{-1} by proceeding as follows.

Steps for Finding f^{-1}

1.   Interchange the symbols x and y in $y = f(x)$ to write $x = f(y)$.
2.   Solve $x = f(y)$ for y to obtain $y = f^{-1}(x)$.

EXAMPLE 4   Find f^{-1} for $f(x) = 4x - 3$.

Solution   First, we write the function in the form

$$y = 4x - 3.$$

Next, we interchange the symbols x and y to obtain

$$x = 4y - 3,$$

which defines f^{-1}. We then solve for y in terms of x to obtain

$$y = \frac{1}{4}(x + 3).$$

Thus, $f^{-1}(x) = \frac{1}{4}(x + 3). \quad ■$

**Graphs of
Inverse
Functions**

Note in Figure 4.23 that the graphs of the ordered pairs (a, b) and (b, a) are symmetric with respect to the graph of $y = x$ (see Exercise 59 on page 191). Next, observe that if the point (a, b) is on the graph of the one-to-one function f, then the point (b, a) is on the graph of f^{-1}. Therefore, the graphs of the functions $y = f^{-1}(x)$ and $y = f(x)$ are reflections of each other about the graph of $y = x$.

Figure 4.24 shows the graphs of the linear function

$$y = f(x) = 4x - 3$$

and its inverse,

$$y = f^{-1}(x) = \frac{1}{4}(x + 3),$$

which we obtained above, together with the graph of $y = x$.

FIGURE 4.23

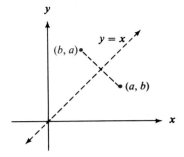

FIGURE 4.24

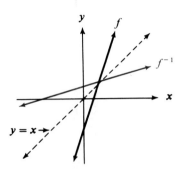

EXAMPLE 5   Graph $f(x) = \frac{1}{3}(6 - x)$ and its inverse on the same set of axes.

Solution   We first write the function as

$$y = \frac{1}{3}(6 - x)$$

and graph this function as shown in the figure. We next find f^{-1} by interchanging the roles of x and y and solving for y. We have

$$x = \frac{1}{3}(6 - y)$$

from which $f^{-1}(x) = y = -3x + 6$. We then graph this function as shown in the figure. ■

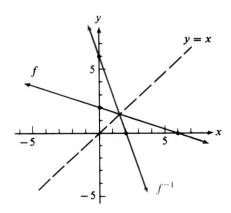

Notice that in Example 5 we could have graphed the equation $x = \frac{1}{3}(6 - y)$ directly without first finding f^{-1}.

$x = f[f^{-1}(x)]$ and $x = f^{-1}[f(x)]$

For any one-to-one function f and any x in the *domain of f^{-1}*, if we write

$$y = f^{-1}(x), \tag{2}$$

then by Condition (1) on page 186, we have

$$x = f(y). \tag{3}$$

Using Equation (2), we substitute $f^{-1}(x)$ for y in Equation (3) to obtain

$$x = f[f^{-1}(x)].$$

Similarly, if x is in the *domain of f*, then by interchanging the roles of x and y and using Condition (1) we can show that

$$x = f^{-1}[f(x)].$$

The details are left as an exercise.

EXAMPLE 6   In Example 4 on page 186, we saw that if $f(x) = 4x - 3$, then $f^{-1}(x) = \frac{1}{4}(x + 3)$. We see that for these functions

$$f^{-1}[f(x)] = \frac{1}{4}[f(x) + 3]$$

$$= \frac{1}{4}[(4x - 3) + 3] = x$$

and

$$f[f^{-1}(x)] = 4[f^{-1}(x)] - 3$$

$$= 4\left[\frac{1}{4}(x + 3)\right] - 3 = x. \qquad ■$$

Restricting Domains

In some cases where f is not a one-to-one function, the domain of f can be restricted, so that on the restricted domain, f is one-to-one. For example, if $f(x) = x^2$, then f is not one-to-one (see Example 1 on page 184). If, however, we restrict the domain so that $x \geq 0$, then f is one-to-one.

EXAMPLE 7   Find f^{-1} if $f(x) = x^2$, $x \geq 0$. Sketch the graphs of f and f^{-1} on the same coordinate axes.

Solution   We first write the function in the form

$$y = x^2, \qquad x \geq 0.$$

Next, we interchange the roles of x and y to obtain

$$x = y^2, \qquad y \geq 0.$$

Then we solve for y in terms of x to obtain

$$y = \sqrt{x}, \qquad y \geq 0.$$

Note that since $y \geq 0$, we are using only the positive square root of x. The graphs are shown in the figure   ■

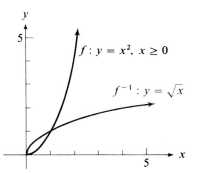

EXERCISE 4.6

A

■   *Determine whether the curve is the graph of a one-to-one function. See Example 2.*

1.

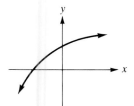

2.

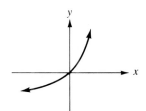

3.

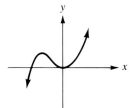

4.

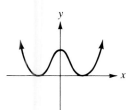

5.

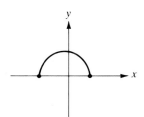

6.

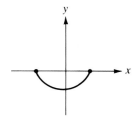

■ *Each of the following functions is one-to-one. Find $f^{-1}(x)$. For Exercises 7–16, see Examples 3, 4, and 7.*

7. $f = \{(4, 1), (2, 3), (1, 5)\}$

8. $f = \{(-2, 2), (-1, 3), (1, 4)\}$

9. $f = \{(-2, 2), (0, 0), (2, -2)\}$

10. $f = \{(-3, -3), (-1, 1), (1, -1)\}$

11. $f(x) = x$

12. $f(x) = -x$

13. $f(x) = 4 - 2x$

14. $f(x) = 3 + 4x$

15. $f(x) = \frac{1}{4}(12 - 3x)$

16. $f(x) = -\frac{1}{2}(4 - x)$

17. $f(x) = x^3 - 1$

18. $f(x) = 8x^3 + 1$

19. $f(x) = \frac{1}{x}$

20. $f(x) = \frac{1}{x^3 + 1}$

21. $f(x) = x^2, \quad x \le 0$

22. $f(x) = x^2 + 1, \quad x \ge 0$

■ *Graph each function and its inverse on the same set of axes. See Examples 5 and 7.*

23. $f = \{(-2, 3), (4, 7), (5, 9)\}$

24. $f = \{(-3, 1), (2, -1), (3, 4)\}$

25. $f = \{(-1, -1), (2, 2), (3, 3)\}$

26. $f = \{(-4, 4), (0, 0), (4, -4)\}$

27. $f(x) = 2x + 6$

28. $f(x) = 3x - 6$

29. $f(x) = 4 - 2x$

30. $f(x) = 6 + 3x$

31. $f(x) = -3 + \frac{3}{4}x$

32. $f(x) = \frac{1}{6}x - 1$

33. $f(x) = 4 - 4x$

34. $f(x) = -4 + \frac{2}{3}x$

35. $f(x) = x^2 + 1, \quad x \ge 0$

36. $f(x) = x^2 + 1, \quad x \le 0$

37. $f(x) = x^2 - 1, \quad x \ge 0$

38. $f(x) = x^2 - 1, \quad x \le 0$

39. $f(x) = x^2 - 4, \quad x \ge 0$

40. $f(x) = x^2 - 4, \quad x \le 0$

41. $f(x) = x^2 - 4x + 4, \quad x \ge 2$

42. $f(x) = x^2 - 2x - 3, \quad x \le 1$

■ *Verify that $x = f[f^{-1}(x)]$ and $x = f^{-1}[f(x)]$. See Example 6.*

43. $f(x) = x - 1; \quad f^{-1}(x) = x + 1$

44. $f(x) = 2x + 3; \quad f^{-1}(x) = \frac{1}{2}(x - 3)$

45. $f(x) = x^3; \quad f^{-1}(x) = \sqrt[3]{x}$

46. $f(x) = x^3 + 1; \quad f^{-1}(x) = \sqrt[3]{x - 1}$

47. $f(x) = \frac{1}{x - 1}; \quad f^{-1}(x) = \frac{x + 1}{x}$

48. $f(x) = \frac{1}{x + 2}; \quad f^{-1}(x) = \frac{1 - 2x}{x}$

■ *Determine whether g is f^{-1}.*

49.

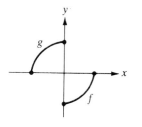

50.

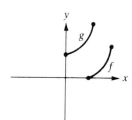

51.

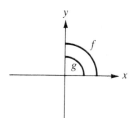

52. 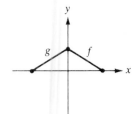   **53.** 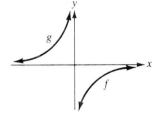   **54.**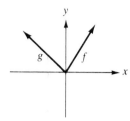

B

55. If $f(2) = 4$, find $f^{-1}(4)$.

56. If $f^{-1}(0) = 1$, find $f(1)$.

57. If $f(3) = -1$, find $f^{-1}(-1)$.

58. If $f^{-1}(-2) = -3$, find $f(-3)$.

59. Show that the line $y = x$ is the perpendicular bisector of the line segment between the points (a, b) and (b, a) for every value of a and b. Notice that this proves the graphs of $y = f(x)$ and $y = f^{-1}(x)$ are reflections of one another in the line $y = x$.

60. Show that if f is a one-to-one function and x is in the domain of f, then $x = f^{-1}[f(x)]$.

■ *Determine whether g is f^{-1}.*

61. $f(x) = \dfrac{1}{x}$

$g(x) = \dfrac{1}{x}$

62. $f(x) = \dfrac{1}{x^2}, x \ge 0$

$g(x) = \dfrac{1}{x^2}, x \ge 0$

63. $f(x) = x^3$

$g(x) = \dfrac{1}{x^3}$

64. $f(x) = x^4, x \le 0$

$g(x) = \dfrac{1}{x^4}, x \le 0$

65. $f(x) = x - 1$

$g(x) = x + 1$

66. $f(x) = 2x + 3$

$g(x) = 2x - 3$

4.7

GRAPHS OF FIRST-DEGREE AND SECOND-DEGREE INEQUALITIES IN TWO VARIABLES

An inequality of the form

$$Ax + By + C \le 0 \qquad \text{or} \qquad Ax + By + C < 0,$$

A and B not both 0, defines a relation.

Graph of an Inequality

The graphs of such relations will be a *region* of the plane. For example, consider the relation

$$2x + y - 3 < 0. \tag{1}$$

When the inequality is rewritten in the equivalent form

$$y < -2x + 3, \tag{2}$$

we see that solutions (x, y) are such that for each x, y is less than $-2x + 3$. The graph of the equation

$$y = -2x + 3 \tag{3}$$

is simply a straight line, as shown in Figure 4.25a. To graph Inequality (1), we need only observe that any point below this line has a y-coordinate that satisfies (2). Consequently, the solution set of (2) corresponds to the entire region below the line.

FIGURE 4.25

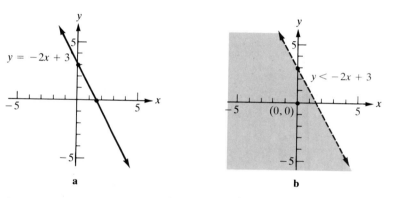

The region is indicated on the graph with shading. That the line itself is not in the graph is shown by means of a broken line, as in Figure 4.25b. Had the inequality been

$$2x + y - 3 \le 0,$$

the line would be a part of the graph and would be shown as a solid line. In general, the graphs of the inequalities

$$Ax + By + C < 0 \qquad \text{or} \qquad Ax + By + C > 0$$

are the points in a half-plane on one side of the graph of the **associated equation**

$$Ax + By + C = 0,$$

depending on the constants and inequality symbols involved.

To determine which half-plane to shade in constructing graphs of first-degree inequalities, one can select any point in either half-plane and test its coordinates in the given inequality to see whether or not the selected point lies in the graph. If the coordinates satisfy the inequality, then the half-plane containing the selected point is shaded; if not, the opposite half-plane is shaded. A very convenient point to use in this process is the origin, provided the origin is not contained in the graph of the associated equation.

EXAMPLE 1 Graph $3x + 4y \geq 12$.

Solution Graph the associated equation

$$3x + 4y = 12.$$

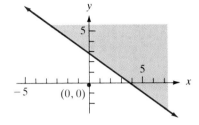

The line is the edge of the half-plane that is the graph of the inequality. Observe that the origin is *not* part of the graph of the inequality because $(0, 0)$ does not satisfy the original inequality:

$$3(0) + 4(0) \not\geq 12.$$

Hence, the region above the line is shaded. Note that the line is included in the graph. ■

If only one variable appears in the inequality, the other variable may take on any value and we proceed as in Example 1.

EXAMPLE 2 Graph $x > 2$ in the plane.

Solution We first graph the associated equation $x = 2$. We next substitute 0 for x and 0 for y in the original inequality (note that since the variable y does not appear in the original inequality, the choice of 0 for y is unimportant) to obtain $0 \not> 2$. Hence, the region not containing $(0, 0)$ is shaded. The line is excluded from the graph since the inequality is $>$ not $\geq$. ■

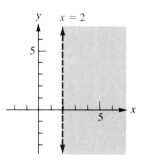

Quadratic Inequalities

Inequalities of the form

$$y < ax^2 + bx + c \qquad \text{and} \qquad y > ax^2 + bx + c, \qquad a \neq 0,$$

can be graphed in the same manner in which we graphed linear inequalities in two variables. We first graph the associated equation, and then shade an appropriate region as required.

EXAMPLE 3   Graph $y < x^2 + 2$.

Solution   Graph the associated equation

$$y = x^2 + 2.$$

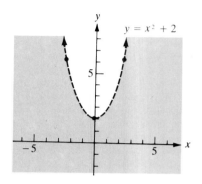

The coordinates of a point not on the graph of the associated equation are used to determine which of the two resulting regions of the plane is the graph of the inequality. Substituting 0 for x and 0 for y in $y < x^2 + 2$ gives $(0) < (0)^2 + 2$, a true statement. Hence, the region containing the origin is shaded as shown in the figure. Since the graph of the equation is not part of the graph of the inequality, a broken curve is used.   ■

EXERCISE 4.7

A

■   *Graph each inequality. For Exercises 1–12, see Example 1.*

1. $y < x$ **2.** $y > x$ **3.** $y \leq x + 2$ **4.** $y \geq x - 2$
5. $x + y < 5$ **6.** $2x + y < 2$ **7.** $x - y < 3$ **8.** $x - 2y < 5$
9. $x \leq 2y - 4$ **10.** $2x \leq y + 1$ **11.** $3 \geq 2x - 2y$ **12.** $0 \geq x + y$

■   *For Exercises 13–20, see Example 2.*

13. $x > 0$ **14.** $y < 0$ **15.** $x < 0$ **16.** $x < -2$
17. $-1 < x < 5$ **18.** $0 \leq y \leq 1$ **19.** $|x| < 3$ **20.** $|y| > 1$

■   *For Exercises 21–28, see Example 3.*

21. $y > x^2$ **22.** $y < x^2$ **23.** $y \geq x^2 + 3$ **24.** $y \leq x^2 + 3$
25. $y < 3x^2 + 2x$ **26.** $y > 3x^2 + 2x$ **27.** $y \leq x^2 + 3x + 2$ **28.** $y \geq x^2 + 3x + 2$

B

29. Graph the set of points whose coordinates satisfy both

$$y \le x - 1 \quad \text{and} \quad y \ge 1 - x.$$

30. Graph the set of points whose coordinates satisfy both

$$y \le 3 - 2x \quad \text{and} \quad y \ge 2x - 3.$$

31. Graph the set of points whose coordinates satisfy both

$$y \le 4 - x^2 \quad \text{and} \quad y \ge x^2 - 4.$$

32. Graph the set of points whose coordinates satisfy both

$$x \ge y^2 - 4 \quad \text{and} \quad y \ge x^2 - 4.$$

■ *Graph each inequality. [Hint: Consider the graphs in each quadrant separately:* $x, y \ge 0; \quad x \le 0, y \ge 0; \quad x, y \le 0; \quad and \ x \ge 0, y \le 0.$]

33. $|x| + |y| \le 1$ **34.** $|x| + |y| \ge 1$ **35.** $|x| - |y| \le 1$ **36.** $|x| - |y| \ge 1$
37. $|x| - |y| \ge 0$ **38.** $|x| - |y| \le 0$ **39.** $|x + y| \ge 1$ **40.** $|x + y| \le 1$

4.8

FUNCTIONS AS MATHEMATICAL MODELS

In applications of mathematics to other sciences, mathematical models are often used to answer questions posed or to make predictions. Mathematical models generally involve one or more functions. In this section we construct several such models and discuss restrictions on the independent variable imposed by limitations on the quantities involved.

Direct Variation

There are two types of functional relationships, widely used in the sciences, to which custom has assigned special names. First, the variable y is said to **vary directly** as the variable x if

$$y = kx \quad (k \text{ a positive constant}). \quad (1)$$

Note that since Equation (1) associates one and only one y with each x, a direct variation defines a function.

EXAMPLE I   (a)   The circumference of a circle varies directly as the radius since

$$c = 2\pi r.$$

(b)   The area of a circle varies directly as the square of the radius since

$$A = \pi r^2. \qquad ■$$

Inverse Variation   The second important type of variation arises from the equation

$$xy = k \qquad (k \text{ a positive constant}), \qquad (2)$$

in which x and y are said to **vary inversely**. When (2) is written in the form

$$y = \frac{k}{x}, \qquad (3)$$

y is said to vary inversely at x.

Since Equation (3) associates only one y with each x ($x \neq 0$), an inverse variation defines a function, with domain $\{x \mid x \neq 0\}$.

EXAMPLE 2   (a)   For an ideal gas at constant absolute temperature (T), the volume (V) and pressure (P) vary inversely since

$$VP = kT \qquad (k, T \text{ constants}),$$

or

$$V = \frac{kT}{P}.$$

(b)   For a right circular cylinder with constant volume (V), the height (h) and the square of the radius (r) vary inversely since

$$V = \pi r^2 h \qquad (V \text{ a constant}),$$

or

$$h = \frac{V}{\pi r^2}. \qquad ■$$

The names "direct" and "inverse," as applied to variation, arise from the fact that in direct variation an assignment of increasing absolute values of x results in an association with increasing absolute values of y, whereas in inverse variation an assignment of increasing absolute values of x results in an association with decreasing absolute values of y.

The constant involved in an equation defining a direct or inverse variation is called the **constant of variation**. If we know that one variable varies directly or inversely as another, and if we have one set of associated values for the variables, we can find the constant of variation involved. We can then use this constant to express one of the variables as a function of the other.

EXAMPLE 3   The speed (v) at which a particle falls in a certain medium varies directly with the time (t) it falls. The particle is falling at a speed of 20 feet per second in 4 seconds. Express v as a function of t.

Solution   Since v varies directly with t, we know there is a positive constant k so that

$$v = kt.$$

Since $v = 20$ when $t = 4$, we have

$$20 = k(4),$$

from which

$$k = 5.$$

Thus, the functional relationship between v and t is given by

$$v = 5t. \qquad ■$$

The following examples illustrate other functional relationships that occur in mathematical models.

EXAMPLE 4   A company that produces computer chips finds that after a start-up cost of $10,000, each new chip costs $200 to manufacture. Express the cost (C) of producing x new chips as a function of x.

Solution   The cost C is the sum of the initial or start-up cost and the cost of manufacturing x

chips. Since the cost of manufacturing x chips is

$$\$200 \qquad \cdot \qquad x$$

[Cost per chip] · [Number of chips]

and the start-up cost is $10,000, we have

$$C = 10,000 + 200x.$$

Since the company cannot produce a negative number of chips, x is always nonnegative, and thus we restrict x to the interval $[0, \infty)$.

EXAMPLE 5 A family has two investments. One investment is more speculative and is paying 12% annually, whereas the other is low risk but is paying only 7% annually. The investment strategy adopted by this family is to invest twice as much in the low-risk investment as in the speculative one. Express their annual income (I) from these investments as a function of the amount invested in the speculative investment (x).

Solution The income from the two investments is the sum of the incomes from each. The income from the speculative investment is

$$0.12 \qquad \cdot \qquad x$$

[Annual rate of return] · [Amount invested]

The amount invested in the low-risk investment is $2x$, twice the amount invested in the speculative investment. Thus, the annual income from this investment is

$$0.07 \cdot 2x.$$

Therefore, the annual income (I) from these investments is

$$I = 0.12x + 0.07(2x)$$
$$= 0.26x.$$

We assume that the family invested some money in each type of investment, so x is always positive and thus we restrict x to the interval $(0, \infty)$. ■

It is sometimes helpful to use a table to organize the data presented in the statement of a problem.

EXAMPLE 6 Let n be the number of liters of a 10% solution of acid that must be added to y liters of a 60% solution of acid to obtain a 50% solution. Express n as a function of y.

Solution   The amount of acid in any mixture is given by

[Concentration (%) of acid] · [Amount of mixture].

We use this relationship to set up the following table.

Mixture	Part of Acid in Mixture	Amount of Mixture	Amount of Acid
10%	0.10	n	$0.10n$
60%	0.60	y	$0.60y$
50%	0.50	$n + y$	$0.50(n + y)$

Note that the amount of acid in the combined solution must be the sum of the amounts of acid in the individual solutions. Thus,

$$0.50(n + y) = 0.10n + 0.60y.$$

We now solve this equation for n, obtaining

$$0.50n + 0.50y = 0.10n + 0.60y,$$

from which

$$-0.10y = -0.40n,$$

$$n = \frac{1}{4}y.$$

Since we must have some 60% solution, y is always positive, and thus we restrict y to the interval $(0, \infty)$.   ■

It is often useful to draw a figure and label parts of the figure using data in the statement of the problem, as shown in the following example. The example also illustrates a case in which the value of the independent variable is restricted to intervals other than $[0, \infty)$.

EXAMPLE 7   A rope 12 feet long is to be cut into two pieces of unequal length, and each piece is to be shaped to enclose a square. Express the total area (A) enclosed as a function of the length (x) of the shorter piece of rope.

Solution   We first draw and label a figure as shown.

Length of short piece: x

Length of long piece: $12 - x$

Smaller area: A_1

Larger area: A_2

Total area enclosed: $A_1 + A_2$

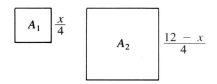

Since the perimeter of the smaller square is x, the side length of this square is $x/4$, as shown in the figure. Similarly, the side length of the larger square is $(12 - x)/4$.
Since the area of a square is the square of the length of its side,

$$A_1 = \left(\frac{x}{4}\right)^2 \quad \text{and} \quad A_2 = \left(\frac{12 - x}{4}\right)^2,$$

and the total area is given by

$$A = \left(\frac{x}{4}\right)^2 + \left(\frac{12 - x}{4}\right)^2.$$

Since x represents a length, it is always positive (we assume a cut is made). Further, x represents the shorter of the two lengths of rope, and hence it is less than one half the total length of the rope. Thus, we restrict x to the interval $(0, 6)$.   ■

It is sometimes helpful to introduce extra variables in constructing models.

EXAMPLE 8   A boat travels downriver twice as fast as it can travel upriver, and its rate in still water is greater than the rate of the current. Express the rate (r_u) of the boat going upriver as a function of the rate (r_c) of the current in the river.

Solution   Let r_s be the rate of the boat in still water and r_d be the rate of the boat downriver. Then we have the following relationships,

$$r_d = r_s + r_c, \tag{4}$$

$$r_u = r_s - r_c, \tag{5}$$

and

$$r_d = 2r_u. \tag{6}$$

Subtracting both members of Equation (5) from the corresponding members of

Equation (4), we have

$$r_d - r_u = 2r_c,$$

from which

$$r_c = \frac{1}{2}(r_d - r_u).$$

Substituting $2r_u$ for r_d from Equation (6), we have

$$r_c = \frac{1}{2}(2r_u - r_u) = \frac{r_u}{2}.$$

Thus,

$$r_u = 2r_c.$$

Since the rate of the current in a river is always nonnegative, $r_c \geq 0$, and thus we restrict r_c to the interval $[0, \infty)$.   ■

In this section we restricted our attention to setting up models in order to emphasize this important skill. Once this has been done, function values can be readily computed for specified domain values.

EXERCISE 4.8

A

■   *In each exercise, give a mathematical model and state any necessary restrictions on the independent variable.*

For Exercises 1–4, see Examples 1–3.

1. The distance (s) a particle falls in a certain medium varies directly with the time (t) it falls. If the particle falls 16 feet in 2 seconds, express s as a function of t.

2. The pressure (P) exerted by a liquid at a given point varies directly as the depth (d) of the point beneath the surface of the liquid. If the liquid exerts a pressure of 40 pounds per square foot at a depth of 10 feet, express P as a function of d.

3. The maximum-safe uniformly distributed load (L) for a horizontal beam with breadth $b = 2$ feet and depth $d = 4$ feet varies inversely with the length (l). If an 8-foot long beam will support up to 750 pounds, express L as a function of l.

4. The resistance (R) of wire 50 feet long varies inversely as the square of its diameter (d). If a wire with diameter 0.012 inch has a resistance of 10 ohms, express R as a function of d.

■ *For Exercises 5–10, see Example 4.*

5. The set-up cost for printing a book is $5000, after which it costs $10 for each book printed. Express the printing costs (C) for a book as a function of the number (n) of books printed.

6. A computer company purchases a shipment of computer chips for $10,000. Express the cost per chip (C) as a function of the number (n) of chips purchased.

7. The cost of fencing material is $7.50 per foot. Express the cost (C) of fencing a square field as a function of the length (s) of the side of the field.

8. The cost of fencing material is $5.50 per foot. Express the cost (C) of fencing a rectangular field with its length twice its width as a function of its width (w).

9. The cost of building a brick wall is $15 per foot, and the cost of fencing material is $7.50 per foot. A rancher is building a square corral with three sides brick wall and one side fencing. Express the cost (C) as a function of the side length (s) of the corral.

10. Assume the same costs as in Exercise 9. A farmer wishes to build a rectangular pen using the wall of his barn as one length of the pen and with the opposite side fencing. The other two sides are to be brick walls. The length is to be twice the width. Express the cost (C) of building the pen as a function of the width (w).

■ *For Exercises 11 and 12, see Example 5.*

11. A woman can invest her retirement funds in stocks paying 12% and in bonds paying 9%. She decides to invest $1\frac{1}{2}$ times as much in stocks as in bonds. Express her annual income (I) from these two investments as a function of the amount (x) invested in bonds.

12. A man directs his company to invest his retirement funds using the rule: invest three times as much in stocks paying 12.5% as in bonds paying 8%. Express the amount (x) he has invested in bonds as a function of the annual income (I) from these two investments.

■ *For Exercises 13–16, see Example 6.*

13. Let n be the number of liters of a 5% solution of acid that must be added to y liters of a 20% solution of acid to obtain a 15% solution. Express y as a function of n.

14. Let n be the number of liters of a 25% solution of acid that must be added to y liters of a 15% solution of acid to obtain a 20% solution. Express n as a function of y.

15. Let n be the number of liters of pure acid that must be added to y liters of a 10% solution of acid to obtain a 55% solution. Express n as a function of y.

16. Let n be the number of liters of water that must be added to y liters of pure acid to obtain a 35% solution of acid. Express y as a function of n.

■ *For Exercises 17–20, see Example 7.*

17. A rope l feet long is to be cut in half, and each half is to be shaped to enclose a square. Express the total area (A) enclosed by the two pieces as a function of l.

18. A rope 200 feet long is to be cut into two pieces. The shorter piece is to be shaped to enclose a square, and the longer piece is to be shaped to enclose a rectangle with length equal to four times the width. Express the total area (A) enclosed by the two pieces as a function of the length (x) of the shorter piece.

19. A rope of length 50 feet is to be cut into two pieces. Each piece is to be shaped to enclose a circle. Express the total area (A) enclosed by the two pieces as a function of the length (x) of the shorter piece. [*Hint:* The formulas for area and circumference of a circle are $A = \pi r^2$ and $c = 2\pi r$, respectively.]

20. A rope of length 100 feet is to be cut into two pieces. The shorter piece is to be shaped to enclose a square, and the longer piece is to be shaped to enclose a circle. Express the total area (A) enclosed by the two pieces as a function of the length (x) of the shorter piece.

21. A double-spaced typed page has approximately 3 lines of type per inch. If pica type is used, there are approximately 10 characters per inch on a line. Express the number (N) of characters on an $8\frac{1}{2} \times 11$-inch double-spaced typed page with uniform margins on the top, bottom, and both sides as a function of the width (x) of the margin. Assume pica type was used.

22. An $8\frac{1}{2} \times 11$-inch double-spaced type page has $\frac{1}{4}$-inch wider margins on the top and bottom than on the sides. Use the information in Exercise 21 to express the number (N) of characters on the page as a function of the width (x) of the top margin. Assume pica type was used.

23. A printer is to leave a uniform margin of 1 inch on top, bottom, and both sides of a square poster. Express the area (A) of printed matter on the poster as a function of the side length (x) of the poster.

24. A printer is to leave a $\frac{1}{4}$-inch smaller margin on the lengths of a poster than on the widths. The width of the poster is one-half the length and the area is 512 square inches. Express the area (A) of printed matter as a function of the margin (x) along the width of the poster.

25. The cost of fencing material is $7.50 per foot. A rancher is to enclose a rectangular area of 100 square feet with this fencing. Express the cost (C) of the fencing material as a function of the width (w) of the rectangle.

26. The cost of fencing material is $10 per foot. A rancher is to enclose a rectangular area of 125 square feet adjacent to his barn. The barn wall will act as one length of the enclosure, and fencing material will be used for the other three sides. Express the cost (C) of the fencing material as a function of the width (w) of the enclosure.

27. A cement sidewalk of uniform width (x) is to enclose a 50-foot- × -50-foot rose garden. The cost of laying cement is $10 per square foot, and the cost of planting roses is $25 per square foot. Express the cost (C) of the garden as a function of the width of the sidewalk (x).

28. A circular disk of radius 10 inches is to have a uniform border of width (x), which is highly polished, and the area inside the border is to be painted. The polishing process costs 15¢ per square inch and paint costs 8¢ per square inch. Express the cost (C) of finishing the disk as a function of the width of the border. [*Hint:* Recall that the area of a circle is given by $A = \pi r^2$.]

29. The telephone company is laying a cable from an island 12 miles offshore to a relay station 18 miles from the point (A) on the shore directly opposite the island, as shown in the figure. It costs $50 per foot to lay cable underwater and $30 per foot to lay cable underground. Express the cost (C) of laying

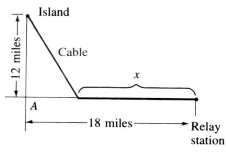

the cable as a function of the distance (x) from the relay station to where the cable leaves the water. Assume that the cable will leave the water at a point at or between A and the relay station.

30. A cable TV company is laying a cable from Island A, which is 12 miles offshore, to a relay station 15 miles from the point on the shore directly opposite the island, and then to Island B, which is 6 miles directly opposite the relay station, as shown in the figure. The cost of laying cable underwater is $50 per foot and the cost of laying cable underground is $30 per foot.

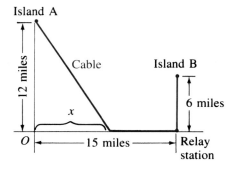

Express the cost (C) of laying the cable as a function of the distance (x) from point O in the figure to the point where the cable leaves the water. Assume the cable will leave the water at a point at or between O and the relay station.

■ *For Exercises 31–34, see Example 8.*

31. A boat travels 25 miles upriver in the same time it travels 45 miles downriver. Express the rate (r_u) of the boat going up the river as a function of the rate (r_c) of the current in the river. [*Hint:* Use Equations (4) and (5) on page 200.]

32. A boat travels 20 miles upriver in the same time that it travels 25 miles downriver. Express the rate (r_d) of the boat going down the river as a function of the rate (r_c) of the current in the river. [*Hint:* Use Equations (4) and (5) on page 200.]

33. An airplane is flying into a headwind of 20 knots (nautical miles per hour) from City A to City B. It makes the round trip from A to B and back in 8 hours. Express the rate (r) at which it flew from City A to City B as a function of the time (t) it takes the plane to fly from City A to City B. [*Hint:* $r = d/t$ is not the correct answer because d is unknown.]

34. For the trip described in Exercise 33, express the distance (d) from City A to City B as a function of the time (t) required for the plane to make the trip from City B back to City A.

35. A shipment of computer chips was purchased by a company for $8000. Four of the chips were damaged in shipment and could not be resold. The remaining chips were sold for a total profit of $1500. Express the profit (P) made on each chip sold as a function of the number of chips purchased (x).

36. A coal supplier purchases a carload of coal for $20,000. Due to inaccurate measurement the car contained 500 pounds more than the amount contracted for. A total profit of $5000 was realized when all of the coal was sold. Express the profit per pound (P) made on the coal that was sold as a function of the amount of coal contracted for (x).

37. A coal company has the following pricing scheme. For every order of between 0 and 3 tons inclusive, the price is $10 per ton or part thereof. For every order of between 3 and 10 tons inclusive a 30¢ per ton discount is given. Express the price (P) per ton of an order as a function of the amount in tons (x) of coal ordered.

38. Assume the coal company in Exercise 37 gives a 35¢ per ton discount for every order between 10 and 15 tons inclusive. Express the price (P) of an order as a function of the amount in tons (x) of coal ordered.

CHAPTER REVIEW

Key Words, Phrases, and Symbols

■   *Define or explain each of the following words, phrases, and symbols.*

1. ordered pair

2. $R \times R$

3. relation

4. domain

5. range

6. function

7. $f(x)$

8. change in f

9. $f(x)|_a^b$

10. composition

11. Cartesian coordinate system

12. linear equation in x and y

13. standard form of a linear equation

14. intercepts of a graph

15. constant function

16. linear function

17. distance between two points

18. slope of a line

19. point-slope form of a linear equation

20. two-point form of a linear equation

21. slope-intercept form of a linear equation

22. intercept form of a linear equation

23. quadratic function

24. vertex of a parabola

25. axis of symmetry of a parabola

26. piecewise-defined function

27. absolute-value function

28. greatest-integer function

29. one-to-one function

30. inverse of a one-to-one function

31. varies directly as

32. varies inversely with

Review Exercises

A

[4.1]   ■   *Specify the domain of each relation.*

1. $\{(4, -1), (2, -4), (3, -5)\}$

2. $\{(1, 2), (1, 3), (1, 6)\}$

3. $y = \dfrac{1}{x + 4}$

4. $y = \sqrt{x - 6}$

■   *Find the indicated range value for the given function.*

5. $f(x) = \dfrac{1}{2x - 1}; \quad f(1)$

6. $f(x) = \dfrac{x}{2x + 1}; \quad f(0)$

7. $f(x) = \sqrt{4 - x^2}; \quad f(0)$

8. $f(x) = \dfrac{1}{\sqrt{x+1}}; \quad f(4)$

■   Find $f(a + h) - f(a)$ for the given function.

9. $f(x) = 3x - 4$

10. $f(x) = -4x + 8$

11. $f(x) = x^2 - 2x$

12. $f(x) = -x^2 + x$

■   Evaluate $f(x)\Big|_2^4$ for each of the following.

13. $f(x) = x^2 + 2$

14. $f(x) = \dfrac{1}{\sqrt{x-1}}$

■   For each pair of functions find the indicated values.

15. $f(x) = x - 5, \quad g(x) = 2x + 3$
    (a) $[f \circ g](1)$ (b) $[g \circ f](1)$

16. $f(x) = x^2 + 2x, \quad g(x) = x + 1$
    (a) $[f \circ g](-2)$ (b) $[g \circ f](-2)$

[4.2]   ■   Graph.

17. $3x - 4y = 4$

18. $2x - y = 1$

19. $x - y = 0$

20. $x + y = 0$

21. $x - 2y = 1$

22. $3x + y = 0$

■   Find the distance between each of the given pairs of points, and find the slope of the line segment joining them.

23. $(2, 0)$ and $(-3, 4)$

24. $(-6, 1)$ and $(-8, 2)$

25. $(1, 2)$ and $(1, -4)$

26. $(-8, 2)$ and $(4, 2)$

[4.3]   ■   Find the equation, in standard form, of the line passing through each of the given points and having the given slope.

27. $(2, -7), \quad m = 4$

28. $(-6, 3), \quad m = \dfrac{1}{2}$

■   Find the equation in standard form of the line passing through the given points.

29. $(3, -2); \quad (2, -1)$

30. $(-4, 1); \quad (2, -4)$

■   Write each equation in slope-intercept form, and specify the slope and the y-intercept of its graph.

31. $4x + y = 6$

32. $3x - 2y = 16$

■   Find the equation in standard form of the line with the given intercepts.

33. $x = -3; \quad y = 2$

34. $x = \dfrac{1}{3}; \quad y = -4$

■   *Find an equation of the line that passes through the given point and (**a**) is parallel to and (**b**) is perpendicular to the graph of the given equation.*

35. $(1, 1);\quad -3x + 2y = 7$ **36.** $(-1, 1);\quad x - y = 0$

[4.4]   ■   *Find the x-intercepts, the axis of symmetry, and the vertex of the graph of each function by algebraic methods.*

37. $f(x) = x^2 - x - 6$ **38.** $f(x) = -x^2 + 7x - 10$

39. Graph the function of Exercise 37. Draw a line segment from $(4, 0)$ to $(4, f(4))$.

40. Graph the function of Exercise 38. Draw a line segment from $(3, 0)$ to $(3, f(3))$.

[4.5]   ■   *Find the indicated range value for the given function.*

41. $f(-1)$ if $f(x) = \begin{cases} -3x + 1 & \text{if } x < 2 \\ x - 2 & \text{if } x \geq 2 \end{cases}$ **42.** $f(2)$ if $f(x) = \begin{cases} x^2 & \text{if } x \leq 1 \\ -2x^2 + 1 & \text{if } x > 1 \end{cases}$

■   *Graph the function.*

43. $f(x) = \begin{cases} x - 1 & \text{if } x \leq 2 \\ \dfrac{1}{2}x & \text{if } x > 2 \end{cases}$ **44.** $f(x) = \begin{cases} x + 4 & \text{if } x \leq -1 \\ -3x & \text{if } x > -1 \end{cases}$

45. $y = \left| \dfrac{x}{5} \right|$ **46.** $y = -\left| x - \dfrac{1}{4} \right|$

47. $y = \left[\dfrac{x + 1}{4} \right], \quad -5 \leq x \leq 5$ **48.** $y = [x] - 1;\quad -5 \leq x \leq 5$

[4.6]   ■   *Find $f^{-1}(x)$.*

49. $f(x) = 2x - 3$ **50.** $f(x) = x^2 + 9,\quad x \leq 0$

■   *Graph f and its inverse f^{-1} using the same set of axes.*

51. $f(x) = 2x - 6$ **52.** $f(x) = \dfrac{3}{4}x - 2$

53. $f(x) = x^2 - 9,\quad x \geq 0$ **54.** $f(x) = x^2 - 4x - 5,\quad x \leq 2$

■   *Verify that $x = f[f^{-1}(x)]$ and $x = f^{-1}[f(x)]$.*

55. $f(x) = 2x - 1;\quad f^{-1}(x) = \dfrac{1}{2}(x + 1)$ **56.** $f(x) = \dfrac{1}{x^3};\quad f^{-1}(x) = \dfrac{1}{\sqrt[3]{x}}$

[4.7]   ■   *Graph the inequality.*

57. $2x - y < 6$ **58.** $2y + x > 0$ **59.** $-2 \leq y < 3$

60. $3 < x \leq 5$ **61.** $y < x^2 - 9$ **62.** $y \geq x^2 + 6x + 5$

[4.8] **63.** If y varies directly as x^2, and $y = 20$ when $x = 2$, express y as a function of x.

64. If r varies directly as x^2 and inversely as z^3, and $r = 4$ when $x = 3$ and $z = 2$, express r as a function of x and z.

65. The number of posts needed to string a telephone line over a given distance varies inversely as the distance between posts. It takes 80 posts separated by 120 feet to string a wire between two points. Express the number (n) of posts required to string a telephone line over a distance (d).

66. The speed of a gear varies directly as the number of teeth it contains. A gear with 10 teeth rotates at 240 revolutions per minute (rpm). Express the rate (r) at which a gear rotates as a function of the number of teeth (n).

B

67. For $f(x) = \dfrac{2}{x + 1}$, find and simplify $\dfrac{f(a + h) - f(a)}{h}$, $h \neq 0$.

68. For $f(x) = \dfrac{2}{\sqrt{x + 1}}$, find and simplify $\dfrac{f(a + h) - f(a)}{h}$, $h \neq 0$.

69. A plane flying at a constant groundspeed is 400 miles from its destination after 3 hours of flying and is 160 miles from its destination after 6 hours of flying. Find a linear equation relating the distance (d) from the plane's destination (in miles) and the amount of time (t) flying (in hours).

70. A runner is 8k from the finish line in a 10k race after 12 minutes from the start of the race. If the runner runs the race at a constant speed, find a linear equation relating the distance run (d) in kilometers and the time (t) in minutes.

71. Find the smallest value of $f(x) = (x^2 - 3x - 4)^2$.

72. Find the smallest value of $f(x) = (x^2 + x + 1)^2$.

73. If $f(3) = 6$, find $f^{-1}(6)$.

74. If $f^{-1}(10) = 4$, find $f(4)$.

75. Graph the solution set of $\frac{1}{4}|x| + \frac{3}{4}|y| \leq 1$.

76. Graph the solution set of $\frac{2}{3}|x| + \frac{1}{3}|y| \leq 1$.

77. City A is 100 miles due south of City B. An airplane leaves from each city at the same time. The airplane leaving City A travels due south at the rate of 150 mph, and the one leaving City B travels due north at the rate of 120 mph. Express the distance (d) in miles between the two planes as a function of time (t) in hours.

78. Assume the plane leaving from City B in Exercise 77 travels due east at the rate of 120 mph. Express the distance (d) in miles between the two planes as a function of time (t) in hours.

5 Polynomial and Rational Functions

In Chapter 4 we studied functions defined by first-degree (linear) and second-degree (quadratic) polynomials. In this chapter we continue the study of polynomial functions and then conclude the chapter with a discussion of rational functions.

In our work with these functions it will sometimes be necessary to obtain values of polynomials and rational expressions. In Section 1.1 we evaluated polynomials for given values of the variable by direct substitution. In Section 5.1 we consider another method of evaluating polynomials that can be helpful when a large number of such values are to be obtained.

5.1

SYNTHETIC DIVISION

The process of dividing one polynomial by another, which was discussed in Section 1.4, can be simplified by a process called synthetic division if the divisor is of the form $x + c$. Consider the example on page 23, where $x^4 + x^2 + 2x - 1$ is divided by $x - 3$. If we write only the coefficients of the terms (not the variables) and use zero for the coefficient of any missing power, we have

$$
\begin{array}{r}
1 + 3 + 10 + 32 \\
1 - 3 \overline{)1 + 0 + \ \ 1 + \ \ 2 - \ \ 1} \\
\underline{1 - 3} \\
3 + (1) \\
\underline{3 - \ \ 9} \\
10 + (2) \\
\underline{10 - 30} \\
32 - (1) \\
\underline{32 - 96} \\
95 \quad \text{(Remainder)}.
\end{array}
$$

Note that the numerals shown in color are the same as the numerals written immediately above and are also the same as the coefficients of the quotient. Also note that the numerals in parentheses, ( ), are the same as the coefficients of the dividend. Therefore, the whole process can be written in compact form as

$$
\begin{array}{rrrrrr}
(1) & \underline{-3}\,| & 1 & 0 & 1 & 2 & -1 \\
(2) & & & -3 & -9 & -30 & -96 \\
(3) & & 1 & 3 & 10 & 32 & 95
\end{array}
\quad \text{(Remainder: 95),}
$$

where the repetitions and 1, the coefficient of x in the divisor, have been omitted.

The entries in line (3) are the coefficients of the variables in the quotient and the remainder. They have been obtained by subtracting the entries in line (2) from the corresponding entries in line (1). We could obtain the same result by replacing -3 with 3 in the divisor and *adding* the entries in line (2) to the corresponding entries in line (1) instead of subtracting at each step. This is what is done in the **synthetic-division** process. The final form is:

$$
\begin{array}{rrrrrr}
(1) & \underline{3}\,| & 1 & 0 & 1 & 2 & -1 \\
(2) & & & 3 & 9 & 30 & 96 \\
(3) & & 1 & 3 & 10 & 32 & 95
\end{array}
\quad \text{(Remainder: 95).}
$$

The entries in line (3) are the coefficients of the polynomial $x^3 + 3x^2 + 10x + 32$, and there is a remainder of 95. These are the quotient and remainder obtained by using long division in the example on page 23.

EXAMPLE 1   Write $\dfrac{3x^3 - 4x - 1}{x + 2}$ in the form $Q(x) + \dfrac{r}{x + 2}$, where $Q(x)$ is a polynomial and r is a constant.

Solution   Note that the divisor is $x - (-2)$ and that the coefficient of the second-degree term in the numerator is 0. Thus, to apply synthetic division, we first write

$$
\begin{array}{rrrrr}
(1) & \underline{-2}\,| & 3 & 0 & -4 & -1 \\
(2) & & & & & \\
(3) & & \multicolumn{4}{c}{\rule{4cm}{0.4pt}}
\end{array}
$$

and proceed as follows:

1. 3 is "brought down" from line (1) to line (3).
2. -6, the product of -2 and 3, is written in the next position on line (2).
3. -6, the sum of 0 and -6, is written on line (3).

4. 12, the product of -2 and -6, is written in the next position on line (2).

5. 8, the sum of -4 and 12, is written on line (3).

6. -16, the product of -2 and 8, is written in the next position on line (2).

7. -17, the sum of -1 and -16, is written on line (3).

These steps result in the following diagram:

$$
\begin{array}{rl|rrrr}
(1) & -2\big| & 3 & 0 & -4 & -1 \\
(2) & & & \searrow -6 & \searrow 12 & \searrow -16 \\
& & & \text{(add)} & \text{(add)} & \text{(add)} \\
(3) & & 3 & -6 & 8 & -17 \\
& & & \underbrace{(-2)(3)} & \underbrace{(-2)(-6)} & \underbrace{(-2)(8)}
\end{array}
$$

We use the first three entries on line (3) as coefficients to write a polynomial of degree one less than the degree of the dividend. This polynomial is the quotient lacking the remainder. The last number is the remainder. Thus,

$$\frac{3x^3 - 4x - 1}{x + 2} = 3x^2 - 6x + 8 - \frac{17}{x + 2} \qquad (x \neq -2).$$

EXAMPLE 2 Find a polynomial $Q(x)$ and a real number r such that for every real number x
$x^3 + 2x^2 + 1 = (x - 1)Q(x) + r$.

Solution Using synthetic division to compute $\dfrac{x^3 + 2x^2 + 1}{x - 1}$, we write

$$
\begin{array}{r|rrrr}
1\big| & 1 & 2 & 0 & 1 \\
& & 1 & 3 & 3 \\
\hline
& 1 & 3 & 3 & 4
\end{array}
$$

Thus,

$$\frac{x^3 + 2x^2 + 1}{x - 1} = x^2 + 3x + 3 + \frac{4}{x - 1} \qquad (x \neq 1).$$

Now, multiplying both members of the preceding equation by $x - 1$ yields

$$x^3 + 2x^2 + 1 = (x^2 + 3x + 3)(x - 1) + 4 \qquad (x \neq 1). \tag{1}$$

Replacing x with 1 in both members of Equation (1), we find that Equation (1) is also satisfied by 1. Hence,

$$x^3 + 2x^2 + 1 = (x^2 + 3x + 3)(x - 1) + 4$$

for every real number x. Thus, $Q(x) = x^2 + 3x + 3$ and $r = 4$. ■

Example 2 illustrates a theorem that we shall state without proof.*

Theorem 5.1

If $P(x)$ is a real polynomial and c is any real number, then there is a unique real polynomial $Q(x)$ and a unique real number r, such that for every real number x

$$P(x) = (x - c)Q(x) + r. \qquad (2)$$

Remainder Theorem

Since Equation (2) of Theorem 5.1 is true for every real number x, it must be true for $x = c$. Thus, we have

$$P(c) = (c - c)Q(c) + r$$
$$= 0 \cdot Q(c) + r = r.$$

This proves the following important result.

Theorem 5.2

If $P(x)$ is a polynomial, then for every real number c there is a unique polynomial $Q(x)$ such that for every real number x

$$P(x) = (x - c)Q(x) + P(c).$$

This theorem is called the **remainder theorem** because it says that the remainder, when $P(x)$ is divided by $x - c$, is $P(c)$. Since synthetic division offers a quick means of obtaining this remainder, we can sometimes find values $P(c)$ more rapidly by synthetic division than by direct substitution.

EXAMPLE 3   Given $P(x) = x^3 - x^2 + 3$, find $P(3)$ by the remainder theorem.

Solution   Synthetically dividing $x^3 - x^2 + 3$ by $x - 3$, we have

$$
\begin{array}{r|rrrr}
3 & 1 & -1 & 0 & 3 \\
  &   & 3 & 6 & 18 \\
\hline
  & 1 & 2 & 6 & 21
\end{array}
$$

and, by inspection, $r = 21$. Thus, by the remainder theorem, $P(3) = 21$.   ■

Factor Theorem

Observe that if $P(r) = 0$, then, by the remainder theorem,

$$P(x) = (x - r)Q(x) + P(r)$$
$$= (x - r)Q(x) + 0,$$

---

* The proofs of several theorems in this chapter are somewhat lengthy and/or require mathematical concepts that are considered in more advanced courses, and hence they are omitted here.

and $(x - r)$ is a factor of $P(x)$. This proves the following result, called the **factor theorem**.

Theorem 5.3   *If $P(x)$ is a polynomial with real-number coefficients and $P(c) = 0$, then $(x - c)$ is a factor of $P(x)$.*

Note that the converse of the factor theorem, that is, if $(x - c)$ is a factor of $P(x)$, then $P(c) = 0$, is also true, since

$$P(c) = (c - c)Q(c)$$
$$= 0 \cdot Q(c) = 0.$$

E X A M P L E   4   Assume $P(3) = 0$ and factor $P(x) = x^3 - 3x^2 - x + 3$ completely.

Solution   Since $P(3) = 0$ we have, from Theorem 5.3, $x - 3$ is a factor of $P(x)$. Using synthetic division to divide $P(x)$ by $x - 3$, we have

$$
\begin{array}{r|rrrr}
3 & 1 & -3 & -1 & 3 \\
  &   & 3 & 0 & -3 \\
\hline
  & 1 & 0 & -1 & 0
\end{array}
$$

and the quotient is $x^2 - 1$. Thus,

$$P(x) = (x - 3)(x^2 - 1) = (x - 3)(x + 1)(x - 1). \qquad ■$$

E X E R C I S E   5.1

A

■ *Use synthetic division to write the quotient* $\dfrac{P(x)}{x - c}$ *in the form $Q(x)$ or in the form $Q(x) + \dfrac{r}{x - c}$, where $Q(x)$ is a polynomial and r is a constant. See Example 1.*

1. $\dfrac{x^2 + 4x - 21}{x - 3}$ **2.** $\dfrac{3x^2 - 4x - 7}{x + 1}$ **3.** $\dfrac{x^2 - 30}{x + 5}$

4. $\dfrac{x^2 + 30}{x - 5}$ **5.** $\dfrac{2x^3 + x + 18}{x + 2}$ **6.** $\dfrac{2x^3 + x + 18}{x - 2}$

7. $\dfrac{3x^4 - 2x^2 + 6}{x - 2}$ **8.** $\dfrac{x^5 + x^3 + 1}{x + 1}$ **9.** $\dfrac{x^6 - 1}{x - 1}$

10. $\dfrac{x^5 + 1}{x + 1}$ **11.** $\dfrac{x^7 + 1}{x + 1}$ **12.** $\dfrac{x^{10} - 1}{x - 1}$

13. $\dfrac{x^4 + x - 14}{x + 3}$ **14.** $\dfrac{x^4 + x - 14}{x - 2}$ **15.** $\dfrac{4x^5 + x^4 - 2x^3 + x^2}{x - 3}$

16. $\dfrac{3x^3 + 4x^2 - 2x + 4}{x + 2}$ **17.** $\dfrac{x^3 - 3x^2 + 3x - 1}{x + 1}$ **18.** $\dfrac{x^3 - 3x^2 + 3x - 1}{x - 1}$

19. $\dfrac{x^5 + 2x^4 + 3x^3 + 9x^2 - 7}{x + 2}$ **20.** $\dfrac{x^4 - 3x^3 + 5x^2 - x + 2}{x - 3}$

■ *Use synthetic division and Theorem 5.2. See Example 3.*

21. $P(x) = x^3 + 2x^2 + x - 1$; find $P(1)$, $P(2)$, and $P(3)$.

22. $P(x) = x^3 - 3x^2 - x + 3$; find $P(1)$, $P(2)$, and $P(3)$.

23. $P(x) = 2x^4 - 3x^3 + x + 2$; find $P(-2)$, $P(2)$, and $P(4)$.

24. $P(x) = 3x^4 + 3x^2 - x + 3$; find $P(-2)$, $P(2)$, and $P(4)$.

25. $P(x) = 3x^5 - x^3 + 2x^2 - 1$; find $P(-3)$, $P(2)$, and $P(3)$.

26. $P(x) = 2x^6 - x^4 + 3x^3 + 1$; find $P(-3)$, $P(2)$, and $P(3)$.

■ *Assume the given number(s) is a solution of $P(x) = 0$ and factor $P(x)$ completely. See Example 4.*

27. 2; $P(x) = x^3 - 3x - 2$ **28.** 3; $P(x) = x^3 - 5x^2 + 7x - 3$

29. -1; $P(x) = x^3 + 2x^2 - 5x - 6$ **30.** -1; $P(x) = x^3 - 7x - 6$

31. $\dfrac{1}{2}$; $P(x) = 2x^3 - x^2 - 8x + 4$ **32.** $\dfrac{2}{3}$; $P(x) = 3x^3 - 2x^2 - 3x + 2$

33. $1, -2$; $P(x) = 2x^4 - 3x^3 - 12x^2 + 7x + 6$ **34.** $-1, 2$; $P(x) = 2x^4 + x^3 - 6x^2 - 7x - 2$

35. $-2, \dfrac{2}{3}$; $P(x) = 6x^5 - x^4 - 17x^3 + 16x^2 - 4x$ **36.** $1, -\dfrac{1}{2}$; $P(x) = 4x^5 + 4x^4 - 23x^3 + 6x^2 + 9x$

B

■ *Use synthetic division to write the quotient $\dfrac{P(x)}{ax - b}$ in the form $Q(x) + \dfrac{r}{ax - b}$, where $Q(x)$ is a polynomial and r is a constant.*

37. $\dfrac{x^2 + x + 1}{2x - 2}$ **38.** $\dfrac{x^2 - x - 4}{2x - 4}$ **39.** $\dfrac{2x^4 - 3x^3 + x - 1}{2x - 2}$

40. $\dfrac{3x^4 + x^2 + 1}{3x + 3}$ **41.** $\dfrac{x^4 - 2x^3 + x + 2}{2x - 1}$ **42.** $\dfrac{x^4 - x^2 - 1}{2x + 1}$

43. $\dfrac{x^3 + x - 2}{3x - 2}$ **44.** $\dfrac{x^4 - x^2 + 2}{3x + 2}$

■   *The remainder theorem (Theorem 5.2) is also true when c is a complex number. Use the remainder theorem with synthetic division to find the indicated value.*

45. $P(x) = x^4 - 3x^3 - 2x - 2$, $P(i)$

46. $P(x) = x^4 - 2x^2 + 3x - 1$, $P(-i)$

47. $P(x) = 4x^2 + 3x + 1$, $P(2 - 3i)$

48. $P(x) = 4x^2 - 2x + 5$, $P(1 + 2i)$

49. $P(x) = x^5 + x^3 - 4$, $P(1 + i)$

50. $P(x) = x^4 - x^2 + 3$, $P(1 - i)$

5.2

ZEROS OF A POLYNOMIAL FUNCTION

In our previous work, we obtained x-intercepts to help us graph linear and quadratic functions. These intercepts are also helpful in graphing polynomial functions. Since the x-intercepts of the graph of a polynomial function are the zeros of the function, we shall first consider some facts that will aid us in finding such zeros.

Bounds on Real Zeros

When searching for the real zeros of a polynomial function, we can narrow our search by finding real numbers r_1 and r_2 with the property that every real zero, x_0, satisfies $r_1 \leq x_0 \leq r_2$. Such values r_1 and r_2 are called **lower** and **upper bounds**, respectively, for the set of real zeros of the polynomial function.

The following theorem enables us to find upper and lower bounds for the real zeros of a polynomial function with real coefficients.

Theorem 5.4

Let $P(x)$ be a polynomial with real coefficients.

   I.   *If $r_2 \geq 0$ and if the third row in the synthetic division of $P(x)$ by $x - r_2$ has only nonnegative values, then $P(x) = 0$ has no solutions greater than r_2.*

   II. *If $r_1 \leq 0$ and if the numbers in the third row of the synthetic division of $P(x)$ by $x - r_1$ alternate in sign (zero being considered $+$ or $-$ as needed), then $P(x) = 0$ has no solution less than r_1.*

We shall verify only the first part of the theorem here.

First recall (Theorem 5.2) that we can write

$$P(x) = (x - r_1)Q(x) + P(r_1).$$

Also, the third row in the synthetic division of $P(x)$ by $x - r_1$ is composed of the coefficients of $Q(x)$ followed by $P(r_1)$. Now for any $x > r_1$, we have $(x - r_1) > 0$. Further, if all the coefficients in $Q(x)$ have the same sign, say positive, then, since $x > r_1 \geq 0$, we have

$$Q(x) > 0 \qquad \text{and} \qquad (x - r_1)Q(x) > 0.$$

Since $P(r_1)$ has the same sign as the coefficients of $Q(x)$, in this case positive, we have

$$P(x) = (x - r_1)Q(x) + P(r_1) > 0.$$

Thus, x is not a solution of $P(x) = 0$.

EXAMPLE 1 Show that 2 and -2 are upper and lower bounds, respectively, for the real zeros of $P(x) = 18x^3 - 12x^2 - 11x + 10$.

Solution Using synthetic division to divide $P(x)$ by $x - 2$, we have

$$
\begin{array}{r|rrrr}
2 & 18 & -12 & -11 & 10 \\
  &   & 36 & 48 & 74 \\
\hline
  & 18 & 24 & 37 & 84
\end{array}
$$

Since all the numbers in the third row are positive, Theorem 5.4-I guarantees that 2 is an upper bound for the set of real zeros of $P(x)$.
Next, dividing $P(x)$ by $x + 2$, we have

$$
\begin{array}{r|rrrr}
-2 & 18 & -12 & -11 & 10 \\
   &   & -36 & 96 & -170 \\
\hline
   & 18 & -48 & 85 & -160
\end{array}
$$

Since the numbers in the third row alternate in sign, Theorem 5.4-II guarantees that -2 is a lower bound for the set of real zeros of $P(x)$.

EXAMPLE 2 Find the least positive integer and the greatest negative integer that are upper and lower bounds, respectively, for the real zeros of

$$P(x) = x^4 - x^3 - 10x^2 - 2x + 12.$$

Solution We shall first seek an upper bound by dividing $P(x)$ successively by $x - 1$, $x - 2$, and so on. Each row after the first in the following array is the bottom row in the respective synthetic division involved.

$$
\begin{array}{r|rrrrr}
  & 1 & -1 & -10 & -2 & 12 \\
\hline
1 & 1 & 0 & -10 & -12 & 0 \\
2 & 1 & 1 & -8 & -18 & -24 \\
3 & 1 & 2 & -4 & -14 & -30 \\
4 & 1 & 3 & 2   & 6   & 36
\end{array}
$$

Since the numbers in the last row are all positive, 4 is an upper bound. Next, we divide by $x + 1$, $x + 2$, and so on, in search of a lower bound.

	1	-1	-10	-2	12
-1	1	-2	-8	6	6
-2	1	-3	-4	6	0
-3	1	-4	2	-8	36

Since the numbers in the row following -3 alternate from positive to negative, and so on, -3 is a lower bound. (Had the numbers in the row been $1, 0, 2, -8, 36$, then the sign "$-$" could arbitrarily have been assigned to 0 to give the desired pattern of alternating signs.)   ■

Variations in Sign

A **variation in sign** occurs in a polynomial with real coefficients if, in the polynomial in standard form, consecutive coefficients are opposite in sign. For example, in the polynomial

$$P(x) = 3x^5 - 2x^4 - 2x^2 + x - 1,$$

there are three variations in sign, and in

$$P(-x) = -3x^5 - 2x^4 - 2x^2 - x - 1,$$

there are no variations in sign.

Descartes' Rule of Signs

The following theorem, called **Descartes' rule of signs**, provides a way to use variation in sign to determine the number of real zeros of a polynomial function. The proof is omitted.

Theorem 5.5   *If $P(x)$ is a polynomial with real coefficients, then:*

   I. *The number of positive solutions of $P(x) = 0$ is either equal to the number of variations in sign in $P(x)$ or less than that number by an even integer.*

  II. *The number of negative solutions of $P(x) = 0$ is either equal to the number of variations in sign in $P(-x)$ or less than that number by an even integer.*

EXAMPLE 3   Find an upper bound on the number of positive real solutions and an upper bound on the number of negative real solutions for the equation

$$3x^4 + 3x^3 - 2x^2 + x + 1 = 0. \tag{1}$$

Solution   Since

$$P(x) = 3x^4 + 3x^3 - 2x^2 + x + 1$$

has two variations in sign, Equation (1) has either 0 or 2 positive solutions. Thus, the equation can have no more than 2 positive solutions. Since

$$P(-x) = 3x^4 - 3x^3 - 2x^2 - x + 1$$

has two variations in sign, the number of negative solutions of (1) is either 0 or 2 and thus can be no greater than 2.

E X A M P L E   4   Find an upper bound on the number of positive real solutions and an upper bound on the number of negative real solutions for the equation

$$x^4 + 3x^3 - x^2 + 2x - 1 = 0.$$

Solution   Since

$$P(x) = x^4 + 3x^3 - x^2 + 2x - 1 \qquad (2)$$

has three variations in sign, the equation has either 1 or 3 positive solutions. Hence, Equation (2) can have no more than 3 positive solutions. Since

$$P(-x) = x^4 - 3x^3 - x^2 - 2x - 1$$

has one variation in sign, Equation (2) has exactly 1 negative solution.   ■

Enlarging the replacement set for x to include complex values allows us to prove some very useful theorems concerning the zeros of polynomial functions.

Remainder and Factor Theorems over C

In Section 5.1 we proved both the remainder theorem and the factor theorem for polynomials with real-number coefficients and replacement sets of real numbers. Since these proofs depend only on the field axioms (see the Preliminary Concepts), both theorems apply to polynomials with complex-number coefficients and replacement sets of complex numbers. (See Exercises 45–50 in Section 5.1.)

Conjugate Complex Zeros

Recalling from Section 2.5 that the conjugate of the complex number $z = a + bi$ is $\bar{z} = a - bi$, where a and b are real numbers, we state an important property of a polynomial. The proof is left as an exercise.

Theorem 5.6   *If $P(z)$ is a polynomial with real-number coefficients and $P(z) = 0$ for some complex number z, then $P(\bar{z}) = 0$.*

This theorem guarantees that the *complex zeros* of a polynomial function with *real coefficients* always occur in *conjugate pairs*. We can use this fact along with the factor theorem to find all the zeros of a polynomial.

EXAMPLE 5   Given that $2 - i$ is a zero of $P(x) = x^3 - 6x^2 + 13x - 10$, find all zeros of P.

Solution   By Theorem 5.6, $\overline{2 - i}$ or $2 + i$ is a zero of P. Thus, by the factor theorem, there is a polynomial $Q(x)$ with

$$P(x) = [x - (2 - i)][x - (2 + i)]Q(x)$$
$$= (x^2 - 4x + 5)Q(x).$$

Hence,

$$Q(x) = \frac{x^3 - 6x^2 + 13x - 10}{x^2 - 4x + 5} = x - 2.$$

Thus,

$$P(x) = [x - (2 - i)][x - (2 + i)](x - 2) = 0,$$

and, by the factor theorem, the solutions of this equation—and hence the zeros of $P(x)$—are $2 - i$, $2 + i$, and 2.

EXAMPLE 6   Given that $2 + 3i$ and $-i$ are zeros of

$$P(x) = x^6 - 5x^5 + 16x^4 - 10x^3 - 11x^2 - 15x - 26,$$

find all zeros of P.

Solution   By Theorem 5.6, $\overline{2 + 3i}$ or $2 - 3i$ and $\overline{-i}$ or i are zeros of P. Thus, by the factor theorem, there is a polynomial $Q(x)$ with

$$P(x) = [x - (2 + 3i)][x - (2 - 3i)][x - (-i)](x - i)Q(x)$$
$$= (x^4 - 4x^3 + 14x^2 - 4x + 13)Q(x).$$

Hence,

$$Q(x) = \frac{x^6 - 5x^5 + 16x^4 - 10x^3 - 11x^2 - 15x - 26}{x^4 - 4x^3 + 14x^2 - 4x + 13}$$
$$= x^2 - x - 2.$$

Thus,

$$P(x) = [x - (2 + 3i)][x - (2 - 3i)][x - (-i)](x - i)(x^2 - x - 2)$$
$$= [x - (2 + 3i)][x - (2 - 3i)][x - (-i)](x - i)(x - 2)(x - (-1)).$$

By the factor theorem, the solutions of $P(x) = 0$—and thus the zeros of $P(x)$—are $2 + 3i$, $2 - 3i$, $-i$, i, 2, and -1.   ■

Fundamental Theorem of Algebra

When Theorem 5.6 is coupled with the following theorem, which is called the **fundamental theorem of algebra**, a great deal of information relative to the zeros of polynomial functions becomes readily available. The proof of this theorem involves concepts beyond those available to us, and is omitted.

Theorem 5.7

Every polynomial function of degree $n \geq 1$ over the complex numbers has at least one real or complex zero.

Notice that in Examples 5 and 6 we were able to factor the polynomials into linear factors. This is true for every polynomial with complex coefficients. We state this fact formally in the following theorem, which can be proved by repeated applications of the fundamental theorem of algebra and the factor theorem. We leave the details as an exercise.

Theorem 5.8

Every polynomial of degree $n \geq 1$ over the complex numbers can be expressed as a product of a constant and n linear factors of the form $x - r_j$, where each r_j is a complex number, that is, there are complex numbers $a, r_1, \ldots, r_n$ so that

$$P(x) = a(x - r_1) \cdots \cdots (x - r_n).$$

An nth-Degree Polynomial Function Has n Zeros

If a factor $(x - r_i)$ occurs k times in such a linear factorization of $P(x)$, then r_i is said to be a zero of **multiplicity k**. With this agreement, Theorem 5.8 shows that every polynomial function defined by a polynomial $P(x)$ of degree n with complex coefficients has exactly n zeros.

E X A M P L E   7   One zero of $P(x) = x^4 + 5x^3 + 8x^2 + 20x + 16$ is $2i$. Factor $P(x)$ over the complex numbers.

Solution   By Theorem 5.6, $\overline{2i} = -2i$ is a zero of P. Thus, by the factor theorem,

$$P(x) = (x - 2i)(x + 2i)Q(x)$$
$$= (x^2 + 4)Q(x)$$

for some polynomial $Q(x)$. Therefore,

$$Q(x) = \frac{x^4 + 5x^3 + 8x^2 + 20x + 16}{x^2 + 4}$$

$$= x^2 + 5x + 4$$

$$= (x + 4)(x + 1).$$

Thus,

$$P(x) = (x - 2i)(x + 2i)(x + 4)(x + 1). \qquad ■$$

Note that any theorem stated in terms of zeros of polynomial functions applies to solutions of polynomial equations and vice versa; a *zero* of

$$P(x) = a_nx^n + a_{n-1}x^{n-1} + \cdots + a_0$$

is a *solution* of $P(x) = 0$.

EXAMPLE 8   (a) $x^3 + 2x - 1$ has three complex zeros.

(b) $x^5 - 2x^3 + x + 1 = 0$ has five complex solutions.   ■

EXERCISE 5.2

A

■ *Find an upper bound and a lower bound for the real zeros of the polynomial function. For Exercises 1–12 see Examples 1 and 2.*

1. $P(x) = x^3 + 2x^2 - 7x - 8$ **2.** $P(x) = x^3 - 8x + 5$

3. $P(x) = x^4 - 2x^3 - 7x^2 + 10x + 10$ **4.** $P(x) = x^3 - 4x^2 - 4x + 12$

5. $P(x) = x^5 - 3x^3 + 24$ **6.** $P(x) = x^5 - 3x^4 - 1$

7. $P(x) = 2x^5 + x^4 - 2x - 1$ **8.** $P(x) = 2x^5 - 2x^2 + x - 2$

9. $P(x) = x^5 - 2x^4 - 19x^3 + 20x^2$ **10.** $P(x) = 4x^5 - 8x^4 - x^3 + 2x^2$

11. $P(x) = 6x^6 - 7x^5 - 4x^4 + 7x^3 - 2x^2$ **12.** $P(x) = 6x^6 - x^5 - 36x^4 + 4x^3 + 48x^2$

■ *Use Descartes' rule of signs to note the number of positive real zeros possible and the number of negative real zeros possible. See Examples 3 and 4.*

13. $P(x) = x^4 - 2x^3 + 2x + 1$ **14.** $P(x) = 3x^4 + 3x^3 + 2x^2 - x + 1$

15. $P(x) = 2x^5 + 3x^3 + 2x + 1$

16. $P(x) = 4x^5 - 2x^3 - 3x - 2$

17. $P(x) = x^4 + x^3 - x^2 + x + 1$

18. $P(x) = x^4 - x^3 - x^2 - x + 1$

■ *One or more zeros are given for each of the polynomial functions; find the other zeros. See Examples 5 and 6.*

19. $P(x) = x^2 + 4$; $2i$ is a zero.

20. $P(x) = 3x^2 + 27$; $-3i$ is a zero.

21. $Q(x) = x^3 - 3x^2 + x - 3$;   3 and i are zeros.

22. $Q(x) = x^3 - 5x^2 + 7x + 13$; -1 and $3 - 2i$ are zeros.

23. $P(x) = 2x^3 - x^2 + 2x - 1$; $-i$ is a zero.

24. $P(x) = 3x^3 - 10x^2 + 7x + 10$; $2 - i$ is a zero.

25. $Q(x) = x^4 + 5x^2 + 4$; $-i$ and $2i$ are zeros.

26. $Q(x) = x^4 + 11x^2 + 18$; $3i$ and $i\sqrt{2}$ are zeros.

27. $P(x) = x^4 + 3x^3 + 4x^2 + 27x - 45$; $-3i$ is a zero.

28. $P(x) = x^4 - 7x^3 + 18x^2 - 22x + 12$; $1 - i$ is a zero.

29. $Q(x) = x^5 - 2x^4 + 8x^3 - 16x^2 + 16x - 32$; $2i$ is a zero of multiplicity two.

30. $Q(x) = 2x^5 - 9x^4 + 20x^3 - 24x^2 + 16x - 4$; $1 + i$ is a zero of multiplicity two.

■ *For Exercises 31–38, see Example 7.*

31. One zero of $P(x) = 2x^3 - 11x^2 + 28x - 24$ is $2 - 2i$. Factor $P(x)$ over the complex numbers.

32. One zero of $Q(x) = 3x^3 - 10x^2 + 7x + 10$ is $2 + i$. Factor $Q(x)$ over the complex numbers.

33. One zero of $P(x) = x^4 - 4x^3 + 14x^2 - 4x + 13$ is $2 + 3i$. Factor $P(x)$ over the complex numbers.

34. One zero of $P(x) = x^4 - 2x^3 + 14x^2 - 18x + 45$ is $1 + 2i$. Factor $P(x)$ over the complex numbers.

35. The polynomial $P(x) = x^5 - 4x^4 + 8x^3 - 32x^2 + 16x - 64$ has $2i$ as a zero of multiplicity two. Factor $P(x)$ over the complex numbers.

36. The polynomial $P(x) = x^5 - 3x^4 + 18x^3 - 54x^2 + 81x - 243$ has $3i$ as a zero of multiplicity two. Factor $P(x)$ over the complex numbers.

37. Factor $P(x) = x^4 + 5x^2 + 4$ over the complex numbers.

38. Factor $P(x) = x^4 + 13x^2 + 36$ over the complex numbers.

39. One solution of $x^4 - 10x^3 + 35x^2 - 50x + 34 = 0$ is $4 - i$. Find the remaining solutions.

40. One solution of $5x^4 + 34x^3 + 40x^2 - 78x + 51 = 0$ is $-4 - i$. Find the remaining solutions.

41. A cubic equation with real coefficients has solutions -2 and $1 + i$. What is the third solution? Write the equation in the form $P(x) = 0$, given that the leading coefficient (the coefficient of the highest power of x) is 1.

42. A cubic equation with real coefficients has solutions 4 and $2 - i$. What is the third solution? Write the equation in the form $P(x) = 0$, given that the leading coefficient is 1.

43. One zero of a polynomial P is $1 + i$. Is $1 - i$ necessarily a zero of P? Why or why not?

44. One zero of a polynomial P is i. Is $-i$ necessarily a zero of P? Why or why not?

B

■ *Find the polynomial of lowest degree with real coefficients that satisfies the given condition.*

45. 1, 2, 3 are zeros; leading coefficient is -1.

46. $-1, -2, -3$ are zeros; leading coefficient is -1.

47. $1 + 2i, 2 + i$ are zeros; constant term is 25.

48. $2 + 3i, 1 + 5i$ are zeros; constant term is 338.

49. $2, -1, 3$ are zeros; $P(1) = 2$.

50. $-1, 0, 1$ are zeros; $P(2) = -2$.

51. $1 + i, 2$ are zeros; $P(-1) = -5$.

52. $1 - i, -1$ are zeros; $P(2) = 1$.

53. Determine the three cube roots of -1.

54. Determine the three cube roots of 1.

55. Argue that every polynomial with real coefficients and of odd degree has at least one real zero.

56. Argue that every polynomial with real coefficients and of even degree has an even number of real zeros when multiplicities are counted.

57. Prove Theorem 5.6 [*Hint:* Show $\overline{P(z)} = P(\bar{z})$.]

58. Fill in the details of the proof of Theorem 5.8.

59. Show that if P and Q are two polynomials of degree n and $P(x) = Q(x)$ for more than n values of x, then $P(x) = Q(x)$ for every x.

60. Show that if $Q(1) = 1$, $Q(2) = -1$, and $Q(3) = 0$ and $Q(x)$ is of degree two, then $Q(x) = P(x)$ where $P(x) = \frac{3}{2}x^2 - \frac{13}{2}x + 6$.

5.3

IRRATIONAL AND RATIONAL ZEROS OF POLYNOMIAL FUNCTIONS

In Section 5.2 we discussed several general properties of polynomials that are useful in the search for zeros of polynomial functions.

In this section we shall develop a method for approximating irrational zeros and a method for finding all of the rational zeros of a polynomial function with integer coefficients.

Location of Zeros

We shall assume that the graph of a polynomial function is a continuous curve in the plane, that is, a curve that has no "jumps" or "breaks." This assumption is shown to be valid in more advanced courses in mathematics. With this assumption, the following theorem aids us in the search for real zeros of such functions.

Theorem 5.9   *Let $P(x)$ be a polynomial with real coefficients. If $x_1 < x_2$ and if $P(x_1)$ and $P(x_2)$ have different signs, then there is at least one value c between x_1 and x_2 such that $P(c) = 0$.*

Although we shall not give a rigorous proof of this theorem, we can certainly make the result seem plausible.

The assumption that $P(x_1)$ and $P(x_2)$ have different signs means that the points $(x_1, P(x_1))$ and $(x_2, P(x_2))$ are on opposite sides of the x-axis. Since the graph of $y = P(x)$ has no jumps or breaks, it must cross the x-axis at some point c between x_1 and x_2 (see Figure 5.1). For this value c, we have $P(c) = 0$.

FIGURE 5.1

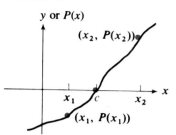

EXAMPLE 1   Show that $P(x) = x^3 + x - 1 = 0$ has a solution between 0 and 1.

Solution   We note that

$$P(0) = -1 \quad \text{and} \quad P(1) = 1.$$

Since $P(0)$ and $P(1)$ have different signs, Theorem 5.9 guarantees that there is a solution between 0 and 1. ■

Approximation of Zeros

Many polynomial functions have one or more irrational zeros. We can never write an exact decimal representation for such a zero. However, we can use Theorem 5.9 with the following procedure to approximate the irrational zeros to any desired degree of accuracy.

Steps to Approximate Zeros of a Polynomial

1. Use Theorem 5.9 to determine consecutive integers a and b with a zero between them.

2. Subdivide the interval $[a, b]$ into ten equal subintervals.

3. Use Theorem 5.9 to find one of these subintervals that contains a zero of $P(x)$.

4. Start over at Step 2, with the endpoints of the subinterval determined in Step 3 taking the place of a and b.

Complete Steps 1–4 k times (eliminating Step 4 the kth time) to approximate a zero to within 10^{-k}.

E X A M P L E 2   Approximate the positive zero of $P(x) = x^2 - 6$ to within 10^{-2}.

Solution   Since we want to approximate the zero to within 10^{-2}, we complete Steps 1–3 two times.

Step 1:   Since $P(2) < 0$ and $P(3) > 0$, by Theorem 5.9, there is a zero between 2 and 3. (See Figure a. Note that the scales on the x-axis and y-axis are different.)

Step 2:   We subdivide the interval $[2, 3]$ into ten equal subintervals, as shown in Figure a.

Step 3:   We calculate the value of $P(x)$ at each subinterval endpoint until Theorem 5.9 applies. We find that $P(2.4) < 0$ while $P(2.5) > 0$. Thus, there is a root between 2.4 and 2.5. (See Figure b. Note that the scales on the x-axis and y-axis are different.)

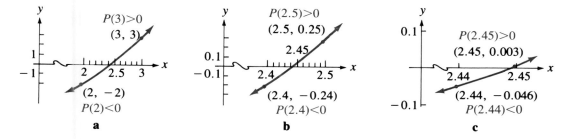

a   b   c

We now return to Step 2 with $a = 2.4$ and $b = 2.5$.

Step 2′:   We subdivide the interval $[2.4, 2.5]$ into ten equal subintervals, as shown in Figure b.

Step 3′:   We calculate the value of $P(x)$ at each subinterval endpoint until Theorem 5.9 applies. We find that $P(2.44) < 0$ and $P(2.45) > 0$. Thus, there is a root between 2.44 and 2.45. (See Figure c.)

This completes two repetitions of the steps, and the zero is located in the interval $[2.44, 2.45]$.

Note that the positive zero of $P(x) = x^2 - 6$ is $\sqrt{6}$, which, to three decimal places, is 2.449. Thus, any value in the interval $[2.44, 2.45]$ is within 10^{-2} of the actual value. ■

The method outlined above for approximating zeros of a polynomial is a variation on a method called the bisection method. There are many methods for approximating zeros of a polynomial function. The advantages of this one are that it *always works* and is easy to program.

Identifying Rational Zeros   The following theorem is useful for identifying rational zeros of polynomial functions with integer coefficients.

Theorem 5.10   *If the rational number p/q, in lowest terms, is a solution of*

$$P(x) = a_n x^n + a_{n-1} x^{n-1} + \cdots + a_0 = 0,$$

where each a_j is an integer, then p is an integer factor of a_0 and q is an integer factor of a_n.

To see that this is true, let p/q be a solution of $P(x) = 0$, with p/q reduced to lowest terms. Then we have

$$a_n \left(\frac{p}{q}\right)^n + a_{n-1} \left(\frac{p}{q}\right)^{n-1} + \cdots + a_0 = 0,$$

and we can multiply each member here by q^n to obtain

$$a_n p^n + a_{n-1} p^{n-1} q + \cdots + a_0 q^n = 0. \tag{1}$$

Adding $-a_0 q^n$ to each member and factoring p from each term in the left-hand member of Equation (1), we have

$$p(a_n p^{n-1} + a_{n-1} p^{n-2} q + \cdots + a_1 q^{n-1}) = -a_0 q^n.$$

Since each a_j is an integer and p and q are both integers, the expression in the parentheses is an integer, say, r. Thus, we have

$$pr = a_0 q^n,$$

where pr is an integer having p as a factor. But we assumed that p/q is reduced to lowest terms. Thus, p and q have no common integer factor. Hence, p and q^n have no common integer factors. This means that p must be an integer factor of a_0.

By initially adding $-a_n p^n$ to both sides of Equation (1) and factoring q from the resulting left-hand member, we can also show that q is a factor of a_n.

EXAMPLE 3   List all possible rational zeros of $P(x) = 2x^3 - 4x^2 + 3x + 9$.

Solution   Rational zeros p/q must, by Theorem 5.10, be such that p is an integer factor of 9 and q is an integer factor of 2. Hence,

$$p \text{ is a member of } \{\pm 1, \pm 3, \pm 9\},$$

$$q \text{ is a member of } \{\pm 1, \pm 2\}.$$

Thus, the set of possible rational zeros is $\left\{ \pm\dfrac{1}{2}, \pm 1, \pm\dfrac{3}{2}, \pm 3, \pm\dfrac{9}{2}, \pm 9 \right\}$.   ■

It is important to observe that Theorem 5.10 does *not* assure us that a polynomial function with integer coefficients has a rational zero. It simply allows us to identify possibilities for rational zeros. These can then be checked by synthetic division.

We can sometimes use the methods of Section 5.2 to reduce the number of possible rational zeros. The ones that remain can then be checked by synthetic division.

EXAMPLE 4   Find all the rational zeros of $P(x) = 2x^3 - 4x^2 + 3x + 9$.

Solution   From Example 3 we know the possible rational zeros are

$$\left\{ \pm\frac{1}{2},\ \pm 1,\ \pm\frac{3}{2},\ \pm 3,\ \pm\frac{9}{2},\ \pm 9 \right\}.$$

From Theorem 5.4 we find that -2 is a lower bound for, and 2 is an upper bound for, the zeros of $P(x)$. Thus, of our original list we need only check

$$\left\{ \pm\frac{1}{2},\ \pm 1,\ \pm\frac{3}{2} \right\}.$$

Starting with the values ± 1 (because they are computationally easy to work with) we find

$$
\begin{array}{r|rrrr}
-1 & 2 & -4 & 3 & 9 \\
   &   & -2 & 6 & -9 \\
\hline
   & 2 & -6 & 9 & 0 \\
\end{array}
$$

Thus, -1 is a zero of $P(x)$ and

$$P(x) = (x + 1)(2x^2 - 6x + 9).$$

To find the remaining zeros of P, we solve the reduced equation

$$2x^2 - 6x + 9 = 0.$$

From the quadratic formula we find that this equation has no real (and hence no rational) solutions. Thus, the only rational zero of $P(x)$ is -1.   ■

Test for Integer Zeros   As a special case of Theorem 5.10, note that if a function $P(x)$ is defined by the equation

$$P(x) = x^n + a_{n-1}x^{n-1} + \cdots + a_0,$$

in which each a_i is an integer and the coefficient of x^n is 1, then any rational zero of $P(x)$ must be an integer and must be an integer factor of a_0.

EXAMPLE 5   Find all rational zeros of $P(x) = x^3 - 4x^2 + x + 6$.

Solution   The only possible rational zeros of P are ± 1, ± 2, ± 3, and ± 6. Using synthetic division, we set up the following array:

$$
\begin{array}{r|rrrr}
 & 1 & -4 & 1 & 6 \\
\hline
1 & 1 & -3 & -2 & 4 \\
\hline
-1 & 1 & -5 & 6 & 0
\end{array}
$$

The third row in this array is the bottom row in division of $x^3 - 4x^2 + x + 6$ by $x + 1$. Thus, $P(-1) = 0$, and -1 is one zero of $P(x)$. Now, by the factor theorem,

$$P(x) = x^3 - 4x^2 + x + 6 = (x + 1)(x^2 - 5x + 6).$$

Hence, the remaining zeros can be found by solving $x^2 - 5x + 6 = 0$. Factoring, we obtain

$$(x - 2)(x - 3) = 0.$$

Thus, the remaining zeros are 2 and 3.   ■

EXERCISE 5.3

A

■ *Use Theorem 5.9 to verify the given statement. See Example 1.*

1. $f(x) = x^3 - 3x + 1$ has a zero between 0 and 1.

2. $f(x) = 2x^3 + 7x^2 + 2x - 6$ has a zero between -2 and -1.

3. $g(x) = x^4 - 2x^2 + 12x - 17$ has a zero between -3 and -2.

4. $g(x) = 2x^4 + 3x^3 - 14x^2 - 15x + 9$ has a zero between -2 and -1.

5. $P(x) = 2x^2 + 4x - 4$ has zeros between -3 and -2 and between 0 and 1.

6. $P(x) = x^3 - x^2 - 2x + 1$ has zeros between -2 and -1, between 0 and 1, and between 1 and 2.

■ *Use Theorem 5.9 to approximate the given zero of the indicated polynomial to the accuracy indicated. See Example 2.*

To within one-tenth:

7. $P(x) = x^2 - 3$,   the zero between 1 and 2.

8. $P(x) = x^2 - 5$,   the zero between 2 and 3.

9. $P(x) = x^2 - 7$,   the zero between -3 and -2.

10. $P(x) = x^2 - 10$,   the zero between -4 and -3.

11. $P(x) = x^3 - 2$, the zero between 1 and 2.

12. $P(x) = x^3 - 3$, the zero between 1 and 2.

13. $P(x) = 2x^4 - 5$, the zero between -2 and -1.

14. $P(x) = 5x^4 - 6$, the zero between -2 and -1.

■ *To within one-hundredth:*

15. $P(x) = x^3 - 4$, the zero between 1 and 2.

16. $P(x) = x^3 - 6$, the zero between 1 and 2.

17. $P(x) = x^3 + 5$, the zero between -1 and -2.

18. $P(x) = x^3 + 7$, the zero between -1 and -2.

19. $P(x) = x^4 - 7$, the zero between 1 and 2.

20. $P(x) = x^4 - 3$, the zero between 1 and 2.

21. $P(x) = 2x^4 - 1$, the zero between -1 and 0.

22. $P(x) = 3x^4 - 2$, the zero between -1 and 0.

■ *Find all integer zeros of the given function. See Examples 3–5.*

23. $f(x) = 3x^3 - 13x^2 + 6x - 8$

24. $f(x) = 5x^3 + 11x^2 - 2x - 8$

25. $f(x) = x^4 + x^3 + 2x - 4$

26. $f(x) = x^4 - x^2 - 4x + 4$

27. $P(x) = 2x^4 - 3x^3 - 8x^2 - 5x - 3$

28. $P(x) = 3x^4 - 40x^3 + 130x^2 - 120x + 27$

29. $P(x) = 6x^4 - 7x^3 - 10x^2 + 14x - 4$

30. $P(x) = 6x^4 + x^3 - 19x^2 - 3x + 3$

■ *Find all rational zeros of the given function. See Examples 3–5.*

31. $f(x) = 2x^3 + 3x^2 - 14x - 21$

32. $P(x) = 2x^3 - 4x^2 + 3x + 9$

33. $P(x) = 4x^4 - 13x^3 - 7x^2 + 41x - 14$

34. $f(x) = 3x^4 - 11x^3 + 9x^2 + 13x - 10$

35. $Q(x) = 2x^3 - 7x^2 + 10x - 6$

36. $Q(x) = x^3 + 3x^2 - 4x - 12$

37. $P(x) = 2x^4 - 5x^3 + 14x^2 - 30x + 12$

38. $P(x) = 4x^4 - 4x^3 + x^2 - 4x - 3$

■ *Find all zeros of the given function. [Hint: First find all rational zeros.] See Examples 3–5.*

39. $P(x) = 3x^3 - 5x^2 - 14x - 4$

40. $P(x) = x^3 + 2x^2 - 7x - 14$

41. $P(x) = 8x^4 + 12x^3 - 4x^2 - 6x + 2$

42. $P(x) = 8x^4 - 22x^3 + 29x^2 - 66x + 15$

43. $P(x) = 12x^4 + 7x^3 - 36x^2 - 14x + 24$

44. $P(x) = 2x^4 + 5x^3 - x^2 - 5x + 2$

45. $P(x) = 6x^6 - 7x^5 - 10x^4 + 14x^3 - 4x^2$

46. $P(x) = 6x^6 + x^5 - 19x^4 - 3x^3 + 3x^2$

B

■ *Find the given zero to the indicated polynomial to within one-thousandth.*

47. $P(x) = x^3 - 4x^2 - 2x + 8$, the zero between -2 and -1.

48. $P(x) = 2x^3 + 2x^2 - x - 1$, the zero between 0 and 1.

49. $P(x) = 2x^4 - x^3 - 2$, the zero between -1 and 0.

50. $P(x) = 3x^4 + 2x - 1$, the zero between 0 and 1.

51. $P(x) = x^5 + x - 1$, the zero between 0 and 1.

52. $P(x) = x^5 - 2x - 3$, the zero between 1 and 2.

53. $P(x) = x^5 + x^4 + 2$, the zero between -2 and -1.

54. $P(x) = x^5 - x^4 + 1$, the zero between -1 and 0.

■ *Find all zeros of the given polynomial to within one-tenth.*

55. $P(x) = 4x^3 - 24x^2 + 21x + 20$ **56.** $P(x) = 4x^3 - 6x^2 - 9x + 8$

57. $P(x) = 3x^4 + 4x^3 - 12x^2 + 4$ **58.** $P(x) = 3x^4 - 4x^3 - 12x^2 + 4$

59. Factor the polynomial $2x^3 + 3x^2 - 2x - 3$.

60. Factor the polynomial $2x^4 - 3x^3 - 6x^2 + 6x + 4$.

61. Show that $\sqrt{3}$ is irrational. [*Hint:* Consider the equation $x^2 - 3 = 0$.]

62. Show that $\sqrt{2}$ is irrational.

63. Show that if p is a prime number, then $\sqrt{p}$ is irrational.

64. Show that if n is any natural number *not* of the form $n = k^2$, where k is a natural number, then $\sqrt{n}$ is irrational.

5.4

GRAPHING POLYNOMIAL FUNCTIONS

In Sections 4.2 and 4.4 we discussed methods for graphing polynomial functions of degree one (linear) and degree two (quadratic), respectively. In this section we consider methods for graphing polynomial functions of higher degree.

The method we employ is to determine certain general properties of the graph and then to use those properties to draw the complete graph. One of these properties is the set of real zeros of a polynomial function, which we discussed in Sections 5.2 and 5.3.

General Form of a Polynomial Graph

Another general property of a polynomial function

$$P(x) = a_n x^n + \cdots + a_1 x + a_0$$

is the behavior of the graph for large values of $|x|$. This behavior is determined by the leading term $a_n x^n$. Table 5.1 shows the behavior for the possible values of n and a_n. The entry in the table states the direction the curve moves for the given situation. The various possibilities are illustrated in Figure 5.2.

Notice that for each graph in Figure 5.2 there are intervals on which the function values increase as x increases and intervals on which the function values

TABLE 5.1

	n Odd		*n* Even	
	To the Left	To the Right	To the Left	To the Right
$a_n > 0$	Down	Up	Up	Up
$a_n < 0$	Up	Down	Down	Down

FIGURE 5.2

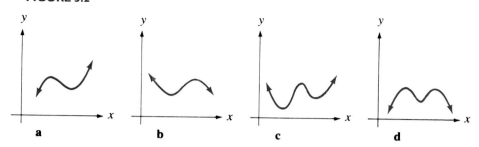

    a   b   c   d

decrease as x increases. A function is said to be **increasing** on an interval if the function values increase as x increases on the given interval. Similarly, a function is said to be **decreasing** on an interval if the function values decrease as x increases on the given interval.

Finding Turning Points

Another property used in graphing a polynomial function is the set of *turning points*, or local maxima and local minima. These are the points where the graph changes from increasing to decreasing or vice versa. For example, in Figure 5.3a, there is one local maximum as well as one local minimum, or a total of two turning points. In Figure 5.3b, there is one local maximum and two local minima, or a total of three turning points. In general, we have the following result.

FIGURE 5.3

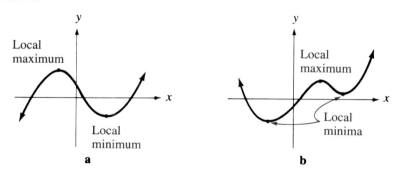

              a   b

Theorem 5.11   *If*

$$P(x) = a_n x^n + a_{n-1} x^{n-1} + \cdots + a_0 \qquad (a_n \neq 0)$$

is a polynomial of degree n with real coefficients, then its graph is a smooth curve that has at most $n - 1$ turning points.

The proof of this theorem, which involves ideas from the calculus, is omitted.

Locating Turning Points

To determine where the turning points of the graph of a polynomial lie, we use the following result. Its proof also involves ideas from the calculus and is omitted.

Theorem 5.12

The first components of the turning points of the graph of

$$P(x) = a_n x^n + a_{n-1} x^{n-1} + \cdots + a_1 x + a_0$$

are solutions of the equation

$$P'(x) = n a_n x^{n-1} + (n-1) a_{n-1} x^{n-2} + \cdots + 2a_2 x + a_1 = 0.$$

Notice that Theorem 5.12 does *not* say that every solution of $P'(x) = 0$ is the first component of a turning point of the graph of $y = P(x)$. It only says the solutions of $P'(x) = 0$ are the *possible* first components of turning points.

To find the turning points of the graph of a polynomial function $y = P(x)$, we first solve $P'(x) = 0$, where $P'(x)$ is as defined in Theorem 5.12. We then substitute these values for x that are real in $P(x)$ to obtain the second coordinates of the points. By plotting these points we can often determine by inspection whether each point is a local maximum or a local minimum or neither. If inspection is not sufficient, then plotting a few more points on either side of the point will enable us to determine its nature.

E X A M P L E   I   Find the turning points of $P(x) = x^3 - 3x^2 - 9x + 15$.

Solution   We first note from Theorem 5.11 that there are at most two turning points. By Theorem 5.12, the x-components of these turning points are solutions of $P'(x) = 0$, where

$$P'(x) = 3(1)x^{3-1} + 2(-3)x^{2-1} + (-9).$$

Then we set $P'(x)$ equal to zero to obtain

$$3x^2 - 6x - 9 = 3(x + 1)(x - 3) = 0,$$

from which $x = -1$ and $x = 3$. Next, we use the remainder theorem and synthetic division to compute $P(-1)$ and $P(3)$. We have

$$
\begin{array}{r|rrrr}
-1 & 1 & -3 & -9 & 15 \\
   &   & -1 & 4 & 5 \\
\hline
   & 1 & -4 & -5 & 20 \\
\end{array}
\qquad
\begin{array}{r|rrrr}
3 & 1 & -3 & -9 & 15 \\
  &   & 3 & 0 & -27 \\
\hline
  & 1 & 0 & -9 & -12 \\
\end{array}
$$

and the only possible turning points are $(-1, 20)$ and $(3, -12)$. Now we plot $(3, -12)$ and $(-1, 20)$. From Table 5.1 we find that the curve ultimately goes down to the left and up to the right. Thus, we see by inspection of the figure that $(-1, 20)$ is a local maximum and $(3, -12)$ is a local minimum.

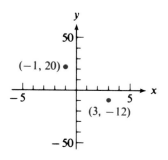

■

Graphing Polynomial Functions

To simplify the graphing of a polynomial function we perform the following steps.

Steps for Graphing a Polynomial Function

1. Locate and plot the intercepts.
2. Determine the general form of the graph using Table 5.1.
3. Locate and plot the possible turning points.
4. Plot a few more points, as needed.
5. Sketch the graph through the points obtained in Steps 1–4.

EXAMPLE 2   Graph $P(x) = 4x^3 + 3x^2 - 6x$.

Solution   We first determine the intercepts. The y-intercept is $P(0) = 0$. To obtain the x-intercepts we solve

$$P(x) = 4x^3 + 3x^2 - 6x = 0.$$

Factoring, we obtain

$$4x^3 + 3x^2 - 6x = x(4x^2 + 3x - 6) = 0.$$

Thus, $x = 0$ is one solution. By using the quadratic formula to solve

$$4x^2 + 3x - 6 = 0,$$

we find that

$$x = \frac{-3 + \sqrt{105}}{8} \approx 0.91 \qquad \text{and} \qquad x = \frac{-3 - \sqrt{105}}{8} \approx -1.66$$

are the other solutions. We then plot the intercepts. (See the figure on page 237.)

Next, we note that the polynomial is of odd degree and has a positive leading coefficient. Thus, by Table 5.1 the graph ultimately goes down to the left and up to the right, as in Figure 5.2a on page 233.

Then we use Theorem 5.12 to find the possible turning points. We solve

$$P'(x) = (3)4x^{3-1} + (2)3x^{2-1} + (-6)$$
$$= 12x^2 + 6x - 6 = 6(2x - 1)(x + 1) = 0,$$

obtaining $x = \frac{1}{2}$ and $x = -1$. Using the remainder theorem and synthetic division to compute $P(-1)$ and $P(\frac{1}{2})$, we have

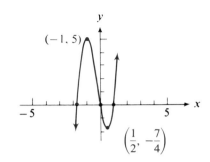

Thus, $(-1, 5)$ and $(\frac{1}{2}, -\frac{7}{4})$ are on the graph and we plot these points. By inspection of the figure we see that $(-1, 5)$ is a local maximum and $(\frac{1}{2}, -\frac{7}{4})$ is a local minimum. Finally, we sketch the graph as shown in the figure.

EXAMPLE 3 Graph $P(x) = x^4 - 10x^2 + 9$.

Solution We first determine the intercepts. The y-intercept is $P(0) = 9$. To obtain the x-intercepts we solve

$$P(x) = x^4 - 10x^2 + 9 = 0.$$

Factoring, we obtain

$$(x + 1)(x - 1)(x + 3)(x - 3) = 0.$$

The x-intercepts are -3, -1, 1, and 3. We then plot the intercepts.

Next, we note that $P(x)$ has a positive leading coefficient and is of even degree. Thus, from Table 5.1 we find that the graph of $y = P(x)$ ultimately goes up to both the left and the right, as in Figure 5.2c.

Then we use Theorem 5.12 to find the possible turning points. We solve

$$P'(x) = 4(1)x^{4-1} + 2(-10)x^{2-1}$$
$$= 4x^3 - 20x$$
$$= 4x(x^2 - 5) = 0,$$

obtaining $x = 0, \sqrt{5}$, and $-\sqrt{5}$. We next compute $P(0) = 9$, $P(\sqrt{5}) = -16$, and $P(-\sqrt{5}) = -16$. Thus, $(0, 9)$, $(\sqrt{5}, -16)$, and $(-\sqrt{5}, -16)$ are on the graph and we plot these points. By inspection we see that $(-\sqrt{5}, -16)$ and $(\sqrt{5}, -16)$ are local minima and $(0, 9)$ is a local maximum. Finally, we sketch the curve as shown in the figure.

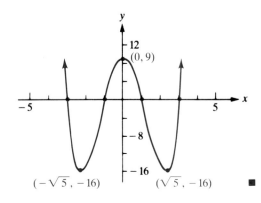

■

Flat Points

In Example 3, all the solutions of the equation $P'(x) = 0$ resulted in turning points. This will not always be the case. If a solution of $P'(x) = 0$ does not result in a turning point, then the graph of the polynomial has a "flat" point instead of a turning point. The proof of this relies on knowledge of the calculus and hence is omitted.

EXAMPLE 4   Graph $P(x) = x^3 - 3x^2 + 3x - 1$.

Solution   We first determine the intercepts. The y-intercept is $P(0) = -1$. We can use the methods described in Section 5.2 to find that the only solution to

$$P(x) = x^3 - 3x^2 + 3x - 1 = 0$$

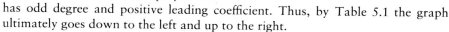

is $x = 1$. Thus, the only x-intercept is 1. We then plot the intercepts.

Next, we note that the polynomial has odd degree and positive leading coefficient. Thus, by Table 5.1 the graph ultimately goes down to the left and up to the right.

Then we use Theorem 5.12 to find the possible turning points. We first obtain

$$P'(x) = 3(1)x^{3-1} + 2(-3)x^{2-1} + 3.$$

Solving

$$P'(x) = 3x^2 - 6x + 3$$
$$= 3(x - 1)^2 = 0,$$

we obtain $x = 1$. Hence, the *only possible* turning point [since 1 is the only solution to $P'(x) = 0$] is $(1, 0)$. We have already determined that the graph ultimately goes down to the left and up to the right. If the point $(1, 0)$ were a turning point, then there would have to be at least one more turning point for the graph to have this form. Hence, the graph has no turning points. The curve has a flat point at $(1, 0)$. We now plot several more points and sketch the curve as shown.   ■

EXERCISE 5.4

A

■   *Find all the possible turning points and determine which are local maxima, which are local minima, and which are neither. See Example 1.*

1. $P(x) = 2x^3 - 9x^2 + 12x + 10$

2. $P(x) = 2x^3 + 3x^2 - 72x + 50$

3. $P(x) = x^4 - 2x^2 + 1$

4. $P(x) = 3x^4 + 4x^3 - 12x^2 + 10$

5. $P(x) = 6x^4 - 8x^3 + 10$

6. $P(x) = 6x^5 - 15x^4 + 10x^3 + 20$

■ *Graph. Include enough of the domain to show all turning points and intercepts. See Examples 2–4.*

7. $P(x) = x^3 - 3x$

8. $P(x) = 2x^3 - 3x^2 - 12x + 18$

9. $P(x) = 4x^3 - 12x^2 + 9x$

10. $P(x) = 2x^3 - 5x^2 + 4x + 1$

11. $P(x) = \dfrac{x^3}{3} - \dfrac{x^2}{2} - 6x$

12. $P(x) = \dfrac{2x^3}{3} + \dfrac{5x^2}{2} - 3x$

13. $P(x) = x^3 - 4x^2 - 3x$

14. $P(x) = 3x^3 + 2x^2 - \dfrac{4x}{3} + 1$

15. $P(x) = -x^3 + 2x^2 - 1$

16. $P(x) = -10x^3 + 9x^2 + 12x$

17. $P(x) = -3x^3 + 9x^2 + 7x + 6$

18. $P(x) = \dfrac{-x^3}{3} + x$

B

■ *Graph. Include enough of the domain to show all turning points and intercepts.*

19. $P(x) = -x^4 + x^2$

20. $P(x) = -x^4 - 4x$

21. $P(x) = x^4 - 10x^2 + 9$

22. $P(x) = x^4 - 5x^2 + 4$

23. $P(x) = x^4 - 4x^3 + 4x^2 + 12$

24. $P(x) = x^4 - x^3 - 2x^2 + 3x - 3$

25. $P(x) = x^3$

26. $P(x) = -x^3 + 8$

27. Use Theorem 5.12 to find coordinates of the vertex of the graph of the equation $P(x) = ax^2 + bx + c$.

28. Find conditions on the real numbers a, b, and c, $a \neq 0$, that guarantee the graph of $P(x) = ax^3 + bx^2 + cx + d$ has no turning points; has one turning point; has two turning points. If no such conditions exist, then so state.

5.5

RATIONAL FUNCTIONS

So far in this chapter we have discussed polynomial functions. In this section we consider techniques for graphing rational functions.

A function defined by an equation of the form

$$y = \frac{P(x)}{Q(x)}, \tag{1}$$

where $P(x)$ and $Q(x)$ are polynomials in x, and $Q(x)$ is not the zero polynomial, is called a **rational function**. Notice, in particular, that $Q(x)$ might be a nonzero constant, say, $Q(x) = 1$, so that polynomial functions are special cases of rational functions. For example,

$$R(x) = \frac{x}{x^2 + 1}, \qquad S(x) = \frac{x^3 + 1}{x^2 - 4}, \qquad \text{and} \qquad T(x) = x^2 - 2$$

are rational functions.

Vertical Asymptotes

If $Q(x_0) = 0$, then $P(x_0)/Q(x_0)$ is not defined. Therefore, there is no point on the graph of Equation (1) with first coordinate x_0. We can, however, consider the graph for values of x as close as we please to x_0 but with $x \neq x_0$ and $Q(x) \neq 0$. This consideration is usually described by saying that x "approaches" x_0 and that $Q(x)$ "approaches" 0. Observe that if $P(x)$ does not approach 0, then the closer $Q(x)$ approaches 0, the larger $|y|$ becomes in Equation (1).

EXAMPLE 1   Consider the behavior of the graph of $y = \dfrac{2}{x - 2}$ as x approaches 2.

Solution   We construct a table showing the values of y for values of x that are very close to 2.

x	$y = \dfrac{2}{x - 2}$
1.9	-20
1.99	-200
1.999	-2000
2.1	20
2.01	200
2.001	2000

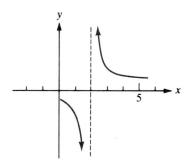

Notice that as x gets closer to 2 from the left (the first three entries of the table), $|y|$ grows large and y is negative; but as x gets closer to 2 from the right (the last three entries in the table), $|y|$ grows large and y is positive. The graph of the function near $x = 2$ is shown in the figure.   ■

Note that in Example 1, the graph of the rational function approaches the vertical line $x = 2$ as x approaches 2. In such a case, we call the vertical line a **vertical asymptote** of the graph. The following theorem gives us a way of determining the vertical asymptotes of the graph of a rational function.

Theorem 5.13   *The graph of the rational function* $y = P(x)/Q(x)$ *has the line* $x = a$ *as a vertical asymptote if* $Q(a) = 0$ *and* $P(a) \neq 0$.

E X A M P L E 2   Find all vertical asymptotes of the graph of

$$y = \frac{1}{x^2 - 3x + 2}.$$

Solution   We first find the values at which the denominator equals 0. Solving the equation

$$x^2 - 3x + 2 = (x - 1)(x - 2) = 0,$$

we have $x = 1$ and $x = 2$. Next, we note that the numerator is never 0. Thus, by Theorem 5.13, the lines $x = 1$ and $x = 2$ are both vertical asymptotes.   ■

Horizontal Asymptotes   For some rational functions, $y = P(x)/Q(x)$, the value of y approaches a fixed value y_0 as $|x|$ grows large.

E X A M P L E 3   Consider the behavior of the graph of

$$y = \frac{2x}{x - 2}$$

as $|x|$ grows large.

Solution   We construct a table showing the values of y for values of x that have large absolute value.

x	$y = \dfrac{2x}{x-2}$
10	2.500
100	2.041
1000	2.004
−10	1.667
−100	1.961
−1000	1.996

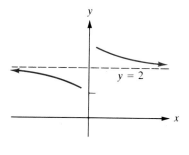

Notice that as $|x|$ grows large, the value of y grows closer to 2. The graph of the function for large values of $|x|$ is shown in the figure.   ■

Note that in Example 3, the graph of the rational function approaches the horizontal line $y = 2$ as $|x|$ grows large. In such a case, we call the horizontal line a **horizontal asymptote** of the graph. The following theorem gives us a way to determine the horizontal asymptote of a rational function if there is one.

Theorem 5.14   *The graph of the rational function defined by*

$$y = R(x) = \frac{a_n x^n + a_{n-1} x^{n-1} + \cdots + a_0}{b_m x^m + b_{m-1} x^{m-1} + \cdots + b_0} \tag{2}$$

where $a_n, b_m \neq 0$, and n, m are nonnegative integers, has

   I.   *a horizontal asymptote at $y = 0$ if the degree of the numerator is less than the degree of the denominator $(n < m)$;*

   II. *a horizontal asymptote at $y = a_n/b_m$ if the degree of the numerator equals the degree of the denominator $(n = m)$;*

   III. *no horizontal asymptotes if the degree of the numerator is greater than the degree of the denominator $(n > m)$.*

Though we shall not give a rigorous proof of Theorem 5.14, we can certainly make the results seem plausible. If $n < m$, we can divide the numerator and denominator of the right-hand member of (2) by x^m to obtain, for $x \neq 0$,

$$y = R(x) = \frac{\dfrac{a_n}{x^{m-n}} + \dfrac{a_{n-1}}{x^{m-n+1}} + \cdots + \dfrac{a_0}{x^m}}{b_m + \dfrac{b_{m-1}}{x} + \cdots + \dfrac{b_0}{x^m}}.$$

Now, as $|x|$ grows greater and greater, each term containing an x in its denominator grows closer and closer to 0. Thus, the expression on the right approaches $0/b_m$, so that y approaches 0. But if, as $|x|$ increases without bound, y grows close to 0, then the graph of $y = R(x)$ approaches the line $y = 0$ asymptotically. A similar argument shows that if $n = m$, then, as $|x|$ increases without bound, the graph of $y = R(x)$ approaches the line $y = a_n/b_m$ asymptotically. Finally, note that if $n > m$, then as $|x|$ becomes greater and greater, so does $|y|$. Thus, the curve has no horizontal asymptote.

EXAMPLE 4   Find all vertical and horizontal asymptotes of the graph of the function

$$y = \frac{2x^3 + x^2 - 1}{3x^3 - 3x^2}.$$

Solution   Since $3x^3 - 3x^2 = 3x^2(x - 1) = 0$ when $x = 0$ and when $x = 1$, we observe that 0 and 1 are zeros of the denominator but not of the numerator. Thus, by Theorem 5.13,

$$x = 0 \qquad \text{and} \qquad x = 1$$

are vertical asymptotes. Further, since the degree of the numerator equals the degree of the denominator, by Theorem 5.14, the line

$$y = \frac{2}{3}$$

is a horizontal asymptote.

EXAMPLE 5   Find all vertical and horizontal asymptotes of the graph of the function

$$y = \frac{x^2 - 4}{x - 1}.$$

Solution   We observe by Theorem 5.13 that $x = 1$ is a vertical asymptote and by Theorem 5.14 that there are no horizontal asymptotes because the degree of the numerator is greater than the degree of the denominator.

EXAMPLE 6   Find all vertical and horizontal asymptotes of the graph of the function

$$y = \frac{x}{x^2 + 1}.$$

Solution   Since the denominator polynomial has no real zeros, the rational function is defined for all real values of x. Thus, the graph has no vertical asymptotes.

We observe that the degree of the denominator is greater than the degree of the numerator. Thus, the line $y = 0$ is a horizontal asymptote.   ■

Oblique Asymptotes   If $R(x)$ is a rational function in which the degree of the numerator is one larger than the degree of the denominator, then by Theorem 5.14 the graph has no horizontal asymptotes. However, the graph approaches an oblique line as $|x|$ grows large. This line is called an **oblique asymptote**. We find an equation for such an asymptote by dividing the numerator by the denominator and observing that the remainder approaches 0 as $|x|$ grows large.

The following example illustrates this concept.

EXAMPLE 7   Find all asymptotes for the graph of the function

$$y = \frac{x^2 - 4}{x - 1}.$$

Solution   This is the same function we investigated in Example 5, where we found the vertical asymptote $x = 1$ and no horizontal asymptotes. If we first rewrite the equation $y = (x^2 - 4)/(x - 1)$ by dividing $x^2 - 4$ by $x - 1$, we obtain

$$y = x + 1 - \frac{3}{x - 1}.$$

Now, as x grows greater and greater, $3/(x - 1)$ becomes closer and closer to 0, and the graph of $y = (x^2 - 4)/(x - 1)$ approaches the graph of $y = x + 1$. Hence the graph of $y = x + 1$, which is an oblique line, is an asymptote to the curve.   ■

Graphing Rational Functions

To sketch the graph of a rational function, we complete the following steps.

Steps for Graphing a Rational Function

1. Find and plot the intercepts.
2. Identify and sketch the vertical, horizontal, and oblique asymptotes.
3. Determine the asymptotic behavior of the curve.
4. Sketch the curve.

EXAMPLE 8   Graph $y = \dfrac{x - 1}{x - 2}$.

Solution   We first observe that the numerator of the right-hand member is equal to 0 when x is equal to 1. Therefore, when $x = 1$ we have $y = 0$, and 1 is an x-intercept. Also, when $x = 0$ we have $y = \frac{1}{2}$, so that $\frac{1}{2}$ is a y-intercept. Thus, we can begin our graph as shown in Figure a.

Next, we observe that for $x = 2$ the denominator is 0 but the numerator is not. Thus, by Theorem 5.13 the line $x = 2$ is the only vertical asymptote. Further, since

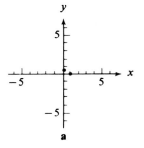

a

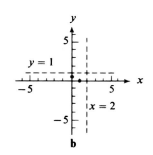

b

the degrees of the numerator and denominator are equal, by Theorem 5.14 we see that the line $y = \frac{1}{1} = 1$ is the only horizontal asymptote. We then plot these asymptotes as dashed lines (Figure b on page 243).

To determine the behavior of the function near the vertical asymptote, we consider one side of the asymptote at a time. On the left side, x is just less than 2, say, $2 - p$, where $0 < p < \frac{1}{10}$, so the denominator $x - 2$ in the expression for y is

$$(2 - p) - 2 = -p,$$

which is barely negative, whereas the numerator

$$x - 1 = 2 - p - 1 = 1 - p$$

is definitely positive. Hence, y is negative and $|y|$ is large. Thus, the graph goes down on the left side of the vertical asymptote (Figure c).

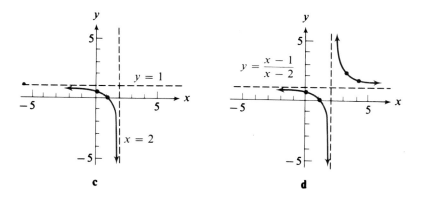

c   d

To the right of the vertical asymptote, we take $x = 2 + p$, where $0 < p < \frac{1}{10}$, to find that the graph goes up on the right side of the vertical asymptote. This, along with the knowledge that $y = 1$ is a horizontal asymptote, leads us to the complete graph of the function $y = (x - 1)/(x - 2)$ as shown in Figure d.

EXAMPLE 9   Graph $y = \dfrac{x^2 - 4}{x - 1}$.

Solution   We first observe that $y = 0$ only when $x^2 - 4 = 0$, or when $x = \pm 2$. Thus, the x-intercepts are 2 and -2. For $x = 0$ we have $y = 4$; thus, the y-intercept is 4. We plot these points in Figure a.

Next, we observe that this is the same function we investigated in Example 7 on pages 242 and 243, for which we found the vertical asymptote $x = 1$ and the oblique asymptote $y = x + 1$. We plot these asymptotes as dashed lines in Figure a.

To determine the behavior of the curve near the line $x = 1$, we note that for values of x near 1 the value of $x^2 - 4$ is near -3 and thus is negative. Hence, y is positive when $x < 1$ and y is negative when $x > 1$. Thus, the curve goes up on the

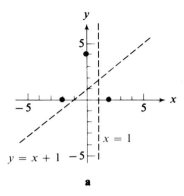

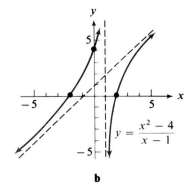

a   b

left side of $x = 1$ and it goes down on the right side of $x = 1$. Further,

$$\frac{x^2 - 4}{x - 1} = x + 1 - \frac{3}{x - 1},$$

and $3/(x - 1)$ is positive for $x > 1$ and negative for $x < 1$. Thus, the graph of the equation $y = (x^2 - 4)/(x - 1)$ is below the graph of $y = x + 1$ for $x > 1$ and above the graph of $y = x + 1$ for $x < 1$. This, along with the fact that we know $y = x + 1$ is an oblique asymptote, leads us to complete the graph as shown in Figure b.   ■

Common Zeros

In some cases the numerator and denominator of a rational function have a common zero. In such a case, the fraction should be reduced to lowest terms and the resulting function graphed with the added restriction that the common zero not be included in the domain.*

EXAMPLE 10   Graph $y = \dfrac{x^2 - 4}{x - 2}$.

Solution   We note that 2 is a zero of both the numerator and the denominator. Factoring the numerator and reducing to lowest terms yields

$$y = \frac{x^2 - 4}{x - 2}$$

$$= \frac{(x - 2)(x + 2)}{x - 2}$$

$$= x + 2, \qquad x \neq 2.$$

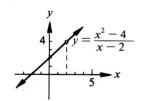

* Often in the definition of rational function, the numerator and denominator are restricted from having common zeros.

Thus, the graph of the function $y = (x^2 - 4)/(x - 2)$ is the same as that of

$$y = x + 2, \qquad x \neq 2.$$

The graph is shown in the figure. Note that there is a "hole" in the graph at $x = 2$ because the original function is not defined for $x = 2$.   ■

EXERCISE 5.5

A

■ *Determine the vertical asymptotes of the graphs of the given functions. For Exercises 1–14, see Examples 2 and 4–6.*

1. $y = \dfrac{1}{x - 3}$ **2.** $y = \dfrac{1}{x + 4}$ **3.** $y = \dfrac{4}{(x + 2)(x - 3)}$

4. $y = \dfrac{8}{(x - 1)(x + 3)}$ **5.** $y = \dfrac{2x - 1}{x^2 + 5x + 4}$ **6.** $y = \dfrac{x + 3}{2x^2 - 5x - 3}$

7. $y = \dfrac{4x}{x^2 + 4}$ **8.** $y = \dfrac{3x^2 + 1}{x^2 + x + 1}$ **9.** $y = \dfrac{2x - 1}{x^2 + 2x + 2}$

10. $y = \dfrac{3x^2 + 1}{x^2 - x + 4}$ **11.** $xy + y = 4$ **12.** $x^2 y + xy = 3$

13. $x^2 y - y = x$ **14.** $x^2 y - 4y = x^2 + 1$

■ *Determine any vertical, horizontal, or oblique asymptotes of the graphs of the given functions. See Examples 4–7.*

15. $y = \dfrac{x}{x^2 - 4}$ **16.** $y = \dfrac{3x - 6}{x^2 + 3x + 2}$ **17.** $y = \dfrac{x^2 - 9}{x - 4}$

18. $y = \dfrac{x^3 - 27}{x^2 - 1}$ **19.** $y = \dfrac{x^2 - 3x + 2}{x^2 - 3x - 4}$ **20.** $y = \dfrac{x^2}{x^2 - x - 6}$

21. $y = \dfrac{x^2 - 1}{x^2 + 1}$ **22.** $y = \dfrac{x^2 - x}{3x^2 + 2}$ **23.** $y = \dfrac{x^3 + 2x^2 + 3x + 1}{x^2 + x + 1}$

24. $y = \dfrac{3x^3 + 4x^2 + 10x - 7}{x^2 + 2x + 4}$ **25.** $y = \dfrac{x^4}{x^2 + 1}$ **26.** $y = \dfrac{x^4 + x^2 + 1}{x^2 + x + 4}$

■ *Graph. Use information concerning the zeros of the function and concerning vertical, horizontal, and oblique asymptotes. See Examples 8 and 9.*

27. $y = \dfrac{1}{x}$ **28.** $y = \dfrac{1}{x + 4}$ **29.** $y = \dfrac{1}{x - 3}$

30. $y = \dfrac{1}{x - 6}$

31. $y = \dfrac{4}{(x + 2)(x - 3)}$

32. $y = \dfrac{8}{(x - 1)(x + 3)}$

33. $y = \dfrac{2}{(x - 3)^2}$

34. $y = \dfrac{1}{(x + 4)^2}$

35. $y = \dfrac{x}{x - 2}$

36. $y = \dfrac{x - 1}{x + 3}$

37. $y = \dfrac{2x - 4}{x^2 - 9}$

38. $y = \dfrac{3x}{x^2 - 5x + 4}$

39. $y = \dfrac{x^2 - 4}{x^3}$

40. $y = \dfrac{x - 2}{x^2}$

41. $y = \dfrac{x + 1}{x(x^2 - 4)}$

42. $y = \dfrac{x^2 + x - 2}{x(x^2 - 9)}$

43. $y = \dfrac{x^2 - 4x + 4}{x - 1}$

44. $y = \dfrac{x^2 + 4}{x - 2}$

45. $y = \dfrac{x^3 + 1}{x^2}$

46. $y = \dfrac{x^3 - x^2 - 2x}{x^2 - 1}$

B

■ Graph. Use information concerning the zeros of the function, and vertical, horizontal, and oblique asymptotes if they exist. For Exercises 47–52, see Example 10.

47. $y = \dfrac{x^2 - 9}{x + 3}$

48. $y = \dfrac{x^2 - 1}{x + 1}$

49. $y = \dfrac{x^3 - 3x^2 + 2x}{x - 1}$

50. $y = \dfrac{-x^3 + 2x^2 - x}{x}$

51. $y = \dfrac{x - 1}{x^2 - x}$

52. $y = \dfrac{x}{x^2 - 4x}$

53. Describe the asymptotic behavior of $y = \dfrac{x^3 + x^2 + 2}{x + 1}$.

54. Sketch the graph of $y = \dfrac{x^3 + x^2 + 2}{x + 1}$.

55. Is it possible for the graph of a rational function to cross a vertical asymptote? Why or why not?

56. Is it possible for the denominator of a rational function to be a zero at some value x_0 but not have $x = x_0$ as a vertical asymptote of the graph?

57. Find the point where the graph of $y = \dfrac{x^2 + 1}{x^2 - 2x - 1}$ crosses its horizontal asymptote.

58. Find the point where the graph of $y = \dfrac{2x^3 + x^2 + 3x}{x^2 + 1}$ crosses its oblique asymptote.

59. Give an example of a rational function with the property that its graph crosses its horizontal asymptote more than once.

60. Give an example of a rational function with the property that its graph crosses its oblique asymptote more than once.

CHAPTER REVIEW

Key Words and Phrases

■ *Define or explain each of the following words and phrases.*

1. remainder theorem

2. factor theorem

3. upper bound for the set of real zeros

4. lower bound for the set of real zeros

5. variation in sign

6. Descartes' rule of signs

7. fundamental theorem of algebra

8. multiplicity k

9. approximation of zeros

10. test for rational zeros

11. general form of the graph of a polynomial function

12. increasing function

13. decreasing function

14. turning points

15. steps in graphing a polynomial function

16. vertical asymptote

17. horizontal asymptote

18. oblique asymptote

19. steps in graphing a rational function

Review Exercises

A

[5.1] ■ *Use synthetic division to write the quotient* $\dfrac{P(x)}{D(x)}$ *in the form* $Q(x)$ *or in the form* $Q(x) + \dfrac{r}{D(x)}$, *where* $Q(x)$ *is a polynomial and r is a real number.*

1. $\dfrac{x^4 - 5x^2 + 7x - 10}{x - 2}$

2. $\dfrac{3x^3 + 6x^2 - 4x - 5}{x + 2}$

■ *Use synthetic division and Theorem 5.2*

3. $P(x) = x^4 - 2x^3 + x - 4$; find $P(-2)$, $P(1)$, and $P(2)$.

4. $P(x) = -x^4 - 3x^3 + x^2 + 2$; find $P(-2)$, $P(1)$, and $P(2)$.

■ *Assume the given number is a solution of* $P(x) = 0$ *and factor* $P(x)$ *completely.*

5. -2; $P(x) = 2x^3 - x^2 - 7x + 6$

6. $\dfrac{1}{2}$; $P(x) = 6x^3 + 13x^2 - 4$

[5.2] ■ *Find an upper bound and a lower bound for the real zeros of the given polynomial function.*

7. $P(x) = 2x^3 + x^2 - 4x - 2$ **8.** $P(x) = x^4 + 2x^3 - 7x^2 - 10x + 10$

■ *Use Descartes' rule of signs to determine the number of positive real zeros possible and the number of negative real zeros possible.*

9. $P(x) = 2x^3 + x^2 - 4x - 2$ **10.** $P(x) = x^4 + 2x^3 - 7x^2 - 10x + 10$

■ *Find all zeros of the given polynomial function.*

11. $P(x) = x^3 - 2x^2 + 4x - 8$; $2i$ is one zero.
12. $Q(x) = 2x^3 - 11x^2 + 28x - 24$; $2 + 2i$ is one zero.

[5.3] ■ *Use the method given in Section 5.3 to approximate the positive zero of the given polynomial function to within one-tenth.*

13. $P(x) = x^2 - 12$ **14.** $P(x) = x^3 - 7$
15. Find all integral zeros of $Q(x) = x^4 - 3x^3 - x^2 - 11x - 4$.
16. Find all rational zeros of $P(x) = 2x^3 - 11x^2 + 12x + 9$.
17. Find all zeros of $P(x) = x^3 + 2x^2 + 2x + 4$.
18. Find all zeros of $P(x) = 18x^4 - 3x^3 - 15x^2 + 2x + 2$.

[5.4] ■ *Find all possible turning points and state which are local maxima, which are local minima, and which are neither.*

19. $P(x) = 2x^3 - 9x^2 + 12x$ **20.** $P(x) = x^3 - x^2$

■ *Graph the function. Indicate all x- and y-intercepts.*

21. $P(x) = 2x^3 - 9x^2 + 12x$ **22.** $P(x) = 2x^3 + 3x^2$
23. $P(x) = x^3 - x^2$ **24.** $P(x) = x^4 - 2x^2 - 3$

[5.5] ■ *Determine all asymptotes and graph each function. Indicate all x- and y-intercepts.*

25. $y = \dfrac{3}{x + 2}$ **26.** $y = \dfrac{2x}{(x + 3)(x - 4)}$

27. $y = \dfrac{x + 2}{x - 3}$ **28.** $y = \dfrac{2x}{x^2 - 3x + 2}$

B

29. Determine the fourth roots of 1.
30. Determine the fourth roots of 16.

31. Find the zero of $P(x) = x^4 - x - 1$ between -1 and 0 to within one-thousandth.

32. Find the zero of $P(x) = x^4 - 2x - 1$ between -1 and 0 to within one-thousandth.

33. Find all zeros of $P(x) = x^3 - 3x^2 + 1$ to within one-tenth.

34. Find all zeros of $P(x) = 4x^3 - 6x^2 + 1$ to within one-tenth.

■ *Graph. Include enough of the domain to show all turning points and intercepts.*

35. $P(x) = x^4 - 2x^2 + 1$ **36.** $P(x) = x^4 - 4x^2 + 4$

37. $P(x) = x^4 + 4x^3 + 4x^2$ **38.** $P(x) = x^4 - 8x^3 + 22x^2 - 24x + 10$

■ *Graph. Use information concerning the zeros of the function and concerning vertical, horizontal, and oblique asymptotes.*

39. $y = \dfrac{x}{x^2 + 1}$ **40.** $y = \dfrac{x}{2x^2 + 1}$

41. $y = \dfrac{x^3 - x}{x^2 - 1}$ **42.** $y = \dfrac{x^4 - 2x^2}{x^2 - 2}$

6

Exponential and Logarithmic Functions

The exponents involved in the polynomial and rational functions considered in Chapters 4 and 5 were constants. In this chapter we shall consider functions in which exponents are variables. We shall also consider the inverses of such functions.

6.1

EXPONENTIAL FUNCTIONS

Properties of powers of the form b^x, where b is a positive real number and x is a *rational number*, were discussed in Chapter 2. We restricted the base b for such powers to positive real numbers to ensure that b^x would be a real number for all rational numbers x.

Powers with Irrational Exponents

We now want to define b^x for any real number x. Hence, we must be able to interpret powers with *irrational exponents*, such as

$$b^{\sqrt{2}}, \qquad b^{-\sqrt{3}}, \qquad \text{and} \qquad b^{\pi}.$$

We shall use the fact that irrational numbers can be approximated by rational numbers to as great a degree of accuracy as desired to interpret expressions of the form b^x as real numbers, where x is a positive irrational number. As an example, we consider the interpretation of $2^{\sqrt{2}}$, where

$$\sqrt{2} \approx 1.4, \qquad \sqrt{2} \approx 1.41, \qquad \text{or} \qquad \sqrt{2} \approx 1.414, \qquad \text{and so on.}$$

Since 2^x is defined for rational values of x, we can use a calculator with a y^x key to construct Table 6.1 on page 254.

The entries in the last column of Table 6.1 suggest that as x and y get closer to $\sqrt{2}$, the value of $2^y - 2^x$ gets closer to zero. It can be shown, although we shall not do so here, that $2^y - 2^x$ can be made as close to zero as we like. This fact guarantees

TABLE 6.1

x	y	2^x	2^y	$2^y - 2^x$
1.4	1.5	2.639	2.828	0.189
1.41	1.42	2.657	2.676	0.019
1.414	1.415	2.665	2.667	0.002

that there is only one real number, which we denote by $2^{\sqrt{2}}$, with the property that

$$2^x < 2^{\sqrt{2}} < 2^y$$

for all rational numbers x and y with $x < \sqrt{2} < y$.

We can use the same idea to obtain b^x for any base $b > 0$ and irrational exponent x. Hence, we shall assume that b^x $(b > 0)$ is defined for all real values of x and that the laws of exponents discussed in Section 2.2 for rational exponents are valid for real exponents.

Exponential Functions

Since for any given $b > 0$ and for each real number x, there is only one value for b^x, the equation

$$f(x) = b^x \qquad (b > 0) \tag{1}$$

defines a function. Because $1^x = 1$ for all real numbers x, (1) defines a constant function if $b = 1$. If $b \neq 1$, we say that (1) defines an **exponential function** with base b.

EXAMPLE 1   Find $f(-2)$ for each exponential function.

(a) $f(x) = 2^x$
(b) $f(x) = \left(\dfrac{1}{3}\right)^x$

Solution   (a) $f(-2) = 2^{-2}$

$$= \frac{1}{2^2} = \frac{1}{4}$$

(b) $f(-2) = \left(\dfrac{1}{3}\right)^{-2}$

$$= \frac{1}{(\frac{1}{3})^2} = 9 \qquad ■$$

10^x and e^x

Two exponential functions are of special interest because they are frequently used in mathematics and other sciences. The base for one of these functions is 10, and the base for the other is an irrational number that we shall denote by e.* The value of e to eight decimal places is 2.71828183.

* We shall consider this number in more detail in Section 6.5.

We can use most scientific calculators to compute an approximation of 10^x and e^x for most decimal values of x. For all irrational values and some rational values of x in fractional form, we can first use a calculator to approximate x and then to approximate 10^x (or e^x). The answers in the examples and exercises were obtained by using a calculator and have been rounded off to four decimal places. Although values obtained in this manner are approximations, for convenience of notation we shall use "$=$" instead of "$\approx$."

EXAMPLE 2   Approximate each power in decimal form.

(a) $10^{\sqrt{2}}$ (b) $e^{-1/3}$

Solution   (a)   We first approximate $\sqrt{2}$ and then approximate the power.

$$10^{\sqrt{2}} = 25.9546$$

(b)   We first approximate $\frac{1}{3}$ and then approximate the power.

$$e^{-1/3} = 0.7165 \qquad ■$$

Values of 10^x and e^x can also be approximated by using tables. Methods for using tables are discussed in Appendix A.

Graphs of Exponential Functions   We consider two typical examples of the graphs of exponential functions. In one case $0 < b < 1$ and in the other $b > 1$.

EXAMPLE 3   Graph each function.

(a) $f(x) = 2^x$ (b) $f(x) = \left(\dfrac{1}{2}\right)^x$

Solution   We first select several values of x and compute the corresponding function values. We then plot the points obtained and sketch a smooth curve through the points. The graphs are shown in Figures a and b on page 256. For convenience, we selected integer values of x. In fact, any real values of x can be used, and the powers obtained by using the y^x key on a calculator.

In Example 3, $f(x) = 2^x$ is an increasing function and $f(x) = (\frac{1}{2})^x$ is a decreasing function, as can be seen in Figures a and b. In general, if $0 < b < 1$, the function $f(x) = b^x$ is decreasing, whereas for $b > 1$, it is increasing.

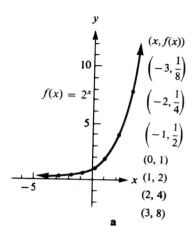

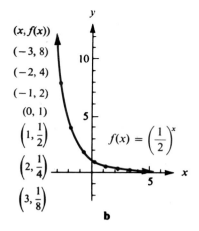

a   b

EXERCISE 6.1

A

■   *Compute the indicated function values. Where appropriate, approximate the value to four decimal places. See Examples 1 and 2.*

1. $f(x) = 3^x$; $f(0), f(1)$

2. $g(x) = 4^x$; $g(0), g(1)$

3. $f(x) = -5^x$; $f(0), f(2)$

4. $g(x) = -2^x$; $g(0), g(2)$

5. $f(x) = \left(\dfrac{1}{2}\right)^x$; $f(-3), f(3)$

6. $g(x) = \left(\dfrac{1}{3}\right)^x$; $g(-3), g(3)$

7. $f(x) = 10^x$; $f(-2), f(1)$

8. $g(x) = 10^{-x}$; $g(-1), g(2)$

9. $f(x) = e^x$; $f(-1), f(1)$

10. $g(x) = e^{-x}$; $g(-1), g(1)$

11. $f(x) = 10^x$; $f(-1.2), f(3.5)$

12. $g(x) = 10^{-x}$; $g(-1.2), g(3.5)$

13. $f(x) = e^x$; $f(-2.3), f(-1.4)$

14. $g(x) = e^{-x}$; $g(-2.3), g(-1.4)$

15. $f(x) = 10^x$; $f\left(\dfrac{1}{3}\right), f(\sqrt{2})$

16. $g(x) = 10^{-x}$; $g\left(\dfrac{1}{9}\right), g(\sqrt{3})$

17. $f(x) = e^x$; $f\left(-\dfrac{1}{7}\right), f(\sqrt{5})$

18. $g(x) = e^{-x}$; $g\left(-\dfrac{5}{6}\right), g(\sqrt{10})$

■   *Graph the function. See Example 3.*

19. $y = 4^x$ **20.** $y = 5^x$ **21.** $y = 10^x$ **22.** $y = 10^{-x}$ **23.** $y = 2^{-x}$

24. $y = 3^{-x}$ **25.** $y = \left(\dfrac{1}{3}\right)^x$ **26.** $y = \left(\dfrac{1}{4}\right)^x$ **27.** $y = e^x$ **28.** $y = e^{-x}$

■   *Solve for x by inspection.*

29. $3^x = 243$ **30.** $5^x = 125$ **31.** $10^x = \dfrac{1}{100}$ **32.** $4^x = \dfrac{1}{256}$

33. $\left(\dfrac{1}{3}\right)^x = 81$ **34.** $\left(\dfrac{1}{2}\right)^x = 16$ **35.** $16^x = 8$ **36.** $8^x = 16$

B

■ *Determine an integer n such that n < x < n + 1.*

37. $3^x = 16.20$ **38.** $4^x = 87.10$ **39.** $10^x = 0.016$ **40.** $2^x = 6$

41. $\left(\dfrac{3}{2}\right)^x = 6$ **42.** $\left(\dfrac{7}{3}\right)^x = 380$ **43.** $\left(\dfrac{7}{8}\right)^x = 1.50$ **44.** $\left(\dfrac{3}{5}\right)^x = 3$

45. Graph $y = b^x$, $b = 2, 3$, and 4, on the same set of coordinate axes. What effect does increasing b have on the curves if $b > 1$?

46. Graph $y = b^x$, $b = \frac{1}{4}, \frac{1}{3}$, and $\frac{1}{2}$, on the same set of coordinate axes. What effect does increasing b have on the curves if $0 < b < 1$?

47. Graph $f: y = 10^x$ and $f^{-1}: x = 10^y$ on the same set of axes.

48. Graph $f: y = 2^x$ and $f^{-1}: x = 2^y$ on the same set of axes.

49. Graph $y = 1^x$ and $x = 1^y$ on the same set of axes.

50. Does the function $f(x) = 1^x$ have an inverse function?

51. Evaluate $f(x) = 3^x$ at $x = -1, -5, -10$, and -100. Express the result in scientific notation. Describe the behavior of $f(x)$ as $|x|$ becomes large with $x < 0$.

52. Evaluate $f(x) = 7^x$ at $x = 1, 5, 10$, and 100. Express the result in scientific notation. Describe the behavior of $f(x)$ as $|x|$ becomes large with $x > 0$.

53. Evaluate $f(x) = \left(\frac{2}{3}\right)^x$ at $x = 1, 5, 10$, and 100. Express the result in scientific notation. Describe the behavior of $f(x)$ as $|x|$ becomes large with $x > 0$.

54. Evaluate $f(x) = \left(\frac{3}{4}\right)^x$ at $x = -1, -5, -10$, and -100. Express the result in scientific notation. Describe the behavior of $f(x)$ as $|x|$ becomes large with $x < 0$.

55. Evaluate each of the following at $x = 1, 5, 10$, and 100. Express your answer to five-decimal-place accuracy.

  a. $f(x) = (0.9)^x$ b. $f(x) = (1.1)^x$

56. Is the following statement true or false? State the reason for your answer. "If $b_1 > 0$ is approximately equal to $b_2 > 0$, then for every value of x, b_1^x is approximately equal to b_2^x."

57. Evaluate each of the following at $x = 30, 40, 50$, and 60.

  a. $f(x) = xe^{-x}$ b. $f(x) = x^4 e^{-x}$ c. $f(x) = x^{10} e^{-x}$

58. Let n be a positive integer. Make a conjecture about the behavior of $f(x) = x^n e^{-x}$ for large values of x. Explain this behavior in terms of the values of x^n and e^{-x} for large values of x.

59. Evaluate each of the following at $x = 1, 5, 10$, and 100.

  a. $f(x) = \dfrac{e^x}{x}$ b. $f(x) = \dfrac{e^x}{x^4}$ c. $f(x) = \dfrac{e^x}{x^{10}}$

60. Let n be a positive integer. Make a conjecture about the behavior of $f(x) = e^x/x^n$ for large values of x. Explain this behavior in terms of the values of e^x and x^n for large values of x.

6.2

LOGARITHMIC FUNCTIONS

In this section we shall consider the inverses of the exponential functions that we considered in Section 6.1.

Inverse of an Exponential Function

Since exponential functions are either increasing (base b greater than 1) or decreasing (base b between 0 and 1), they are one-to-one functions. Hence, the exponential function has an inverse function defined by $x = b^y$ (see Section 4.6). This function is called the **logarithmic function with base b**. The function values are denoted $\log_b x$.

Thus, for $b > 0$, $b \neq 1$, and $x > 0$,

$$y = \log_b x \qquad \text{if and only if} \qquad x = b^y. \tag{1}$$

$\text{Log}_b x$ is read "**logarithm of x with the base b**."

Graphs of $y = \log_b x$

Since $y = b^x$ and $y = \log_b x$ are inverse functions, their graphs are symmetric with respect to the line $y = x$, as illustrated in Figure 6.1a for $b = 2$ and in Figure 6.1b for $b = \frac{1}{2}$.

FIGURE 6.1

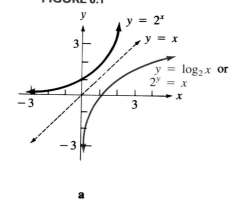

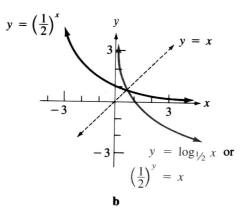

a   b

Basic Properties of $y = \log_b x$

We can observe from Figures 6.1a and 6.1b that the domain of both $y = \log_2 x$ and $y = \log_{1/2} x$ is the set of positive real numbers; the range of both functions is the set of real numbers; the x-intercept of both functions is 1; and $y = \log_2 x$ is increasing whereas $y = \log_{1/2} x$ is decreasing. These properties are summarized in the following box.

For any $b > 0$ the function $f(x) = \log_b x$ has the following properties:

1. The domain of f is the set of *positive* real numbers.
2. The range of f is the set of real numbers.
3. $f(1) = \log_b 1 = 0$
4. The function f is increasing if $b > 1$ and is decreasing if $0 < b < 1$.

EXAMPLE 1   (a) $\log_{10} 1 = 0$ (b) $\log_e 1 = 0$

(c) $\log_{10} 0$ is undefined.   (d) $\log_e 0$ is undefined.

(e) $\log_{10}(-4)$ is undefined.   (f) $\log_e(-5)$ is undefined.   ■

Logarithmic and Exponential Statements

Note that the equations $y = \log_b x$ and $x = b^y$ are equivalent equations that determine the same function. Thus, logarithmic statements may be written in exponential form and exponential statements may be written in logarithmic form.

EXAMPLE 2   (a) $\log_{10} 100 = 2$ can be written as $10^2 = 100.$

(b) $\log_3 81 = 4$ can be written as $3^4 = 81.$

(c) $\log_2 \dfrac{1}{2} = -1$ can be written as $2^{-1} = \dfrac{1}{2}.$

EXAMPLE 3   (a) $5^2 = 25$ can be written as $\log_5 25 = 2.$

(b) $8^{1/3} = 2$ can be written as $\log_8 2 = \dfrac{1}{3}.$

(c) $3^{-3} = \dfrac{1}{27}$ can be written as $\log_3 \dfrac{1}{27} = -3.$ ■

Notice that in each of these examples the logarithm is the exponent. In fact, $\log_b x$ is the *exponent y* such that the power b^y is equal to x; that is,

$$b^{\log_b x} = x. \qquad\qquad (2)$$

EXAMPLE 4 (a) $10^{\log_{10} 1000} = 1000$ (b) $10^{\log_{10} 2x} = 2x$ (c) $5^{\log_5 x^2} = x^2$

EXAMPLE 5 Find the value of each of the following.

(a) $\log_8 64$ (b) $\log_{10} 0.001$

Solution (a) We note that $\log_8 64$ is the exponent placed on 8 to obtain $64 = 8^2$. Thus, by inspection $\log_8 64 = 2$.

(b) We note that $\log_{10} 0.001$ is the exponent placed on 10 to obtain $0.001 = 10^{-3}$. Thus, by inspection $\log_{10} 0.001 = -3$. ■

We can sometimes solve simple equations involving logarithms by first rewriting the logarithmic statement in exponential form and solving the resulting equation by inspection.

EXAMPLE 6 Solve each equation.

(a) $\log_2 x = 3$ (b) $\log_b 2 = \dfrac{1}{2}$ (c) $\log_{1/4} 16 = y$

Solution In each case we first write the equation equivalently in exponential form and then solve the resulting equation by inspection.

(a) $2^3 = x$ (b) $b^{1/2} = 2$ (c) $\left(\dfrac{1}{4}\right)^y = 16$

$\quad\ \ x = 8$ $\quad (b^{1/2})^2 = (2)^2$ $\qquad\ 4^{-y} = 4^2$

$\qquad\qquad\qquad\quad\ b = 4$ $\qquad\quad y = -2$ ■

Common and Natural Logarithms

In most applications, the base that is used for $\log_b x$ is either 10 or e. The values $\log_{10} x$ are called **common logarithms**, and the values of $\log_e x$ are called **natural logarithms**. We frequently denote $\log_e x$ by **ln x**.

Most values for $\log_{10} x$ and $\ln x$, $x > 0$, are irrational. A hand calculator can be used to obtain approximations for such values. In the examples and exercises, the approximations are rounded off to four decimal places.

EXAMPLE 7 (a) $\log_{10} 1.61 = 0.2068$ (b) $\log_{10} 0.52 = -0.2840$

(c) $\ln 10.75 = 2.3749$ (d) $\ln 0.01 = -4.6052$ ■

Tables can also be used to find approximations of $\log_{10} x$ and $\ln x$. Methods for using tables are discussed in Appendix A.

Graphs of
$y = \log_b x$

We can graph $y = \log_b x$ for a given value of b by using the fact that it is the inverse of the exponential function $y = b^x$ as in Figure 6.1 on page 258. However, it is sometimes convenient to use the same method of plotting points that we used in Section 6.1 to graph exponential functions.

EXAMPLE 8 Graph each function.

(a) $f(x) = \log_{10} x$ (b) $f(x) = \log_{1/2} x$

Solution We first select some arbitrary values of x and compute the corresponding values of $f(x)$. We then plot the points and sketch a smooth curve through the points to obtain the figures shown. The values of x that we selected did not require the use of a calculator to find values of $f(x)$. We could have obtained additional values by using a calculator.

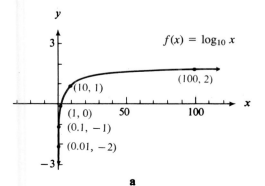

a

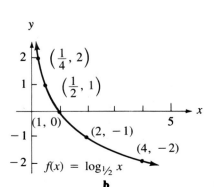

b   ■

EXERCISE 6.2

A

■ *Express in exponential notation. See Examples 1 and 2.*

1. $\log_2 64 = 6$ **2.** $\log_5 25 = 2$ **3.** $\log_3 9 = 2$ **4.** $\log_{16} 256 = 2$

5. $\log_{1/3} 9 = -2$ **6.** $\log_{1/2} 8 = -3$ **7.** $\log_{10} 1000 = 3$ **8.** $\log_{10} 1 = 0$

9. $\log_{10} 0.1 = -1$ **10.** $\log_{10} 0.01 = -2$ **11.** $\ln e^2 = 2$ **12.** $\ln e^{-3} = -3$

■ *Express in logarithmic notation. See Example 3.*

13. $4^2 = 16$ **14.** $5^3 = 125$ **15.** $3^3 = 27$ **16.** $8^2 = 64$

17. $\left(\dfrac{1}{2}\right)^2 = \dfrac{1}{4}$ **18.** $\left(\dfrac{1}{3}\right)^2 = \dfrac{1}{9}$ **19.** $8^{-1/3} = \dfrac{1}{2}$ **20.** $64^{-1/6} = \dfrac{1}{2}$

21. $10^2 = 100$ **22.** $10^0 = 1$ **23.** $10^{-1} = 0.1$ **24.** $10^{-2} = 0.01$

■ *Find the value of each expression. See Examples 4 and 5.*

25. $\log_7 49$ **26.** $\log_2 32$ **27.** $\log_4 64$ **28.** $\log_3 \dfrac{1}{3}$

29. $\log_5 \dfrac{1}{5}$ **30.** $\log_3 3$ **31.** $\log_2 2$ **32.** $\log_{10} 10$

33. $\log_{10} 100$ **34.** $\log_{10} 1$ **35.** $\log_{10} 0.1$ **36.** $\log_{10} 0.01$

■ *Solve for x, y, or b. See Example 6.*

37. $\log_3 9 = y$ **38.** $\log_5 125 = y$ **39.** $\log_b 8 = 3$

40. $\log_b 625 = 4$ **41.** $\log_4 x = 3$ **42.** $\log_{1/2} x = -5$

■ *Use a hand calculator to approximate each value to four decimal places. See Example 7.*

43. $\log_{10} 1.54$ **44.** $\log_{10} 2.68$ **45.** $\log_{10} 0.51$ **46.** $\log_{10} 0.66$

47. $\ln 3.67$ **48.** $\ln 2.00$ **49.** $\ln 0.50$ **50.** $\ln 0.75$

■ *Graph the function. See Example 8.*

51. $y = \log_2 x$ **52.** $y = \log_3 x$ **53.** $y = \log_2 4x$ **54.** $y = \log_3 9x$

55. $y = \log_2(x + 3)$ **56.** $y = \log_3(x + 4)$ **57.** $y = \log_{10} x^3$ **58.** $y = \log_{10} x^5$

B

■ *Evaluate each of the following functions at x = 0.001, 0.0001, 0.00001, and 0.000001.*

59. $f(x) = x \ln x$ **60.** $f(x) = x^{1/2} \ln x$ **61.** $f(x) = \dfrac{x}{\ln x}$ **62.** $f(x) = \dfrac{x}{(\ln x)^2}$

■ *Evaluate each of the following functions at $x = 10^3, 10^6, 10^9, 10^{12}$, and 10^{15}.*

63. $f(x) = \dfrac{\ln x}{x}$ **64.** $f(x) = \dfrac{\ln x}{x^{1/2}}$ **65.** $f(x) = \dfrac{\ln x}{x^{1/4}}$ **66.** $f(x) = \dfrac{\ln x}{x^{1/8}}$

67. Graph $y = \log_b x$ for $b = 2, 4, 8$ on the same set of axes. What effect does increasing b have on the curves if $b > 1$?

68. Graph $y = \log_b x$ for $b = \frac{1}{2}, \frac{1}{4}, \frac{1}{8}$ on the same set of axes. For $0 < b < 1$, where is $\log_b x < 0$? Where is $\log_b x > 0$?

■ *Find a positive integer n so that $n < x < n + 1$.*

69. $\log_{10} x = 0.75$ **70.** $\log_{10} x = 1.25$ **71.** $\ln x = 2.31$

72. $\ln x = 4.20$ **73.** $\log_{10}(\log_{10} x) = 0.125$ **74.** $\log_{10}(\log_{10} x) = 0.50$

75. $\ln(\ln x) = 0.95$ **76.** $\ln(\ln x) = 2.50$

77. Does the graph of $y = \ln x$ have a horizontal asymptote, that is, does y get close to a particular value as x grows very large? [*Hint:* Consider the graph of $y = \ln x$ as the graph of the inverse of $y = e^x$.] What can you conclude about the behavior of $y = \ln x$ as x gets very large?

78. Compute the value of $f(x) = \ln x$ at $x = 1, 10, 10^{20}, 10^{50}$, and 10^{90}. What can you say about the rate at which the natural logarithm increases with x? [*Hint:* Use the slope of the line joining $(1, \ln 1)$ and $(10^{90}, \ln 10^{90})$ as a measure of the average rate of growth.]

79. Does the graph of $y = \ln(\ln x)$ have a horizontal asymptote?

80. Compute the value of $f(x) = \ln(\ln x)$ at $x = 10, 10^5, 10^{10}$, and 10^{90}. What can you say about the rate at which the natural logarithm of the natural logarithm increases with x?

6.3

PROPERTIES OF LOGARITHMS; LOGARITHMIC EQUATIONS

In Section 6.2 we defined logarithmic functions and considered the graphs of such functions. In this section we consider several useful properties of logarithms.

Properties of Logarithms

Since a logarithm is an exponent, the following theorem follows directly from properties of powers with real exponents.

Theorem 6.1   *If x_1, x_2, and b are positive real numbers with $b \neq 1$ and m any real number, then*

$$\text{I.} \quad \log_b(x_1 x_2) = \log_b x_1 + \log_b x_2,$$

$$\text{II.} \quad \log_b \frac{x_2}{x_1} = \log_b x_2 - \log_b x_1,$$

$$\text{III.} \quad \log_b(x_1)^m = m \log_b x_1.$$

We justify Property I as follows: Since

$$x_1' = b^{\log_b x_1}$$

and

$$x_2 = b^{\log_b x_2},$$

it follows that

$$x_1 x_2 = b^{\log_b x_1} \cdot b^{\log_b x_2}$$
$$= b^{\log_b x_1 + \log_b x_2},$$

and, by the definition of a logarithm,

$$\log_b(x_1 x_2) = \log_b x_1 + \log_b x_2.$$

The validity of II and III can be established in a similar way.

EXAMPLE I   (a) $\log_{10} 100 = \log_{10}(10 \cdot 10)$
$\qquad\qquad\qquad = \log_{10} 10 + \log_{10} 10$
$\qquad\qquad\qquad = 1 + 1 = 2$

(b) $\log_2 \dfrac{1}{4} = \log_2 1 - \log_2 4$
$\qquad\qquad = 0 - 2 = -2$

(c) $\log_3 81 = \log_3 3^4$
$\qquad\qquad = 4 \log_3 3$
$\qquad\qquad = 4 \cdot 1 = 4$

(d) $\log_2 4^{-8} = -8 \log_2 4$
$\qquad\qquad = -8 \cdot 2$
$\qquad\qquad = -16$ ■

Theorem 6.1 can be used to write expressions involving more than one variable in equivalent forms.

EXAMPLE 2   Express $\log_{10}(xy^2/z)$ as the sum and difference of simpler logarithmic quantities.

Solution   By Theorem 6.1-II, we have

$$\log_{10}\left(\frac{xy^2}{z}\right) = \log_{10}(xy^2) - \log_{10} z.$$

By Theorem 6.1-I and III,

$$\log_{10}\left(\frac{xy^2}{z}\right) = \log_{10} x + 2 \log_{10} y - \log_{10} z.$$

EXAMPLE 3 Express $\ln\left[\dfrac{(x-1)^2(2x-1)^{-3}}{\sqrt{x+1}}\right]$ as the sum and difference of simpler logarithmic quantities.

Solution By Theorem 6.1-II, we have

$$\ln\left[\frac{(x-1)^2(2x-1)^{-3}}{\sqrt{x+1}}\right] = \ln[(x-1)^2(2x-1)^{-3}] - \ln\sqrt{x+1}.$$

By Theorem 6.1-I and III, the right-hand member of the preceding equation is equivalent to

$$2\ln(x-1) + (-3)\ln(2x-1) - \ln\sqrt{x+1}.$$

Rewriting $\sqrt{x+1}$ as $(x+1)^{1/2}$ and using Theorem 6.1-III, we have

$$\ln\left[\frac{(x-1)^2(2x-1)^{-3}}{\sqrt{x+1}}\right] = 2\ln(x-1) - 3\ln(2x-1) - \frac{1}{2}\ln(x+1). \qquad ■$$

The properties stated in Theorem 6.1 can also be used to compute unknown logarithms in terms of known values.

EXAMPLE 4 Given that $\log_b 2 = 0.6931$, $\log_b 3 = 1.0986$, and $\log_b 5 = 1.6094$, find the value of each of the following.

(a) $\log_b \dfrac{1}{9}$ (b) $\log_b \sqrt{2}$

Solution (a) $\log_b \dfrac{1}{9} = \log_b 3^{-2}$ (b) $\log_b \sqrt{2} = \log_b 2^{1/2}$

$\qquad\qquad = -2\log_b 3$ (by 6.1-III) $\qquad\qquad = \dfrac{1}{2}\log_b 2$ (by 6.1-III)

$\qquad\qquad = -2(1.0986)$ $\qquad\qquad\qquad = \dfrac{1}{2}(0.6931)$

$\qquad\qquad = -2.1972$ $\qquad\qquad\qquad = 0.3466$ ■

In Examples 1–4 we view the properties in Theorem 6.1 from left to right, that is, to express one logarithm as the sum or difference of several others. It is sometimes useful to view the properties from right to left, that is, to express a sum or difference of several logarithms as a single logarithm.

E X A M P L E 5   Express $\frac{1}{2}(\log_b x - \log_b y)$ as a single logarithm with coefficient 1.

Solution   We first use Theorem 6.1-II to write

$$\frac{1}{2}(\log_b x - \log_b y) = \frac{1}{2}\log_b\left(\frac{x}{y}\right).$$

We then use Theorem 6.1-III to obtain

$$\frac{1}{2}(\log_b x - \log_b y) = \log_b\left(\frac{x}{y}\right)^{1/2}. \qquad ■$$

Logarithmic Equations   An equation of the form

$$\log_b f(x) = a, \tag{1}$$

where a is a real number, can be solved by first writing the equation in exponential form.

E X A M P L E 6   Solve $\ln x^2 = 4$.

Solution   We first rewrite the equation in exponential form to obtain

$$x^2 = e^4,$$

from which

$$x = \pm e^2 = \pm 7.3891.$$

E X A M P L E 7   Solve $\ln(x^2 - 1) = -2$.

Solution   We rewrite the given equation in exponential form to obtain

$$x^2 - 1 = e^{-2}.$$

Solving this equation, we obtain

$$x = \pm\sqrt{1 + e^{-2}} = \pm 1.0655. \qquad ■$$

An equation that involves logarithms can sometimes be solved by using Theorem 6.1 to rewrite the equation equivalently in the form of Equation (1) and then writing the equation in exponential notation.

E X A M P L E 8   Solve $\log_{10}(x + 6) = 1 - \log_{10}(x - 3)$.

Solution   We first rewrite the given equation as

$$\log_{10}(x + 6) + \log_{10}(x - 3) = 1.$$

We then use Theorem 6.1-I to obtain

$$\log_{10}[(x + 6)(x - 3)] = 1.$$

Next, we rewrite this equation in exponential form to obtain

$$(x + 6)(x - 3) = 10^1.$$

Solving this equation for x, we obtain $x = -7$ and $x = 4$. Neither $\log_{10}(-7 + 6)$ nor $\log_{10}(-7 - 3)$ is defined. Thus, -7 does not satisfy the original equation. Since 4 satisfies the equation, the solution is $x = 4$.

EXAMPLE 9   Solve $\ln x - \ln(x + 1) = -2$.

Solution   We use Theorem 6.1-II to rewrite the equation as

$$\ln\left(\frac{x}{x + 1}\right) = -2.$$

Next, we rewrite this equation in exponential form to obtain

$$\frac{x}{x + 1} = e^{-2}.$$

Solving this equation for x, we obtain

$$x = e^{-2}x + e^{-2}$$
$$x - e^{-2}x = e^{-2}$$
$$x(1 - e^{-2}) = e^{-2}$$

$$x = \frac{e^{-2}}{1 - e^{-2}} = 0.1565.   ■$$

EXERCISE 6.3

A

■ *Express as the sum or difference of simpler logarithmic quantities. See Examples 1–3.*

1. $\log_b(xy)$ **2.** $\log_b(xyz)$ **3.** $\log_b\left(\dfrac{x}{y}\right)$ **4.** $\log_b\left(\dfrac{xy}{z}\right)$

5. $\log_b x^5$ **6.** $\log_b x^{1/2}$ **7.** $\log_b \sqrt[3]{x}$ **8.** $\log_b \sqrt[3]{x^2}$

9. $\log_b \dfrac{\sqrt[3]{xy^2}}{z^3}$ **10.** $\log_b \dfrac{\sqrt{x^3 y}}{2z}$ **11.** $\log_b \sqrt{\dfrac{x^3 y}{3z}}$ **12.** $\log_b \sqrt[3]{\dfrac{x^2 yz^2}{3}}$

■ *Given that* $\log_b 2 = 0.6931$, $\log_b 3 = 1.0986$, *and* $\log_b 5 = 1.6094$, *find the value of each expression. See Example 4.*

13. $\log_b 6$ **14.** $\log_b 10$ **15.** $\log_b \dfrac{2}{5}$ **16.** $\log_b \dfrac{3}{2}$

17. $\log_b 9$ **18.** $\log_b 25$ **19.** $\log_b \dfrac{\sqrt{5}}{2}$ **20.** $\log_b \dfrac{1}{\sqrt[3]{5}}$

21. $\log_b \dfrac{10}{\sqrt{6}}$ **22.** $\log_b \dfrac{\sqrt{10}}{50}$ **23.** $\log_b 0.06$ **24.** $\log_b \sqrt{0.10}$

■ *Express as a single logarithm with coefficient* 1. *See Example 5.*

25. $\log_b 2x + 3 \log_b y$ **26.** $3 \log_b x - \log_b 2y$

27. $\dfrac{1}{2} \log_b x + \dfrac{2}{3} \log_b y$ **28.** $\dfrac{1}{4} \log_b x - \dfrac{3}{4} \log_b y$

29. $3 \log_b x + \log_b y - 2 \log_b z$ **30.** $\dfrac{1}{3} (\log_b x + \log_b y - 2 \log_b z)$

31. $4 \left(\dfrac{3}{2} \log_b x + \dfrac{1}{3} \log_b y^2 \right)$ **32.** $\dfrac{4}{3} \left(\dfrac{1}{3} \log_b 3x + \dfrac{3}{2} \log_b 2y \right)$

33. $\dfrac{1}{2} (\log_b x + \log_b y) + \dfrac{1}{3} \log_b 2x$ **34.** $\dfrac{2}{3} (\log_b 2x - \log_b y) - \dfrac{1}{3} \log_b x$

35. $\sqrt{2} (\log_b x - \log_b y) + \sqrt{3} (\log_b 2y - \log_b 2x)$ **36.** $\sqrt{5} (\log_b 3x + \log_b y) - \sqrt{2} (\log_b 2y + \log_b x)$

■ *Solve. See Examples 6 and 7.*

37. $\log_2 3x = 4$ **38.** $\log_2 (-2x) = -1$ **39.** $\log_2 (2x^2) = 1$

40. $\log_2 (3x^3) = -3$ **41.** $\log_{10} 4x = 2$ **42.** $\log_{10} (-3x) = -1$

43. $\log_{10} \dfrac{1}{2} x^2 = -2$ **44.** $\log_{10} x^3 = 2$ **45.** $\ln 3x = 2$

46. $\ln 2x = 8$ **47.** $\ln 2x^2 = -1$ **48.** $\ln 3x^2 = 6$

■ *Solve. See Examples 8 and 9.*

49. $\log_{10} x + \log_{10} 6 = 5$ **50.** $\log_{10} (x + 3) - \log_{10} 3 = 4$

51. $\log_{10} x + \log_{10} 2 = 3$ **52.** $\log_{10} (x - 1) - \log_{10} 4 = 2$

53. $\log_{10} x + \log_{10} (x + 21) = 2$ **54.** $\log_{10} (x + 3) + \log_{10} x = 1$

55. $\log_{10} (x + 2) + \log_{10} (x - 1) = 1$ **56.** $\log_{10} (x - 3) - \log_{10} (x + 1) = 1$

57. $\log_{10} x = 1 + \log_{10} \sqrt{x}$ **58.** $\log_{10} x^2 = 1 + \log_{10} \sqrt[3]{x}$

B

■ *Solve.*

59. $\log_{10}(\ln x) = 1.05$

60. $\log_{10}(\log_{10} x) = 0.75$

61. $\ln(\ln x) = 2$

62. $\ln(\log_{10} x) = -1$

■ *Express as the sum or difference of simpler logarithmic quantities.*

63. $\ln \dfrac{(x-1)^2(x^2+2)^{1/2}}{x^2-3}$

64. $\ln \dfrac{\sqrt{x^2+1}\,(2x-1)}{x^2(x+1)}$

65. $\ln \dfrac{\sqrt{x^4-1}}{2x}$

66. $\ln \dfrac{x^2-1}{\sqrt{x^3-1}}$

67. $\ln \sqrt{\dfrac{x^3-1}{x^2+1}}$

68. $\ln \sqrt{\dfrac{x^2-3x+2}{x^2-4}}$

69. $\ln \sqrt{\dfrac{x}{\sqrt{x+1}}}$

70. $\ln \sqrt{\dfrac{\sqrt{x^2+6}}{x-3}}$

6.4

SOLVING EXPONENTIAL EQUATIONS

When solving an exponential equation, $y = b^{f(x)}$, we are interested in finding the value for y, given a value of x, or in finding the value(s) for x, given a value of y.

When using a calculator to find unknown values in $y = b^{f(x)}$, we can either round off each approximation and use that value in further calculations or carry all approximations internally on the calculator. These two procedures sometimes yield slightly different values. Since the latter is generally more accurate, we shall use it for all examples and exercises.

Finding Powers

As noted in Section 6.1, a scientific calculator can be used to find y or $b^{f(x)}$ for a given value of x.

EXAMPLE 1   Find each power.

(a) $y = (2.51)^{(0.2)^2}$

(b) $y = (3.21)^{-1/0.2}$

Solution   (a) $(0.2)^2 = 0.04$;

$y = (2.51)^{0.04} = 1.0375$

(b) $-\dfrac{1}{0.2} = -5$;

$y = (3.21)^{-5} = 0.0029$

EXAMPLE 2   Solve $y = k(2^{-5t})$; for y given that $k = 0.2$, $t = 0.5$.

Solution   We substitute 0.2 for k and 0.5 for t to obtain

$$y = 0.2[2^{-5(0.5)}].$$
$$= 0.2[2^{-2.5}] = 0.0354. \qquad ■$$

Finding Exponents   For a given value of y, in order to solve an exponential equation $y = b^{f(x)}$ for x, it is necessary first to rewrite the equation so that the variable no longer appears in the exponent. In the cases when the base is either 10 or e, we simply rewrite the equation using logarithmic notation with the appropriate base. The following examples illustrate the procedure.

EXAMPLE 3   Solve $10^{2x} = 6$.

Solution   Rewriting the equation in logarithmic notation with base 10, we have

$$2x = \log_{10} 6 = 0.7782,$$

from which

$$x = \frac{\log_{10} 6}{2} = 0.3891.$$

EXAMPLE 4   Solve $96.3 = 2.4(10^{2-3.1t})$.

Solution   We first divide both members by 2.4 to obtain

$$\frac{96.3}{2.4} = 10^{2-3.1t}.$$

Rewriting this equation in logarithmic notation with base 10, we have

$$2 - 3.1t = \log_{10}\left(\frac{96.3}{2.4}\right),$$

from which

$$t = \frac{-2 + \log_{10}(96.3/2.4)}{-3.1} = 0.1279.$$

EXAMPLE 5   Solve $e^{-4x} = 2$.

Solution   Rewriting the equation in logarithmic notation with base e, we obtain

$$-4x = \ln 2 = 0.6931.$$

Thus,

$$x = \frac{\ln 2}{-4} = -0.1733.$$

EXAMPLE 6   Solve $12.4 = 1.3e^{1-0.04t}$.

Solution   We first divide both members by 1.3 to obtain

$$\frac{12.4}{1.3} = e^{1-0.04t}.$$

Rewriting this equation in logarithmic notation with base e, we have

$$1 - 0.04t = \ln\frac{12.4}{1.3},$$

from which

$$t = \frac{-1 + \ln(12.4/1.3)}{-0.04} = -31.38. \qquad ■$$

Most scientific calculators only have log keys for the base 10 or e. Hence, when the base is not 10 or e we cannot apply the method used in the foregoing examples. However, we can use Theorem 6.1-III and the fact that

$$a = b \qquad \text{implies} \qquad \log_{10} a = \log_{10} b$$

(because $y = \log_{10} x$ is a function) to rewrite the equation so that the variable does not appear in the exponent.

EXAMPLE 7   Find the solution of $5^x = 7$.

Solution   Taking the logarithm to base 10 of each member, we have

$$\log_{10} 5^x = \log_{10} 7;$$

and from Theorem 6.1-III,

$$x \log_{10} 5 = \log_{10} 7.$$

Dividing each member by $\log_{10} 5$, we obtain

$$x = \frac{\log_{10} 7}{\log_{10} 5} = 1.2091.$$

EXAMPLE 8   Find the solution of $3^{x-2} = 12$.

Solution   Taking the logarithm to base 10 of each member, we have

$$\log_{10} 3^{x-2} = \log_{10} 12.$$

From Theorem 6.1-III, we obtain

$$(x - 2) \log_{10} 3 = \log_{10} 12.$$

Solving for x yields

$$x = \frac{\log_{10} 12}{\log_{10} 3} + 2$$

$$= 4.2619. \qquad ■$$

The technique used in the preceding examples can be used to solve an exponential equation that involves any positive base b.

Literal Equations   Exponential equations involving more than one variable can be solved for any one of the variables in terms of the others.

EXAMPLE 9   Solve $y = A(10^{-x+t})$ for t.

Solution   We first rewrite the equation as

$$\frac{y}{A} = 10^{-x+t}.$$

We then rewrite this equation in logarithmic notation with base 10 to obtain

$$\log_{10} \frac{y}{A} = -x + t.$$

Solving for t, we have

$$t = x + \log_{10} \frac{y}{A}.$$

EXAMPLE 10   Solve $y = Ce^{kt}$ for k.

Solution   We first rewrite the equation as

$$\frac{y}{C} = e^{kt}.$$

We then rewrite this equation in logarithmic notation with base e to obtain

$$\ln\left(\frac{y}{C}\right) = kt.$$

Solving for k, we have

$$k = \frac{1}{t}\ln\left(\frac{y}{C}\right). \qquad ■$$

As noted in Examples 7 and 8 on pages 271 and 272, if the base of the power is not 10 or e, we can take the $\log_{10}$ of both sides and apply Theorem 6.1-III.

EXAMPLE 11   Solve $y = (1 + r)^n$ for n in terms of y and r, using logarithms to base 10.

Solution   We have

$$\log_{10} y = \log_{10}(1 + r)^n;$$

and from Theorem 6.1-III,

$$\log_{10} y = n \log_{10}(1 + r).$$

Thus,

$$n = \frac{\log_{10} y}{\log_{10}(1 + r)}. \qquad ■$$

EXERCISE 6.4

A

■ *Solve for the indicated variable. See Examples 1 and 2.*

1. $y = k10^t$;   for y given $k = 5, t = 0.2455$

2. $y = k10^t$;   for y given $k = 5, t = 2.2455$

3. $y = ke^{2t}$;   for y given $k = 0.01, t = 5$

4. $y = ke^{-4t}$;   for y given $k = 10, t = 0.5$

5. $A = (1 + r)^n$;   for A given $r = 0.01, n = 8$

6. $A = (1 + r)^n$;   for A given $r = 0.0025, n = 20$

7. $V = (1 - r)^n$;   for V given $r = 0.07, n = 5$

8. $V = (1 - r)^n$;   for V given $r = 0.05, n = 5$

■ *Solve. Give the solution in logarithmic form using base* 10 *and also as an approximation in decimal form. See Examples 3 and 4.*

9. $10^{3x} = 6$ **10.** $10^{-2x} = 15$ **11.** $10^{3x+1} = 9$

12. $10^{-2x+3} = 20$ **13.** $10^{x^2} = 150$ **14.** $10^{2x^2+1} = 200$

15. $10^{\sqrt{x}} = 25$ **16.** $10^{-\sqrt{x}} = 0.50$ **17.** $10^{\sqrt{x+1}} = 2$

18. $10^{-\sqrt{x+3}} = 0.5$ **19.** $1.4(10^{1-0.1t}) = 2.3$ **20.** $2.6(10^{3-0.1t}) = 2.8$

■ *Solve. Give the solution in logarithmic form with base e and also as an approximation in decimal form. See Examples 5 and 6.*

21. $e^{3x} = 5$ **22.** $e^{-2x} = 10$ **23.** $e^{2x+1} = 25$

24. $e^{-x+2} = 8$ **25.** $e^{x^2/2} = 15$ **26.** $e^{x^2+1} = 4$

27. $e^{\sqrt{x}} = 20$ **28.** $e^{\sqrt{x+1}} = 100$ **29.** $e^{\sqrt{x+2}} = 2e$

30. $e^{-1} = 4e^{-\sqrt{x+2}}$ **31.** $1.4e^{1+0.04t} = 2.5$ **32.** $1.8e^{1+0.2t} = 6.5$

■ *Solve. Give the solution in logarithmic form with base* 10 *and also as an approximation in decimal form. See Examples 7 and 8.*

33. $2^x = 7$ **34.** $3^x = 4$ **35.** $7^{2x-1} = 3$ **36.** $3^{x+2} = 10$

37. $4^{x^2} = 16$ **38.** $8^{x^2+1} = 64$ **39.** $8 = 4^{\sqrt{x}}$ **40.** $2.5 = \left(\dfrac{1}{8}\right)^{-x^2}$

41. $6^{x^2-2} = 14$ **42.** $3^{2x^2-1} = 12$ **43.** $5^{x^2-3} = 0.45$ **44.** $7^{x^2-5} = 0.25$

■ *Solve for the indicated variable. Leave the result in the form of an equation equivalent to the given equation. Use logarithms with the indicated base b. See Examples 9–11.*

45. $A = 10^{kt}$;   for t, $b = 10$ **46.** $A = B10^{-kt}$;   for t, $b = 10$

47. $y = e^{kt}$;   for t, $b = e$ **48.** $y = Ce^{-kt}$;   for t, $b = e$

49. $y = x^n$;   for n, $b = 10$ **50.** $y = Cx^{-n}$;   for n, $b = 10$

51. $y = -2 + 3e^{-4t}$;   for t, $b = e$ **52.** $y = 3 + 2e^{2t}$;   for t, $b = e$

53. $2y + 4 = 3 - 2e^{-2t+1}$;   for t, $b = e$ **54.** $3y - 6 = 2 + 6e^{-3t+4}$;   for t, $b = e$

55. $V = P(1 + r)^n$;   for n, $b = 10$ **56.** $V = P(1 - r)^n$;   for n, $b = 10$

B

■ *Evaluate each of the following functions at* $x = 0.1, 0.01, 0.001,$ *and* $0.0001.$ *Show six-decimal-place accuracy.*

57. $f(x) = 1 - x^x$ **58.** $f(x) = (x + 1)^x - 1$

59. $f(x) = (1 + x)^{1/x} - e$ **60.** $f(x) = (1 + 2x)^{2/x}$

■ *Solve. Leave the result in the form of an equation equivalent to the given equation. Use common logarithms.*

61. $V = P\left(1 + \dfrac{r}{100n}\right)^{nk}$;   for k **62.** $p = S\left[\dfrac{r/12}{1 - \left(1 + \dfrac{r}{12}\right)^{-n}}\right]$;   for n

63. $P = S\left(1 + \dfrac{r}{n}\right)^{-nk}$; for k

64. $p = S\left[\dfrac{r/n}{\left(1 + \dfrac{r}{n}\right)^{nk} - 1}\right]$; for k

6.5

APPLICATIONS

In this section we shall apply the principles discussed in the preceding sections to problems that arise from the physical, biological, and social sciences. First we shall consider how the number e, which occurs in many models for applied problems, arises.

Definition of e
Let us consider a quantity Q, which is growing at the average rate of 100% per year. If we start out with 1 unit of this quantity, and if we assume that this growth is compounded n times yearly, then after one compounding period we have the original 1 unit plus $1/n$th part of a full year's growth, or

$$Q_1 = 1 + \frac{1}{n}$$

units. After two compounding periods we have Q_1 units plus the amount that Q_1 grows over $1/n$th of a year, or

$$Q_2 = Q_1 + \frac{1}{n}Q_1$$
$$= \left(1 + \frac{1}{n}\right)Q_1 = \left(1 + \frac{1}{n}\right)^2$$

units. Similarly, after three compounding periods we have

$$Q_3 = Q_2 + \frac{1}{n}Q_2$$
$$= \left(1 + \frac{1}{n}\right)Q_2 = \left(1 + \frac{1}{n}\right)^3$$

units. In general, after k compounding periods we have

$$Q_k = \left(1 + \frac{1}{n}\right)^k$$

units. Thus, after the n compounding periods in 1 year, we have precisely

$$Q_n = \left(1 + \frac{1}{n}\right)^n$$

units. If the number of compounding periods per year is very large, then

$$Q_n \approx 2.71828 \approx e,$$

as Table 6.2 suggests.

TABLE 6.2

n	$\left(1 + \dfrac{1}{n}\right)^n$	e (approx.)
10	2.59374	2.71828
100	2.70481	2.71828
1000	2.71692	2.71828
10,000	2.71815	2.71828
100,000	2.71827	2.71828

Thus, for very large values of n, we have

$$e \approx \left(1 + \frac{1}{n}\right)^n.$$

Growth and Decay Models

In attempting to mathematically model the growth or decay of a quantity or a population, assumptions must be made about the population and its environment. One such assumption is that the **relative growth rate**—the rate of change of the population size [denoted by $r(Q)$] divided by the population size (denoted by Q)—depends *only* on the size of the population.

These models are called **density-dependent** models and are characterized by the fact that

$$\frac{r(Q)}{Q} = f(Q) \qquad (2)$$

for some function f. We shall consider two types of density-dependent models.

Exponential (or Malthusian) Growth Model

For this model we shall assume that the function f in Equation (2) is a constant. In this case the relative rate of change is given by

$$\frac{r(Q)}{Q} = \alpha, \qquad \alpha \text{ a constant.}$$

Hence, the rate of change of Q is proportional to Q. With this assumption and using methods from the calculus, it can be shown that the amount of the quantity present at any time t, denoted by $Q(t)$, is

$$Q(t) = Q_0\, e^{\alpha t}, \qquad t \geq 0, \tag{3}$$

where Q_0 is the amount present initially, t is the time measured in the appropriate unit, and α is the constant of proportionality.

If the quantity is increasing, then $\alpha > 0$ and the model given by Equation (3) is called an **exponential growth model**. This model is also called the **Malthusian growth model** after the British economist Thomas Malthus who published the model in his *Essay on the Principle of Population* in 1798. If the quantity is decreasing, then $\alpha < 0$ and the model given by Equation (3) is called an **exponential decay model**.

EXAMPLE 1   The population of a city in the sun belt is growing according to the formula

$$Q(t) = 100{,}000e^{0.14t},$$

where t is measured in years starting from January 1, 1988. In what year will the population reach 150,000?

Solution   We are seeking the value of t so that $Q(t) = 150{,}000$. Thus, we must solve the equation

$$150{,}000 = 100{,}000e^{0.14t}$$

or, equivalently,

$$\frac{3}{2} = e^{0.14t}.$$

Taking the natural logarithm of both members yields

$$\ln\left(\frac{3}{2}\right) = 0.14t,$$

from which we have

$$t = \frac{1}{0.14}\ln\left(\frac{3}{2}\right) = 2.90.$$

Thus, it will take approximately 2.90 years for the population to reach 150,000. This will occur late in 1990.   ■

We next consider an example involving exponential decay.

EXAMPLE 2 A sample of a radioactive element is initially 25 grams. If the amount y (measured in grams) remaining at any time t is given by $y = 25e^{-0.5t}$, where t is in seconds, how much of the element (to the nearest tenth of a gram) is remaining after 3 seconds?

Solution Substituting 3 for t, we have

$$y = 25e^{-0.5(3)}$$
$$= 25e^{-1.5} = 5.58.$$

Thus, there are approximately 5.6 grams of radioactive material remaining in the sample.   ■

Logistic Growth Model The second density-dependent growth model we shall consider is called the **logistic growth model**. It is characterized by the assumption that the relative rate at which the population size Q is changing is nonnegative and linear with negative slope, that is,

$$\frac{r(Q)}{Q} = -\frac{A}{Q_M}(Q - Q_M),$$

where A and Q_M are positive constants. With this assumption and using methods from the calculus, it can be shown that the population size at time t, denoted by $Q(t)$, is

$$Q(t) = \frac{Q_M}{1 + \dfrac{Q_M - Q_0}{Q_0}e^{-At}}, \qquad t \geq 0, \tag{4}$$

where Q_0 is the initial population size and Q_M is the maximum population size.

EXAMPLE 3 A herd of antelope is growing according to the formula

$$Q(t) = \frac{750}{1 + \frac{1}{2}e^{-0.3t}},$$

where t is measured in years starting from 1985. What will the size of the herd be, to the nearest 10 antelopes, in 1990?

Solution   We want to know the size of the herd when $t = 5$. Thus, we substitute 5 for t and obtain

$$Q(5) = \frac{750}{1 + \frac{1}{2}e^{(-0.3)(5)}}$$

$$= \frac{750}{1 + \frac{1}{2}e^{-1.5}}$$

$$= 674.7$$

Hence, to the nearest 10, there will be 670 antelopes in the herd in 1990.   ■

Gompertz Growth Model

We shall now consider a model of population growth that is not density-dependent: the **Gompertz growth model**. In this model we assume that the relative rate of change of Q decays exponentially with time, that is,

$$\frac{r(Q)}{Q} = ke^{-\alpha t},$$

where k and α are positive constants. With this assumption and using methods from the calculus, it can be shown that

$$Q(t) = Q_M e^{-\frac{k}{\alpha}e^{-\alpha t}}, \qquad t \geq 0,$$

where Q_M is the maximum population size.

EXAMPLE 4   A population grows in size according to the formula

$$Q(t) = 10{,}000{,}000e^{-10e^{-0.1t}},$$

where t is time measured in years. What is the initial population size?

Solution   We are seeking the value $Q(0)$. Thus, we substitute 0 for t to obtain

$$Q(0) = 10{,}000{,}000e^{-10e^{(-0.1)(0)}}$$

$$= 10{,}000{,}000e^{-10}$$

$$= 454.0.$$

Thus, the initial population size is 454.   ■

Exponential functions with bases other than e are sometimes used as mathematical models of quantities such as compound interest, annuities, or population size. For example, if the annual interest rate on a savings account is $r\%$ and the interest is compounded n times annually, then the value V of the account after k years is

$$V = P\left(1 + \frac{r}{100n}\right)^{nk}, \qquad (4)$$

where P is the amount initially deposited.

EXAMPLE 5   One dollar compounded annually for 12 years yields $1.90. What is the rate of interest paid to the nearest $\frac{1}{2}\%$?

Solution   From Equation (4) we are seeking the value of r so that

$$\left(1 + \frac{r}{100}\right)^{12} = 1.90.$$

Equating $\log_{10}$ of each member of this equation and applying Theorem 6.1-III, we have

$$12 \log_{10}\left(1 + \frac{r}{100}\right) = \log_{10} 1.90$$

or, equivalently,

$$\log_{10}\left(1 + \frac{r}{100}\right) = \frac{1}{12}(\log_{10} 1.90).$$

Rewriting this equation in exponential notation, we obtain

$$1 + \frac{r}{100} = 10^{(1/12)(\log_{10} 1.90)}$$

$$= 1.0549.$$

Solving for r, we have

$$r = 5.49.$$

Thus, to the nearest $\frac{1}{2}\%$ the rate of interest paid is $5\frac{1}{2}\%$.   ■

Logarithmic Models   The relationship between some quantities can be modeled using logarithmic functions.

EXAMPLE 6   The chemist defines the pH (hydrogen potential) of a solution by

$$pH = -\log_{10}[H^+],$$

where $[H^+]$ is a numerical value for the concentration of hydrogen ions in aqueous solution in moles per liter. Calculate the hydrogen ion concentration of a solution with pH = 5.4.

Solution   We first substitute 5.4 for pH in the given formula to obtain

$$5.4 = -\log_{10}[H^+].$$

Hence,

$$[H^+] = 10^{-5.4} = 3.98 \times 10^{-6}.$$

EXAMPLE 7   The work W done on a gas piston system in an isothermal expansion is given by the formula

$$W = -nRT \ln\left(\frac{V_f}{V_0}\right),$$

where n is the number of moles of the gas, T is the temperature in degrees Kelvin, V_f is the final volume, V_0 is the initial volume, and $R = 8.314$ joules per (mole)(degree K). How much work (to the nearest tenth of a joule) is done on a system containing 1 mole of gas at 0°C (273 K) if the volume is halved?

Solution   We first note that since the volume is halved we have

$$V_f = \frac{1}{2}V_0$$

or, equivalently,

$$\frac{V_f}{V_0} = \frac{1}{2}.$$

Then substituting 1 for n, 8.314 for R, 273 for T, and $\frac{1}{2}$ for V_f/V_0 we obtain

$$W = -1(8.314)(273) \ln\left(\frac{1}{2}\right)$$

$$= 1573.25.$$

Thus, the work done on the system is 1573.25 joules.   ■

Logarithmic Scales

Scientists and engineers often use logarithmic scales to measure phenomena that vary over a wide range of values.

One example of a logarithmic scale is the Richter scale, proposed by C. F. Richter in 1935 to measure earthquake magnitude. The Richter scale is defined by

$$M = \log_{10}\left(\frac{A}{A_0}\right), \tag{5}$$

where A is the seismographic trace amplitude measured in microns and A_0 is the amplitude generated by the smallest quake detectable by good seismographs.

EXAMPLE 8   The 1971 Los Angeles earthquake measured 6.7 on the Richter scale. In 1986 an earthquake in Los Angeles measured 3.5 on the Richter scale. How much stronger was the 1971 quake than the 1986 quake?

Solution   We denote the amplitude of the 1971 quake by A_{1971} and of the 1986 quake by A_{1986}. Then from Equation (5) we have

$$6.7 = \log_{10}\left(\frac{A_{1971}}{A_0}\right) \quad \text{and} \quad 3.5 = \log_{10}\left(\frac{A_{1986}}{A_0}\right).$$

Rewriting these equations in exponential form, we obtain

$$\frac{A_{1971}}{A_0} = 10^{6.7} \quad \text{and} \quad \frac{A_{1986}}{A_0} = 10^{3.5}.$$

Dividing corresponding members of these equations yields

$$\frac{A_{1971}}{A_{1986}} = 10^{6.7-3.5}$$

$$= 10^{3.2} = 1584.89$$

or, equivalently,

$$A_{1971} = 1584.89 A_{1986}.$$

Thus, the 1971 quake was approximately 1585 times as strong as the 1986 quake.   ■

Intensity levels of sound are measured in decibels (db), which is another logarithmic scale. Intensity level is denoted by

$$\beta = 10 \log_{10}\left(\frac{I}{I_0}\right), \tag{6}$$

where I is the intensity of the sound wave measured in watts/m^2 and I_0 is the threshold of audibility ($I_0 \approx 10^{-12}$ watts/m^2 at a frequency of 1000 cycles/sec).

EXAMPLE 9   A whisper has an intensity level of about 20 db, and the threshold of pain for the human ear is about 90 db. How much more intense is a painful noise than a whisper?

Solution   We denote the intensity of the whisper by I_w and that of the painful noise by I_P. Then from Equation (6) we have

$$20 = 10 \log_{10}\left(\frac{I_w}{I_0}\right) \quad \text{and} \quad 90 = 10 \log_{10}\left(\frac{I_P}{I_0}\right).$$

Dividing both members of each equation by 10 and rewriting in exponential form, we have

$$\frac{I_w}{I_0} = 10^2 \quad \text{and} \quad \frac{I_P}{I_0} = 10^9.$$

Dividing corresponding members of these equations yields

$$\frac{I_w}{I_P} = 10^{2-9}$$

$$= 10^{-7}$$

or, equivalently,

$$I_P = 10^7 I_w.$$

Thus, the painful noise is 10^7 (ten million) times more intense than the whisper.   ■

Other applications of exponential and logarithmic functions are given in the exercises.

EXERCISE 6.5

A

■   *For Exercises 1–8, see Example 1.*

The number of bacteria present in a culture is given by the formula

$$N = N_0 e^{\alpha t},$$

where N_0 is the number of bacteria present at $t = 0$, t is time in hours, and α is a constant.

1. If a certain culture has an initial size of 10,000 bacteria and $\alpha = 0.01$, how many bacteria (to the nearest hundred) will be present 48 hours after the beginning of an experiment?

2. If a certain culture has an initial size of 1000 bacteria and $\alpha = 0.04$, how many bacteria (to the nearest hundred) will be present 2 hours after the beginning of an experiment?

3. If a certain culture had 1000 bacteria present initially and 10,000 are present after 4 hours, what is the value of α (to the nearest hundredth) for that culture?

4. If a certain culture had 2000 bacteria present initially and 10,000,000 are present after 10 hours, what is the value of α (to the nearest hundredth) for that culture?

■ *The population of a city is growing according to the formula*

$$Q(t) = Q_0 e^{\alpha t},$$

where Q_0 is the initial population and t is time in years measured from January 1, 1980.

5. If the initial population is 500 and $\alpha = 0.1$, in what year will the population reach 5000?

6. If the initial population is 5000 and $\alpha = 0.1$, in what year will the population reach 25,000?

7. If the initial population is 500 and after 25 years it is 50,000, what is the relative growth rate α (to the nearest hundredth)?

8. If the initial population is 1000 and after 10 years it is 1200, what is the relative growth rate α (to the nearest hundredth)?

■ *For Exercises 9–16, see Example 2.*

The amount of a radioactive element in a given sample remaining at any time t is given by

$$y = y_0 e^{\alpha t},$$

where y is measured in grams, t in seconds, and α is a constant.

9. How much of the material is present (to the nearest tenth of a gram) after 600 seconds if there was 1 kilogram present initially ($y_0 = 1000$) and $\alpha = -0.001$?

10. How much of the material is present (to the nearest tenth of a gram) after 1200 seconds under the same conditions as Exercise 9?

11. If $y_0 = 50$ and $\alpha = -0.05$, how long (to the nearest hundredth of a second) does it take for the amount of the radioactive element to be reduced to 25 grams?

12. If $y_0 = 50$ and it takes 3.15×10^9 seconds (approximately 100 years) for the sample to be reduced to 25 grams, find the value of α. Show the answer in scientific notation.

■ *Newton's law of cooling states that the rate of change of the temperature difference between a body and its surrounding medium is proportional to that difference. Equation (3) then gives the formula*

$$D(t) = D_0 e^{-\alpha t},$$

which relates the temperature difference $D(t)$ at time t to the initial difference D_0 and the proportionality constant $\alpha > 0$.

13. Suppose an object at 100°C is placed in a surrounding medium of 0°C. What is the temperature difference (to the nearest tenth of a degree) between the object and the surrounding medium 10 seconds later if the proportionality constant for this object and medium is 0.01?

14. Using the information in Exercise 13, determine the temperature difference (to the nearest tenth of a degree) after 600 seconds.

15. Using the information in Exercise 13, determine the amount of time (to the nearest tenth of a second) required for the temperature difference to drop to 0.00001°C.

16. Suppose an object at 100°C is placed in a surrounding medium of 0°C and 5 seconds later the temperature difference between the surrounding medium is measured to be 0.5°C. Determine the constant of proportionality (to the nearest tenth) for this object and this medium.

■ *For Exercises 17–20, see Example 3. A herd of elk grows in size according to the formula*

$$Q(t) = \frac{Q_M}{1 + Ke^{-At}},$$

where Q_M is the maximum size of the herd, $K = \dfrac{Q_M - Q_0}{Q_0}$ with Q_0 being the initial size of the herd, and t is time measured in years.

17. If $Q_M = 500$, $Q_0 = 300$, and $A = 0.05$, what will be the size (to the nearest 10) of the herd in 15 years?

18. If $Q_M = 500$, $Q_0 = 300$, and $A = 0.5$, what will be the size (to the nearest 10) of the herd in 15 years?

19. If the environment will support a herd of size 1000 (that is, $Q_M = 1000$) and we place 300 elk in the herd initially, what is the value of A (to the nearest thousandth) if we have 600 elk after 20 years?

20. If the environment will support a herd of size 1000, we initially place 600 elk in the herd, and we determine the value of A to be 0.063, how long (to the nearest year) will it take the herd to reach size 900?

■ *For Exercises 21–26, see Example 4. A population grows according to the formula*

$$Q(t) = Q_M e^{-\frac{k}{\alpha}e^{-\alpha t}},$$

where Q_M is the maximum population and k and α are positive constants and t is time measured in years.

21. If the maximum population is 100,000, $k = 2$, and $\alpha = 0.5$, what is the initial population size to the nearest unit?

22. If the maximum population is 100,000, $k = 2$, and $\alpha = 0.9$, what is the initial population size to the nearest unit?

23. Suppose the initial population size is 5000, $k = 1$, and $\alpha = 0.35$. What is the maximum population size to the nearest unit?

24. Suppose the initial population size is 500, $k = 1$, and $\alpha = 0.15$. What is the maximum population size to the nearest unit?

25. Suppose the maximum population size is 10,000 with $k = 1$ and $\alpha = 0.25$. How long (to the nearest year) will it take for the population to reach 7500?

26. Suppose the maximum population size is 10,000 with $k = 1$ and $\alpha = 0.25$. How long (to the nearest year) will it take for the population to reach 9900?

■ *For Exercises 27–32, see Equation (4) and Example 5.*

27. One dollar compounded annually for 10 years yields $1.95. What is the rate of interest to the nearest $\frac{1}{2}\%$?

28. How many years (to the nearest year) would it take for $1.00 to yield $2.19 if compounded annually at 9%?

29. Find the compounded amount of $5000 invested at $9\frac{1}{2}\%$ for 10 years when compounded annually; when compounded semiannually.

30. Two men, A and B, each invested $10,000 at $9\frac{1}{2}\%$ for 20 years with a bank that computed interest quarterly. A withdrew his interest at the end of each 3-month period, but B let his investment be compounded. How much more than A did B earn over the period of 20 years?

31. Institution A pays 9% compounded quarterly and Institution B pays 7.75% compounded daily. Which institution gives the best return on an investment of P dollars and by how much over a 1-year period?

32. Institution A pays 7% compounded quarterly and Institution B pays 6.5% compounded daily. Which institution gives the best return on an investment of P dollars and by how much over a 1-year period?

■ *For Exercises 33–36, see Example 6.*

33. Calculate the pH of a solution whose hydrogen ion concentration $[H^+]$ is 2.0×10^{-8}.

34. Calculate the pH of a solution whose hydrogen ion concentration is 6.3×10^{-7}.

35. Calculate the hydrogen ion concentration of a solution whose pH is 5.6.

36. Calculate the hydrogen ion concentration of a solution whose pH is 7.2.

■ *For Exercises 37–40, see Example 7.*

37. How much work (to the nearest tenth of a joule) is done on a system containing 1 mole of gas at $0°C$ (273 K) if the volume doubles?

38. How much work (to the nearest tenth of a joule) is done on a system containing 2 moles of gas at $100°C$ (373 K) if the volume is reduced by a factor of $\frac{1}{2}$?

39. A system containing 1 mole of gas at $0°C$ has -3413.8 joules of work performed on it. Find the ratio of the final volume to the initial volume (to the nearest hundredth).

40. A system containing 1 mole of gas at $0°C$ has 2493.5 joules of work performed on it. Find the ratio of the final volume to the initial volume (to the nearest hundredth).

■ *For Exercises 41–44, see Example 8.*

41. In 1964 an earthquake in Alaska measured 8.4 on the Richter scale. An earthquake of this magnitude will collapse masonry structures, put underground pipelines out of service, destroy bridges, and bend rails. An earthquake measuring 4.0 on the Richter scale is considered small and causes little or no damage. How much stronger (to the nearest unit) was the Alaska quake than one measuring 4.0 on the Richter scale?

42. In 1946 the Nankai, Japan earthquake measured 8.1 on the Richter scale. How much stronger (to the nearest unit) was the Alaska quake of 1964?

43. If earthquake X measures 3.0 on the Richter scale and earthquake Y is 3 times as strong as earthquake X, then what does Y measure (to the nearest tenth) on the Richter scale?

44. If earthquake X measures 5.0 on the Richter scale and earthquake Y is 200 times as strong as X, then what does Y measure (to the nearest tenth) on the Richter scale?

■ *For Examples 45–48, see Example 9.*

45. What is the effect on intensity level of doubling the intensity of sound?

46. By what factor must the intensity of sound be increased to double intensity level?

47. The intensity level 50 meters from an English pop music group was recorded at 120 db. How much more intense was this sound than a noise at 90 db (the threshold of pain for the human ear)?

48. A sound registering 210 db has been produced in a laboratory by a horn. The energy from such a sound is sufficient to bore holes in solid material. A whisper is recorded at 20 db. How much more intense is the sound of the horn than the whisper?

B

49. Institution A pays 9% annual interest compounded annually. Institution B wishes to compound interest daily at a rate that will produce the same yearly return that Institution A gives. To the nearest hundredth of a percent, what rate should Institution B give?

50. If in Exercise 49 Institution A compounds interest quarterly, what rate should Institution B give (to the nearest hundredth of a percent)?

51. Institution A pays 7% annual interest compounded daily. Institution B wishes to compound interest quarterly. In order to pay the same return as Institution A, what rate (to the nearest hundredth of a percent) should Institution B give?

52. If in Exercise 51 Institution B wishes to compound semiannually, what rate should Institution B give (to the nearest hundredth of a percent)?

■ *If S dollars is borrowed at an annual interest rate r (simple interest) and is to be repaid in equal monthly payments over a period of n months, then each monthly payment p is given by*

$$p = S\left[\frac{r/12}{1 - \left(1 + \dfrac{r}{12}\right)^{-n}}\right].$$

In Exercises 53–56, assume simple interest.

53. A family buying a house costing $95,000 has $19,000 to use as a down payment. If they borrow the remaining amount at an annual rate of $16\frac{3}{4}\%$ for a 30-year period, what is their monthly payment?

54. A person wishes to borrow $100,000. One loan company will lend it at $16\frac{1}{2}\%$ annual interest for a period of 30 years. A second loan company will lend it at 16% annual interest but only for a period of 25 years. Which loan results in the lower monthly payment, and how much difference is there in the monthly payments?

55. A family wishes to buy a house costing $125,000. They can obtain a loan at 15% annual interest. They want to pay off the loan in 30 years. If they want to keep their monthly house payment at $1000, how much of the $125,000 do they have to use as a down payment?

56. Suppose in Exercise 55 the family had $50,000 to use as a down payment. How long would the term of the loan have to be (to the nearest month) to keep the monthly payments at $1000?

57. Does the graph of the function describing logistic growth have a horizontal asymptote? If so, what is the asymptote? If not, explain why.

58. Does the graph of the function describing Gompertz growth have a horizontal asymptote? If so, what is the asymptote? If not, explain why.

59. Does the graph of the function describing exponential growth have a horizontal asymptote? If so, what is the asymptote? If not, explain why. Comment on this function as a model for a population whose environment limits its growth potential.

60. Does the graph of the function describing exponential decay have a horizontal asymptote? If so, what is the asymptote? If not, explain why.

C H A P T E R   R E V I E W

Key Words, Phrases, and Symbols

■ *Define or explain each of the following words, phrases, and symbols.*

1. real-number powers with irrational exponents
2. exponential function with base b
3. e
4. logarithmic function with base b
5. properties of logarithmic functions
6. common logarithms
7. natural logarithms
8. exponential growth model
9. exponential decay model
10. logistic growth model
11. Gompertz growth model
12. logarithmic models
13. logarithmic scales

Review Exercises

A

[6.1] ■ *Find the power.*

1. $10^{1/2}$
2. $10^{1/3}$
3. $10^{\sqrt{2}}$
4. $10^{-\sqrt{3}}$

5. $e^{1/4}$ **6.** $e^{1/6}$ **7.** $e^{\sqrt{5}}$ **8.** $e^{-\sqrt{5}}$

9. Graph $y = 3^x$.   **10.** Graph $y = e^{5x}$.

[6.2] ■ *Express in logarithmic notation.*

11. $16^{-1/2} = \dfrac{1}{4}$ **12.** $7^3 = 343$

■ *Express in exponential notation.*

13. $\log_2 8 = 3$ **14.** $\log_{10} 0.0001 = -4$

■ *Solve.*

15. $\log_2 16 = y$ **16.** $\log_{10} x = 3$

■ *Compute.*

17. $\log_{10} 42$ **18.** $\log_{10} 0.00314$ **19.** $\log_{10} 682$ **20.** $\log_{10} 0.0414$

21. $\ln 7$ **22.** $\ln 23$ **23.** $\ln 241$ **24.** $\ln 510$

■ *Graph the function.*

25. $y = \log_4 x$ **26.** $y = \ln 2x$

[6.3] ■ *Express as the sum or difference of simpler logarithmic quantities.*

27. $\log_{10} \sqrt[3]{xy^2}$ **28.** $\log_{10} \dfrac{2R^3}{\sqrt{PQ}}$

■ *Express as a single logarithm with coefficient 1.*

29. $2 \log_b x - \dfrac{1}{3} \log_b y$ **30.** $\dfrac{1}{3}(2 \log_{10} x + \log_{10} y) - 3 \log_{10} z$

■ *Solve.*

31. $\ln 2x^2 = 4$ **32.** $\ln x^3 = -1$

33. $\log_{10}(x + 1) - \log_{10} x = 1$ **34.** $\log_{10}(x + 1) + \log_{10}(x - 1) = 1$

[6.4] ■ *Solve. Give the solution in logarithmic form using base 10 and also as an approximation in decimal form.*

35. $10^{-2x} = 8$ **36.** $10^{-x-2} = 25$

■ *Solve. Give the solution in logarithmic form with base e and also as an approximation in decimal form.*

37. $e^{4x} = 8$ **38.** $e^{-2x+2} = 16$

■ *Solve. Give the solution in logarithmic form with base 10 and also as an approximation in decimal form.*

39. $3^x = 2$ **40.** $3^{x+1} = 80$

■ *Solve. Leave the result in the form of an equation equivalent to the given equation.*

41. $y = A + ke^{-t}$;   for t **42.** $y = A + ke^{-(t+c)}$;   for t

[6.5] ■ *Solve. In Exercises 45–48, use an appropriate formula from Section 6.5.*

43. The number of bacteria in a given culture is given by $N = 10{,}000e^{0.05t}$, where t is measured in seconds. How many bacteria (to the nearest thousand) are present after 1 minute?

44. The amount of a radioactive element in a given sample remaining at any time t is given by $y = 100e^{-10t}$, where y is measured in grams and t in seconds. How long (to the nearest tenth of a second) will it take for the amount to be reduced to 0.1 gram?

45. A herd of buffalo is exhibiting logistic growth with a maximum herd size of 500 and an initial size of 313. If $A = 0.15$, how large will the herd be in 30 years?

46. A population exhibits growth according to the formula $Q(t) = 10e^{-9.1e^{-1.1t}}$ where t is measured in years. How many years (to the nearest year) will it take the population to reach size 5?

47. What annual rate of interest (to the nearest $\frac{1}{4}$%) must a bank pay to triple your investment in 20 years if the account is compounded quarterly?

48. How many years (to the nearest quarter) will it take to triple your money in an account paying 7.25% annually and compounded quarterly?

49. Calculate the pH (to the nearest hundredth) of a solution whose hydrogen ion concentration is 5.4×10^{-6}.

50. How much work (to the nearest tenth of a joule) is done on a system containing 1 mole of gas at 0°C (273 K) if the volume triples?

51. How many times stronger (to the nearest unit) than the smallest detectable quake is an earthquake that measures 6.5 on the Richter scale?

52. How much more intense (to the nearest watt/m²) than a whisper is a sound that has an intensity level of 55 db?

B

■ *Determine an integer n such that $n < x < n + 1$.*

53. $(0.95)^x = 0.915$ **54.** $(2.3)^x = 1727$

55. $\log_{10}(\ln x) = 0.70$ **56.** $\ln(\log_{10} x) = 0.70$

■ *Solve. Show the solution to the nearest tenth.*

57. $\ln(\log_{10}(\ln x)) = 0.25$ **58.** $\log_{10}(\ln(\log_{10} x)) = 0.25$

59. Does the graph of $f(x) = (0.95)^x$ have a horizontal asymptote? Explain your answer.

60. Does the graph of $f(x) = (1.05)^x$ have a horizontal asymptote? Explain your answer.

61. A person borrows \$25,000 at 18% simple annual interest. If the loan is to be repaid in 5 years, what are the monthly payments?

62. A person borrows \$30,000 at 15% simple annual interest. If the monthly payments are \$400, how long is the term of the loan?

63. Does the graph of

$$Q(t) = \frac{100}{1 + \frac{1}{2}e^{2t}}, \qquad t \text{ any real number}$$

have a horizontal asymptote? If so, what is it? If not, explain why.

64. Does the graph of

$$Q(t) = 100e^{-10e^{10t}}, \qquad t \text{ any real number}$$

have a horizontal asymptote? If so, what is it? If not, explain why.

7

Systems of Equations and Inequalities

For a given collection of equations or inequalities in two variables, it is often necessary to determine the ordered pairs that satisfy all the equations or inequalities in the collection. In such a case, the collection of equations or inequalities is called a **system in two variables**. Any ordered pair that is a solution of all the equations or inequalities in a system is called a **solution of the system**, and the set of all solutions of a system is called the **solution set of the system**. Systems in more than two variables are defined similarly.

SYSTEMS OF LINEAR EQUATIONS IN TWO VARIABLES

We shall begin by considering the system

$$a_1 x + b_1 y + c_1 = 0 \qquad (a_1, b_1 \text{ not both } 0) \tag{1}$$

$$a_2 x + b_2 y + c_2 = 0 \qquad (a_2, b_2 \text{ not both } 0). \tag{2}$$

Linear Dependence

The left-hand members of Equations (1) and (2) are said to be **linearly dependent** if one of them can be obtained from the other through multiplication by a constant. If the left-hand members of Equations (1) and (2) are *not* linearly dependent, then they are said to be **linearly independent**.

E X A M P L E   I   (a)   The expressions $2x + 4y - 8$ and $6x + 12y - 24$ are linearly dependent since

$$3(2x + 4y - 8) = 6x + 12y - 24.$$

(b)   The expressions $2x + 4y - 8$ and $6x + 12y - 23$ are linearly independent because there is no constant c so that $c(2x + 4y - 8) = 6x + 12y - 23$.

■

293

Consistency   A system is said to be **consistent** if its solution set is nonempty. A system that is not consistent is **inconsistent**.

E X A M P L E 2   (a)   The system

$$x - 2 = 0$$
$$y + 3 = 0$$

is consistent. The solution set is $\{(2, -3)\}$.

(b)   The system

$$x + y = 0$$
$$x + y - 1 = 0$$

is inconsistent. The solution set is empty.   ■

Geometric Interpretation   In a geometric sense, because the graphs of both Equations (1) and (2) are straight lines, we are confronted with three possibilities. One possibility is that the graphs of the equations are the same line, as illustrated in Figure 7.1. In this case, the solution set of the system is the same as the solution set of each equation. The left-hand members of the two equations in standard form are linearly dependent and the system is consistent.

FIGURE 7.1

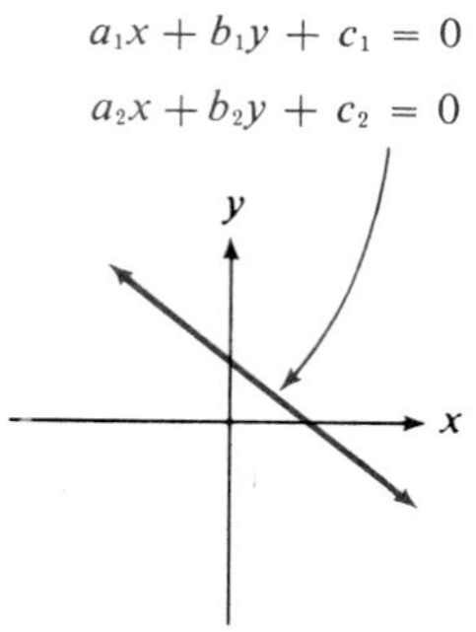

$$a_1x + b_1y + c_1 = 0$$
$$a_2x + b_2y + c_2 = 0$$

FIGURE 7.2

$$a_1x + b_1y + c_1 = 0$$
$$a_2x + b_2y + c_2 = 0$$

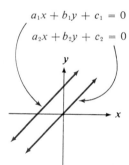

FIGURE 7.3

$$a_1x + b_1y + c_1 = 0$$
$$a_2x + b_2y + c_2 = 0$$

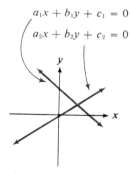

A second possibility is that the graphs of the equations are parallel lines, as illustrated in Figure 7.2. Since the lines have no point in common, there is no ordered pair that satisfies both equations. Hence, the solution set of the system is empty. In this case, the left-hand members are linearly independent and the system is inconsistent.

The last possibility is that the graphs of the equations intersect in a single point, as illustrated in Figure 7.3. Since the lines have exactly one point in common, there is exactly one ordered pair that satisfies both equations. Hence, the solution set of the system contains one ordered pair. In this case, the left-hand members are linearly independent and the system is consistent.

Equivalent Systems   You will recall that equivalent equations have the same solution set. Systems may be said to be equivalent in a similar sense.

Definition 7.1

*If the solution set of one system is equal to (the same as) the solution set of another system, then the systems are **equivalent**.*

In seeking the solution set of a system of equations, our procedure will be to generate equivalent systems until we arrive at a system for which the solution set is obvious. One way to obtain an equivalent system is to replace one equation by a certain *linear combination* of the equations in the system. If $f(x, y)$ and $g(x, y)$ are polynomials, the polynomial $af(x, y) + bg(x, y)$, where a and b are not both zero, is said to be a linear combination of $f(x, y)$ and $g(x, y)$. We shall then refer to the equation

$$af(x, y) + bg(x, y) = 0$$

as a **linear combination** of the equations $f(x, y) = 0$ and $g(x, y) = 0$.

We now compare the two systems:

$$f(x, y) = 0$$
$$g(x, y) = 0 \tag{3}$$

and

$$af(x, y) + bg(x, y) = 0$$
$$g(x, y) = 0, \tag{4}$$

where $a \neq 0$. If (x_1, y_1) is a solution to (3), then $f(x_1, y_1) = 0$ and $g(x_1, y_1) = 0$, so

$$af(x_1, y_1) + bg(x_1, y_1) = 0,$$

and (x_1, y_1) is a solution to (4). On the other hand, suppose (x_2, y_2) is a solution to (4). Then $g(x_2, y_2) = 0$ and

$$af(x_2, y_2) + bg(x_2, y_2) = 0.$$

Thus, $af(x_2, y_2) = 0$ and since $a \neq 0$, $f(x_2, y_2) = 0$, so that (x_2, y_2) is a solution to (3). Therefore, (3) and (4) have the same solution set. This proves the following theorem.

Theorem 7.1

If either equation in the system

$$f(x, y) = 0$$
$$g(x, y) = 0$$

is replaced by a linear combination of the two equations (with nonzero coefficients), the result is an equivalent system.

Theorem 7.1 is useful in solving linear systems of the form

$$a_1 x + b_1 y + c_1 = 0 \qquad (a_1, b_1 \text{ not both } 0) \tag{5}$$
$$a_2 x + b_2 y + c_2 = 0 \qquad (a_2, b_2 \text{ not both } 0). \tag{6}$$

By appropriate choice of multipliers a and b, the linear combination

$$a(a_1x + b_1y + c_1) + b(a_2x + b_2y + c_2) = 0 \qquad (7)$$

will be free of one variable; that is, the coefficient of one variable will be 0. We can then form an equivalent system by substituting Equation (7) for either (5) or (6).

E X A M P L E 3   Solve

$$\begin{aligned} x - 3y + 5 &= 0 \\ 2x + y - 4 &= 0. \end{aligned} \qquad (8)$$

We first form the linear combination

$$1(x - 3y + 5) + 3(2x + y - 4) = 0,$$

or

$$7x - 7 = 0,$$

from which

$$x - 1 = 0.$$

Replacing $x - 3y + 5 = 0$ with $x - 1 = 0$, we then have the equivalent system

$$\begin{aligned} x - 1 &= 0 \\ 2x + y - 4 &= 0. \end{aligned} \qquad (9)$$

Next, we can replace the second equation in (9) with the linear combination

$$-2(x - 1) + 1(2x + y - 4) = 0,$$

or

$$y - 2 = 0,$$

to obtain the equivalent system

$$\begin{aligned} x - 1 &= 0 \\ y - 2 &= 0, \end{aligned} \qquad (10)$$

from which, by inspection, we can obtain the unique solution $(1, 2)$. Hence, the

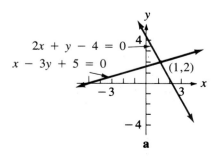

a

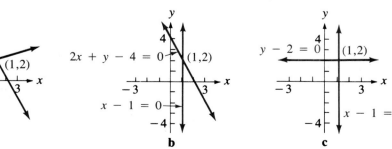

b

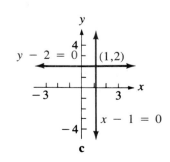

c

solution set of (8) is $\{(1, 2)\}$. Graphs of the systems (8), (9), and (10) appear in Figures a, b, and c, respectively. The point of intersection in each case has coordinates (1, 2). ■

Choice of Multipliers for Linear Combinations

Observe that the multipliers used in Example 3 were first 1 and 3 and later -2 and 1. These were chosen because they produced coefficients that were additive inverses, first for the terms in y and later for the terms in x. In general, the linear combination

$$b_2 (a_1 x + b_1 y + c_1) - b_1 (a_2 x + b_2 y + c_2) = 0$$

will always be free of y, and

$$a_2 (a_1 x + b_1 y + c_1) - a_1 (a_2 x + b_2 y + c_2) = 0$$

will be free of x.

E X A M P L E 4   Form a linear combination of the equations in the system

$$2x + 6y + 7 = 0$$
$$5x + 4y + 3 = 0$$

to obtain an equation that is free of:

(a)   The variable x.   (b)   The variable y.

Solution   (a) $5 (2x + 6y + 7) - 2 (5x + 4y + 3) = 0$

$$22y + 29 = 0$$

(b) $4 (2x + 6y + 7) - 6 (5x + 4y + 3) = 0$

$$-22x + 10 = 0$$ ■

Solution of a System by Substitution

Note that in Example 3 we could have proceeded somewhat differently from (9), as follows. A solution of system (9) must be of the form $(1, y)$ (because any such ordered pair is a solution of $x - 1 = 0$); if x is replaced by 1 in $2x + y - 4 = 0$, then $y = 2$; thus, the only ordered pair that satisfies both of the equations in (9) is $(1, 2)$. Hence, again the solution set of (8) is $\{(1, 2)\}$. In this latter way of proceeding from (9) to the solution of (8), we have used what is known as the **method of substitution**.

E X A M P L E 5   Solve

$$4x + 6y = 31 \qquad\qquad (11)$$

$$3x + 4y = 22 \qquad\qquad (12)$$

by the method of substitution.

Solution   We first replace (11) by 3 times itself plus -4 times (12) to obtain

$$3(4x + 6y) + (-4)(3x + 4y) = (3)(31) + (-4)(22)$$

$$3x + 4y = 22$$

or, equivalently,

$$2y = 5 \qquad\qquad (13)$$

$$3x + 4y = 22. \qquad\qquad (14)$$

Next, we solve (13) for y to obtain

$$y = \frac{5}{2}.$$

We then substitute $\frac{5}{2}$ for y in (14) and solve for x to obtain

$$x = 4.$$

Thus, the solution set of the system is $\{(4, \frac{5}{2})\}$.   ■

Systems as Models

Systems of linear equations are quite useful in expressing relationships in practical applications. By assigning separate variables to represent separate physical quantities, we can usually decrease the difficulty in symbolically representing these relationships.

E X A M P L E 6   A company offers split-rail fence for sale in two prepackaged options. One option consists of 4 posts and 6 rails for $31; the other consists of 3 posts and 4 rails for $22. What are the individual values of posts and rails?

Solution   Let x represent the value of a post in dollars and y represent the value of a rail in dollars. Then

$$4x + 6y = 31$$

$$3x + 4y = 22.$$

This system is solved in Example 5, from which we have $x = 4$ and $y = \frac{5}{2}$. Thus, the value of a post is $4, and the value of a rail is $2.50.   ■

Criteria for Dependent or Inconsistent Equations

If the coefficients of the variables in one equation in a system are proportional to the corresponding coefficients in the other equation, then the equations might be either dependent or inconsistent. The equations in the system

$$a_1 x + b_1 y + c_1 = 0$$

$$a_2 x + b_2 y + c_2 = 0 \qquad (a_2, b_2, c_2 \neq 0)$$

are dependent if

$$\frac{a_1}{a_2} = \frac{b_1}{b_2} = \frac{c_1}{c_2}$$

and inconsistent if

$$\frac{a_1}{a_2} = \frac{b_1}{b_2} \neq \frac{c_1}{c_2}.$$

(See Exercises 7.1–37 and 7.1–38.)

If

$$\frac{a_1}{a_2} \neq \frac{b_1}{b_2},$$

then the system has a unique solution.

E X E R C I S E   7.1

A

■ *Solve each system in Exercises 1–18. See Examples 3–6.*

1. $x - y = 1$
$x + y = 5$

2. $2x - 3y = 6$
$2x + 3y = 3$

3. $3x + y = 7$
$2x - 5y = -1$

4. $2x - y = 7$
$3x + 2y = 14$

5. $5x - y = -29$
$2x + 3y = 2$

6. $6x + 4y = 12$
$3x + 2y = 12$

7. $5x + 2y = 3$
$x = 0$

8. $2x - y = 0$
$x = -3$

9. $3x - 2y = 4$
$y = -1$

10. $x + 2y = 6$
$x = 2$

11. $\frac{1}{4}x - \frac{1}{3}y = -\frac{5}{12}$
$\frac{1}{10}x + \frac{1}{5}y = \frac{1}{2}$

12. $\frac{2}{3}x - y = 4$
$x - \frac{3}{4}y = 6$

13. $\frac{1}{7}x - \frac{3}{7}y = 1$
$2x - y = -4$

14. $\frac{1}{3}x - \frac{2}{3}y = \frac{4}{3}$
$x - 3y = 5$

15. $6x + 4y = 12$
$3x + 2y = 6$

16. $\frac{1}{3}x - \frac{2}{3}y = 2$
$x - 2y = 6$

17. $3x - y = 4$
$18x - 6y = -24$

18. $\frac{1}{2}x - \frac{3}{4}y = 1$
$x - \frac{3}{2}y = 6$

19. A woman has $2 in quarters and nickels. She has two more quarters than nickels. How many quarters and nickels does she have?

20. A man has $1.80 in nickels and dimes, with three more dimes than nickels. How many dimes and nickels does he have?

21. How many pounds of an alloy containing 45% silver must be melted with an alloy containing 60% silver to obtain 40 pounds of a 48% silver alloy?

22. How many pounds of an alloy containing 20% silver must be melted with an alloy containing 15% silver to obtain 100 pounds of a 17% silver alloy?

23. A woman has $1000 more invested at 9% than she has invested at 11%. If her combined annual income from the two investments is $698, how much does she have invested at each rate?

24. A man has three times as much money invested in 9% bonds as he has in stocks paying 8%. How much does he have invested in each if his yearly income from the investments is $1680?

25. Two cars start together and travel in the same direction, one going twice as fast as the other. At the end of 3 hours, they are 96 miles apart. How fast is each traveling?

26. An airplane travels 1260 miles in the same time that an automobile travels 420 miles. If the rate of the airplane is 120 miles per hour greater than the rate of the automobile, find the rate of each.

27. Find a and b so that the graph of $ax + by + 3 = 0$ passes through the points $(-1, 2)$ and $(-3, 0)$.

28. Find a and b so that the graph of $ax + by - 4 = 0$ passes through the points $(0, 2)$ and $(2, -4)$.

29. Find a and b so that the solution set of the system below is $\{(1, 1)\}$.

$$ax + by = 2$$
$$bx - ay = 2$$

30. Find a and b so that the solution set of the system below is $\{(1, 2)\}$.

$$ax + by = 4$$
$$bx - ay = -3$$

31. Recall that the slope-intercept form of the equation of a straight line is $y = mx + b$. Find an equation of the line that passes through the points $(0, 2)$ and $(3, -8)$.

32. Find an equation of the line that passes through the points $(-6, 2)$ and $(4, 1)$.

B

■ Solve each system. Hint: Using substitutions, change the given system to one of the form

$$a_1 u + b_1 v = c_1$$
$$a_2 u + b_2 v = c_2.$$

33. $\dfrac{1}{x} + \dfrac{1}{y} = 2$

$\dfrac{1}{x} - \dfrac{2}{y} = -1$

34. $\dfrac{4}{x} - \dfrac{3}{y} = -7$

$\dfrac{-1}{x} - \dfrac{2}{y} = -1$

35.
$$\frac{1}{x+2} - \frac{2}{y+3} = 0$$
$$\frac{2}{x+2} - \frac{1}{y+3} = \frac{3}{4}$$

36.
$$\frac{3}{x+4} + \frac{1}{y-4} = \frac{1}{6}$$
$$\frac{2}{x+4} + \frac{1}{y-4} = 0$$

37. Show that the left-hand members of

$$a_1 x + b_1 y + c_1 = 0 \qquad (a_1, b_1, c_1 \neq 0)$$
$$a_2 x + b_2 y + c_2 = 0 \qquad (a_2, b_2, c_2 \neq 0)$$

are dependent if and only if

$$\frac{a_1}{a_2} = \frac{b_1}{b_2} = \frac{c_1}{c_2}.$$

[*Hint:* Write the equations in slope-intercept form.]

38. Show that the equations in the system

$$a_1 x + b_1 y + c_1 = 0$$
$$a_2 x + b_2 y + c_2 = 0 \qquad (a_2, b_2, c_2 \neq 0)$$

are inconsistent if and only if

$$\frac{a_1}{a_2} = \frac{b_1}{b_2} \neq \frac{c_1}{c_2}.$$

[*Hint:* Write the equations in slope-intercept form.]

39. Find a and b so that the system below is inconsistent.

$$ax + by = -1$$
$$bx - ay = 3$$

40. Find a and b so that the system below is inconsistent.

$$ax + by = 2$$
$$bx + ay = -4$$

7.2

SYSTEMS OF LINEAR EQUATIONS IN THREE VARIABLES

Solutions of an Equation in Three Variables

A solution of an equation in three variables, such as

$$x + 2y - 3z + 4 = 0, \tag{1}$$

is an ordered triple of numbers (x, y, z), because all three of the variables must be replaced before we can decide whether or not the statement is true. Thus, $(0, -2, 0)$ and $(-1, 0, 1)$ are solutions of Equation (1), whereas $(1, 1, 1)$ is not. There are, of course, infinitely many members in the solution set of such an equation.

Solution of a System

The solution set of a system of linear (first-degree) equations in three variables is the intersection of the solution sets of the separate equations in the system. We are primarily interested in systems involving three equations, such as

$$x + 2y - 3z + 4 = 0$$
$$2x - y + z - 3 = 0$$
$$3x + 2y + z - 10 = 0.$$

We can find the members of this set by methods analogous to those used in Section 7.1.

Theorem 7.2

If any equation in the system

$$a_1x + b_1y + c_1z + d_1 = 0$$
$$a_2x + b_2y + c_2z + d_2 = 0$$
$$a_3x + b_3y + c_3z + d_3 = 0$$

is replaced by a linear combination, with nonzero coefficients, of itself and any one of the other equations in the system, then the result is an equivalent system.

The proof of this theorem is similar to that of Theorem 7.1 and is omitted here.

EXAMPLE 1 Find the solution set of the system

$$x + 2y - 3z + 4 = 0 \tag{1}$$
$$2x - y + z - 3 = 0 \tag{2}$$
$$3x + 2y + z - 10 = 0. \tag{3}$$

Solution We begin by replacing Equation (2) with the linear combination formed by multiplying Equation (1) by -2 and Equation (2) by 1, that is, with

$$-2(x + 2y - 3z + 4) + 1(2x - y + z - 3) = 0,$$

or

$$-5y + 7z - 11 = 0.$$

We obtain the equivalent system

$$x + 2y - 3z + \;4 = 0 \tag{1}$$

$$-\,5y + 7z - 11 = 0 \tag{2$'$}$$

$$3x + 2y + \;z - 10 = 0, \tag{3}$$

where (2$'$) is free of x. Next, if we replace (3) by the sum of -3 times (1) and 1 times (3), we have

$$x + 2y - \;\;3z + \;4 = 0 \tag{1}$$

$$-\,5y + \;\;7z - 11 = 0 \tag{2$'$}$$

$$-\,4y + 10z - 22 = 0, \tag{3$'$}$$

where both (2$'$) and (3$'$) are free of x. Now, if (3$'$) is replaced by the sum of 4 times (2$'$) and -5 times (3$'$), we have

$$x + 2y - \;\;3z + \;4 = 0 \tag{1}$$

$$-\,5y + \;\;7z - 11 = 0 \tag{2$'$}$$

$$-\,22z + 66 = 0, \tag{3$''$}$$

which are equivalent to the original (1), (2), and (3). But, from (3$''$), we see that for any solution of (1), (2$'$), and (3$''$), $z = 3$; that is, the solution will be of the form $(x, y, 3)$. If 3 is substituted for z in (2$'$), we have

$$-\,5y + 7(3) - 11 = 0, \qquad \text{or} \qquad y = 2,$$

and any solution of the system must be of the form $(x, 2, 3)$. Substituting 2 for y and 3 for z in (1) gives

$$x + 2(2) - 3(3) + 4 = 0, \qquad \text{or} \qquad x = 1,$$

so that the single member of the solution set of (1), (2$'$), and (3$''$) is $(1, 2, 3)$. Therefore, the solution set of the original system is $\{(1, 2, 3)\}$.   ■

The foregoing process of solving a system of linear equations can be reduced to a series of mechanical procedures as shown in the following example.

E X A M P L E   2   Find the solution set of the system

$$x + 2y - \;z + 1 = 0 \tag{1}$$

$$x - 3y + \;z - 2 = 0 \tag{2}$$

$$2x + \;y + 2z - 6 = 0. \tag{3}$$

Solution We first find linear combinations which eliminate the x-terms.

We first multiply (1) by -1 and add the result to 1 times (2). We then multiply (1) by -2 and add the result to 1 times (3). In each linear combination the x-terms vanish.

$$-5y + 2z - 3 = 0 \tag{4}$$

$$-3y + 4z - 8 = 0 \tag{5}$$

Next, we eliminate the y-terms. To do this, we multiply (4) by -3 and add the result to 5 times (5). In this linear combination the y-terms vanish.

$$14z - 31 = 0 \tag{6}$$

Now (1), (4), and (6) constitute a set of equations equivalent to (1), (2), and (3).

$$x + 2y - \quad z + \quad 1 = 0 \tag{1}$$

$$-5y + \quad 2z - \quad 3 = 0 \tag{4}$$

$$14z - 31 = 0 \tag{6}$$

Solving for z in (6), we obtain

$$z = \frac{31}{14}.$$

We then substitute $\frac{31}{14}$ for z in (4) and solve for y to obtain

$$-5y + 2\left(\frac{31}{14}\right) - 3 = 0$$

$$y = \frac{2}{7}.$$

Substituting $\frac{2}{7}$ for y and $\frac{31}{14}$ for z in either (1), (2), or (3), say (1), and solving for x, we obtain

$$x + 2\left(\frac{2}{7}\right) - \left(\frac{31}{14}\right) + 1 = 0$$

$$x = \frac{9}{14}.$$

Thus, the solution set of (1), (2), and (3) is $\left\{\left(\frac{9}{14}, \frac{2}{7}, \frac{31}{14}\right)\right\}$. ■

If at any step in the procedure used in the preceding examples, the resulting linear combination vanishes or yields a contradiction, the system contains linearly dependent left-hand members or else inconsistent equations, or both, and it either has an infinite number of solutions or else has no member in its solution set.

Applications   Many applied problems require the solution of systems of linear equations.

EXAMPLE 3   A projectile is moving along a parabolic arc. Three measurements are taken and it is observed to pass through the points $(1, 3)$, $(3, 5)$, and $(4, 9)$. Find the equation of the curve along which the projectile is traveling.

Solution   Since the projectile is traveling along a parabolic arc, we shall assume the equation of the curve along which it travels is of the form

$$y = ax^2 + bx + c.$$

We must determine the value of a, b, and c. Since $(1, 3)$ is on the path, we must have

$$3 = a(1)^2 + b(1) + c$$

or, equivalently,

$$a + b + c = 3. \tag{1}$$

From the other two points, we obtain the equations

$$9a + 3b + c = 5 \tag{2}$$

$$16a + 4b + c = 9. \tag{3}$$

To solve this system we first replace (2) with itself plus -1 times (1) and replace (3) with itself plus -1 times (1) to obtain

$$a + b + c = 3 \tag{1'}$$

$$8a + 2b \quad = 2 \tag{2'}$$

$$15a + 3b \quad = 6. \tag{3'}$$

We now replace (3') with itself plus $-\frac{3}{2}$ times (2') to obtain

$$a + b + c = 3 \tag{1''}$$

$$8a + 2b \quad = 2 \tag{2''}$$

$$3a \quad = 3. \tag{3''}$$

From (3'') we have $a = 1$. Now, substituting 1 for a in (2'') and solving for b, we obtain $b = -3$. Finally, substituting 1 for a and -3 for b in (1'') and solving for c, we obtain $c = 5$. Thus, the equation of the path traveled is

$$y = x^2 - 3x + 5. \quad ■$$

Graphs in Three Dimensions

By establishing a three-dimensional Cartesian coordinate system as shown in Figure 7.4, a one-to-one correspondence can be established between the points in a three-dimensional space and ordered triples of real numbers. If this is done, it can be shown that the graph of a linear equation in three variables is a plane. For example, the equation

FIGURE 7.4

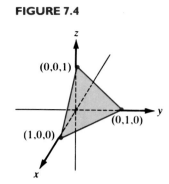

$$x + y + z = 1$$

represents the plane through the points (1, 0, 0), (0, 1, 0), and (0, 0, 1), as illustrated in Figure 7.4. Hence, the solution set of a system of three linear equations in three variables consists of the coordinates of the common intersection of three planes. Figure 7.5 shows the possibilities for the relative positions of the plane graphs of three linear equations in three variables.

FIGURE 7.5

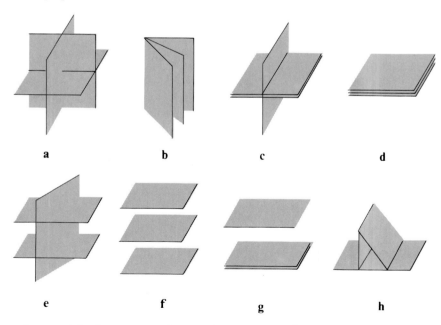

In case (a), the common intersection consists of a single point; hence, the solution set of the corresponding system of three equations contains a single member. In cases (b), (c), and (d), the intersection is a line or a plane, and the solution set of the corresponding system has infinitely many members. In cases (e), (f), (g), and (h), the three planes have no common intersection, and the solution set of the corresponding system is the null set. Linear equations corresponding to cases (a), (b), (c), and (d) are consistent; the others are inconsistent.

Systems with
$n > 3$

The use of linear combinations to solve systems of linear equations can be extended to cover cases of n equations in n variables. Clearly, as n grows larger, the time and effort necessary to find the solution set of a system increase correspondingly. Fortunately, modern high-speed computers can handle such computations in stride for fairly large values of n. Analytic methods other than those exhibited here become preferable for very large systems.

E X E R C I S E   7.2

A

■   *Solve each system. See Examples 1–3.*

1. $x + y + z = 2$
$2x - y + z = -1$
$x - y - z = 0$

2. $x + y + z = 1$
$2x - y + 3z = 2$
$2x - y - z = 2$

3. $x + y + 2z = 0$
$2x - 2y + z = 8$
$3x + 2y + z = 2$

4. $2x - 3y + z = 3$
$x - y - 3z = -1$
$-x + 2y - 3z = -4$

5. $x - 2y + z = -1$
$2x + y - 3z = 3$
$3x + 3y - 2z = 10$

6. $x - 2y + 4z = -3$
$3x + y - 2z = 12$
$2x + y - 3z = 11$

7. $4x - 2y + 3z = 4$
$2x - y + z = 1$
$3x - 3y + 4z = 5$

8. $x + 5y - z = 2$
$3x - 9y + 3z = 6$
$x - 3y + z = 4$

9. $x + z = 5$
$y - z = -4$
$x + y = 1$

10. $5y - 8z = -19$
$5x - 8z = 6$
$3x - 2y = 12$

11. $x - \frac{1}{2}y - \frac{1}{2}z = 4$
$x - \frac{3}{2}y - 2z = 3$
$\frac{1}{4}x + \frac{1}{4}y - \frac{1}{4}z = 0$

12. $x + 2y + \frac{1}{2}z = 0$
$x + \frac{3}{5}y - \frac{2}{5}z = \frac{1}{5}$
$4x - 7y - 7z = 6$

13. The sum of three numbers is 15. The second equals two times the first and the third equals the second. Find the numbers.

14. The sum of three numbers is 2. The first number is equal to the sum of the other two, and the third number is the result of subtracting the first from the second. Find the numbers.

15. A box contains $6.25 in nickels, dimes, and quarters. There are 85 coins in all, with three times as many nickels as dimes. How many coins of each kind are there?

16. A man has $446 in ten-dollar, five-dollar, and one-dollar bills. There are 94 bills in all and 10 more five-dollar bills than ten-dollar bills. How many bills of each kind does he have?

17. The perimeter of a triangle is 155 centimeters. Side x is 20 centimeters shorter than side y, and side y is 5 centimeters longer than side z. Find the lengths of the sides of the triangle.

18. The perimeter of a triangle is 120 centimeters. Side x is the same length as side y and 10 centimeters shorter than side z. Find the length of each side.

19. Find values for a, b, and c so that the graph of $x^2 + y^2 + ax + by + c = 0$ contains the points $(0, 0)$, $(6, 0)$, and $(0, 8)$.

20. The equation for a circle can be written $x^2 + y^2 + ax + by + c = 0$. Find the equation of the circle so that the graph contains the points $(2, 3)$, $(3, 2)$, and $(-4, -5)$.

21. Find values for a, b, and c so that the graph of $y = ax^2 + bx + c$ contains the points $(-1, 0)$, $(2, 12)$, and $(-2, 8)$.

22. Find values for a, b, and c so that the graph of $y = ax^2 + bx + c$ contains the points $(-1, 2)$, $(1, 6)$, and $(2, 11)$.

23. Three solutions of the equation $ax + by + cz = 1$ are $(0, 2, 1)$, $(6, -1, 2)$, and $(0, 2, 0)$. Find the coefficients a, b, and c.

24. Three solutions of the equation $ax + by + cz = 1$ are $(2, 1, 0)$, $(-1, 3, 2)$, and $(3, 0, 0)$. Find the coefficients a, b, and c.

B

■ *Solve each system. Hint: Using substitutions, change the given system to one of the form*

$$a_1 u + b_1 v + c_1 w = d_1$$

$$a_2 u + b_2 v + c_2 w = d_2$$

$$a_3 u + b_3 v + c_3 w = d_3.$$

25.
$$\frac{1}{x} + \frac{1}{y} - \frac{1}{z} = 1$$
$$\frac{2}{x} - \frac{2}{y} + \frac{1}{z} = 1$$
$$\frac{-3}{x} + \frac{1}{y} - \frac{1}{z} = -3$$

26.
$$\frac{4}{x} - \frac{2}{y} + \frac{1}{z} = 4$$
$$\frac{3}{x} - \frac{1}{y} + \frac{2}{z} = 0$$
$$\frac{-1}{x} + \frac{3}{y} - \frac{2}{z} = 0$$

27.
$$\frac{2}{x-1} + \frac{1}{y+2} - \frac{1}{z-3} = \frac{-2}{3}$$
$$\frac{-3}{x-1} + \frac{1}{y+2} - \frac{2}{z-3} = \frac{16}{3}$$
$$\frac{2}{x-1} + \frac{3}{y+2} - \frac{1}{z-3} = 0$$

28.
$$\frac{2}{x+1} - \frac{1}{y-3} + \frac{1}{z} = -3$$
$$\frac{-3}{x+1} + \frac{2}{y-3} - \frac{2}{z} = \frac{13}{2}$$
$$\frac{-2}{x+1} + \frac{4}{y-3} - \frac{1}{z} = 9$$

29. Show that the system

$$x + y + 2z = 2$$

$$2x - y - z = 3$$

has an infinite number of members in its solution set. List two ordered triples that are solutions. [*Hint:* Express x in terms of z alone, and express y in terms of z alone.]

30. Give a geometric argument to show that any system of two consistent linear equations in three variables has an infinite number of solutions.

7.3

SYSTEMS OF NONLINEAR EQUATIONS

In Sections 7.1 and 7.2 we discussed systems of *linear* equations. In this section we introduce methods that can sometimes be used to solve systems containing nonlinear equations.

Use of Graphs

In solving a system containing nonlinear equations, it is frequently helpful to sketch the graphs of all the equations in the system. By studying the graphs, we can approximate the intersection points of the curves. This usually tells us how many solutions with real components the system has, and it gives us approximate values for the solutions, which we can use as a rough check on any solutions we find algebraically.

E X A M P L E I   Determine the number of solutions with real components of the system

$$x^2 + y^2 = 25$$

$$x + y = 1$$

by graphical methods and approximate the solutions.

Solution   We sketch the graphs of $x^2 + y^2 = 25$*
and $x + y = 1$ on the same set of co-
ordinate axes. Observe from the figure
that there are two points of intersection with coordinates approximately
$(-3, 4)$ and $(4, -3)$. Thus, the system has two solutions with real components, and they are approximately
$(-3, 4)$ and $(4, -3)$.   ■

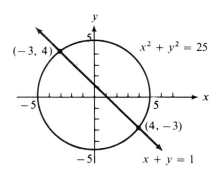

Substitution   The substitution method is sometimes a convenient means of finding solution sets of systems in which nonlinear equations are present.

E X A M P L E 2   Solve.

$$x^2 + y^2 = 25 \tag{1}$$

$$x + y = 1 \tag{2}$$

* The graph of the equation $x^2 + y^2 = r^2$ is a circle of radius r centered at the origin. This equation is treated in more detail in Chapter 11.

Solution   In Example 1, we found that the system has two solutions with real components. We now use the method of substitution to determine the solutions algebraically. Equation (2) can be written equivalently in the form

$$y = 1 - x, \tag{3}$$

and we can replace y in (1) by $(1 - x)$ from (3). This produces

$$x^2 + (1 - x)^2 = 25, \tag{4}$$

which has as a solution set those values of x for which the ordered pair (x, y) is a common solution of (1) and (2). We can now find the solution set of (4):

$$x^2 + 1 - 2x + x^2 = 25,$$

$$2x^2 - 2x - 24 = 0,$$

$$2(x + 3)(x - 4) = 0,$$

which is satisfied if x is either -3 or 4. Now, by replacing x in the *first-degree equation* (3) by each of these numbers in turn, we obtain

$$y = 1 - (-3) = 4$$

and

$$y = 1 - 4 = -3,$$

so that the solution set of the system (1) and (2) is $\{(-3, 4), (4, -3)\}$.   ■

If, in Example 2, we had substituted the values obtained for x (-3 and 4) in the *second-degree equation* (1), we would have obtained some extraneous solutions that do not satisfy Equation (2).

EXAMPLE 3   Solve.

$$y = x^2 + 2x + 1 \qquad (1)$$

$$x = y - 3 \qquad (2)$$

Solution   We first sketch the graphs of Equations (1) and (2) on the same set of coordinate axes. Observe from the figure that there are two points of intersection with coordinates approximately $(-2, 1)$ and $(1, 4)$. Next, we use the method of substitution to solve the system. We solve Equation (2) explicitly for y to

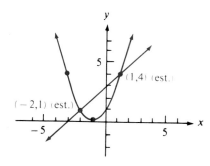

obtain

$$y = x + 3. \qquad (2')$$

Next, we substitute $x + 3$ for y in (1), obtaining

$$x + 3 = x^2 + 2x + 1. \qquad (3)$$

We then solve (3) for x to obtain

$$x^2 + x - 2 = 0$$
$$(x + 2)(x - 1) = 0$$
$$x = -2, \qquad x = 1.$$

Finally, we substitute each of these values in turn in (2′) and determine values for y.

$$\text{If } x = -2, \quad \text{then} \quad y = 1.$$
$$\text{If } x = 1, \quad \text{then} \quad y = 4.$$

Thus, the solution set is $\{(-2, 1), (1, 4)\}$.   ■

Linear Combination

If both of the equations in a system of two equations in two variables are of the second degree in both variables, the use of linear combinations of the equations often provides a simpler means of solution than does substitution.

EXAMPLE 4   Solve.

$$3x^2 - 7y^2 + 15 = 0 \qquad (1)$$
$$3x^2 - 4y^2 - 12 = 0 \qquad (2)$$

Solution   By forming the linear combination of -1 times (1) and 1 times (2), we obtain

$$3y^2 - 27 = 0,$$
$$y^2 = 9,$$

from which

$$y = 3 \qquad \text{or} \qquad y = -3,$$

and we have the y-components of the members of the solution set of Equations (1) and (2). Substituting 3 for y in either (1) or (2), say (1), we have

$$3x^2 - 7(3)^2 + 15 = 0,$$
$$x^2 = 16,$$

from which

$$x = 4 \qquad \text{or} \qquad x = -4.$$

Thus, the ordered pairs $(4, 3)$ and $(-4, 3)$ are solutions of the system. Substituting -3 for y in (1) or (2) [this time we shall use (2)] gives us

$$3x^2 - 4(-3)^2 - 12 = 0,$$

$$x^2 = 16,$$

so that

$$x = 4 \qquad \text{or} \qquad x = -4.$$

Thus, the ordered pairs $(4, -3)$ and $(-4, -3)$ are solutions of the system, and the complete solution set is $\{(4, 3), (4, -3), (-4, 3), (-4, -3)\}$.   ■

Linear Combination and Substitution

The solution of some systems requires the application of both linear combinations *and* substitution.

EXAMPLE 5   Solve.

$$x^2 + y^2 = 5 \qquad (1)$$

$$x^2 - 2xy + y^2 = 1 \qquad (2)$$

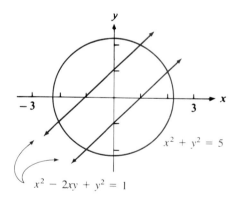

Solution   We graph the two equations on the same set of coordinate axes. Note that Equation (2) is equivalent to $(x - y)^2 = 1$, or $x - y = \pm 1$. Thus, the graph of this equation consists of the two parallel lines $x - y = 1$ and $x - y = -1$. Observe that there are four solutions with real components.

By forming the linear combination of 1 times (1) and -1 times (2), we obtain

$$2xy = 4,$$

$$xy = 2. \qquad (3)$$

The system (1) and (2) is equivalent to the system (1) and (3). This latter system can be solved by substitution. From (3), we have

$$y = \frac{2}{x}.$$

Replacing y in (1) by $2/x$, we obtain

$$x^2 + \left(\frac{2}{x}\right)^2 = 5,$$

$$x^2 + \frac{4}{x^2} = 5, \tag{4}$$

$$x^4 + 4 = 5x^2,$$
$$x^4 - 5x^2 + 4 = 0, \tag{5}$$

which is a quadratic in x^2. The left-hand member of (5), when factored, yields

$$(x^2 - 1)(x^2 - 4) = 0,$$

from which we obtain

$$x^2 - 1 = 0 \quad \text{or} \quad x^2 - 4 = 0,$$

so that

$$x = 1, \quad x = -1, \quad x = 2, \quad x = -2.$$

Since the step from (4) to (5) was not an elementary transformation, we are careful to note that these values of x all satisfy (4). Substituting 1, -1, 2, and -2 in turn for x in (3), we obtain the corresponding values for y. Thus, the solution set of the system (1) and (2) is $\{(1, 2), (-1, -2), (2, 1), (-2, -1)\}$.   ■

There are other techniques involving substitution in conjunction with linear combinations that are useful in handling systems of higher-degree equations, but they all bear similarity to those illustrated.

Imaginary Solutions

In the foregoing examples, each solution is a member of $R \times R$ and their graphs are the points of intersection of the graphs of each equation. If one or more of the components of the solutions are complex numbers, we find these solutions in

$$C \times C = \{(x, y) \mid x \in C \quad \text{and} \quad y \in C\}.$$

However, the graphs in the real plane of the equations do *not* have points of intersection corresponding to these solutions.

EXAMPLE 6   Solve.

$$x^2 + y^2 = 26 \tag{1}$$
$$x + y = 8 \tag{2}$$

Solution   We graph the two equations on the same set of coordinate axes and observe that there are no solutions with real components. We use substitution to find the solutions with complex components.

Equation (2) can be written equivalently as

$$y = 8 - x. \qquad (3)$$

Substituting $8 - x$ for y in (1) and simplifying yields

$$x^2 + (8 - x)^2 = 26$$
$$x^2 + 64 - 16x + x^2 = 26$$
$$2x^2 - 16x + 38 = 0$$
$$x^2 - 8x + 19 = 0.$$

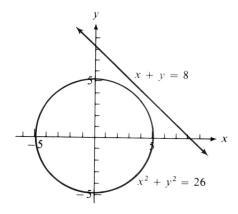

Using the quadratic formula to solve for x, we obtain

$$x = \frac{8 \pm \sqrt{64 - 76}}{2(1)} = \frac{8 \pm \sqrt{-12}}{2}$$
$$= \frac{8 \pm 2i\sqrt{3}}{2} = \frac{2(4 \pm i\sqrt{3})}{2} = 4 \pm i\sqrt{3}.$$

Then, substituting $4 + i\sqrt{3}$ for x in (3) gives $y = 4 - i\sqrt{3}$, and substituting $4 - i\sqrt{3}$ for x in (3) gives $y = 4 + i\sqrt{3}$. Hence, the solution set is

$$\{(4 + i\sqrt{3}, 4 - i\sqrt{3}), (4 - i\sqrt{3}, 4 + i\sqrt{3})\}.$$

EXAMPLE 7   Solve.

$$x^2 - 3xy + 2y^2 = 0 \qquad (1)$$
$$2x^2 - xy - 3y^2 = 12 \qquad (2)$$

Solution   We first factor the left-hand member of (1) to obtain

$$(x - y)(x - 2y) = 0.$$

Since the product is 0, we have

$$x - y = 0 \qquad \text{or} \qquad x - 2y = 0.$$

Thus,

$$x = y, \qquad (1')$$

or

$$x = 2y. \qquad (1'')$$

Next, we substitute y for x in (2) to obtain

$$2y^2 - y^2 - 3y^2 = 12,$$
$$-2y^2 = 12,$$
$$y = i\sqrt{6} \qquad \text{or} \qquad y = -i\sqrt{6}.$$

Finally, we substitute each of these values in turn in (1') and determine values for x.

$$\text{If} \quad y = i\sqrt{6}, \quad \text{then} \quad x = i\sqrt{6}.$$
$$\text{If} \quad y = -i\sqrt{6}, \quad \text{then} \quad x = -i\sqrt{6}.$$

Thus, $(i\sqrt{6}, i\sqrt{6})$ and $(-i\sqrt{6}, -i\sqrt{6})$ are solutions of the system. We now substitute $2y$ for x in (2) and solve for y to obtain

$$8y^2 - 2y^2 - 3y^2 = 12,$$
$$3y^2 = 12,$$
$$y = 2 \qquad \text{or} \qquad y = -2.$$

Finally, we substitute each of these values in turn in (1'') and determine values for x.

$$\text{If} \quad y = 2, \quad \text{then} \quad x = 2(2) = 4.$$
$$\text{If} \quad y = -2, \quad \text{then} \quad x = 2(-2) = -4.$$

Thus, the solution set is $\{(4, 2), (-4, -2), (i\sqrt{6}, i\sqrt{6}), (-i\sqrt{6}, -i\sqrt{6})\}$.   ■

Approximations to Real Solutions

When the degree of any equation in a set of equations is greater than two, or when the left-hand member of one of the equations of the form $f(x, y) = 0$ is not a polynomial, it may be very difficult or impossible to find common solutions analytically. In this case, we can at least obtain approximations to any real solutions by graphical methods.

EXAMPLE 8   Approximate the solutions to the system

$$y = 2^{-x}$$
$$y = \sqrt{0.01x}.$$

Solution   We first graph these equations as shown on page 316.

Examining Figure b, an enlargement of part of Figure a, we observe that the curves intersect at approximately $(2.6, 0.16)$. From geometric considerations we conclude that this ordered pair approximates the only member in the solution set of the system.

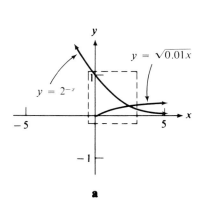

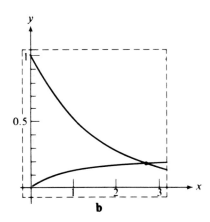

Example 8 shows, in particular, that the equation

$$2^{-x} - \sqrt{0.01x} = 0$$

has just one solution, approximately 2.6, and suggests a method of approximating solutions of similar equations.

EXERCISE 7.3

A

■ *Solve by the method of substitution. Check the solutions by sketching the graphs of the equations and estimating the coordinates of any points of intersection. See Examples 1–3.*

1. $y = x^2 - 5$
$y = 4x$

2. $y = x^2 - 2x + 1$
$y + x = 3$

3. $x^2 + y^2 = 13$
$x + y = 5$

4. $x^2 + y^2 = 4$
$2x - y = 2$

5. $x + y = 1$
$xy = -12$

6. $2x - y = 9$
$xy = -4$

■ *Solve using linear combinations. See Examples 4 and 5.*

7. $x^2 + y^2 = 10$
$9x^2 + y^2 = 18$

8. $x^2 + 4y^2 = 52$
$x^2 + y^2 = 25$

9. $x^2 - y^2 = 7$
$2x^2 + 3y^2 = 24$

10. $x^2 + 4y^2 = 25$
$4x^2 + y^2 = 25$

11. $4x^2 - 9y^2 + 132 = 0$
$x^2 + 4y^2 - 67 = 0$

12. $16y^2 + 5x^2 - 26 = 0$
$25y^2 - 4x^2 - 17 = 0$

■ *Solve. See Examples 5–7.*

13. $x^2 - xy + y^2 = 7$
$x^2 + y^2 = 5$

14. $3x^2 - 2xy + 3y^2 = 34$
$x^2 + y^2 = 17$

15. $3x^2 + 3xy - y^2 = 35$
$x^2 - xy - 6y^2 = 0$

16. $x^2 - xy + y^2 = 21$
$x^2 + 2xy - 8y^2 = 0$

17. $2x^2 - xy - 6y^2 = 0$
$x^2 + 3xy + 2y^2 = 4$

18. $2x^2 + xy - y^2 = 0$
$6x^2 + xy - y^2 = 1$

19. $x^2 + y^2 = 9$
$y = 4$

20. $x^2 + 2y^2 = 6$
$x + y = 10$

21. $2x^2 + xy - 4y^2 = 12$
$x^2 - 2y^2 = 4$

22. $x^2 + 3xy - y^2 = -3$
$x^2 - xy - y^2 = 1$

23. $y - 10^x = 0$
$\log_{10} y - 2x = 1$

24. $y - 10^x = 0$
$\log_{10} y + 3x = 4$

25. $10^y = \dfrac{1}{x}$
$\log_{10} x = y^2$

26. $10^y = \sqrt{x}$
$\log_{10} x^2 = y^2$

■ *Solve by graphing. Approximate components of solutions to the nearest half-unit. See Example 8.*

27. $y = 10^x$
$x + y = 2$

28. $y = 2^x$
$y - x = 2$

29. $y = \log_{10} x$
$y = x^2 - 2x + 1$

30. $y = 10^x$
$y = x^2$

31. The sum of the squares of two positive numbers is 13. If twice the first number is added to the second, the sum is 7. Find the numbers.

32. The sum of two numbers is 6 and their product is 35/4. Find the numbers.

33. The annual income from an investment is $32. If the amount invested were $200 more and the rate $\frac{1}{2}\%$ less, the annual income would be $35. What are the amount and rate of the investment?

34. At a constant temperature, the pressure P and volume V of a gas are related by the equation $PV = K$. The product of the pressure (in pounds per square inch) and the volume (in cubic inches) of a certain gas is 30 inch-pounds. If the temperature remains constant as the pressure is increased 4 pounds per square inch, the volume is decreased by 2 cubic inches. Find the original pressure and volume of the gas.

B

35. Consider the system

$$x^2 + y^2 = 8 \tag{1}$$

$$xy = 4. \tag{2}$$

We can solve this system by substituting $4/x$ for y in (1) to obtain

$$x^2 + \frac{16}{x^2} = 8,$$

from which we obtain $x = 2$ or $x = -2$. Now if we obtain the y-components of the solution from (2), we find that for $x = 2$ we have $y = 2$, and that for $x = -2$ we have $y = -2$. But if we seek y-components from (1), for $x = 2$ we have $y = \pm 2$, and for $x = -2$, $y = \pm 2$. Discuss the fact that we seem to obtain two more solutions from (1) than from (2). What is the solution set of the system?

36. What relationships must exist between the numbers a and b so that the solution set of the system

$$x^2 + y^2 = 25$$

$$y = ax + b$$

has two ordered pairs of real numbers? One ordered pair of real numbers? No ordered pairs of real numbers? [*Hint:* Use substitution and consider the nature of the roots of the resulting quadratic equation.]

7.4

SYSTEMS OF INEQUALITIES

Graphs of Systems of Inequalities

In Section 4.7, we observed that the graph of the solution set of an inequality in two variables might be a region in the plane. The graph of the solution set of a system of inequalities in two variables, which consists of the intersection of the graphs of the inequalities in the system, might also be a plane region.

EXAMPLE 1 Graph the solution set of the system

$$x + 2y \leq 6$$

$$2x - 3y \geq 12.$$

Solution Graphing each inequality by the method discussed in Section 4.7, we obtain the figure shown, where the doubly shaded region is the graph of the solution set of the system.

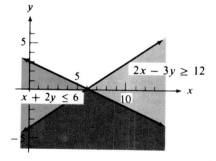

EXAMPLE 2 Graph the solution set of the system

$$y \leq x + 2$$

$$y \geq x^2.$$

Solution The graph of each inequality is shaded as shown in the figure. The doubly shaded region constitutes the graph of the solution set of the system.

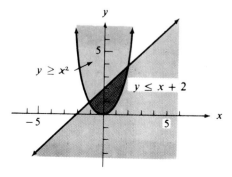

EXAMPLE 3   Graph the solution set of the system

$$y > 2$$

$$x > -2$$

$$x + y > 1.$$

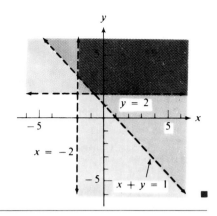

Solution   The triply shaded region in the figure constitutes the graph of the solution set of the system. ■

EXERCISE 7.4

A

■   *By double or triple shading, indicate the region in the plane representing the solution set of each system. For Exercises 1–6, see Example 1.*

1. $y \geq x + 1$
$y \geq 5 - x$

2. $y \geq 4$
$x \geq 2$

3. $y - x \geq 0$
$y + x \geq 0$

4. $2y - x \geq 1$
$x < -3$

5. $y + 3x < 6$
$y > 2$

6. $x > 3$
$x + y \geq 1$

■   *For Exercises 7–12, see Example 2.*

7. $y > x^2 + 1$
$x + y > 4$

8. $y < x^2 + 4$
$x - y \leq 4$

9. $x^2 + y^2 < 25$
$y > 3$

10. $x^2 + y^2 \leq 4$
$x < -1$

11. $x^2 + y^2 \geq 9$
$y < x^2$

12. $x^2 + y^2 \geq 25$
$y > x^2$

■   *For Exercises 13–16, see Example 3.*

13. $y \leq 3$
$x \leq 2$
$y < x$

14. $y \geq 2$
$x \geq 2$
$y \geq x$

15. $x^2 + y^2 \leq 9$
$x \leq 2$
$y \leq -2$

16. $x^2 + y^2 \leq 36$
$x \geq 3$
$y \geq 3$

17. $x = 3$
$x + y < 4$

18. $y = 2$
$2y + x < 3$

19. $x - 3y < 6$
$y + x = 1$

20. $2x + y \geq 4$
$x - y = -2$

B

21. $x^2 + y^2 \leq 25$
$y \geq x^2 - 4$
$y \leq -x^2 + 4$

22. $y \leq \log_{10} x$
$y \geq x - 1$

23. $y \leq 10^x$
$y \leq 2 - x^2$

24. $9 \leq x^2 + y^2 \leq 16$
$-1 \leq x - y \leq 1$

25. $9 \leq x^2 + y^2 \leq 16$
$x^2 + 1 \leq y \leq x^2 + 3$

26. $x^2 - 2 \leq y \leq 2 - x^2$
$|x| \geq 1$

CHAPTER REVIEW

Key Words and Phrases

■ *Define or explain each of the following words and phrases.*

1. system of equations (inequalities)
2. solution of a system
3. solution set of a system
4. linearly dependent expressions
5. linearly independent expressions

6. consistent system
7. inconsistent system
8. equivalent systems
9. linear combination of equations
10. method of substitution

Review Exercises

A

[7.1] ■ *Solve each system.*

1. $x + 5y = 18$
  $x - y = -3$

2. $x + 5y = 11$
  $2x + 3y = 8$

3. $2x - 3y = 8$
  $3x + 2y = 7$

4. $3x - y = -7$
  $x - 2y = -4$

5. $2x - 3y = 1$
  $4x - 6y = 2$

6. $9x + 12y = 1$
  $6x + 8y = 0$

7. Find values for a and b so that the graph of $ax + by = 19$ passes through the points $(2, 3)$ and $(-3, 5)$.

8. Find a and b so that the solution set of the system below is $\{(2, 1)\}$.

$$ax + by = 3$$
$$2ax - by = 1$$

[7.2] ■ *Solve each system.*

9. $x + 3y - z = 3$
  $2x - y + 3z = 1$
  $3x + 2y + z = 5$

10. $x + y + z = 2$
  $3x - y + z = 4$
  $2x + y + 2z = 3$

11. $2x + 3y - z = -2$
  $x - y + z = 6$
  $3x - y + z = 10$

12. Find values for a, b, and c so that the graph of $y = ax^2 + bx + c$ contains the points $(-1, 9)$, $(0, 4)$, and $(1, 3)$.

[7.3] ■ *Solve each system.*

13. $x^2 + y = 3$
  $5x + y = 7$

14. $x^2 - xy + y^2 = 1$
  $2x^2 - xy + 2y^2 = 3$

15. $2x^2 + 5y^2 - 53 = 0$
  $4x^2 + 3y^2 - 43 = 0$

16. $x^2 + 3xy + 2y^2 = 0$
  $x^2 - 3xy + y^2 = 1$

17. $y - 10^{2x} = 0$
  $\log_{10} y^2 - x + 1 = 0$

18. $10^y - 2x = 0$
  $\log_{10} x + \log_{10} 2 = -1$

19. Approximate the solution of the system below by graphical methods.

$$y = 3^x$$

$$y = 3 - x$$

[7.4] ■ *By double or triple shading, indicate the region representing the solution set of each system.*

20. $y > x^2 - 4$
$\quad\ y < 2 - x$

21. $y + x^2 < 0$
$\qquad y + 3 > 0$
$\qquad y < x - 3$

22. $y - x^2 < 4$
$\qquad y - 3 > 0$
$\qquad y - x > 0$

B

■ *Solve each system.*

23. $\dfrac{1}{x} + \dfrac{2}{y} = 4$

$\quad\ \dfrac{2}{x} - \dfrac{3}{y} = 2$

24. $\dfrac{3}{x + 2} + \dfrac{4}{y - 1} = 1$

$\qquad \dfrac{5}{x + 2} + \dfrac{7}{y - 1} = 2$

25. $\dfrac{2}{x} - \dfrac{1}{y} + \dfrac{1}{z} = 5$

$\quad\ \dfrac{1}{x} - \dfrac{2}{y} + \dfrac{3}{z} = 9$

$\quad\ \dfrac{1}{x} \quad\ + \dfrac{2}{z} = 5$

26. $\dfrac{3}{x - 1} - \dfrac{2}{y + 3} + \dfrac{1}{z - 5} = -15$

$\qquad \dfrac{-2}{x - 1} + \dfrac{1}{y + 3} + \dfrac{1}{z - 5} = 4$

$\qquad \dfrac{1}{x - 1} + \dfrac{1}{y + 3} + \dfrac{1}{z - 5} = -2$

27. What relationship must exist between real numbers a and b so that the solution set of the system

$$2x^2 - y = 2$$

$$y = ax + b$$

has two ordered pairs of real numbers? One ordered pair of real numbers? No ordered pair of real numbers?

28. Answer the same questions as in Exercise 27 for the system

$$x^2 - y^2 = 1$$

$$y = ax + b.$$

■ *By double or triple shading, indicate the region representing the solution set of each system.*

29. $\qquad\quad y \geq \ln x$
$\qquad\qquad y \leq e^x$
$\qquad\ x^2 + y^2 \leq 4$

30. $1 \leq x^2 + y^2 \leq 4$
$\qquad y \leq x^2$

8 Matrices and Determinants

Matrices, as introduced in this chapter, are frequently used in mathematics and engineering, and also in the physical, social, and life sciences.

8.1

DEFINITIONS; MATRIX ADDITION

A **matrix** is a rectangular array of numbers (or other suitable entities), which are called the **entries** or **elements** of the matrix. In this book, we shall consider only real numbers as entries. A matrix is customarily displayed in a pair of brackets or parentheses (we shall use brackets). Thus,

$$\begin{bmatrix} 1 & 2 & 3 \\ 4 & 5 & 6 \end{bmatrix} \quad \text{and} \quad \begin{bmatrix} 2 \\ 1 \end{bmatrix}$$

are matrices. The **order**, or **dimension**, of a matrix is the ordered pair having as first component the number of (horizontal) **rows** and as second component the number of (vertical) **columns** in the matrix.

EXAMPLE 1

$$\begin{bmatrix} 1 & 2 & 3 \\ 4 & 5 & 6 \end{bmatrix}, \quad \begin{bmatrix} 1 \\ 2 \\ 3 \end{bmatrix}, \quad \begin{bmatrix} a_1 & a_2 & a_3 & a_4 \\ b_1 & b_2 & b_3 & b_4 \\ c_1 & c_2 & c_3 & c_4 \\ d_1 & d_2 & d_3 & d_4 \end{bmatrix}$$

are 2×3 (read "two by three"), 3×1 (read "three by one"), and 4×4 (read "four by four") matrices, respectively. Note that the number of *rows* is given first and then the number of *columns*. ■

A matrix consisting of a single row is called a **row matrix** or a **row vector**, whereas a matrix consisting of a single column is called a **column matrix** or a **column vector**.

Matrices are frequently represented by capital letters. Thus, we might want to talk about the matrices A and B, where

$$A = \begin{bmatrix} a_1 & a_2 \\ b_1 & b_2 \end{bmatrix} \quad \text{and} \quad B = [b_1 \quad b_2].$$

To indicate that A is a 2×2 matrix, we can write $A_{2 \times 2}$. Similarly, $B_{1 \times 2}$ is a matrix with one row and two columns.

To represent the entries of a matrix, double-subscript notation can be used. A single letter, say a, is used to denote an entry in a matrix, and then *two* subscripts are appended, the first subscript telling in which *row* the entry occurs and the second telling in which *column*. Thus, we write

$$A = \begin{bmatrix} a_{11} & a_{12} & a_{13} \\ a_{21} & a_{22} & a_{23} \\ a_{31} & a_{32} & a_{33} \end{bmatrix},$$

where a_{21} is the element in the second *row* and first *column*, a_{33} is the element in the third *row* and third *column*, and, in general, a_{ij} is the element in the *i*th *row* and *j*th *column*.

Definition 8.1   *Two matrices, A and B, are **equal** if and only if both matrices are of the same order and $a_{ij} = b_{ij}$ for each i, j.*

E X A M P L E 2   (a) $\begin{bmatrix} 2 & 1 \\ 3 & 0 \end{bmatrix} = \begin{bmatrix} \frac{4}{2} & 2-1 \\ \sqrt{9} & 0 \end{bmatrix}$ (b) $\begin{bmatrix} 2 & 1 \\ 3 & 0 \end{bmatrix} \neq \begin{bmatrix} 2 & 3 \\ 1 & 0 \end{bmatrix}$ (c) $\begin{bmatrix} 1 \\ 2 \end{bmatrix} \neq [1, \quad 2]$ ■

Definition 8.2   *The **transpose** of a matrix A, denoted by A^t, is the matrix in which the rows are the columns of A and the columns are the rows of A.*

Thus, if a_{ij} and b_{ij} represent the entries in the *i*th row and *j*th column of A and A^t, respectively, then we have

$$b_{ij} = a_{ji}$$

for each i and j.

E X A M P L E 3   (a) $\begin{bmatrix} 2 & 1 \\ 3 & 0 \end{bmatrix}^t = \begin{bmatrix} 2 & 3 \\ 1 & 0 \end{bmatrix}$ (b) $\begin{bmatrix} 1 & 2 & 3 \\ 4 & 5 & 6 \end{bmatrix}^t = \begin{bmatrix} 1 & 4 \\ 2 & 5 \\ 3 & 6 \end{bmatrix}$ (c) $\begin{bmatrix} 1 \\ 2 \end{bmatrix}^t = [1 \quad 2]$ ■

Definition 8.3 The **sum** of two matrices of the same order, $A_{m \times n}$ and $B_{m \times n}$, is the matrix $(A + B)_{m \times n}$, in which the entry in the ith row and jth column is $a_{ij} + b_{ij}$, for $i = 1, 2, 3, \ldots, m$ and $j = 1, 2, 3, \ldots, n$.

Thus, the sum of two matrices of the same order is obtained by adding the corresponding entries; the sum of two matrices of different orders is not defined.

EXAMPLE 4 (a) $\begin{bmatrix} 3 & 1 & 2 \\ 2 & 1 & 4 \end{bmatrix} + \begin{bmatrix} 1 & 0 & 2 \\ -1 & 3 & 0 \end{bmatrix} = \begin{bmatrix} 3+1 & 1+0 & 2+2 \\ 2+(-1) & 1+3 & 4+0 \end{bmatrix} = \begin{bmatrix} 4 & 1 & 4 \\ 1 & 4 & 4 \end{bmatrix}$

(b) $\begin{bmatrix} 1 & 2 \\ 1 & -3 \end{bmatrix} + \begin{bmatrix} 1 & -2 \\ 2 & 3 \end{bmatrix} = \begin{bmatrix} 1+1 & 2+(-2) \\ 1+2 & -3+3 \end{bmatrix} = \begin{bmatrix} 2 & 0 \\ 3 & 0 \end{bmatrix}$ ■

Definition 8.4 A *matrix with each entry equal to 0 is a **zero matrix**.*

Zero matrices are generally denoted by the symbol 0. This distinguishes the zero matrix from the real number 0. For example,

$$0_{2 \times 4} = \begin{bmatrix} 0 & 0 & 0 & 0 \\ 0 & 0 & 0 & 0 \end{bmatrix}$$

is the 2×4 zero matrix.

Definition 8.5 The **negative of a matrix** $A_{m \times n}$, denoted by $-A_{m \times n}$, is formed by replacing each entry a_{ij} in the matrix $A_{m \times n}$ with $-a_{ij}$.

For example, if

$$A_{3 \times 2} = \begin{bmatrix} 3 & -1 \\ 2 & -2 \\ -4 & 5 \end{bmatrix}, \quad \text{then} \quad -A_{3 \times 2} = \begin{bmatrix} -3 & 1 \\ -2 & 2 \\ 4 & -5 \end{bmatrix}.$$

The sum $B_{m \times n} + (-A_{m \times n})$ is called the **difference** of $B_{m \times n}$ and $A_{m \times n}$ and is denoted $B_{m \times n} - A_{m \times n}$.

Properties of Sums At this point, we are able to establish the following facts concerning sums of matrices with real-number entries. We shall prove Parts III and IV; the rest are similar.

Theorem 8.1 If A, B, and C are $m \times n$ matrices with real-number entries, then:

I. $(A + B)_{m \times n}$ is a matrix with real-number entries.   **Closure law for addition**

continued

II. $(A + B) + C = A + (B + C)$ Associative law for
addition

III. *The matrix* $0_{m \times n}$ *has the property that for*   Additive-identity law
every matrix $A_{m \times n}$,

$$A + 0 = A \quad and \quad 0 + A = A.$$

IV.   *For every matrix* $A_{m \times n}$, *the matrix*   Additive-inverse law
$- A_{m \times n}$ *has the property that*

$$A + (-A) = 0 \quad and \quad (-A) + A = 0.$$

V. $A + B = B + A$ Commutative law for
addition

To prove Theorem 8.1-III, observe that since each entry of the zero matrix is 0, the entries of $A_{m \times n} + 0_{m \times n}$ are $a_{ij} + 0 = a_{ij}$ and the entries of $0_{m \times n} + A_{m \times n}$ are $0 + a_{ij} = a_{ij}$. Thus, $A + 0 = 0 + A = A$.

EXAMPLE 5   (a) $\begin{bmatrix} a_{11} & a_{12} \\ a_{21} & a_{22} \end{bmatrix} + \begin{bmatrix} 0 & 0 \\ 0 & 0 \end{bmatrix} = \begin{bmatrix} a_{11} & a_{12} \\ a_{21} & a_{22} \end{bmatrix}$ (b) $\begin{bmatrix} a_1 \\ a_2 \\ a_3 \end{bmatrix} + \begin{bmatrix} 0 \\ 0 \\ 0 \end{bmatrix} = \begin{bmatrix} a_1 \\ a_2 \\ a_3 \end{bmatrix}$ ■

To see that Theorem 8.1-IV is true, let the entries of $A_{m \times n}$ and $- A_{m \times n}$ be a_{ij} and $-a_{ij}$, respectively. Since each entry of $A + (-A)$ is $a_{ij} + (-a_{ij})$, or 0, we have $A + (-A) = 0$. Similarly, $(-A) + A = 0$.

EXAMPLE 6   If

$$A = \begin{bmatrix} 1 & -1 & 2 \\ 3 & -1 & 1 \end{bmatrix},$$

then

$$A + (-A) = \begin{bmatrix} 1 & -1 & 2 \\ 3 & -1 & 1 \end{bmatrix} + \begin{bmatrix} -1 & 1 & -2 \\ -3 & 1 & -1 \end{bmatrix} = \begin{bmatrix} 0 & 0 & 0 \\ 0 & 0 & 0 \end{bmatrix} = 0.$$

EXAMPLE 7   As an example of Theorem 8.1-II, consider the matrices

$$A = \begin{bmatrix} 1 & 3 \\ -1 & 2 \end{bmatrix}, \quad B = \begin{bmatrix} 2 & 1 \\ -3 & 2 \end{bmatrix}, \quad and \quad C = \begin{bmatrix} -1 & 1 \\ 0 & 2 \end{bmatrix}.$$

Then

$$(A + B) + C = \left(\begin{bmatrix} 1 & 3 \\ -1 & 2 \end{bmatrix} + \begin{bmatrix} 2 & 1 \\ -3 & 2 \end{bmatrix} \right) + \begin{bmatrix} -1 & 1 \\ 0 & 2 \end{bmatrix}$$

$$= \begin{bmatrix} 3 & 4 \\ -4 & 4 \end{bmatrix} + \begin{bmatrix} -1 & 1 \\ 0 & 2 \end{bmatrix} = \begin{bmatrix} 2 & 5 \\ -4 & 6 \end{bmatrix}$$

and

$$A + (B + C) = \begin{bmatrix} 1 & 3 \\ -1 & 2 \end{bmatrix} + \left(\begin{bmatrix} 2 & 1 \\ -3 & 2 \end{bmatrix} + \begin{bmatrix} -1 & 1 \\ 0 & 2 \end{bmatrix} \right)$$

$$= \begin{bmatrix} 1 & 3 \\ -1 & 2 \end{bmatrix} + \begin{bmatrix} 1 & 2 \\ -3 & 4 \end{bmatrix} = \begin{bmatrix} 2 & 5 \\ -4 & 6 \end{bmatrix}.$$

Hence,

$$A + (B + C) = (A + B) + C.$$

EXAMPLE 8   As an example of Theorem 8.1-V, consider the matrices

$$A = \begin{bmatrix} 1 & -1 \\ 2 & 3 \end{bmatrix} \quad \text{and} \quad B = \begin{bmatrix} 3 & 4 \\ -2 & -4 \end{bmatrix}.$$

Then

$$A + B = \begin{bmatrix} 1 & -1 \\ 2 & 3 \end{bmatrix} + \begin{bmatrix} 3 & 4 \\ -2 & -4 \end{bmatrix} = \begin{bmatrix} 4 & 3 \\ 0 & -1 \end{bmatrix}$$

and

$$B + A = \begin{bmatrix} 3 & 4 \\ -2 & -4 \end{bmatrix} + \begin{bmatrix} 1 & -1 \\ 2 & 3 \end{bmatrix} = \begin{bmatrix} 4 & 3 \\ 0 & -1 \end{bmatrix}.$$

Hence,

$$A + B = B + A. \quad ■$$

Solution of Matrix Equations

Theorem 8.1 can be used to solve matrix equations of the form $X + A = B$ as follows.

$$(X + A) + (-A) = B + (-A)$$

$$X + (A + (-A)) = B + (-A) \qquad \text{By Theorem 8.1-II}$$

$$X + 0 = B + (-A) \qquad \text{By Theorem 8.1-IV}$$

$$X = B + (-A) \qquad \text{By Theorem 8.1-III}$$

$$= B - A$$

E X A M P L E 9   Solve $X + \begin{bmatrix} -3 & 1 \\ 2 & -1 \end{bmatrix} = \begin{bmatrix} 1 & -1 \\ 2 & -2 \end{bmatrix}$.

Solution $X = \begin{bmatrix} 1 & -1 \\ 2 & -2 \end{bmatrix} - \begin{bmatrix} -3 & 1 \\ 2 & -1 \end{bmatrix} = \begin{bmatrix} 4 & -2 \\ 0 & -1 \end{bmatrix}$ ■

Applications   Matrices can be used to model a variety of problems in science and engineering. For example, if we have a collection of people (or objects) and some relationship exists between the members of the collection, we can use a matrix of 0's and 1's to represent the relationship by assigning 1 to the i, jth entry if member i is related to member j and assigning a 0 to the i, jth entry if the members are not related. Such a matrix is called an **incidence matrix** of the relation.

E X A M P L E 10   Suppose a communications network has four transmitter-decoders. The decoders will decode messages only from certain transmitters, as shown in the following table.

Transmitter	Possible Decoder
1	3, 4
2	1, 2, 3, 4
3	4
4	1, 2

This network can be modeled by the incidence matrix

$$\begin{bmatrix} 0 & 0 & 1 & 1 \\ 1 & 1 & 1 & 1 \\ 0 & 0 & 0 & 1 \\ 1 & 1 & 0 & 0 \end{bmatrix}$$

where the row number is the transmitter being used and the column number is the decoder.   ■

EXERCISE 8.I

A

■ State the order and find the transpose of each matrix. See Examples 1–3.

1. $\begin{bmatrix} 6 & -1 \\ 2 & 3 \end{bmatrix}$

2. $\begin{bmatrix} 4 & 1 \\ 0 & -2 \end{bmatrix}$

3. $\begin{bmatrix} 2 & -7 & 3 \\ 1 & 4 & 0 \end{bmatrix}$

4. $\begin{bmatrix} -3 & 1 \\ 6 & 0 \\ 0 & 2 \end{bmatrix}$

5. $\begin{bmatrix} 2 & 3 & -1 \\ 4 & 0 & 1 \\ -2 & 3 & 1 \end{bmatrix}$

6. $\begin{bmatrix} 4 & -1 & -2 \\ 3 & 0 & 0 \\ 2 & 1 & 1 \end{bmatrix}$

7. $\begin{bmatrix} 4 & -3 & -1 & 0 \\ 2 & 1 & 1 & 6 \end{bmatrix}$

8. $\begin{bmatrix} -2 & 1 & 3 & 2 \\ 4 & 0 & 0 & -2 \\ -1 & 3 & 2 & 4 \end{bmatrix}$

■ Write each sum or difference as a single matrix. See Example 4.

9. $\begin{bmatrix} 2 & 3 \\ 1 & 6 \end{bmatrix} + \begin{bmatrix} 1 & -2 \\ 2 & 3 \end{bmatrix}$

10. $\begin{bmatrix} 4 & -1 & 3 \\ 2 & 1 & 0 \end{bmatrix} + \begin{bmatrix} 3 & -1 & 0 \\ 4 & 0 & -2 \end{bmatrix}$

11. $\begin{bmatrix} 3 & 0 & -1 \\ 2 & 1 & 2 \end{bmatrix} + \begin{bmatrix} 6 & -1 & 0 \\ 0 & 2 & 4 \end{bmatrix}$

12. $[1 \quad 3 \quad 5] + [0 \quad -2 \quad 1]$

13. $\begin{bmatrix} 4 & -3 \\ 2 & 1 \end{bmatrix} - \begin{bmatrix} 6 & 0 \\ -2 & 1 \end{bmatrix}$

14. $\begin{bmatrix} 4 & -1 & 2 \\ 3 & 1 & -4 \end{bmatrix} - \begin{bmatrix} -1 & -1 & 2 \\ 3 & 1 & 4 \end{bmatrix}$

15. $\begin{bmatrix} 10 & 3 & 2 \\ 5 & 1 & 7 \\ 6 & 1 & 9 \end{bmatrix} - \begin{bmatrix} 8 & 12 & 15 \\ -2 & 5 & 6 \\ -3 & 1 & 9 \end{bmatrix}$

16. $\begin{bmatrix} 3 & -1 & 2 \\ 4 & -2 & 1 \\ 6 & 3 & 2 \end{bmatrix} - \begin{bmatrix} 2 & -1 & 2 \\ 4 & -1 & 1 \\ 6 & 3 & 1 \end{bmatrix}$

17. $\begin{bmatrix} 4 \\ 3 \\ -1 \end{bmatrix} + \begin{bmatrix} 6 \\ 0 \\ -2 \end{bmatrix}$

18. $\begin{bmatrix} 2 & 3 \\ 1 & 0 \\ -1 & 2 \end{bmatrix} + \begin{bmatrix} -2 & 0 \\ -3 & 0 \\ 4 & -1 \end{bmatrix}$

19. $\begin{bmatrix} 2 & 3 & 4 \\ -1 & 6 & 2 \\ 1 & 0 & 3 \end{bmatrix} + \begin{bmatrix} 0 & 0 & 0 \\ 0 & 0 & 0 \\ 0 & 0 & 0 \end{bmatrix}$

20. $\begin{bmatrix} 2 & -3 \\ 4 & -1 \\ -2 & 1 \end{bmatrix} + \begin{bmatrix} -2 & 3 \\ -4 & 1 \\ 2 & -1 \end{bmatrix}$

■ Solve each of the following matrix equations. See Example 9.

21. $X - \begin{bmatrix} -1 & 0 \\ 0 & 0 \end{bmatrix} = \begin{bmatrix} 3 & -1 \\ 2 & 1 \end{bmatrix}$

22. $X + \begin{bmatrix} 3 & -1 \\ 2 & 1 \end{bmatrix} = \begin{bmatrix} 5 & 1 \\ -3 & 5 \end{bmatrix}$

23. $\begin{bmatrix} 1 & 3 \\ -1 & 0 \end{bmatrix}^t = \begin{bmatrix} 2 & -2 \\ -1 & 3 \end{bmatrix}^t - X$

24. $X + \begin{bmatrix} 3 & 2 \\ -1 & 4 \end{bmatrix}^t = \begin{bmatrix} 2 & 6 \\ 1 & 5 \end{bmatrix}^t$

■ For Exercises 25 and 26, see Example 10.

25. A basketball team has 8 players. The coach has arranged the following scheme for

transmitting messages: Player 1 calls Players 2 and 3, Player 2 calls Players 4 and 5, Player 3 calls Players 6 and 7, and Player 7 calls Player 8. Write an incidence matrix that models this communication scheme.

26. Suppose 9 computers are networked in such a way that Computer 1 can communicate with Computers 2 through 4, Computer 2 can communicate with Computers 1, 5, and 6, and Computer 3 can communicate with Computers 1 and 7 through 9. All communications are two-way. Assume each computer communicates with itself. Write an incidence matrix that models this network.

B

■   For a two-person zero sum game, the **payoff matrix** for a given player is the matrix whose i, jth entry is the amount won (losses are negative wins) by the player if he chooses alternative i and his opponent chooses alternative j. Write the payoff matrix for Player A in each of the following games.

27. Player A and Player B each secretly choose either 1 or −1. They then reveal their choices simultaneously. If the sum of the choices is not zero, Player A wins 1 dollar; otherwise, Player A loses 2 dollars.

28. Player A and Player B each secretly choose a 1, 2, or 3. They reveal their choices simultaneously. If the sum of the choices is even, Player A wins the sum of the choices; otherwise, Player A loses the sum plus the smaller of the two choices.

29. Show that $[A_{2 \times 2}^t]^t = A_{2 \times 2}$. Does an analogous result seem valid for $n \times n$ matrices? For $m \times n$ matrices?

30. Show that $[A_{2 \times 2} + B_{2 \times 2}]^t = A_{2 \times 2}^t + B_{2 \times 2}^t$. Does an analogous result seem valid for $n \times n$ matrices?

8.2

MATRIX MULTIPLICATION

We shall be interested in two kinds of products involving matrices: (1) the product of a matrix and a real number, and (2) the product of two matrices.

Products of Real Numbers and Matrices   **Definition 8.6**   *The **product** of a real number c and an m × n matrix A with entries a_{ij} is the matrix cA with corresponding entries ca_{ij}, where $i = 1, 2, 3, \ldots, m$ and $j = 1, 2, 3, \ldots, n$.*

EXAMPLE 1   (a) $3\begin{bmatrix} 2 & 1 \\ 0 & 5 \end{bmatrix} = \begin{bmatrix} 3(2) & 3(1) \\ 3(0) & 3(5) \end{bmatrix} = \begin{bmatrix} 6 & 3 \\ 0 & 15 \end{bmatrix}$

(b) $-4\begin{bmatrix} 1 \\ 2 \\ -1 \end{bmatrix} = \begin{bmatrix} -4(1) \\ -4(2) \\ -4(-1) \end{bmatrix} = \begin{bmatrix} -4 \\ -8 \\ 4 \end{bmatrix}$ ■

The following theorem states some simple algebraic laws for the multiplication of matrices by real numbers.

Theorem 8.2   *If A and B are m × n matrices and c and d are real numbers, then*

I.	cA is an $m \times n$ matrix,	V.	$1A = A$,
II.	$c(dA) = (cd)A$,	VI.	$(-1)A = -A$,
III.	$(c + d)A = cA + dA$,	VII.	$0A = 0$,
IV.	$c(A + B) = cA + cB$,	VIII.	$c0 = 0$.

We shall prove only Property IV. The proofs of the remaining properties are similar. Since the elements of $A + B$ are of the form $a_{ij} + b_{ij}$, it follows, by definition, that the elements of $c(A + B)$ are of the form $c(a_{ij} + b_{ij})$. But, since a_{ij}, b_{ij}, and c denote real numbers,

$$c(a_{ij} + b_{ij}) = ca_{ij} + cb_{ij}.$$

Now, the elements of cA are of the form ca_{ij}, and those of cB are of the form cb_{ij}, so that the elements of $cA + cB$ are of the form $ca_{ij} + cb_{ij}$ and Property IV is proved.

Some of the properties of Theorems 8.1 and 8.2 can be used to solve simple equations involving unknowns that are matrix-valued.

E X A M P L E   2   Solve $3X + 2\begin{bmatrix} -1 & 2 \\ 3 & 1 \end{bmatrix} = \begin{bmatrix} 1 & 0 \\ 0 & 1 \end{bmatrix}$ for X.

Solution   We use the properties of Theorems 8.1 and 8.2 to generate the following sequence of equations.

$$3X + 2\begin{bmatrix} -1 & 2 \\ 3 & 1 \end{bmatrix} = \begin{bmatrix} 1 & 0 \\ 0 & 1 \end{bmatrix}$$

$$3X = \begin{bmatrix} 1 & 0 \\ 0 & 1 \end{bmatrix} - 2\begin{bmatrix} -1 & 2 \\ 3 & 1 \end{bmatrix}$$

$$X = \frac{1}{3}\left(\begin{bmatrix} 1 & 0 \\ 0 & 1 \end{bmatrix} - 2\begin{bmatrix} -1 & 2 \\ 3 & 1 \end{bmatrix}\right)$$

$$= \frac{1}{3}\begin{bmatrix} 1 & 0 \\ 0 & 1 \end{bmatrix} - \frac{2}{3}\begin{bmatrix} -1 & 2 \\ 3 & 1 \end{bmatrix}$$

$$= \begin{bmatrix} 1 & -\frac{4}{3} \\ -2 & -\frac{1}{3} \end{bmatrix} \quad ■$$

Products of Matrices

Turning now to the *product of two matrices*, we first define the product of a row matrix and a column matrix.

Definition 8.7   For $A_{1 \times p} = [a_1 \cdots a_p]$ and $B_{p \times 1} = \begin{bmatrix} b_1 \\ \vdots \\ b_p \end{bmatrix}$, *the product is*

$$A_{1 \times p} B_{p \times 1} = a_1 b_1 + a_2 b_2 + \cdots + a_p b_p.$$

EXAMPLE 3   (a) $[1 \quad -2] \begin{bmatrix} 3 \\ -4 \end{bmatrix} = 1(3) + (-2)(-4) = 11$

   (b) $[1 \quad 0 \quad -1 \quad 2] \begin{bmatrix} 3 \\ 1 \\ -2 \\ 4 \end{bmatrix} = 1(3) + 0(1) + (-1)(-2) + 2(4) = 13$ ■

Note from Definition 8.7 that the product $A_{1 \times p} B_{p \times 1}$ is always a real number. However, the product of matrices with more than one row or column is a matrix.

Definition 8.8   *The **product** of matrices $A_{m \times p}$ and $B_{p \times n}$ is the $m \times n$ matrix whose i, jth entry is the product of the ith row of A and the jth column of B. This product is denoted AB or $A \cdot B$.*

EXAMPLE 4   Multiply $\begin{bmatrix} 3 & 0 & 1 \\ 0 & 1 & 2 \end{bmatrix} \begin{bmatrix} 1 & -2 \\ -1 & 2 \\ 1 & 1 \end{bmatrix}$.

Solution   Since the matrix on the left is 2×3 and the one on the right is 3×2, the product will be 2×2.

The 1, 1 entry is $[3 \quad 0 \quad 1] \begin{bmatrix} 1 \\ -1 \\ 1 \end{bmatrix} = 3 \cdot 1 + 0(-1) + 1 \cdot 1 = 4.$

The 1, 2 entry is $[3 \quad 0 \quad 1] \begin{bmatrix} -2 \\ 2 \\ 1 \end{bmatrix} = 3(-2) + 0 \cdot 2 + 1 \cdot 1 = -5.$

The 2, 1 entry is $\begin{bmatrix} 0 & 1 & 2 \end{bmatrix} \begin{bmatrix} 1 \\ -1 \\ 1 \end{bmatrix} = 0 \cdot 1 + 1(-1) + 2 \cdot 1 = 1.$

The 2, 2 entry is $\begin{bmatrix} 0 & 1 & 2 \end{bmatrix} \begin{bmatrix} -2 \\ 2 \\ 1 \end{bmatrix} = 0(-2) + 1 \cdot 2 + 2 \cdot 1 = 4.$

Therefore, $\begin{bmatrix} 3 & 0 & 1 \\ 0 & 1 & 2 \end{bmatrix} \begin{bmatrix} 1 & -2 \\ -1 & 2 \\ -1 & 1 \end{bmatrix} = \begin{bmatrix} 4 & -5 \\ 1 & 4 \end{bmatrix}.$

EXAMPLE 5   If $A = \begin{bmatrix} 1 & 2 \\ -1 & 3 \end{bmatrix}$ and $B = \begin{bmatrix} 2 & 1 \\ 1 & 1 \end{bmatrix}$, find AB and BA.

Solution $AB = \begin{bmatrix} 1 & 2 \\ -1 & 3 \end{bmatrix} \begin{bmatrix} 2 & 1 \\ 1 & 1 \end{bmatrix} = \begin{bmatrix} 2+2 & 1+2 \\ -2+3 & -1+3 \end{bmatrix} = \begin{bmatrix} 4 & 3 \\ 1 & 2 \end{bmatrix}$

$BA = \begin{bmatrix} 2 & 1 \\ 1 & 1 \end{bmatrix} \begin{bmatrix} 1 & 2 \\ -1 & 3 \end{bmatrix} = \begin{bmatrix} 2-1 & 4+3 \\ 1-1 & 2+3 \end{bmatrix} = \begin{bmatrix} 1 & 7 \\ 0 & 5 \end{bmatrix}$ ■

Properties of Products

Example 5 shows very clearly that the multiplication of matrices, in general, is *not commutative*. Thus, when discussing products of matrices, we must specify the *order* in which the matrices are to be considered as factors. For the product AB, we say that A is *right-multiplied* by B and that B is *left-multiplied* by A.

Note that the definition of the product of two matrices, A and B, requires that the matrix A have the same number of *columns* as B has *rows*; the result, AB, then has the same number of rows as A and the same number of columns as B. Such matrices A and B are said to be **compatible** for multiplication. The fact that two matrices are conformable in the order AB, however, does not mean that they necessarily are compatible in the order BA.

EXAMPLE 6   If $A = \begin{bmatrix} 3 & 1 \\ 1 & 0 \\ 2 & 1 \end{bmatrix}$ and $B = \begin{bmatrix} 1 & -1 \\ 2 & 1 \end{bmatrix}$, find AB.

Solution   Since A is a 3×2 matrix and B is a 2×2 matrix, they are compatible for multiplication in the order AB. We have

$$AB = \begin{bmatrix} 3 & 1 \\ 1 & 0 \\ 2 & 1 \end{bmatrix} \begin{bmatrix} 1 & -1 \\ 2 & 1 \end{bmatrix} = \begin{bmatrix} 3+2 & -3+1 \\ 1+0 & -1+0 \\ 2+2 & -2+1 \end{bmatrix} = \begin{bmatrix} 5 & -2 \\ 1 & -1 \\ 4 & -1 \end{bmatrix}.$$ ■

Note that the matrices A and B in Example 6 are not compatible in the order BA.

In much of the matrix work in this book, we shall focus our attention on matrices having the same number of rows as columns. For brevity, a matrix of order $n \times n$ is often called a **square matrix** of order n. Although many of the ideas we shall discuss are applicable to compatible matrices of any order, we shall apply the notions only to square matrices.

Theorem 8.3   *If A, B, and C are $n \times n$ square matrices, then*

$$(AB)C = A(BC).$$

If A is a square matrix, then A^2, A^3, etc., denote AA, $(AA)A$, etc.

Theorem 8.4   *If A, B, and C are $n \times n$ square matrices, then*

$$A(B + C) = AB + AC$$

and

$$(B + C)A = BA + CA.$$

The proofs of these theorems involve some complicated symbolism and are omitted here. Observe that, because matrix multiplication is not, in general, commutative, we must establish both the left-hand and the right-hand distributive property.

Definition 8.9   *The **principal diagonal** of a square matrix is the ordered set of entries a_{jj}, extending from the upper left-hand corner to the lower right-hand corner of the matrix. Thus, the principal diagonal contains a_{11}, a_{22}, a_{33}, etc.*

For example, the principal diagonal of

$$\begin{bmatrix} 1 & 3 & -1 \\ 5 & 2 & 3 \\ 6 & 4 & 0 \end{bmatrix}$$

consists of 1, 2, and 0, in that order.

Definition 8.10   *A **diagonal matrix** is a square matrix in which all entries not in the principal diagonal are 0.*

Thus,

$$\begin{bmatrix} 4 & 0 \\ 0 & 2 \end{bmatrix} \quad \text{and} \quad \begin{bmatrix} 1 & 0 & 0 \\ 0 & 1 & 0 \\ 0 & 0 & 0 \end{bmatrix}$$

are diagonal matrices.

Definition 8.11 $I_{n \times n}$ *denotes the diagonal matrix having 1's for entries on the principal diagonal.*

For example,

$$I_{2 \times 2} = \begin{bmatrix} 1 & 0 \\ 0 & 1 \end{bmatrix} \quad \text{and} \quad I_{4 \times 4} = \begin{bmatrix} 1 & 0 & 0 & 0 \\ 0 & 1 & 0 & 0 \\ 0 & 0 & 1 & 0 \\ 0 & 0 & 0 & 1 \end{bmatrix}.$$

The following properties are consequences of the definitions we have adopted.

Theorem 8.5   *For each matrix $A_{n \times n}$,*

$$A_{n \times n} I_{n \times n} = I_{n \times n} A_{n \times n} = A_{n \times n}.$$

Furthermore, $I_{n \times n}$ is the unique matrix having this property for all matrices $A_{n \times n}$.

Accordingly, $I_{n \times n}$ is the **identity element for multiplication** in the set of $n \times n$ square matrices. The proof of this theorem, for the illustrative case $n = 2$, is left as an exercise.

The following result relates the order in which matrices can be multiplied by real numbers and by other matrices.

Theorem 8.6   *If A and B are $n \times n$ square matrices and a is a real number, then*

$$a(AB) = (aA)B = A(aB).$$

EXAMPLE 7   Let

$$A = \begin{bmatrix} 1 & 3 \\ 2 & -4 \end{bmatrix} \quad \text{and} \quad B = \begin{bmatrix} 4 & 1 \\ -2 & -1 \end{bmatrix}.$$

Then

$$4(AB) = 4\left(\begin{bmatrix} 1 & 3 \\ 2 & -4 \end{bmatrix} \begin{bmatrix} 4 & 1 \\ -2 & -1 \end{bmatrix} \right) = 4 \begin{bmatrix} -2 & -2 \\ 16 & 6 \end{bmatrix} = \begin{bmatrix} -8 & -8 \\ 64 & 24 \end{bmatrix},$$

$$(4A)B = \begin{bmatrix} 4 & 12 \\ 8 & -16 \end{bmatrix} \begin{bmatrix} 4 & 1 \\ -2 & -1 \end{bmatrix} = \begin{bmatrix} -8 & -8 \\ 64 & 24 \end{bmatrix},$$

and

$$A(4B) = \begin{bmatrix} 1 & 3 \\ 2 & -4 \end{bmatrix} \begin{bmatrix} 16 & 4 \\ -8 & -4 \end{bmatrix} = \begin{bmatrix} -8 & -8 \\ 64 & 24 \end{bmatrix}.$$

Thus, $4(AB) = (4A)B = A(4B)$.   ■

Applications   Matrix multiplication is used to solve problems in many areas of science and engineering. For example, if A is an incidence matrix for a communications network, then the i, jth entry of A^2 is the number of two-stage communications (a communication through 1 intermediate) from station i to station j. Similarly, A^3 gives the number of three-stage communications, and so on.

EXAMPLE 8   Let A be the incidence matrix of Example 8.1-10, namely

$$A = \begin{bmatrix} 0 & 0 & 1 & 1 \\ 1 & 1 & 1 & 1 \\ 0 & 0 & 0 & 1 \\ 1 & 1 & 0 & 0 \end{bmatrix}.$$

Show that there are no two-stage communications from Transmitter-Decoder 1 to Transmitter-Decoder 3.

Solution   The number of two-stage communications from 1 to 3 is the $(1, 3)$ entry of the matrix

$$A^2 = \begin{bmatrix} 1 & 1 & 0 & 1 \\ 2 & 2 & 2 & 3 \\ 1 & 1 & 0 & 0 \\ 1 & 1 & 2 & 2 \end{bmatrix}.$$

Since $a_{13} = 0$, there are no two-stage communications from 1 to 3.   ■

We see another example of an application of matrix multiplication in game theory. A **strategy** for a player in a two-person zero sum game is a row matrix whose entries are all nonnegative and which add to 1. The jth entry is the relative frequency with which the player chooses strategy j. If S is a strategy for Player A and P is a payoff matrix (see Exercises 27–28 of Section 8.1) for Player A, then the jth entry of the matrix SP is the average amount Player A will "win" if Player B uses alternative j.

EXAMPLE 9   What are the average outcomes for Player A if she uses strategy $S = \begin{bmatrix} \frac{1}{4} & \frac{3}{4} \end{bmatrix}$ in a game where her payoff matrix is

$$P = \begin{bmatrix} 1 & 0 \\ -1 & 1 \end{bmatrix}$$

and where the entries in P are in dollars?

Solution   We compute the 1×2 matrix

$$SP = [\tfrac{1}{4} \quad \tfrac{3}{4}] \begin{bmatrix} 1 & 0 \\ -1 & 1 \end{bmatrix} = [-\tfrac{1}{2} \quad \tfrac{3}{4}].$$

Thus, if Player B chooses alternative 1, then Player A will average losing $\tfrac{1}{2}$ dollar. If Player B chooses alternative 2, then Player A will average winning $\tfrac{3}{4}$ dollar.   ■

Finally, one of the most common applications of matrix multiplication is to solve systems of equations. This will be discussed in Section 8.7.

EXERCISE 8.2

A

■ *Write each product as a single matrix. See Example 1.*

1. $2\begin{bmatrix} 1 & 1 \\ 0 & 2 \end{bmatrix}$

2. $4\begin{bmatrix} 1 & 0 & 1 & 0 \\ 1 & 1 & 0 & 0 \\ 0 & 0 & 1 & 1 \end{bmatrix}$

3. $-5\begin{bmatrix} 0 & 1 & -1 \\ 3 & -1 & 2 \end{bmatrix}$

4. $2\begin{bmatrix} 2 & 1 & 3 & -2 \\ 4 & 2 & 0 & -1 \\ 0 & 0 & -1 & 2 \end{bmatrix}$

■ *Find a matrix X satisfying each matrix equation. See Example 2.*

5. $3X + \begin{bmatrix} 1 & 0 \\ 2 & 1 \end{bmatrix} = \begin{bmatrix} -2 & 3 \\ -1 & -2 \end{bmatrix}$

6. $2X + 3\begin{bmatrix} 1 & 1 \\ 0 & 1 \end{bmatrix} = \begin{bmatrix} 7 & -1 \\ 3 & -5 \end{bmatrix}$

7. $X + 2I = \begin{bmatrix} 3 & -1 \\ 1 & 2 \end{bmatrix}$

8. $3X - 2I = \begin{bmatrix} 7 & 3 \\ 6 & 4 \end{bmatrix}$

■ *Write each product as a real number. See Example 3.*

9. $[2 \quad 3] \cdot \begin{bmatrix} 1 \\ 1 \end{bmatrix}$

10. $[1 \quad 1 \quad 1] \cdot \begin{bmatrix} 0 \\ 1 \\ 0 \end{bmatrix}$

11. $[1 \quad -2] \cdot \begin{bmatrix} 3 \\ 2 \end{bmatrix}$

12. $[3 \quad -2 \quad 2] \cdot \begin{bmatrix} 1 \\ 0 \\ -2 \end{bmatrix}$

■ *For Exercises 13–22, see Examples 4–6.*

13. $\begin{bmatrix} 3 & -1 \\ 2 & 1 \end{bmatrix} \cdot \begin{bmatrix} 1 & -4 \\ 2 & 1 \end{bmatrix}$

14. $\begin{bmatrix} 1 & -5 \\ 0 & 2 \end{bmatrix} \cdot \begin{bmatrix} 3 & 1 \\ -1 & 2 \end{bmatrix}$

15. $\begin{bmatrix} 4 & -5 \\ 7 & 3 \end{bmatrix} \cdot \begin{bmatrix} 5 & -1 \\ -2 & 7 \end{bmatrix}$

16. $\begin{bmatrix} 1 & -2 \\ -3 & 1 \end{bmatrix} \cdot \begin{bmatrix} 5 & 1 \\ 0 & 2 \end{bmatrix}$

17. $\begin{bmatrix} -3 & 1 & 0 \\ 2 & 1 & 1 \end{bmatrix} \cdot \begin{bmatrix} 2 & 0 \\ 1 & -1 \\ 3 & 0 \end{bmatrix}$

18. $\begin{bmatrix} 1 & -1 & 0 \\ 2 & 1 & 3 \end{bmatrix} \cdot \begin{bmatrix} 4 & -1 \\ 2 & 0 \\ 1 & 1 \end{bmatrix}$

19. $\begin{bmatrix} -1 & 0 & 1 \\ 2 & 1 & 0 \\ 1 & 0 & 0 \end{bmatrix} \cdot \begin{bmatrix} 0 & 1 & 3 \\ 1 & 0 & 2 \\ -1 & 1 & 1 \end{bmatrix}$

20. $\begin{bmatrix} 2 & -3 & 1 \\ 0 & 1 & -1 \\ 2 & 0 & 0 \end{bmatrix} \cdot \begin{bmatrix} 1 & 0 & 0 \\ 0 & 1 & 0 \\ 0 & 0 & 1 \end{bmatrix}$

21. $\begin{bmatrix} 2 & -2 & -1 \\ 1 & 1 & -2 \\ 1 & 0 & -1 \end{bmatrix} \cdot \begin{bmatrix} -1 & -2 & 5 \\ -1 & -1 & 3 \\ -1 & -2 & 4 \end{bmatrix}$

22. $\begin{bmatrix} -1 & -2 & 5 \\ -1 & -1 & 3 \\ -1 & -2 & 4 \end{bmatrix} \cdot \begin{bmatrix} 2 & -2 & -1 \\ 1 & 1 & -2 \\ 1 & 0 & -1 \end{bmatrix}$

■ Let $A = \begin{bmatrix} 1 & -2 \\ 1 & 0 \end{bmatrix}$ and $B = \begin{bmatrix} -1 & 2 \\ -1 & 1 \end{bmatrix}$. Compute each of the following products. See Examples 4–7.

23. AB 24. BA 25. $(AB)A$ 26. $(BA)B$ 27. A^tB 28. AB^t

■ Compute the number of two-stage communications between all possible pairs of stations in a communications network with the following incidence matrices. See Example 8.

29. $\begin{bmatrix} 1 & 1 & 1 & 1 \\ 0 & 1 & 1 & 1 \\ 0 & 0 & 1 & 1 \\ 0 & 0 & 0 & 1 \end{bmatrix}$ 30. $\begin{bmatrix} 1 & 0 & 1 & 0 \\ 0 & 1 & 0 & 1 \\ 1 & 0 & 1 & 0 \\ 0 & 1 & 0 & 1 \end{bmatrix}$ 31. $\begin{bmatrix} 1 & 1 & 0 & 0 \\ 1 & 1 & 0 & 0 \\ 0 & 0 & 1 & 1 \\ 0 & 0 & 1 & 1 \end{bmatrix}$ 32. $\begin{bmatrix} 1 & 0 & 0 & 0 \\ 0 & 1 & 1 & 0 \\ 0 & 0 & 1 & 1 \\ 0 & 0 & 0 & 1 \end{bmatrix}$

B

■ Player A's payoff matrix in a two-person zero sum game is

$$P = \begin{bmatrix} 2 & -4 & 4 \\ -4 & 4 & -7 \\ 4 & -7 & 6 \end{bmatrix}.$$

Compute all possible average wins for Player A if he adopts the strategies given. See Example 9.

33. $[1 \quad 0 \quad 0]$ 34. $[0 \quad 1 \quad 0]$ 35. $\begin{bmatrix} \dfrac{1}{3} & \dfrac{1}{2} & \dfrac{1}{6} \end{bmatrix}$ 36. $[0.3 \quad 0.2 \quad 0.5]$

37. Show that if $A = \begin{bmatrix} -1 & 2 \\ 0 & 1 \end{bmatrix}$ and $B = \begin{bmatrix} 1 & 0 \\ -1 & 2 \end{bmatrix}$, then

    a. $(A + B)(A + B) \neq A^2 + 2AB + B^2$.

    b. $(A + B)(A - B) \neq A^2 - B^2$.

38. What conditions can be placed on the matrices A and B to guarantee that $(A + B)^2 = A^2 + 2AB + B^2$ and $A^2 - B^2 = (A + B)(A - B)$?

39. Show that for each matrix $A_{2 \times 2}$,

$$A_{2 \times 2} \cdot I_{2 \times 2} = I_{2 \times 2} \cdot A_{2 \times 2} = A_{2 \times 2}.$$

40. Show that $\begin{bmatrix} 0 & a \\ a & 0 \end{bmatrix}^2 = a^2 I$.

41. Show that $(A_{2 \times 2} \cdot B_{2 \times 2})^t = B_{2 \times 2}^t \cdot A_{2 \times 2}^t$.

42. Show that $A_{2 \times 2}^2 = (-A_{2 \times 2})^2$.

8.3

SOLUTION OF LINEAR SYSTEMS BY USING ROW-EQUIVALENT MATRICES

Elementary Transformations

An **elementary transformation** of a matrix $A_{n \times m}$ is one of the following three operations upon the rows of the matrix.

1. Multiply the entries of any row of $A_{n \times m}$ by k, where k is a nonzero real number.
2. Interchange any two rows of $A_{n \times m}$.
3. Multiply the entries of any row of $A_{n \times m}$ by k, where k is a real number, and add to the corresponding entries of any other row.

EXAMPLE 1 Let $A = \begin{bmatrix} 1 & 3 \\ -2 & 4 \end{bmatrix}$.

(a) The matrix $B = \begin{bmatrix} 1 & 3 \\ -1 & 2 \end{bmatrix}$ is obtained from A by multiplying each entry of Row 2 by $\frac{1}{2}$.

(b) The matrix $C = \begin{bmatrix} -2 & 4 \\ 1 & 3 \end{bmatrix}$ is obtained from A by interchanging Rows 1 and 2.

(c) The matrix $D = \begin{bmatrix} 0 & 5 \\ -2 & 4 \end{bmatrix}$ is obtained from A by multiplying each entry of Row 2 by $\frac{1}{2}$ and adding the result to the corresponding entry of Row 1. ■

Notice that the inverse of an elementary transformation is an elementary transformation. That is, you can undo an elementary transformation by means of an elementary transformation. For example, if A is transformed into B by interchanging two rows, then you can regain A from B by again interchanging the same two rows. It follows that if B results from performing a *succession* of elementary transformations on A, then A can similarly be obtained from B by performing the inverse operations in reverse order.

Row-Equivalent Matrices

Definition 8.12 *If B is a matrix resulting from a succession of a finite number of elementary transformations on a matrix A, then A and B are **row-equivalent** matrices. This is expressed by writing A ∼ B or B ∼ A.*

EXAMPLE 2 Show that $\begin{bmatrix} 1 & -2 & -1 \\ 1 & 0 & 2 \\ -4 & 3 & 1 \end{bmatrix} \sim \begin{bmatrix} 3 & -6 & -3 \\ 1 & 0 & 2 \\ -4 & 3 & 1 \end{bmatrix}$.

Solution Multiplying each entry of the first row of the left-hand matrix by 3, we obtain the right-hand matrix.

EXAMPLE 3 Show that $\begin{bmatrix} 1 & -2 & -1 \\ 1 & 0 & 2 \\ -4 & 3 & 1 \end{bmatrix} \sim \begin{bmatrix} -4 & 3 & 1 \\ 1 & 0 & 2 \\ 1 & -2 & -1 \end{bmatrix}$.

Solution Interchanging the first and third rows of the left-hand matrix, we obtain the right-hand matrix. ■

Sometimes it is convenient to make several elementary transformations on the same matrix.

EXAMPLE 4 Show that $\begin{bmatrix} 1 & -2 & -1 \\ 1 & 0 & 2 \\ -4 & 3 & 1 \end{bmatrix} \sim \begin{bmatrix} 1 & -2 & -1 \\ 0 & 2 & 3 \\ 0 & -5 & -3 \end{bmatrix}$.

Solution Multiplying the entries of the first row of the left-hand matrix by -1 and adding these products to the corresponding entries of the second row, and then multiplying the entries of the first row by 4 and adding these products to the corresponding entries of the third row, we obtain the right-hand matrix. ■

The $n \times n$ matrices that are row-equivalent to $I_{n \times n}$ are called **nonsingular** matrices. If $n \times n$ matrix is not row-equivalent to $I_{n \times n}$, it is called **singular**.

EXAMPLE 5 Show that $A = \begin{bmatrix} 1 & 2 & 1 \\ -1 & 1 & 0 \\ 1 & 0 & 1 \end{bmatrix}$ is a nonsingular matrix.

Solution We first make appropriate transformations to obtain "0" elements in each entry (except for the principal diagonal) in Columns 1, 2, and 3. Each reference to a row indicates the row of the preceding matrix.

$$\begin{bmatrix} 1 & 2 & 1 \\ -1 & 1 & 0 \\ 1 & 0 & 1 \end{bmatrix} \sim \begin{bmatrix} 1 & 2 & 1 \\ 0 & 3 & 1 \\ 0 & -2 & 0 \end{bmatrix} \quad \begin{array}{l} \text{Row 2 + Row 1} \\ \text{Row 3 + [−1 × Row 1]} \end{array}$$

$$\sim \begin{bmatrix} 1 & 0 & 1 \\ 0 & 3 & 1 \\ 0 & 0 & \frac{2}{3} \end{bmatrix} \quad \begin{array}{l} \text{Row 1 + Row 3} \\[1em] \text{Row 3 + [}\frac{2}{3}\text{ × Row 2]} \end{array}$$

$$\sim \begin{bmatrix} 1 & 0 & 0 \\ 0 & 3 & 0 \\ 0 & 0 & \frac{2}{3} \end{bmatrix} \quad \begin{array}{l} \text{Row 1 + [−}\frac{3}{2}\text{ × Row 3]} \\ \text{Row 2 + [−}\frac{3}{2}\text{ × Row 3]} \end{array}$$

$$\sim \begin{bmatrix} 1 & 0 & 0 \\ 0 & 1 & 0 \\ 0 & 0 & 1 \end{bmatrix} \quad \begin{array}{l} \frac{1}{3}\text{ × Row 2} \\ \frac{3}{2}\text{ × Row 3} \end{array}$$

Since the last operation yields the matrix $I_{3 \times 3}$, A is nonsingular.   ■

Notice that in Example 5 each elementary transformation is noted next to the appropriate row. This provides a convenient record that is helpful when checking the work. It is advisable to keep such a record when doing problems.

Matrix Solution of Linear Systems

In a linear system of the form

$$a_{11}x + a_{12}y + a_{13}z = c_1$$
$$a_{21}x + a_{22}y + a_{23}z = c_2$$
$$a_{31}x + a_{32}y + a_{33}z = c_3,$$

the matrices

$$\begin{bmatrix} a_{11} & a_{12} & a_{13} \\ a_{21} & a_{22} & a_{23} \\ a_{31} & a_{32} & a_{33} \end{bmatrix} \quad \text{and} \quad \begin{bmatrix} a_{11} & a_{12} & a_{13} & \vdots & c_1 \\ a_{21} & a_{22} & a_{23} & \vdots & c_2 \\ a_{31} & a_{32} & a_{33} & \vdots & c_3 \end{bmatrix}$$

are called the **coefficient matrix** and the **augmented matrix**, respectively. Similar definitions hold for a system of n linear equations.

Starting with the augmented matrix of a linear system and generating a sequence of row-equivalent matrices, we can obtain a matrix from which the solution set of the system is evident simply by inspection. The validity of the method, which is illustrated by the next example, follows from the fact that performing elementary transformations on the augmented matrix of a system corresponds to performing the same sorts of operations on the equations of the system itself. Neither multiplying an equation by a nonzero constant nor interchanging two equations has any effect on the solution set of the system. Multiplying one equation by a constant and adding it to another equation in effect replaces the other by a linear combination of equations; a generalization of Theorems 7.1 and 7.2 ensures that the solution set remains the same.

EXAMPLE 6   Solve the system

$$x + 2y - 3z = -4$$
$$2x - y + z = 3$$
$$3x + 2y + z = 10.$$

Solution   The augmented matrix of the system is

$$\begin{bmatrix} 1 & 2 & -3 & \vdots & -4 \\ 2 & -1 & 1 & \vdots & 3 \\ 3 & 2 & 1 & \vdots & 10 \end{bmatrix}.$$

We perform elementary transformations on this matrix as follows.

$$\begin{bmatrix} 1 & 2 & -3 & \vdots & -4 \\ 0 & -5 & 7 & \vdots & 11 \\ 3 & 2 & 1 & \vdots & 10 \end{bmatrix} \quad \textbf{Row 2} + [-2 \times \textbf{Row 1}]$$

$$\begin{bmatrix} 1 & 2 & -3 & \vdots & -4 \\ 0 & -5 & 7 & \vdots & 11 \\ 0 & -4 & 10 & \vdots & 22 \end{bmatrix} \quad \textbf{Row 3} + [-3 \times \textbf{Row 1}]$$

$$\begin{bmatrix} 1 & 2 & -3 & \vdots & -4 \\ 0 & -5 & 7 & \vdots & 11 \\ 0 & 0 & -22 & \vdots & -66 \end{bmatrix} \quad \textbf{-5} \times \textbf{Row 3, then Row 3} + [4 \times \textbf{Row 2}]$$

The resulting system is

$$x + 2y - 3z = -4$$
$$-5y + 7z = 11$$
$$-22z = -66,$$

which may be solved by reverse substitution. This example was solved in Section 7.2, and the augmented matrices here correspond exactly to the equivalent systems obtained there. Matrix notation, however, so facilitates the operations on the equations (rows) that it is easy to obtain even simpler equivalent systems. We can rework the example as follows; the system of equations corresponding to each successive matrix is shown on the right of the matrix in the following solution.

$$
\begin{array}{ll}
& \begin{bmatrix} 1 & 2 & -3 & \vdots & -4 \\ 0 & -5 & 7 & \vdots & 11 \\ 0 & -4 & 10 & \vdots & 22 \end{bmatrix} & \begin{array}{l} x + 2y - 3z = -4 \\ 0x - 5y + 7z = 11 \\ 0x - 4y + 10z = 22 \end{array}
\end{array}
$$

Row 2 + [−2 × Row 1]
Row 3 + [−3 × Row 1]

$$
\begin{array}{ll}
& \begin{bmatrix} 1 & 0 & -\frac{1}{5} & \vdots & \frac{2}{5} \\ 0 & -5 & 7 & \vdots & 11 \\ 0 & 0 & \frac{22}{5} & \vdots & \frac{66}{5} \end{bmatrix} & \begin{array}{l} x + 0y - \frac{1}{5}z = \frac{2}{5} \\ 0x - 5y + 7z = 11 \\ 0x + 0y + \frac{22}{5}z = \frac{66}{5} \end{array}
\end{array}
$$

Row 1 + [$\frac{2}{5}$ × Row 2]

Row 3 + [$-\frac{4}{5}$ × Row 2]

$$
\begin{array}{ll}
& \begin{bmatrix} 5 & 0 & -1 & \vdots & 2 \\ 0 & -5 & 7 & \vdots & 11 \\ 0 & 0 & 1 & \vdots & 3 \end{bmatrix} & \begin{array}{l} 5x + 0y - z = 2 \\ 0x - 5y + 7z = 11 \\ 0x + 0y + z = 3 \end{array}
\end{array}
$$

5 × Row 1

$\frac{5}{22}$ × Row 3

$$
\begin{array}{ll}
& \begin{bmatrix} 5 & 0 & 0 & \vdots & 5 \\ 0 & -5 & 0 & \vdots & -10 \\ 0 & 0 & 1 & \vdots & 3 \end{bmatrix} & \begin{array}{l} 5x + 0y + 0z = 5 \\ 0x - 5y + 0z = -10 \\ 0x + 0y + z = 3 \end{array}
\end{array}
$$

Row 1 + Row 3
Row 2 + [−7 × Row 3]

$$
\begin{array}{ll}
& \begin{bmatrix} 1 & 0 & 0 & \vdots & 1 \\ 0 & 1 & 0 & \vdots & 2 \\ 0 & 0 & 1 & \vdots & 3 \end{bmatrix} & \begin{array}{l} x + 0y + 0z = 1 \\ 0x + y + 0z = 2 \\ 0x + 0y + z = 3 \end{array}
\end{array}
$$

$\frac{1}{5}$ × Row 1

$-\frac{1}{5}$ × Row 2

The last system is equivalent to

$$x = 1$$
$$y = 2$$
$$z = 3.$$

From this, the solution set, $\{(1, 2, 3)\}$, for the given system is evident by inspection.   ■

For any given $n \times n$ linear system with a *nonsingular* coefficient matrix, there are many sequences of row operations that will transform the augmented matrix of a system equivalently to one of the form

$$
\begin{bmatrix}
1 & 0 & 0 & \cdots & 0 & \vdots & x_1 \\
0 & 1 & 0 & \cdots & 0 & \vdots & x_2 \\
0 & 0 & 1 & \cdots & 0 & \vdots & x_3 \\
\vdots & \vdots & \vdots & & \vdots & \vdots & \vdots \\
0 & 0 & 0 & \cdots & 1 & \vdots & x_n
\end{bmatrix},
$$

from which the solution set, $\{(x_1, \ldots, x_n)\}$, of the original system is evident by inspection. Finding the most efficient sequence depends on experience and insight, but several good systematic procedures exist. The procedure used above was, briefly, the following one: obtain a 1 on the diagonal using type 1 or 2 transformations if necessary; next, "clear out" the remainder of the column using type 3 transformations to obtain 0's; then proceed to the next column.

Systems with Singular Coefficient Matrices

If the row-reduction process yields a row of the form

$$0 \quad 0 \quad \cdots \quad 0 \quad a,$$

then that row in the system corresponds to an equation of the form

$$0 \cdot x_1 + 0 \cdot x_2 + \cdots + 0 \cdot x_n = a.$$

This equation has a solution only if $a = 0$, and in that case has an infinite number of solutions. Thus, the system will have no solution when the row-reduction process yields a row of the form

$$0 \quad 0 \quad \cdots \quad 0 \quad a,$$

where $a \neq 0$. The system will have an infinite number of solutions when the process does not yield such a row but does yield a row of zeros.

E X A M P L E 7   Solve the system

$$2x + y = 1$$
$$4x + 2y = 0.$$

Solution   The augmented matrix of the system is

$$\begin{bmatrix} 2 & 1 & \vdots & 1 \\ 4 & 2 & \vdots & 0 \end{bmatrix}.$$

Thus, we have

$$\begin{bmatrix} 2 & 1 & \vdots & 1 \\ 4 & 2 & \vdots & 0 \end{bmatrix} \sim \begin{bmatrix} 2 & 1 & \vdots & 1 \\ 0 & 0 & \vdots & -2 \end{bmatrix} \quad \textbf{Row 2 + [−2 × Row 1]}$$

Since the last row in the matrix on the right is of the form

$$0 \quad 0 \quad a,$$

with $a = -2$, the system has no solution.

E X A M P L E 8   Solve the system

$$2x + y = 1$$
$$4x + 2y = 2.$$

Solution   The augmented matrix of this system is

$$\begin{bmatrix} 2 & 1 & \vdots & 1 \\ 4 & 2 & \vdots & 2 \end{bmatrix}.$$

Thus, we have

$$\begin{bmatrix} 2 & 1 & | & 1 \\ 4 & 2 & | & 2 \end{bmatrix} \sim \begin{bmatrix} 2 & 1 & | & 1 \\ 0 & 0 & | & 0 \end{bmatrix}$$ Row 2 + [−2 × Row 1]

Since the last row in the matrix on the right is of the form

$$0 \quad 0 \quad a,$$

with $a = 0$ and there are no rows of this form with $a \neq 0$, the system has an infinite number of solutions.

The system corresponding to the matrix

$$\begin{bmatrix} 2 & 1 & | & 1 \\ 0 & 0 & | & 0 \end{bmatrix}$$

is

$$2x + y = 1$$

$$0x + 0y = 0.$$

Since the second equation is satisfied by any ordered pair, the solution set of the system is the same as that of the first equation. When this occurs, we usually write the general form of the solution by expressing one variable in terms of the other. In this case, we have $y = 1 - 2x$ and we write the solution to this system as $(x, 1 - 2x)$, $x \in R$.

EXAMPLE 9   Solve the system

$$x - y + 2z = 3$$

$$2x + 3y - 6z = 1$$

$$4x + y - 2z = 7.$$

Solution   We have the following sequence of equivalent matrices.

$$\begin{bmatrix} 1 & -1 & 2 & | & 3 \\ 2 & 3 & -6 & | & 1 \\ 4 & 1 & -2 & | & 7 \end{bmatrix}$$ (Augmented matrix)

$$\sim \begin{bmatrix} 1 & -1 & 2 & | & 3 \\ 0 & 5 & -10 & | & -5 \\ 0 & 5 & -10 & | & -5 \end{bmatrix}$$ Row 2 + [−2 × Row 1]
    Row 3 + [−4 × Row 1]

$$\sim \begin{bmatrix} 1 & -1 & 2 & | & 3 \\ 0 & 1 & -2 & | & -1 \\ 0 & 5 & -10 & | & -5 \end{bmatrix}$$ $\frac{1}{5}$ × Row 2

$$\sim \begin{bmatrix} 1 & 0 & 0 & | & 2 \\ 0 & 1 & -2 & | & -1 \\ 0 & 0 & 0 & | & 0 \end{bmatrix}$$ Row 1 + Row 2

    Row 3 + [−5 × Row 2]

Since the third row is of the form

$$0 \quad 0 \quad 0 \quad a$$

with $a = 0$ and there are no rows of this form with $a \neq 0$, the system has an infinite number of solutions. The final equivalent system is

$$
\begin{array}{ccc}
x = 2 & & x = 2 \\
& \text{or} & \\
y - 2z = -1 & & y = 2z - 1 \\
0x + 0y + 0z = 0 & &
\end{array}
$$

For any real number z, the triple $(2, 2z - 1, z)$ is a solution of the system, and conversely. Thus, the solution set may be described as the set of all triples of the form $(2, 2z - 1, z)$, $z \in R$.   ■

Applications

In Sections 7.1 and 7.2, we observed that solving linear systems of equations is useful in the solution of many problems in science and engineering.

EXAMPLE 10 A pesticide is made up of three chemicals. There is four times as much Chemical A as Chemical C, and the combined amount of Chemicals B and C is the same as the amount of Chemical A. The manufacturer wants to mix an 8-pound bag of the pesticide. How much of each chemical should he use?

Solution We first let

$$x = \text{amount of Chemical A}$$
$$y = \text{amount of Chemical B}$$
$$z = \text{amount of Chemical C.}$$

Then from the problem statement we have the system

$$
\begin{aligned}
x &= 4z \\
y + z &= x \\
x + y + z &= 8
\end{aligned}
$$

or, equivalently,

$$
\begin{aligned}
x \quad\quad - 4z &= 0 \\
-x + y + \quad z &= 0 \\
x + y + \quad z &= 8.
\end{aligned}
$$

We solve this system using row-equivalent matrices as follows.

$$\begin{bmatrix} 1 & 0 & -4 & \vdots & 0 \\ -1 & 1 & 1 & \vdots & 0 \\ 1 & 1 & 1 & \vdots & 8 \end{bmatrix}$$ (Augmented matrix)

$$\sim \begin{bmatrix} 1 & 0 & -4 & \vdots & 0 \\ 0 & 1 & -3 & \vdots & 0 \\ 0 & 1 & 5 & \vdots & 8 \end{bmatrix}$$ Row 1 + Row 2
Row 3 + [−1 × Row 1]

$$\sim \begin{bmatrix} 1 & 0 & -4 & \vdots & 0 \\ 0 & 1 & -3 & \vdots & 0 \\ 0 & 0 & 8 & \vdots & 8 \end{bmatrix}$$ Row 3 + [−1 × Row 2]

$$\sim \begin{bmatrix} 1 & 0 & 0 & \vdots & 4 \\ 0 & 1 & 0 & \vdots & 3 \\ 0 & 0 & 1 & \vdots & 1 \end{bmatrix}$$ Row 1 + [$\frac{4}{8}$ × Row 3]
Row 2 + [$\frac{3}{8}$ × Row 3]
$\frac{1}{8}$ × Row 3

Thus, the solution of the given system is $x = 4$, $y = 3$, and $z = 1$. The manufacturer must use 4 pounds of Chemical A, 3 pounds of Chemical B, and 1 pound of Chemical C.   ■

The use of row-equivalent matrices is one method of solving linear systems that is easily programmed on either a hand calculator or a computer.

EXERCISE 8.3

A

■ *Use row transformations of the augmented matrix to solve each system of equations. Note the elementary transformations that are used. (Each coefficient matrix is nonsingular.) See Examples 6 and 10.*

1. $x - 2y = 4$
  $x + 3y = -1$

2. $x + y = -1$
  $x - 4y = -14$

3. $3x - 2y = 13$
  $4x - y = 19$

4. $4x - 3y = 16$
  $2x + y = 8$

5. $x - 2y = 6$
  $3x + y = 25$

6. $x - y = -8$
  $x + 2y = 9$

7. $x + y - z = 0$
  $2x - y + z = -6$
  $x + 2y - 3z = 2$

8. $2x - y + 3z = 1$
  $x + 2y - z = -1$
  $3x + y + z = 2$

9. $2x - y = 0$
  $3y + z = 7$
  $2x + 3z = 1$

10. $3x - z = 7$
  $2x + y = 6$
  $3y - z = 7$

11. $2x - 5y + 3z = -1$
  $-3x - y + 2z = 11$
  $-2x + 7y + 5z = 9$

12. $2x + y + z = 4$
  $3x \quad - z = 3$
  $2x \quad + 3z = 13$

■ *Use row transformations of the augmented matrix to solve each system of equations. Note the elementary transformations that are used. (The coefficient matrix may be singular.) See Example 7–9.*

13. $x + 2y + 2z = 3$
$2x + 5y + 5z = 7$
$y + z = 1$

14. $x - y + z = -1$
$3x - 2y + 2z = 1$
$2x - 4y + 4z = -10$

15. $x + 3y + 7z = 5$
$-x + y + z = -1$
$x + 11y + 22z = 14$

16. $x + 2y = -1$
$3x + y - 5z = 5$
$2x + 9y + 5z = -10$

17. $x + 2y + 2z = 5$
$x + y + z = 4$
$2x + 3y + 3z = 8$

18. $x - y + z = 4$
$2x + 3y - z = 5$
$x - 6y + 4z = 6$

19. $x + 2y - z = 1$
$3x + y + z = 3$
$2x - y + 2z = -1$

20. $x + 3y + z = 1$
$2x + 4y - z = -1$
$2y + 3z = 3$

■ *For Exercises 21–26, see Example 10.*

21. The sum of three numbers is 3. The sum of the first and second is four times the third, and the sum of the second and third is twice the first. Find the three numbers.

22. The sum of three numbers is 2. The difference between the first and second is twice the third. The sum of the second and third is three times the first. Find the three numbers.

23. A caterer is mixing a punch made from fruit juice, wine, and brandy. The alcohol content of the wine is 12% and that of the brandy is 40%. There are equal amounts of fruit juice and brandy in the punch. How much of each ingredient should be present to make 10 quarts of punch that is 16% alcohol?

24. A nut company sells mixed nuts. The manager wants to market a 10-pound bag of mixed nuts consisting of pecans, cashews, and peanuts. She wants to have half as many peanuts as pecans and cashews combined and twice as many cashews as pecans. How many pounds of each type of nut should be placed in each bag?

25. Find values of a, b, and c so that the graph of $x^2 + y^2 + ax + by + c = 0$ passes through the points $(1, 1)$, $(2, 1)$, and $(0, -2)$.

26. Find values of a, b, and c so that the graph of $ax^2 + by^2 + x + y + c = 0$ passes through the points $(1, -5)$, $(-2, -2)$, and $(-5, 1)$.

B

■ *Show that each product is a matrix that is row-equivalent to $\begin{bmatrix} a & b \\ c & d \end{bmatrix}$ if $k \neq 0$.*

27. $\begin{bmatrix} k & 0 \\ 0 & 1 \end{bmatrix} \begin{bmatrix} a & b \\ c & d \end{bmatrix}$

28. $\begin{bmatrix} 1 & 0 \\ 0 & k \end{bmatrix} \begin{bmatrix} a & b \\ c & d \end{bmatrix}$

29. $\begin{bmatrix} 1 & 0 \\ k & 1 \end{bmatrix} \begin{bmatrix} a & b \\ c & d \end{bmatrix}$

30. $\begin{bmatrix} 1 & k \\ 0 & 1 \end{bmatrix} \begin{bmatrix} a & b \\ c & d \end{bmatrix}$

8.4

THE DETERMINANT FUNCTION

Associated with each square matrix A having real-number entries is a real number called the **determinant** of A and denoted by δA or $\delta(A)$ (read "the determinant of

A''). Thus, we have a function, δ (delta), with domain the set of all square matrices having real-number entries and with range the set of all real numbers; $\delta(A_{n \times n})$ is called a determinant of **order** n.

Let us begin by examining δ over the set of 2×2 matrices.

Definition 8.13   *The **determinant** of the matrix*

$$\begin{bmatrix} a_{11} & a_{12} \\ a_{21} & a_{22} \end{bmatrix}$$

is the number $a_{11}a_{22} - a_{12}a_{21}$.

The determinant of a square matrix is customarily displayed in the same form as the matrix, but with vertical bars in lieu of brackets. Thus,

$$\delta \begin{bmatrix} a_{11} & a_{12} \\ a_{21} & a_{22} \end{bmatrix} = \begin{vmatrix} a_{11} & a_{12} \\ a_{21} & a_{22} \end{vmatrix} = a_{11}a_{22} - a_{12}a_{21}.$$

EXAMPLE 1   Compute the determinant of each matrix.

(a) $\begin{bmatrix} 2 & 4 \\ 1 & 0 \end{bmatrix}$ (b) $\begin{bmatrix} 4 & -3 \\ 1 & 2 \end{bmatrix}$

Solution   (a) $\begin{vmatrix} 2 & 4 \\ 1 & 0 \end{vmatrix} = (2)(0) - (4)(1)$ (b) $\begin{vmatrix} 4 & -3 \\ 1 & 2 \end{vmatrix} = (4)(2) - (-3)(1)$

$= -4$ $= 11$ ■

Minors and Cofactors

Before we define the determinant of a higher-order matrix, we need to define two terms. Throughout the following discussion, A is the $n \times n$ matrix

$$A = \begin{bmatrix} a_{11} & \cdots & a_{1n} \\ \vdots & & \vdots \\ a_{n1} & \cdots & a_{nn} \end{bmatrix}.$$

Definition 8.14   *The **minor** M_{ij} of a_{ij} is the determinant of the $(n-1) \times (n-1)$ matrix obtained by deleting the ith row and jth column of the matrix A.*

EXAMPLE 2   Let $A = \begin{bmatrix} 1 & 4 & -2 \\ 2 & 3 & -4 \\ 0 & 2 & 3 \end{bmatrix}$. Compute (a) M_{23},   (b) M_{31}.

Solution   (a)   Write the matrix A and cross out the second row and third column.

$$M_{23} = \delta \begin{bmatrix} 1 & 4 & -2 \\ 2 & 3 & 4 \\ 0 & 2 & 3 \end{bmatrix} = \begin{vmatrix} 1 & 4 \\ 0 & 2 \end{vmatrix} = (1)(2) - (4)(0) = 2$$

(b)   Write the matrix A and cross out the third row and first column.

$$M_{31} = \delta \begin{bmatrix} 1 & 4 & -2 \\ 2 & 3 & -4 \\ 0 & 2 & 3 \end{bmatrix} = \begin{vmatrix} 4 & -2 \\ 3 & -4 \end{vmatrix} = (4)(-4) - (-2)(3) = -10 \quad ■$$

Definition 8.15   *The **cofactor** A_{ij} of the entry a_{ij} is*

$$A_{ij} = (-1)^{i+j} M_{ij}.$$

E X A M P L E 3   Let $A = \begin{bmatrix} 1 & 4 & -2 \\ 2 & 3 & -4 \\ 0 & 2 & 3 \end{bmatrix}$. Compute (a) A_{23},   (b) A_{31}.

Solution   The matrix A is the same one that was used in Example 2. Thus, from that example, we have $M_{23} = 2$ and $M_{31} = -10$. Hence, by Definition 8.15:

(a) $A_{23} = (-1)^{2+3} M_{23}$ (b) $A_{31} = (-1)^{3+1} M_{31}$

          $= (-1)(2) = -2$ $= (1)(-10) = -10$ ■

Determinants of $n \times n$ Matrices

We can use the cofactors of the entries of a square matrix A with order greater than 2×2 to define the determinant of A.

Definition 8.16   *The **determinant** of the square matrix*

$$\begin{bmatrix} a_{11} & a_{12} & \cdots & a_{1n} \\ a_{21} & a_{22} & \cdots & a_{2n} \\ \vdots & \vdots & & \vdots \\ a_{n1} & a_{n2} & \cdots & a_{nn} \end{bmatrix}$$

is the sum of the n products formed by multiplying each entry in any single row (or any single column) by its cofactor.

In applying Definition 8.16, the determinant is said to be **expanded** about whatever row (or column) is chosen.

It can be shown, although we shall not do so here, that the value of the determinant is independent of the row or column about which the determinant is expanded.

EXAMPLE 4   Let $A = \begin{bmatrix} 1 & 0 & 1 \\ 2 & 3 & 0 \\ 3 & 0 & 4 \end{bmatrix}$. Compute $\delta(A)$.

Solution   We shall expand about Column 2.

$$\delta(A) = (0)A_{12} + (3)A_{22} + (0)A_{32}$$

$$= 3(-1)^{2+2}\begin{vmatrix} 1 & 1 \\ 3 & 4 \end{vmatrix} = 3(1)(1)$$

$$= 3 \qquad ■$$

Choosing a Row or Column

Note that each term in the sum used to compute the determinant of a matrix is of the form $a_{ij}A_{ij}$. Thus, if $a_{ij} = 0$, then that term does not contribute to the sum and therefore the cofactor of that entry need not be computed. Therefore, to minimize the number of operations necessary to compute the determinant of a matrix, we expand about the row or column with the most zero entries. Thus, in the preceding example we expanded about the second column, which has the largest number of zero entries.

EXAMPLE 5   Let $A = \begin{bmatrix} 1 & 0 & 1 & 1 \\ 4 & -2 & 0 & 0 \\ -3 & 1 & 1 & 0 \\ 0 & 2 & -4 & 1 \end{bmatrix}$. Compute $\delta(A)$.

Solution   Observe that no row or column has more than two zero entries and both Row 2 and Column 4 have two zero entries. We therefore expand the determinant about either Row 2 or Column 4. We shall use Row 2.

$$\delta(A) = 4A_{21} + (-2)A_{22} + (0)A_{23} + (0)A_{24}$$

$$= 4(-1)^{2+1}\begin{vmatrix} 0 & 1 & 1 \\ 1 & 1 & 0 \\ 2 & -4 & 1 \end{vmatrix} + (-2)(-1)^{2+2}\begin{vmatrix} 1 & 1 & 1 \\ -3 & 1 & 0 \\ 0 & -4 & 1 \end{vmatrix}$$

Now, using cofactors we compute the two 3×3 determinants in the right-hand member. We shall expand each of the determinants about the first column:

$$\begin{vmatrix} 0 & 1 & 1 \\ 1 & 1 & 0 \\ 2 & -4 & 1 \end{vmatrix} = (0)\begin{vmatrix} 1 & 0 \\ -4 & 1 \end{vmatrix} + 1(-1)\begin{vmatrix} 1 & 1 \\ -4 & 1 \end{vmatrix} + (2)\begin{vmatrix} 1 & 1 \\ 1 & 0 \end{vmatrix}$$

$$= 0(1) + (-1)(5) + 2(-1) = -7$$

and

$$\begin{vmatrix} 1 & 1 & 1 \\ -3 & 1 & 0 \\ 0 & -4 & 1 \end{vmatrix} = (1)\begin{vmatrix} 1 & 0 \\ -4 & 1 \end{vmatrix} + (-3)(-1)\begin{vmatrix} 1 & 1 \\ -4 & 1 \end{vmatrix} + (0)\begin{vmatrix} 1 & 1 \\ 1 & 0 \end{vmatrix}$$

$$= 1(1) + 3(5) + 0(-1) = 16.$$

Thus,

$$\delta(A) = 4(-1)(-7) + (-2)(1)(16) = -4. \qquad ■$$

EXERCISE 8.4

A

■ *Compute the determinant of each matrix. See Example 1.*

1. $\begin{bmatrix} 3 & 0 \\ 0 & 1 \end{bmatrix}$ **2.** $\begin{bmatrix} 4 & 0 \\ 0 & -\frac{1}{2} \end{bmatrix}$ **3.** $\begin{bmatrix} 2 & -1 \\ 0 & 2 \end{bmatrix}$ **4.** $\begin{bmatrix} 3 & -5 \\ 0 & 10 \end{bmatrix}$

5. $\begin{bmatrix} -2 & 4 \\ -1 & 2 \end{bmatrix}$ **6.** $\begin{bmatrix} \frac{1}{2} & 1 \\ 4 & 8 \end{bmatrix}$ **7.** $\begin{bmatrix} 3 & 1 \\ 2 & -3 \end{bmatrix}$ **8.** $\begin{bmatrix} 4 & -2 \\ 5 & -4 \end{bmatrix}$

■ *Specify the cofactor (in determinant form) of the indicated entry of*

$$A = \begin{bmatrix} 2 & 1 & -2 & 0 \\ 1 & 0 & 3 & -1 \\ -2 & 1 & 2 & 2 \\ 1 & -1 & 3 & 1 \end{bmatrix}.$$

See Examples 2 and 3.

9. a_{11} **10.** a_{13} **11.** a_{23} **12.** a_{41}

13. a_{31} **14.** a_{33} **15.** a_{44} **16.** a_{14}

■ *Compute the determinant of the given matrix. See Examples 4 and 5.*

17. $\begin{bmatrix} 2 & 1 & 3 \\ 0 & 4 & 1 \\ 0 & 0 & 2 \end{bmatrix}$ **18.** $\begin{bmatrix} -3 & 0 & 0 \\ -1 & 4 & 0 \\ 5 & 6 & -\frac{1}{4} \end{bmatrix}$ **19.** $\begin{bmatrix} 0 & 1 & 1 \\ 1 & 2 & 1 \\ 3 & -1 & 0 \end{bmatrix}$

20. $\begin{bmatrix} 2 & -4 & 1 \\ 1 & 1 & 0 \\ 0 & 2 & 0 \end{bmatrix}$ **21.** $\begin{bmatrix} 2 & 1 & 1 \\ 3 & -4 & 2 \\ 4 & 2 & 2 \end{bmatrix}$ **22.** $\begin{bmatrix} 1 & -4 & 2 \\ 2 & -1 & 4 \\ -3 & 2 & -6 \end{bmatrix}$

23. $\begin{bmatrix} 1 & 0 & 0 & 0 \\ 0 & 3 & 0 & 0 \\ 0 & 0 & -2 & 0 \\ 0 & 0 & 0 & 4 \end{bmatrix}$ **24.** $\begin{bmatrix} 1 & 0 & 2 & 0 \\ 0 & 1 & 3 & 1 \\ 0 & 0 & 2 & -1 \\ 0 & 0 & 0 & 1 \end{bmatrix}$

25. $\begin{bmatrix} 0 & 1 & 0 & 2 \\ 1 & 2 & 1 & 0 \\ 3 & -4 & 0 & 0 \\ 2 & 0 & 1 & 0 \end{bmatrix}$ **26.** $\begin{bmatrix} 1 & 4 & 1 & 1 \\ 0 & 2 & -3 & -4 \\ 2 & 0 & 0 & 4 \\ 1 & 1 & 1 & 2 \end{bmatrix}$

B

27. Show that $\begin{vmatrix} x & y & 1 \\ x_1 & y_1 & 1 \\ x_2 & y_2 & 1 \end{vmatrix} = 0$ represents an equation of the line through the points (x_1, y_1) and (x_2, y_2).

28. Use the results in Exercise 27 to find an equation of the line through the points $(3, -1)$ and $(-2, 5)$.

29. Show that if A is an $n \times n$ matrix with one row (or column) identically 0, then $\delta(A) = 0$.

30. In accordance with Definitions 8.14, 8.15, and 8.16, the determinant of an $n \times n$ matrix is the sum of a certain number of products of the entries. What is this number for $n = 2$? For $n = 3$? For $n = 4$?

31. Show that for any 2×2 matrix A, $\delta(aA) = a^2\delta(A)$.

32. Show that for any 2×2 matrix A, $\delta(A^t) = \delta(A)$.

33. Show that for any 2×2 matrices A and B, $\delta(AB) = \delta(A)\delta(B)$.

8.5

PROPERTIES OF DETERMINANTS

Determinants have some properties that are useful by virtue of the fact that they permit us to generate equivalent determinants (name the same number) with different and simpler configurations of entries. This, in turn, helps us find values for determinants. We shall state these properties without proof.

Theorem 8.7   *If each entry in any row, or each entry in any column, of a determinant is 0, then the determinant is equal to 0.*

E X A M P L E   1 (a) $\begin{vmatrix} 0 & 0 \\ 1 & 2 \end{vmatrix} = 0$ (b) $\begin{vmatrix} 1 & 1 & 0 \\ 3 & 5 & 0 \\ 2 & 7 & 0 \end{vmatrix} = 0$ (c) $\begin{vmatrix} 0 & 1 & 0 & 0 \\ 1 & 0 & 0 & 0 \\ 0 & 0 & 0 & 1 \\ 0 & 0 & 0 & 1 \end{vmatrix} = 0$ ■

Theorem 8.8   *If any two rows (or any two columns) of a determinant are interchanged, then the resulting determinant is the negative of the original determinant.*

E X A M P L E   2 (a) $\begin{vmatrix} 1 & 2 \\ 3 & 4 \end{vmatrix} = -\begin{vmatrix} 3 & 4 \\ 1 & 2 \end{vmatrix}$ (b) $\begin{vmatrix} 1 & 2 & 3 \\ 4 & 5 & 6 \\ 7 & 8 & 9 \end{vmatrix} = -\begin{vmatrix} 3 & 2 & 1 \\ 6 & 5 & 4 \\ 9 & 8 & 7 \end{vmatrix}$

In (a), Rows 1 and 2 were interchanged. In (b), Columns 1 and 3 were interchanged.   ■

Theorem 8.9   *If two rows (or two columns) in a determinant have corresponding entries that are equal, the determinant is equal to 0.*

E X A M P L E   3 (a) $\begin{vmatrix} 1 & 1 \\ 3 & 3 \end{vmatrix} = 0$ (b) $\begin{vmatrix} 1 & 2 & 1 \\ 3 & 1 & 0 \\ 1 & 2 & 1 \end{vmatrix} = 0$ (c) $\begin{vmatrix} 1 & 2 & 3 & 4 \\ 5 & 6 & 7 & 8 \\ 0 & 0 & 1 & 0 \\ 1 & 2 & 3 & 4 \end{vmatrix} = 0$

In (a), Columns 1 and 2 are identical. In (b), Rows 1 and 3 are identical. In (c), Rows 1 and 4 are identical.   ■

Theorem 8.10   *If each of the entries of one row (or column) of a determinant is multiplied by k, the determinant is multiplied by k.*

E X A M P L E   4 (a) $\begin{vmatrix} 1 & 0 & 0 \\ 2 & 1 & 3 \\ 1 \times 2 & 3 \times 2 & 4 \times 2 \end{vmatrix} = 2\begin{vmatrix} 1 & 0 & 0 \\ 2 & 1 & 3 \\ 1 & 3 & 4 \end{vmatrix}$ (b) $\begin{vmatrix} 4 & 5 & 8 \\ 1 & 1 & 2 \\ 3 & 1 & 6 \end{vmatrix} = 2\begin{vmatrix} 4 & 5 & 4 \\ 1 & 1 & 1 \\ 3 & 1 & 3 \end{vmatrix}$ ■

Note that this process is different from that of the multiplication of a matrix by a real number. In the latter, each entry in the matrix is multiplied by the real number, rather than, as here, only the entries in a single row or column being so multiplied.

Theorem 8.11

If each entry of the row (or column) of a determinant is multiplied by a real number k and the resulting product is added to the corresponding entry in another row (or column, respectively) in the determinant, the resulting determinant is equal to the original determinant.

EXAMPLE 5 (a) $\begin{vmatrix} 1 & 1 \\ 2 & 1 \end{vmatrix} = \begin{vmatrix} 1 & 1 \\ 2 + 3(1) & 1 + 3(1) \end{vmatrix} = \begin{vmatrix} 1 & 1 \\ 5 & 4 \end{vmatrix}$

(b) $\begin{vmatrix} 1 & 2 & 3 \\ 4 & 5 & 6 \\ 7 & 8 & 9 \end{vmatrix} = \begin{vmatrix} 1 + 2(3) & 2 & 3 \\ 4 + 2(6) & 5 & 6 \\ 7 + 2(9) & 8 & 9 \end{vmatrix} = \begin{vmatrix} 7 & 2 & 3 \\ 16 & 5 & 6 \\ 25 & 8 & 9 \end{vmatrix}$

EXAMPLE 6 Use Theorem 8.11 to find the value of the missing entry that makes the statement true.

$$\begin{vmatrix} 1 & 5 \\ 4 & 3 \end{vmatrix} = \begin{vmatrix} 1 & 5 \\ 0 & \end{vmatrix}$$

Solution To obtain the 0 in the 2, 1 entry, we added -4 times Row 1 to Row 2. Performing that operation in the second column gives

$$3 + (-4)(5) = -17$$

as the missing entry.   ■

Evaluation of Determinants

The preceding theorems can be used to write sequences of equal determinants, leading from one form of a determinant to another and more useful form.

EXAMPLE 7 Evaluate

$$D = \begin{vmatrix} 2 & -1 & 1 & -3 \\ 1 & 3 & -4 & 2 \\ 1 & 0 & -2 & 1 \\ 3 & -1 & 5 & 2 \end{vmatrix}.$$

Solution   As a step toward evaluating the determinant, we shall use Theorem 8.11 to produce an equal determinant with a row or a column containing zero entries in all but one place. Let us select the second column for this role, because one entry is already zero. Multiplying a_{1j} by 3 and adding the result to a_{2j}, we obtain

$$D = \begin{vmatrix} 2 & -1 & 1 & -3 \\ 1+3(2) & 3+3(-1) & -4+3(1) & 2+3(-3) \\ 1 & 0 & -2 & 1 \\ 3 & -1 & 5 & 2 \end{vmatrix} = \begin{vmatrix} 2 & -1 & 1 & -3 \\ 7 & 0 & -1 & -7 \\ 1 & 0 & -2 & 1 \\ 3 & -1 & 5 & 2 \end{vmatrix}.$$

Next, multiplying a_{1j} by -1 and adding the result to a_{4j}, we find that

$$D = \begin{vmatrix} 2 & -1 & 1 & -3 \\ 7 & 0 & -1 & -7 \\ 1 & 0 & -2 & 1 \\ 3-1(2) & -1-1(-1) & 5-1(1) & 2-1(-3) \end{vmatrix} = \begin{vmatrix} 2 & -1 & 1 & -3 \\ 7 & 0 & -1 & -7 \\ 1 & 0 & -2 & 1 \\ 1 & 0 & 4 & 5 \end{vmatrix}.$$

If we now expand the determinant about the second column, we have

$$D = \begin{vmatrix} 2 & -1 & 1 & -3 \\ 7 & 0 & -1 & -7 \\ 1 & 0 & -2 & 1 \\ 1 & 0 & 4 & 5 \end{vmatrix} = -(-1)\begin{vmatrix} 7 & -1 & -7 \\ 1 & -2 & 1 \\ 1 & 4 & 5 \end{vmatrix} + (0)\,A_{22} + (0)\,A_{32} + (0)\,A_{42}.$$

From this point, we can reduce the third-order determinant to a second-order determinant by a similar procedure or, alternatively, expand directly about the elements in any row or column. Expanding about the elements of the first row, we obtain

$$D = \begin{vmatrix} 7 & -1 & -7 \\ 1 & -2 & 1 \\ 1 & 4 & 5 \end{vmatrix} = 7\begin{vmatrix} -2 & 1 \\ 4 & 5 \end{vmatrix} - (-1)\begin{vmatrix} 1 & 1 \\ 1 & 5 \end{vmatrix} + (-7)\begin{vmatrix} 1 & -2 \\ 1 & 4 \end{vmatrix},$$

from which

$$D = 7(-14) + (4) - 7(6)$$
$$= -98 + 4 - 42 = -136. \quad ■$$

EXERCISE 8.5

A

■ *Without evaluating, state why each statement is true. Verify selected examples by expansion. See Examples 1–5.*

1. $\begin{vmatrix} 2 & 3 & 1 \\ 0 & 0 & 0 \\ -1 & 2 & 0 \end{vmatrix} = 0$

2. $\begin{vmatrix} 3 & 1 & 3 \\ 0 & 1 & 0 \\ 1 & 2 & 1 \end{vmatrix} = 0$

3. $\begin{vmatrix} 2 & 3 & 1 & 1 \\ 2 & 0 & 1 & 2 \\ 2 & 3 & 1 & 1 \\ 0 & 1 & 2 & 0 \end{vmatrix} = 0$

4. $\begin{vmatrix} 7 & 3 & 2 & 0 \\ 2 & 1 & 2 & 0 \\ 4 & 1 & 1 & 0 \\ 0 & 2 & 1 & 0 \end{vmatrix} = 0$

5. $\begin{vmatrix} 4 & 2 & 1 \\ 0 & -1 & -2 \\ 1 & 0 & 2 \end{vmatrix} = -\begin{vmatrix} 4 & 2 & 1 \\ 0 & 1 & 2 \\ 1 & 0 & 2 \end{vmatrix}$

6. $\begin{vmatrix} -2 & 3 & 1 \\ -1 & 0 & 1 \\ -2 & 1 & 0 \end{vmatrix} = -\begin{vmatrix} 2 & 3 & 1 \\ 1 & 0 & 1 \\ 2 & 1 & 0 \end{vmatrix}$

7. $2\begin{vmatrix} 1 & 0 & 2 \\ -1 & 2 & 0 \\ 1 & 1 & 1 \end{vmatrix} = \begin{vmatrix} 1 & 0 & 2 \\ -1 & 2 & 0 \\ 2 & 2 & 2 \end{vmatrix}$

8. $\begin{vmatrix} 3 & -4 & 2 \\ 1 & -2 & 0 \\ 0 & 8 & 1 \end{vmatrix} = -2\begin{vmatrix} 3 & 2 & 2 \\ 1 & 1 & 0 \\ 0 & -4 & 1 \end{vmatrix}$

9. $\begin{vmatrix} 1 & 2 \\ 3 & 4 \end{vmatrix} = \begin{vmatrix} 1+2 & 2 \\ 3+4 & 4 \end{vmatrix}$

10. $\begin{vmatrix} 1 & 2 \\ 3 & 4 \end{vmatrix} = \begin{vmatrix} 1+4 & 2 \\ 3+8 & 4 \end{vmatrix}$

11. $\begin{vmatrix} 1 & 2 & 1 \\ 0 & 2 & 3 \\ 2 & -1 & 2 \end{vmatrix} = \begin{vmatrix} 1 & 2 & 1 \\ 0 & 2 & 3 \\ 0 & -5 & 0 \end{vmatrix}$

12. $\begin{vmatrix} -1 & 1 & 0 \\ 2 & 3 & -1 \\ 2 & 1 & 2 \end{vmatrix} = \begin{vmatrix} 0 & 1 & 0 \\ 5 & 3 & -1 \\ 3 & 1 & 2 \end{vmatrix}$

■ *Theorem 8.11 was used on the left-hand member of each of the following equalities to produce the elements in the right-hand member. Complete the entries. See Example 6.*

13. $\begin{vmatrix} 1 & 3 \\ 2 & 2 \end{vmatrix} = \begin{vmatrix} 1 & 3 \\ 0 & \end{vmatrix}$

14. $\begin{vmatrix} 2 & -1 \\ 3 & 1 \end{vmatrix} = \begin{vmatrix} & 0 \\ 3 & 1 \end{vmatrix}$

15. $\begin{vmatrix} 1 & -2 & 1 \\ 3 & 1 & 4 \\ 0 & 2 & 1 \end{vmatrix} = \begin{vmatrix} 1 & -2 & 1 \\ 0 & 7 & \\ 0 & 2 & 1 \end{vmatrix}$

16. $\begin{vmatrix} 3 & -1 & 0 \\ 1 & 2 & 1 \\ 2 & 3 & 1 \end{vmatrix} = \begin{vmatrix} 3 & -1 & 0 \\ 1 & 2 & 1 \\ 1 & & 0 \end{vmatrix}$

17. $\begin{vmatrix} 2 & 3 & 1 & 4 \\ 0 & 2 & 1 & 2 \\ 1 & 1 & 2 & 3 \\ 0 & 1 & 1 & 1 \end{vmatrix} = \begin{vmatrix} 0 & 1 & & -2 \\ 0 & 2 & 1 & 2 \\ 1 & 1 & 2 & 3 \\ 0 & 1 & 1 & 1 \end{vmatrix}$

18. $\begin{vmatrix} 2 & 1 & 1 & 0 \\ 1 & 2 & 0 & 2 \\ 3 & 1 & 0 & 3 \\ 2 & 1 & 4 & 2 \end{vmatrix} = \begin{vmatrix} 2 & 1 & 1 & 0 \\ 1 & 2 & 0 & 2 \\ 3 & 1 & 0 & 3 \\ & -3 & 0 & 2 \end{vmatrix}$

■ *First reduce each determinant to an equal 2 × 2 determinant and then evaluate. See Example 7.*

19. $\begin{vmatrix} 2 & 1 & 0 \\ 3 & 2 & 1 \\ -1 & 2 & 0 \end{vmatrix}$

20. $\begin{vmatrix} 1 & 2 & 1 \\ 2 & -1 & 2 \\ 0 & 1 & 0 \end{vmatrix}$

21. $\begin{vmatrix} 1 & 0 & 3 \\ 2 & -1 & 1 \\ 1 & 2 & 1 \end{vmatrix}$

22. $\begin{vmatrix} 1 & 2 & -1 \\ 2 & 1 & 3 \\ 0 & 1 & 2 \end{vmatrix}$

23. $\begin{vmatrix} 1 & 2 & 1 \\ -1 & 2 & 3 \\ 2 & -1 & 1 \end{vmatrix}$

24. $\begin{vmatrix} 3 & -1 & 2 \\ 1 & 2 & 1 \\ -2 & 1 & 3 \end{vmatrix}$

25. $\begin{vmatrix} 0 & 0 & 1 & 2 \\ 6 & 0 & 0 & 1 \\ 6 & 1 & 0 & -1 \\ 6 & 1 & 0 & 2 \end{vmatrix}$

26. $\begin{vmatrix} 4 & 2 & 0 & 2 \\ -1 & 0 & 2 & 1 \\ 3 & 0 & -1 & 1 \\ 0 & 0 & 2 & 1 \end{vmatrix}$

27. $\begin{vmatrix} 0 & 1 & 0 & 2 \\ 0 & 2 & 0 & 3 \\ 2 & -1 & 1 & 0 \\ 0 & 0 & 8 & 8 \end{vmatrix}$

28. $\begin{vmatrix} 0 & 2 & -1 & 3 \\ 0 & 0 & 2 & 1 \\ 3 & 0 & 1 & 0 \\ -6 & 6 & 0 & 0 \end{vmatrix}$ **29.** $\begin{vmatrix} 1 & 2 & 3 & -1 \\ 0 & 4 & 8 & 4 \\ -2 & 0 & 1 & 1 \\ 2 & 1 & 0 & 1 \end{vmatrix}$ **30.** $\begin{vmatrix} 1 & 2 & 1 & 1 \\ 2 & -1 & 0 & 1 \\ 0 & 6 & 3 & 9 \\ 2 & 0 & -1 & 1 \end{vmatrix}$

B

31. Show that $\begin{vmatrix} 1 & a & a^2 \\ 1 & b & b^2 \\ 1 & c & c^2 \end{vmatrix} = (b-c)(c-a)(a-b).$

32. Show that $\begin{vmatrix} a_{11} & a_{12} & a_{13} & a_{14} \\ a_{21} & a_{22} & a_{23} & a_{24} \\ 0 & 0 & a_{33} & a_{34} \\ 0 & 0 & a_{43} & a_{44} \end{vmatrix} = \begin{vmatrix} a_{11} & a_{12} \\ a_{21} & a_{22} \end{vmatrix} \cdot \begin{vmatrix} a_{33} & a_{34} \\ a_{43} & a_{44} \end{vmatrix}.$

33. Prove Theorem 8.9.

34. Prove Theorem 8.10.

8.6

THE INVERSE OF A SQUARE MATRIX

In the field of real numbers, every element a except 0 has a multiplicative inverse $1/a$ with the property that $a \cdot 1/a = 1$. The question should (and does) arise, "Does every square matrix A have a multiplicative inverse A^{-1}?"

Definition 8.17   *For a given square matrix A of order n, if there is a square matrix A^{-1} of order n such that*

$$AA^{-1} = I \quad and \quad A^{-1}A = I,$$

*where I is the multiplicative identity matrix of order n, then A^{-1} is the **multiplicative inverse** of A.*

Inverse of a 2 × 2 Matrix

To answer the question about the existence of a multiplicative inverse for a matrix, we shall begin by considering the simple case of 2 × 2 matrices. If we let

$$A = \begin{bmatrix} a_{11} & a_{12} \\ a_{21} & a_{22} \end{bmatrix},$$

we must see whether or not there exists a 2 × 2 matrix A^{-1} such that $AA^{-1} = I$. If so, let $A^{-1} = \begin{bmatrix} b & c \\ d & e \end{bmatrix}$. We wish to have

$$\begin{bmatrix} a_{11} & a_{12} \\ a_{21} & a_{22} \end{bmatrix}\begin{bmatrix} b & c \\ d & e \end{bmatrix} = \begin{bmatrix} 1 & 0 \\ 0 & 1 \end{bmatrix}.$$

This leads to

$$\begin{bmatrix} a_{11}b + a_{12}d & a_{11}c + a_{12}e \\ a_{21}b + a_{22}d & a_{21}c + a_{22}e \end{bmatrix} = \begin{bmatrix} 1 & 0 \\ 0 & 1 \end{bmatrix},$$

which is true if and only if

$$a_{11}b + a_{12}d = 1, \qquad a_{11}c + a_{12}e = 0,$$

$$a_{21}b + a_{22}d = 0, \qquad a_{21}c + a_{22}e = 1.$$

(1)

Solving these equations for b, c, d, and e, we have

$$(a_{11}a_{22} - a_{12}a_{21})b = a_{22}, \qquad (a_{11}a_{22} - a_{12}a_{21})c = -a_{12},$$

$$(a_{11}a_{22} - a_{12}a_{21})d = -a_{21}, \qquad (a_{11}a_{22} - a_{12}a_{21})e = -a_{11},$$

from which

$$b = \frac{a_{22}}{a_{11}a_{22} - a_{12}a_{21}}, \qquad c = \frac{-a_{12}}{a_{11}a_{22} - a_{12}a_{21}},$$

$$d = \frac{-a_{21}}{a_{11}a_{22} - a_{12}a_{21}}, \qquad e = \frac{a_{11}}{a_{11}a_{22} - a_{12}a_{21}},$$

provided $a_{11}a_{22} - a_{12}a_{21} \neq 0$. Now the denominator of each of these fractions is just $\delta(A)$, so that

$$A^{-1} = \begin{bmatrix} b & c \\ d & e \end{bmatrix} = \begin{bmatrix} \dfrac{a_{22}}{\delta(A)} & \dfrac{-a_{12}}{\delta(A)} \\ \dfrac{-a_{21}}{\delta(A)} & \dfrac{a_{11}}{\delta(A)} \end{bmatrix} = \frac{1}{\delta(A)} \begin{bmatrix} a_{22} & -a_{12} \\ -a_{21} & a_{11} \end{bmatrix}.$$

Hence,

$$A^{-1} = \frac{1}{\delta(A)} \begin{bmatrix} a_{22} & -a_{12} \\ -a_{21} & a_{11} \end{bmatrix}.$$

By direct multiplication, it can be verified not only that

$$AA^{-1} = I,$$

but also (surprisingly, since matrix multiplication is not always commutative) that

$$A^{-1}A = I.$$

Thus, to write the inverse of a 2×2 square matrix A for which $\delta(A) \neq 0$, we interchange the entries on the principal diagonal, replace each of the other two entries with its negative, and multiply the result by $1/\delta(A)$.

EXAMPLE 1   If $A = \begin{bmatrix} 1 & 3 \\ 2 & -1 \end{bmatrix}$, find A^{-1}.

Solution   We first observe that $\delta(A) = (1)(-1) - (3)(2) = -7$. Hence,

$$A^{-1} = -\frac{1}{7}\begin{bmatrix} -1 & -3 \\ -2 & 1 \end{bmatrix} = \begin{bmatrix} \frac{1}{7} & \frac{3}{7} \\ \frac{2}{7} & -\frac{1}{7} \end{bmatrix}.$$

It is always a good idea to check the result when finding A^{-1}, because there is much room for blundering in the process of determining the inverse. In the present example, we have

$$A^{-1}A = -\frac{1}{7}\begin{bmatrix} -1 & -3 \\ -2 & 1 \end{bmatrix}\begin{bmatrix} 1 & 3 \\ 2 & -1 \end{bmatrix} = -\frac{1}{7}\begin{bmatrix} -7 & 0 \\ 0 & -7 \end{bmatrix} = \begin{bmatrix} 1 & 0 \\ 0 & 1 \end{bmatrix}. \quad ■$$

Matrices with No Inverse   We have now arrived at a position where we can answer the question, "Does every 2×2 square matrix A have an inverse?" The answer is "No," for if $\delta(A)$ is 0, then the foregoing Equations (1) for b, c, d, e would have no solution.

EXAMPLE 2   The matrix $\begin{bmatrix} 3 & 5 \\ 6 & 10 \end{bmatrix}$ has no inverse because

$$\delta(A) = 3(10) - 6(5) = 0. \quad ■$$

Inverse of an $n \times n$ Matrix   More generally, and without proving it, we have the following result.

Theorem 8.12   *If*

$$A = \begin{bmatrix} a_{11} & a_{12} & \cdots & a_{1n} \\ a_{21} & a_{22} & \cdots & a_{2n} \\ \vdots & \vdots & & \vdots \\ a_{n1} & a_{n2} & \cdots & a_{nn} \end{bmatrix}$$

and if $\delta(A) \neq 0$, then A has an inverse A^{-1} given by

$$A^{-1} = \frac{1}{\delta(A)}\begin{bmatrix} A_{11} & A_{21} & \cdots & A_{n1} \\ A_{12} & A_{22} & \cdots & A_{n2} \\ \vdots & \vdots & & \vdots \\ A_{1n} & A_{2n} & \cdots & A_{nn} \end{bmatrix},$$

where A_{ij} is the cofactor of a_{ij} in A. If $\delta(A) = 0$, then A has no inverse.

Square matrices A for which $\delta(A) \neq 0$ are *nonsingular* (see page 340), for it can be shown that A is row-equivalent to the identity if and only if $\delta(A) \neq 0$. Thus, by Theorem 8.12, A has an inverse if and only if A is nonsingular.

Observe that A^{-1} is $1/\delta(A)$ times the transpose of the matrix obtained by replacing each entry of A with its cofactor.

EXAMPLE 3   If $A = \begin{bmatrix} 1 & 0 & 1 \\ 2 & 1 & 0 \\ 1 & -1 & 1 \end{bmatrix}$, find A^{-1}.

Solution   We first obtain $\delta(A) = -2$, and since $\delta(A)$ is not zero, A has an inverse. Next, replacing each entry in A with its cofactor, we obtain the matrix

$$\begin{bmatrix} 1 & -2 & -3 \\ -1 & 0 & 1 \\ -1 & 2 & 1 \end{bmatrix}, \qquad \text{whose transpose is} \qquad \begin{bmatrix} 1 & -1 & -1 \\ -2 & 0 & 2 \\ -3 & 1 & 1 \end{bmatrix},$$

so that

$$A^{-1} = -\frac{1}{2} \begin{bmatrix} 1 & -1 & -1 \\ -2 & 0 & 2 \\ -3 & 1 & 1 \end{bmatrix}.$$

As a check, we have

$$A^{-1}A = -\frac{1}{2} \begin{bmatrix} 1 & -1 & -1 \\ -2 & 0 & 2 \\ -3 & 1 & 1 \end{bmatrix} \begin{bmatrix} 1 & 0 & 1 \\ 2 & 1 & 0 \\ 1 & -1 & 1 \end{bmatrix}$$

$$= -\frac{1}{2} \begin{bmatrix} -2 & 0 & 0 \\ 0 & -2 & 0 \\ 0 & 0 & -2 \end{bmatrix} = \begin{bmatrix} 1 & 0 & 0 \\ 0 & 1 & 0 \\ 0 & 0 & 1 \end{bmatrix}. \qquad ■$$

Theorem 8.12 is applicable to $n \times n$ square matrices, although, clearly, the process of actually determining A^{-1} by the formula given in that theorem becomes very laborious for matrices much larger than 3×3.

Elementary Transformations   If the inverse of an $n \times n$ matrix A exists, it can be obtained by using elementary transformations. This is the method applied when using a computer to find the inverse of a matrix. To see this, we need the following result, which we state without proof.

Theorem 8.13   *If A is an $n \times n$ nonsingular matrix and if $[A \mid I]$ is the $n \times 2n$ matrix obtained by adjoining the $n \times n$ identity matrix to A, then*

$$[A \ I] \sim [I \mid A^{-1}].$$

Theorem 8.13 is used to compute A^{-1} by using elementary transformations to obtain $[I \mid A^{-1}]$ from $[A \mid I]$.

EXAMPLE 4 If $A = \begin{bmatrix} 1 & 2 \\ 2 & 0 \end{bmatrix}$, find A^{-1}.

Solution We first observe that $\delta(A) = -4$, and since $\delta(A)$ is not zero, A has an inverse.

$$[A \mid I] = \begin{bmatrix} 1 & 2 & \vdots & 1 & 0 \\ 2 & 0 & \vdots & 0 & 1 \end{bmatrix}$$

$$\sim \begin{bmatrix} 0 & 2 & \vdots & 1 & -\frac{1}{2} \\ 2 & 0 & \vdots & 0 & 1 \end{bmatrix} \qquad \text{Row } 1 + [-\tfrac{1}{2} \times \text{Row 2}]$$

$$\sim \begin{bmatrix} 0 & 1 & \vdots & \frac{1}{2} & -\frac{1}{4} \\ 1 & 0 & \vdots & 0 & \frac{1}{2} \end{bmatrix} \qquad \begin{matrix} \tfrac{1}{2} \times \text{Row 1} \\ \tfrac{1}{2} \times \text{Row 2} \end{matrix}$$

$$\sim \begin{bmatrix} 1 & 0 & \vdots & 0 & \frac{1}{2} \\ 0 & 1 & \vdots & \frac{1}{2} & -\frac{1}{4} \end{bmatrix} \qquad \begin{matrix} \text{Interchange} \\ \text{Rows 1 and 2} \end{matrix}$$

$$= [I \mid A^{-1}]$$

Thus, $A^{-1} = \begin{bmatrix} 0 & \frac{1}{2} \\ \frac{1}{2} & -\frac{1}{4} \end{bmatrix}$.

As a check, we have

$$A^{-1}A = \begin{bmatrix} 0 & \frac{1}{2} \\ \frac{1}{2} & -\frac{1}{4} \end{bmatrix} \begin{bmatrix} 1 & 2 \\ 2 & 0 \end{bmatrix} = \begin{bmatrix} 1 & 0 \\ 0 & 1 \end{bmatrix}.$$

EXAMPLE 5 Find B^{-1}, if $B = \begin{bmatrix} 1 & 0 & -1 \\ 1 & 3 & 1 \\ 0 & 1 & 2 \end{bmatrix}$.

Solution We first observe that $\delta(B) = 4$, and since $\delta(B) \neq 0$, B has an inverse.

$$[B \mid I] = \begin{bmatrix} 1 & 0 & -1 & \vdots & 1 & 0 & 0 \\ 1 & 3 & 1 & \vdots & 0 & 1 & 0 \\ 0 & 1 & 2 & \vdots & 0 & 0 & 1 \end{bmatrix}$$

$$\sim \begin{bmatrix} 1 & 0 & -1 & 1 & 0 & 0 \\ 0 & 3 & 2 & -1 & 1 & 0 \\ 0 & 1 & 2 & 0 & 0 & 1 \end{bmatrix} \quad \text{Row 2} + [-1 \times \text{Row 1}]$$

$$\sim \begin{bmatrix} 1 & 0 & -1 & 1 & 0 & 0 \\ 0 & 3 & 2 & -1 & 1 & 0 \\ 0 & 0 & \frac{4}{3} & \frac{1}{3} & -\frac{1}{3} & 1 \end{bmatrix} \quad \text{Row 3} + [-\frac{1}{3} \times \text{Row 2}]$$

$$\sim \begin{bmatrix} 1 & 0 & 0 & \frac{5}{4} & -\frac{1}{4} & \frac{3}{4} \\ 0 & 3 & 0 & -\frac{3}{2} & \frac{3}{2} & -\frac{3}{2} \\ 0 & 0 & \frac{4}{3} & \frac{1}{3} & -\frac{1}{3} & 1 \end{bmatrix} \quad \begin{matrix} \text{Row 1} + [\frac{3}{4} \times \text{Row 3}] \\ \text{Row 2} + [\frac{3}{2} \times \text{Row 3}] \end{matrix}$$

$$\sim \begin{bmatrix} 1 & 0 & 0 & \frac{5}{4} & -\frac{1}{4} & \frac{3}{4} \\ 0 & 1 & 0 & -\frac{1}{2} & \frac{1}{2} & -\frac{1}{2} \\ 0 & 0 & 1 & \frac{1}{4} & -\frac{1}{4} & \frac{3}{4} \end{bmatrix} \quad \begin{matrix} \frac{1}{3} \times \text{Row 2} \\ \frac{3}{4} \times \text{Row 3} \end{matrix}$$

Thus, $B^{-1} = \begin{bmatrix} \frac{5}{4} & -\frac{1}{4} & \frac{3}{4} \\ -\frac{1}{2} & \frac{1}{2} & -\frac{1}{2} \\ \frac{1}{4} & -\frac{1}{4} & \frac{3}{4} \end{bmatrix}.$

As a check, we have

$$\begin{bmatrix} \frac{5}{4} & -\frac{1}{4} & \frac{3}{4} \\ -\frac{1}{2} & \frac{1}{2} & -\frac{1}{2} \\ \frac{1}{4} & -\frac{1}{4} & \frac{3}{4} \end{bmatrix} \begin{bmatrix} 1 & 0 & -1 \\ 1 & 3 & 1 \\ 0 & 1 & 2 \end{bmatrix} = \begin{bmatrix} 1 & 0 & 0 \\ 0 & 1 & 0 \\ 0 & 0 & 1 \end{bmatrix}. \quad ■$$

Properties of Matrices and Their Inverses

A number of useful properties are associated with matrices and their inverses. For example, let A and B be $n \times n$ nonsingular matrices. If we right-multiply AB by $B^{-1}A^{-1}$ and apply Theorem 8.3, we have

$$AB \cdot B^{-1}A^{-1} = A \cdot I \cdot A^{-1} = A \cdot A^{-1} = I.$$

Moreover, if we left-multiply AB by $B^{-1}A^{-1}$, we have

$$B^{-1}A^{-1} \cdot AB = B^{-1} \cdot I \cdot B = B^{-1} \cdot B = I.$$

Thus, since $(AB)(B^{-1}A^{-1})(AB) = (B^{-1}A^{-1})(AB) = I$, by the definition of the inverse of a matrix, we have

$$(AB)^{-1} = B^{-1}A^{-1}.$$

This proves the following theorem.

Theorem 8.14   *If A and B are $n \times n$ nonsingular square matrices, then AB has an inverse, namely*

$$(AB)^{-1} = B^{-1}A^{-1}.$$

Theorem 8.14 can be used to find the inverse of products of any number of nonsingular matrices. For example, if there are three factors, A, B, and C, in a product,

$$(ABC)^{-1} = [(AB)C]^{-1} = C^{-1}(AB)^{-1} = C^{-1}B^{-1}A^{-1}.$$

EXERCISE 8.6

A

■ *Find the inverse of each matrix if one exists. For Exercises 1–12, see Examples 1, 2, and 4.*

1. $\begin{bmatrix} 1 & 2 \\ 1 & 3 \end{bmatrix}$ **2.** $\begin{bmatrix} 3 & 1 \\ 2 & -1 \end{bmatrix}$ **3.** $\begin{bmatrix} 2 & -3 \\ 1 & 1 \end{bmatrix}$ **4.** $\begin{bmatrix} 3 & -2 \\ 2 & 1 \end{bmatrix}$

5. $\begin{bmatrix} -2 & -1 \\ 4 & 2 \end{bmatrix}$ **6.** $\begin{bmatrix} 3 & 1 \\ 9 & 3 \end{bmatrix}$ **7.** $\begin{bmatrix} 5 & 7 \\ 3 & 4 \end{bmatrix}$ **8.** $\begin{bmatrix} 5 & -4 \\ 4 & -3 \end{bmatrix}$

9. $\begin{bmatrix} 7 & 4 \\ -4 & -2 \end{bmatrix}$ **10.** $\begin{bmatrix} -9 & 5 \\ -4 & 2 \end{bmatrix}$ **11.** $\begin{bmatrix} -2 & -6 \\ -3 & -9 \end{bmatrix}$ **12.** $\begin{bmatrix} 21 & 7 \\ 9 & 3 \end{bmatrix}$

■ *For Exercises 13–24, see Examples 3 and 5.*

13. $\begin{bmatrix} 1 & -1 & 2 \\ 2 & 1 & 3 \\ 0 & 0 & 2 \end{bmatrix}$ **14.** $\begin{bmatrix} 0 & 4 & 2 \\ 1 & 0 & 2 \\ 0 & -1 & 1 \end{bmatrix}$ **15.** $\begin{bmatrix} 2 & -1 & 1 \\ 3 & 0 & 1 \\ 2 & 2 & 1 \end{bmatrix}$ **16.** $\begin{bmatrix} 1 & 2 & 1 \\ 0 & 2 & 1 \\ -2 & 2 & 3 \end{bmatrix}$

17. $\begin{bmatrix} 2 & 1 & 1 \\ 1 & 0 & 2 \\ 4 & 2 & 2 \end{bmatrix}$ **18.** $\begin{bmatrix} -3 & 1 & -6 \\ 2 & 1 & 4 \\ 2 & 0 & 4 \end{bmatrix}$ **19.** $\begin{bmatrix} 1 & 2 & -3 \\ 3 & -1 & 0 \\ 5 & 3 & -6 \end{bmatrix}$ **20.** $\begin{bmatrix} 2 & 4 & -1 \\ 1 & 6 & 2 \\ 5 & 14 & 0 \end{bmatrix}$

21. $\begin{bmatrix} 2 & -1 & -5 \\ 1 & 3 & 4 \\ 0 & 1 & 2 \end{bmatrix}$ **22.** $\begin{bmatrix} 2 & 1 & -8 \\ 1 & 1 & -2 \\ 1 & 2 & 3 \end{bmatrix}$ **23.** $\begin{bmatrix} 0 & 0 & 1 \\ 0 & 1 & 0 \\ 1 & 0 & 0 \end{bmatrix}$ **24.** $\begin{bmatrix} 1 & 0 & 1 \\ 0 & 1 & 0 \\ 1 & 0 & 0 \end{bmatrix}$

25. Verify that

$$\left(\begin{bmatrix} 2 & 3 \\ 1 & -1 \end{bmatrix} \cdot \begin{bmatrix} 0 & 1 \\ 3 & 1 \end{bmatrix} \right)^{-1} = \begin{bmatrix} 0 & 1 \\ 3 & 1 \end{bmatrix}^{-1} \cdot \begin{bmatrix} 2 & 3 \\ 1 & -1 \end{bmatrix}^{-1}.$$

26. Verify that

$$\left(\begin{bmatrix} 1 & 2 \\ -1 & 0 \end{bmatrix} \cdot \begin{bmatrix} 1 & 1 \\ 2 & 0 \end{bmatrix} \cdot \begin{bmatrix} 2 & -1 \\ 0 & 1 \end{bmatrix} \right)^{-1} = \begin{bmatrix} 2 & -1 \\ 0 & 1 \end{bmatrix}^{-1} \cdot \begin{bmatrix} 1 & 1 \\ 2 & 0 \end{bmatrix}^{-1} \cdot \begin{bmatrix} 1 & 2 \\ -1 & 0 \end{bmatrix}^{-1}.$$

27. Verify that

$$\left(\begin{bmatrix} 3 & 0 & 1 \\ 2 & 1 & 0 \\ 0 & 1 & 2 \end{bmatrix} \cdot \begin{bmatrix} 2 & 1 & 0 \\ 1 & 1 & 2 \\ 0 & 1 & 0 \end{bmatrix} \right)^{-1} = \begin{bmatrix} 2 & 1 & 0 \\ 1 & 1 & 2 \\ 0 & 1 & 0 \end{bmatrix}^{-1} \cdot \begin{bmatrix} 3 & 0 & 1 \\ 2 & 1 & 0 \\ 0 & 1 & 2 \end{bmatrix}^{-1}.$$

B

28. Show that $[A^t]^{-1} = [A^{-1}]^t$ for each nonsingular 2×2 matrix.

29. Show that $\delta(A^{-1}) = 1/\delta(A)$ for each nonsingular 2×2 matrix.

30. Prove that if a and b are real numbers, then $\delta(aA^2 + bA) = \delta(aA + bI) \times \delta(A)$ for all 2×2 matrices A.

31. Prove that $\delta(B^{-1}AB) = \delta(A)$ for all nonsingular 2×2 matrices A and B.

32. Prove that if A is a 2×2 matrix, if a, b, and c are real numbers, with $c \neq 0$, and if $aA^2 + bA + cI = 0$ then A has an inverse.

8.7

SOLUTION OF LINEAR SYSTEMS USING MATRIX INVERSES

In Section 8.3 we solved linear systems using row-equivalent matrices. The solution for a linear system can also be found by using the inverse of a matrix.

We first verify the matrix product equation

$$\begin{bmatrix} a_{11} & a_{12} & \cdots & a_{1n} \\ \vdots & \vdots & & \vdots \\ a_{n1} & a_{n2} & \cdots & a_{nn} \end{bmatrix} \begin{bmatrix} x_1 \\ \vdots \\ x_n \end{bmatrix} = \begin{bmatrix} a_{11}x_1 + a_{12}x_2 + \cdots + a_{1n}x_n \\ \vdots & \vdots & & \vdots \\ a_{n1}x_1 + a_{n2}x_2 + \cdots + a_{nn}x_n \end{bmatrix},$$

and hence note that the linear system

$$\begin{aligned} a_{11}x_1 + a_{12}x_2 + \cdots + a_{1n}x_n &= c_1 \\ a_{21}x_1 + a_{22}x_2 + \cdots + a_{2n}x_n &= c_2 \\ \vdots \qquad \vdots \qquad \qquad \vdots \qquad \vdots \\ a_{n1}x_1 + a_{n2}x_2 + \cdots + a_{nn}x_n &= c_n \end{aligned} \qquad (1)$$

can be written as the matrix equation

$$\begin{bmatrix} a_{11} & a_{12} & \cdots & a_{1n} \\ \vdots & \vdots & & \vdots \\ a_{n1} & a_{n2} & \cdots & a_{nn} \end{bmatrix} \begin{bmatrix} x_1 \\ \vdots \\ x_n \end{bmatrix} = \begin{bmatrix} c_1 \\ \vdots \\ c_n \end{bmatrix},$$

where the first factor in the left-hand member is the coefficient matrix for the system. In more concise notation, this latter equation can be written

$$AX = C,$$

where A is an $n \times n$ square matrix and X and C are $n \times 1$ column matrices.

Solution of Systems

If A in the foregoing equation is nonsingular, we can left-multiply both members of this equation by A^{-1} to obtain the equivalent matrices

$$A^{-1}AX = A^{-1}C,$$
$$IX = A^{-1}C,$$
$$X = A^{-1}C,$$

where $A^{-1}C$ is an $n \times 1$ column matrix. Since X and $A^{-1}C$ are equal, each entry in X is equal to the corresponding entry in $A^{-1}C$, and hence these latter entries constitute the components of the solution of the given linear system. If A is a singular matrix, then of course it has no inverse, and either the system has no solution or the solution is not unique.

EXAMPLE 1   Use a matrix equation to find the solution set of

$$2x + y + z = 1$$
$$x - 2y - 3z = 1$$
$$3x + 2y + 4z = 5.$$

Solution   We first write this as a matrix equation of the form $AX = C$, thus:

$$\begin{bmatrix} 2 & 1 & 1 \\ 1 & -2 & -3 \\ 3 & 2 & 4 \end{bmatrix} \begin{bmatrix} x \\ y \\ z \end{bmatrix} = \begin{bmatrix} 1 \\ 1 \\ 5 \end{bmatrix}.$$

We next determine $\delta(A)$, obtaining

$$\delta \begin{bmatrix} 2 & 1 & 1 \\ 1 & -2 & -3 \\ 3 & 2 & 4 \end{bmatrix} = 2(-2) - 1(13) + 1(8) = -9.$$

Observing that A is nonsingular, we then find A^{-1} by either of the methods of Section 8.6. We shall use the first one discussed.

$$A^{-1} = \begin{bmatrix} 2 & 1 & 1 \\ 1 & -2 & -3 \\ 3 & 2 & 4 \end{bmatrix}^{-1} = -\frac{1}{9} \begin{bmatrix} -2 & -2 & -1 \\ -13 & 5 & 7 \\ 8 & -1 & -5 \end{bmatrix}$$

As a matter of routine, we check the latter by verifying that $A^{-1}A = I$.

$$A^{-1}A = -\frac{1}{9}\begin{bmatrix} -2 & -2 & -1 \\ -13 & 5 & 7 \\ 8 & -1 & -5 \end{bmatrix}\begin{bmatrix} 2 & 1 & 1 \\ 1 & -2 & -3 \\ 3 & 2 & 4 \end{bmatrix}$$

$$= -\frac{1}{9}\begin{bmatrix} -9 & 0 & 0 \\ 0 & -9 & 0 \\ 0 & 0 & -9 \end{bmatrix} = \begin{bmatrix} 1 & 0 & 0 \\ 0 & 1 & 0 \\ 0 & 0 & 1 \end{bmatrix}$$

Now, since $X = A^{-1}C$, we have

$$\begin{bmatrix} x \\ y \\ z \end{bmatrix} = -\frac{1}{9}\begin{bmatrix} -2 & -2 & -1 \\ -13 & 5 & 7 \\ 8 & -1 & -5 \end{bmatrix}\begin{bmatrix} 1 \\ 1 \\ 5 \end{bmatrix} = -\frac{1}{9}\begin{bmatrix} -9 \\ 27 \\ -18 \end{bmatrix} = \begin{bmatrix} 1 \\ -3 \\ 2 \end{bmatrix}.$$

Hence, $x = 1$, $y = -3$, and $z = 2$, and the solution set is $\{(1, -3, 2)\}$.   ■

The computation of A^{-1} is laborious when A is a square matrix containing many rows and columns. The foregoing method is not always the easiest to use in solving systems. It is, however, very useful when solving several systems of the form $AX = C$ having the same coefficient matrix A because we need only compute A^{-1} one time and we get the solution of all the systems.

EXAMPLE 2   Solve.

(a) $\begin{bmatrix} 2 & 1 & 1 \\ 1 & -2 & -3 \\ 3 & 2 & 4 \end{bmatrix}\begin{bmatrix} x \\ y \\ z \end{bmatrix} = \begin{bmatrix} 8 \\ 5 \\ 10 \end{bmatrix}$ (b) $\begin{bmatrix} 2 & 1 & 1 \\ 1 & -2 & -3 \\ 3 & 2 & 4 \end{bmatrix}\begin{bmatrix} x \\ y \\ z \end{bmatrix} = \begin{bmatrix} 0 \\ 10 \\ -11 \end{bmatrix}$

Solution   In both parts of the example we have $A = \begin{bmatrix} 2 & 1 & 1 \\ 1 & -2 & -3 \\ 3 & 2 & 4 \end{bmatrix}$, and from Example 1,

$$A^{-1} = -\frac{1}{9}\begin{bmatrix} -2 & -2 & -1 \\ -13 & 5 & 7 \\ 8 & -1 & -5 \end{bmatrix}.$$

(a)   The solution is

$$\begin{bmatrix} x \\ y \\ z \end{bmatrix} = A^{-1}\begin{bmatrix} 8 \\ 5 \\ 10 \end{bmatrix} = \begin{bmatrix} 4 \\ 1 \\ -1 \end{bmatrix}.$$

(b)   The solution is

$$\begin{bmatrix} x \\ y \\ z \end{bmatrix} = A^{-1} \begin{bmatrix} 0 \\ 10 \\ -11 \end{bmatrix} = \begin{bmatrix} 1 \\ 3 \\ -5 \end{bmatrix}. \qquad ■$$

E X E R C I S E   8.7

■   *Find the solution set of the given system by using matrices. If the system has no unique solution, so state. See Example 1.*

1. $2x - 3y = -1$
   $x + 4y = 5$

2. $3x - 4y = -2$
   $x - 2y = 0$

3. $3x + 6y = -2$
   $6x + 12y = 36$

4. $2x - 4y = 7$
   $x - 2y = 1$

5. $2x - 3y = 0$
   $2x + y = 16$

6. $2x + 3y = 3$
   $3x - 4y = 0$

7. $3x + y = -5$
   $2x - 4y = -16$

8. $2x - 3y = -8$
   $x - 4y = -9$

9. $3x + 9y = 2$
   $6x + 18y = 4$

10. $2x + y = 1$
   $4x + 2y = 2$

11. $x - 4y = -6$
   $4x - y = 6$

12. $2x + 3y = 3$
   $3x - 2y = -2$

13. $x + y = 2$
   $2x - z = 1$
   $2y - 3z = -1$

14. $2x - 6y + 3z = -12$
   $3x - 2y + 5z = -4$
   $4x + 5y - 2z = 10$

15. $x - 2y + z = -1$
   $3x + y - 2z = 4$
   $y - z = 1$

16. $2x + 5z = 9$
   $4x + 3y = -1$
   $3y - 4z = -13$

17. $2x + 2y + z = 1$
   $x - y + 6z = 21$
   $3x + 2y - z = -4$

18. $4x + 8y + z = -6$
   $2x - 3y + 2z = 0$
   $x + 7y - 3z = -8$

19. $x + y + z = 0$
   $2x - y - 4z = 15$
   $x - 2y - z = 7$

20. $x + y - 2z = 3$
   $3x - y + z = 5$
   $3x + 3y - 6z = 9$

■   *For Exercises 21 and 22, see Example 2.*

21. Find the solution set of each system. [*Hint:* Use the result of Exercise 19.]

a. $x + y + z = 3$
   $2x - y - 4z = 4$
   $x - 2y - z = -1$

b. $x + y + z = -2$
   $2x - y - 4z = 1$
   $x - 2y - z = 0$

c. $x + y + z = 1$
   $2x - y - 4z = 0$
   $x - 2y - z = 1$

22. Find the solution set of each system. [*Hint:* Use the result of Exercise 16.]

a. $2x + 5z = 1$
   $4x + 3y = 1$
   $3y - 4z = 1$

b. $2x + 5z = 2$
   $4x + 3y = -1$
   $3y - 4z = 1$

c. $2x + 5z = 0$
   $4x + 3y = 2$
   $3y - 4z = 1$

8.8

CRAMER'S RULE

In Section 8.7 we obtained the solution set of the linear system (1) on page 365 with nonsingular coefficient matrix by first expressing the system in the matrix form $AX = C$ and then left-multiplying both members of the equation by A^{-1} to obtain

$$A^{-1}AX = X = A^{-1}C.$$

If, now, this technique is viewed in terms of determinants, we arrive at a general solution for such systems.

If the coefficient matrix A is nonsingular, then its inverse A^{-1}, is

$$A^{-1} = \frac{1}{\delta(A)} \begin{bmatrix} A_{11} & A_{21} & \cdots & A_{n1} \\ \vdots & \vdots & & \vdots \\ A_{1n} & A_{2n} & \cdots & A_{nn} \end{bmatrix}.$$

Now, since $C = \begin{bmatrix} c_1 \\ c_2 \\ \vdots \\ c_n \end{bmatrix}$, we have

$$X = A^{-1}C = \frac{1}{\delta(A)} \begin{bmatrix} c_1 A_{11} + c_2 A_{21} + \cdots + c_n A_{n1} \\ c_1 A_{12} + c_2 A_{22} + \cdots + c_n A_{n2} \\ \vdots & \vdots & \vdots \\ c_1 A_{1n} + c_2 A_{2n} + \cdots + c_n A_{nn} \end{bmatrix}.$$

Each entry in $X = A^{-1}C$ can be seen to be of the form

$$\frac{c_1 A_{1j} + c_2 A_{2j} + \cdots + c_n A_{nj}}{\delta(A)}.$$

But $c_1 A_{1j} + c_2 A_{2j} + \cdots + c_n A_{nj}$ is just the expansion of the determinant

$$\begin{array}{c} j\text{th} \\ \text{column} \\ \downarrow \end{array}$$

$$\begin{vmatrix} a_{11} & a_{12} & \cdots & c_1 & \cdots & a_{1n} \\ a_{21} & a_{22} & \cdots & c_2 & \cdots & a_{2n} \\ \vdots & \vdots & & \vdots & & \vdots \\ a_{n1} & a_{n2} & \cdots & c_n & \cdots & a_{nn} \end{vmatrix}$$

about the jth column, which has entries $c_1, c_2, \ldots, c_n$ in place of $a_{1j}, a_{2j}, \ldots, a_{nj}$.

Thus, each entry x_j in the matrix $X = \begin{bmatrix} x_1 \\ x_2 \\ \vdots \\ x_n \end{bmatrix} = A^{-1}C$ is

$$x_j = \frac{\delta(A_j)}{\delta(A)} = \frac{\begin{vmatrix} a_{11} & a_{12} & \cdots & c_1 & \cdots & a_{1n} \\ a_{21} & a_{22} & \cdots & c_2 & \cdots & a_{2n} \\ \vdots & \vdots & & \vdots & & \vdots \\ a_{n1} & a_{n2} & \cdots & c_n & \cdots & a_{nn} \end{vmatrix}}{\begin{vmatrix} a_{11} & a_{12} & \cdots & & & a_{1n} \\ a_{21} & a_{22} & \cdots & & & a_{2n} \\ \vdots & \vdots & & & & \vdots \\ a_{n1} & a_{n2} & \cdots & & & a_{nn} \end{vmatrix}}.$$

jth column ↓

Application of Cramer's Rule

This relationship expresses **Cramer's rule.** Cramer's rule is the assertion that if the determinant of the coefficient matrix of an $n \times n$ linear system is *not* 0, then the equations are consistent (the system has a solution) and have a unique solution that can be found as follows.

Steps for Using Cramer's Rule

To find x_j in solving the matrix equation $AX = C$:

1. Find the determinant of the coefficient matrix for the system.
2. Replace each entry in the *j*th column of the coefficient matrix A with the corresponding entry from the column matrix C, and find the determinant of the resulting matrix.
3. Divide the result in Step 2 by the result in Step 1.

Hence, in a nonsingular 3×3 system:

$$x = \frac{\delta(A_x)}{\delta(A)}, \qquad y = \frac{\delta(A_y)}{\delta(A)}, \qquad \text{and} \qquad z = \frac{\delta(A_z)}{\delta(A)}.$$

EXAMPLE 1   Use Cramer's rule to solve the system

$$-4x + 2y - 9z = 2 \qquad 3x + 4y + z = 5 \qquad x - 3y + 2z = 8.$$

Solution   By inspection,

$$\delta(A) = \begin{vmatrix} -4 & 2 & -9 \\ 3 & 4 & 1 \\ 1 & -3 & 2 \end{vmatrix}$$

$$= -4\,(11) - 2\,(5) - 9\,(-13)$$

$$= -44 - 10 + 117 = 63.$$

Replacing the entries in the first column of A with corresponding constants 2, 5, and 8, we have

$$\delta(A_x) = \begin{vmatrix} 2 & 2 & -9 \\ 5 & 4 & 1 \\ 8 & -3 & 2 \end{vmatrix}$$

$$= 2\,(11) - 2\,(2) - 9\,(-47)$$

$$= 22 - 4 + 423 = 441.$$

Hence, by Cramer's rule,

$$x = \frac{\delta(A_x)}{\delta(A)} = \frac{441}{63} = 7.$$

Similarly, by replacing, in turn, the entries of the second and third columns of A with the corresponding constants 2, 5, and 8, we have

$$\delta(A_y) = \begin{vmatrix} -4 & 2 & -9 \\ 3 & 5 & 1 \\ 1 & 8 & 2 \end{vmatrix} \quad \text{and} \quad \delta(A_z) = \begin{vmatrix} -4 & 2 & 2 \\ 3 & 4 & 5 \\ 1 & -3 & 8 \end{vmatrix}.$$

Now,

$$\delta(A_y) = -4\,(2) - 2\,(5) - 9\,(19)$$

$$= -8 - 10 - 171 = -189$$

and

$$\delta(A_z) = -4\,(47) - 2\,(19) + 2\,(-13)$$

$$= -188 - 38 - 26 = -252,$$

so that

$$y = \frac{\delta(A_y)}{\delta(A)} = \frac{-189}{63} = -3$$

and

$$z = \frac{\delta(A_z)}{\delta(A)} = \frac{-252}{63} = -4.$$

The solution set of the system is $\{(7, -3, -4)\}$.   ■

If $\delta(A) = 0$ for a linear system, then the system either has infinitely many solutions (the equations are consistent and one of them can be obtained from the others by linear combinations) or has no solutions (the equations are inconsistent). The distinction can be determined using methods from Section 8.3.

E X E R C I S E   8.8

A

■ *Find the solution set of each of the following systems by Cramer's rule. If $\delta(A) = 0$ in any of the systems, use the methods in Section 8.3 to determine whether or not the equations in the system are consistent.*

1. $\quad x - y = 2$
$\quad\quad x + 4y = 5$

2. $\quad x + y = 4$
$\quad\quad x - 2y = 0$

3. $\quad 3x - 4y = -2$
$\quad\quad x + y = 6$

4. $\quad 2x - 4y = 7$
$\quad\quad x - 2y = 1$

5. $\quad \frac{1}{3}x - \frac{1}{2}y = 0$
$\quad\quad \frac{1}{2}x + \frac{1}{4}y = 4$

6. $\quad \frac{2}{3}x + y = 1$
$\quad\quad x - \frac{4}{3}y = 0$

7. $\quad x - 2y = 6$
$\quad\quad \frac{2}{3}x - \frac{4}{3}y = 6$

8. $\quad \frac{1}{2}x + y = 3$
$\quad\quad -\frac{1}{4}x - y = -3$

9. $\quad x - 3y = 1$
$\quad\quad y = 1$

10. $\quad 2x - 3y = 12$
$\quad\quad x = 4$

11. $\quad ax + by = 1$
$\quad\quad bx + ay = 1$

12. $\quad x + y = a$
$\quad\quad x - y = b$

13. $\quad x - 2y + z = -1$
$\quad\quad 3x + y - 2z = 4$
$\quad\quad y - z = 1$

14. $\quad 2x + 5z = 9$
$\quad\quad 4x + 3y = -1$
$\quad\quad 3y - 4z = -13$

15. $\quad 2x + 2y + z = 1$
$\quad\quad x - y + 6z = 21$
$\quad\quad 3x + 2y - z = -4$

16. $\quad 4x + 8y + z = -6$
$\quad\quad 2x - 3y + 2z = 0$
$\quad\quad x + 7y - 3z = -8$

17. $\quad x + y + z = 0$
$\quad\quad 2x - y - 4z = 15$
$\quad\quad x - 2y - z = 7$

18. $\quad x + y - 2z = 2$
$\quad\quad 3x - y + z = 5$
$\quad\quad 3x + 3y - 6z = 6$

19. $\quad x - 2y - 2z = 3$
$\quad\quad 2x - 4y + 4z = 1$
$\quad\quad 3x - 3y - 3z = 4$

20. $\quad 3x - 2y + 5z = 6$
$\quad\quad 4x - 4y + 3z = 0$
$\quad\quad 5x - 4y + z = -5$

21. $\quad x - 4z = -1$
$\quad\quad 3x + 3y = 2$
$\quad\quad 3x + 4z = 5$

22. $\quad 2x - \frac{2}{3}y + z = 2$
$\quad\quad 6x - 4y - 3z = 0$
$\quad\quad 4x + 5y - 3z = -1$

23. $\quad x + y + z = 0$
$\quad\quad w + 2y - z = 4$
$\quad\quad 2w - y + 2z = 3$
$\quad\quad -2w + 2y - z = -2$

24. $\quad x + y + z = 0$
$\quad\quad x + z + w = 0$
$\quad\quad x + y + w = 0$
$\quad\quad y + z + w = 0$

B

25. For the system

$$a_1x + b_1y + c_1 = 0$$

$$a_2x + b_2y + c_2 = 0,$$

show that if both $\delta(A_y) = 0$ and $\delta(A_x) = 0$ and if c_1 and c_2 are not both 0, then $\delta(A) = 0$ and the equations are consistent. [*Hint:* Show that the first two determinant equations imply that $a_1 c_2 = a_2 c_1$ and $b_1 c_2 = b_2 c_1$ and that the rest follows from the formation of a proportion with these equations.]

26. Show that if $\delta(A) = 0$ and $\delta(A_x) = 0$ and if a_1 and a_2 are not both 0, then $\delta(A_y) = 0$, where $\delta(A)$ is the determinant of the coefficient matrix of the system in Exercise 25.

CHAPTER REVIEW

Key Words and Phrases

■ *Define or explain each of the following words and phrases.*

1. matrix
2. entries (or elements) of a matrix
3. order (or dimension) of a matrix
4. rows of a matrix
5. columns of a matrix
6. row matrix (or vector)
7. column matrix (or vector)
8. equal matrices
9. transpose of a matrix
10. sum of two matrices
11. zero matrix
12. negative of a matrix
13. product of real number and a matrix
14. product of two matrices
15. square matrices

16. principal diagonal
17. diagonal matrix
18. $n \times n$ identity matrix
19. elementary transformation of a matrix
20. row-equivalent matrices
21. nonsingular matrix
22. singular matrix
23. coefficient matrix
24. augmented matrix
25. determinant of a 2×2 matrix
26. minor of the i, jth entry
27. cofactor of the i, jth entry
28. determinant of an $n \times n$ matrix
29. inverse of a square matrix
30. Cramer's rule

Review Exercises

A

[8.1] ■ *Write each sum or difference as a single matrix.*

1. $\begin{bmatrix} 4 & -7 \\ 2 & 1 \end{bmatrix} + \begin{bmatrix} -3 & 6 \\ -1 & 0 \end{bmatrix}$

2. $\begin{bmatrix} 3 & -1 & 7 \\ 6 & 2 & 5 \end{bmatrix} + \begin{bmatrix} -1 & 6 & -9 \\ 8 & -3 & 7 \end{bmatrix}$

3. $\begin{bmatrix} 2 & -4 & 3 \\ 6 & 1 & 7 \\ 2 & 8 & 0 \end{bmatrix} - \begin{bmatrix} 4 & -1 & 2 \\ 3 & 8 & 1 \\ 7 & 6 & -5 \end{bmatrix}$

4. $\begin{bmatrix} -11 & 2 & -6 \\ 7 & 1 & 2 \\ -3 & 4 & 8 \end{bmatrix} - \begin{bmatrix} -3 & 5 & 0 \\ 1 & 4 & 2 \\ 6 & -1 & 3 \end{bmatrix}$

[8.2] ■ *Write each product as a single matrix.*

5. $-7\begin{bmatrix} 3 & -1 \\ 2 & 0 \\ 1 & 1 \end{bmatrix}$

6. $\begin{bmatrix} 3 & -1 & 2 \end{bmatrix} \cdot \begin{bmatrix} 4 \\ -1 \\ 0 \end{bmatrix}$

7. $\begin{bmatrix} 3 & -1 \\ 6 & 5 \end{bmatrix} \cdot \begin{bmatrix} -4 & 2 \\ 1 & 3 \end{bmatrix}$

8. $\begin{bmatrix} -1 & 7 & 6 \\ 3 & 1 & 2 \\ 1 & 0 & 1 \end{bmatrix} \cdot \begin{bmatrix} 1 & -1 & 2 \\ 1 & 0 & 3 \\ 2 & 1 & 1 \end{bmatrix}$

[8.3] ■ *Use row transformations on the augmented matrix to solve each system of equations.*

9. $2x - y = 5$
$x + 3y = -1$

10. $2x - y + z = 4$
$x + 3y - z = 4$
$x + 2y + z = 5$

[8.4] ■ *Evaluate each determinant.*

11. $\begin{vmatrix} -3 & 0 \\ 2 & 1 \end{vmatrix}$

12. $\begin{vmatrix} 1 & 5 \\ -1 & 2 \end{vmatrix}$

13. $\begin{vmatrix} 3 & 1 & 0 \\ 2 & 0 & 1 \\ 1 & 2 & -1 \end{vmatrix}$

14. $\begin{vmatrix} 3 & 1 & -2 \\ -1 & 2 & 1 \\ 1 & -2 & 1 \end{vmatrix}$

15. $\begin{vmatrix} 3 & 0 & 1 & 1 \\ 0 & 2 & -1 & 0 \\ 0 & 1 & 0 & 2 \\ 1 & 0 & 0 & 1 \end{vmatrix}$

16. $\begin{vmatrix} -2 & 1 & 4 & 0 \\ 2 & 0 & 4 & 1 \\ 1 & 1 & 0 & 0 \\ 2 & -1 & 3 & 1 \end{vmatrix}$

[8.5] ■ *Reduce each determinant to an equal 2 × 2 determinant and evaluate.*

17. $\begin{vmatrix} 3 & -1 & 2 \\ 1 & -2 & 0 \\ 2 & 1 & -1 \end{vmatrix}$

18. $\begin{vmatrix} 1 & 7 & -11 \\ 12 & 10 & 15 \\ 5 & 11 & 14 \end{vmatrix}$

[8.6] ■ *Find the inverse of each nonsingular matrix.*

19. $\begin{bmatrix} -4 & 2 \\ 11 & 3 \end{bmatrix}$

20. $\begin{bmatrix} 1 & -1 & 2 \\ 3 & 1 & 0 \\ 2 & 1 & 1 \end{bmatrix}$

[8.7] ■ *Use matrices to solve each system.*

21. $x - y = -3$
$2x + 3y = -1$

22. $2x + z = 7$
$y + 2z = 1$
$3x + y + z = 9$

23. a. $x - y = 1$
$3x + y = 1$
   b. $x - y = 2$
$3x + y = -1$
   c. $x - y = 3$
$3x + y = 1$

24. a. $x + 2y - z = 4$
$3x - y + 2z = 0$
$x - y - z = 1$
   b. $x + 2y - z = 1$
$3x - y + 2z = 6$
$x - y - z = -2$
   c. $x + 2y - z = 0$
$3x - y + 2z = -3$
$x - y - z = 3$

[8.8] ■   *Use Cramer's rule to solve each system.*

25. $3x - y = -5$
$x + 2y = -6$

26. $x - y + 2z = 3$
$2x + y - z = 3$
$x - 2y + 2z = 4$

B

■   *A matrix A is **symmetric** if $A^t = A$. For Exercises 27–29, assume that A and B are symmetric 2×2 matrices.*

27. Show that $A + B$ is a symmetric matrix.

28. Show that if r is any real number, then rA is a symmetric matrix.

29. Under what condition is AB a symmetric matrix? Prove your answer.

30. Let A be any 2×2 matrix. Show that $(A^t)^2 = (A^2)^t$. Does an analogous result hold if A is an $n \times n$ matrix?

31. Show that for any $n \times n$ matrix A and any real number c, $\delta(cA) = c^n\delta(A)$.

32. Show that if A is a nonsingular 2×2 matrix, then A^2 is a nonsingular matrix.

9 Sequences and Series

SEQUENCES

Let us consider a class of functions in which each function has as its domain either the set N of positive integers or a subset of successive members of N.

Definition 9.1 *A **sequence** is a function having as its domain the set N of positive integers $1, 2, 3, \ldots$. A **finite sequence** has as its domain the set of positive integers $1, 2, 3, \ldots, n$, for some fixed n.*

For example, the function defined by

$$s(n) = n + 3, \tag{1}$$

where n is a positive integer, is a sequence. The elements in the range of a sequence are usually considered in the order

$$s(1), s(2), s(3), s(4), \ldots.$$

In this case, $s(i)$ is called the **ith term** of the sequence.

For example, the first four terms of the sequence defined by Equation (1) are found by successively substituting the numbers 1, 2, 3, 4 for n:

$$s(1) = 1 + 3 = 4,$$
$$s(2) = 2 + 3 = 5,$$
$$s(3) = 3 + 3 = 6,$$
$$s(4) = 4 + 3 = 7.$$

Thus, the first four terms of Equation (1) are 4, 5, 6, and 7. The nth term, commonly called the **general term**, is $n + 3$.

E X A M P L E I   Find the first three terms and the twenty-fifth term in the given sequence.

(a) $s(n) = n^2$ (b) $s(n) = \dfrac{1}{n}$

Solution   (a) $s(1) = 1^2 = 1$ (b) $s(1) = \dfrac{1}{1} = 1$

$s(2) = 2^2 = 4$

$s(3) = 3^2 = 9$ $s(2) = \dfrac{1}{2}$

$s(25) = 25^2 = 625$

$s(3) = \dfrac{1}{3}$

$s(25) = \dfrac{1}{25}$ ■

Given several terms in a sequence, we are often able to construct an expression for a general term of a sequence to which they belong. Such a sequence is not unique. Thus, if the first three terms in a sequence are 2, 4, 6, . . . , we may *surmise* that the general term is $s(n) = 2n$. Note, however, that both of the sequences

$$s(n) = 2n$$

and

$$s(n) = 2n + (n - 1)(n - 2)(n - 3)$$

start with 2, 4, 6, but the two sequences differ for terms following the third.

Sequence Notation   The notation ordinarily used for the terms in a sequence is not function notation as such. It is customary to denote a term in a sequence by means of a subscript. Thus, the sequence $s(1), s(2), s(3), s(4), \ldots$ would appear as $s_1, s_2, s_3, s_4, \ldots$. For example, instead of writing $s(n) = 1/n$, we write $s_n = 1/n$.

E X A M P L E 2   Write s_1, s_2, s_3, and s_4 for the sequence given.

(a) $s_n = \dfrac{1}{n}$ (b) $s_n = 2^n$ (c) $s_n = \dfrac{n\pi}{2}$

Solution   (a) $s_1 = \dfrac{1}{1} = 1$ (b) $s_1 = 2^1 = 2$ (c) $s_1 = \dfrac{\pi}{2}$

$s_2 = 2^2 = 4$

$s_2 = \dfrac{1}{2}$ $s_3 = 2^3 = 8$ $s_2 = \pi$

$s_4 = 2^4 = 16$ $s_3 = \dfrac{3\pi}{2}$

$s_3 = \dfrac{1}{3}$

$s_4 = 2\pi$ ■

$s_4 = \dfrac{1}{4}$

Arithmetic Progressions

Let us next consider two special kinds of sequences that have many applications. The first kind can be defined as follows.

Definition 9.2

An **arithmetic progression** is a sequence defined by equations of the form

$$s_1 = a, \qquad s_{n+1} = s_n + d,$$

where n is a positive integer and a and d are real numbers.

Since each term in such a sequence is obtained from the preceding term by adding d, d is called the **common difference**.

EXAMPLE 3   Find the first four terms in the arithmetic progression.

(a) $s_1 = 1, \quad d = 3$ (b) $s_1 = -3, \quad d = 2$

Solution   (a) $s_1 = 1$ (b) $s_1 = -3$

$s_2 = s_1 + 3 = 4$ $s_2 = s_1 + 2 = -1$

$s_3 = s_2 + 3 = 7$ $s_3 = s_2 + 2 = 1$

$s_4 = s_3 + 3 = 10$ $s_4 = s_3 + 2 = 3$

1, 4, 7, 10 $-3, -1, 1, 3$

EXAMPLE 4   Find the next three terms in the arithmetic progression.

(a) $5, 9, \ldots$ (b) $x, x - a, \ldots$

Solution   We first find the common difference and then continue the sequence.

(a) $d = 9 - 5 = 4;$ (b) $d = (x - a) - x = -a;$

$9 + 4 = 13$ $x - a + (-a) = x - 2a$

$13 + 4 = 17$ $x - 2a + (-a) = x - 3a$

$17 + 4 = 21$ $x - 3a + (-a) = x - 4a$

$13, 17, 21$ $x - 2a, x - 3a, x - 4a$ ■

To obtain the general term of an arithmetic progression with $s_1 = a$, observe that since the sequence progresses from term to term by adding the common difference, any term is obtained by adding an appropriate number of differences to a. The second term requires one difference, the third requires two differences, and, in general, s_n requires $n - 1$ differences added to a. This indicates that the following theorem is true. A complete proof requires the technique of mathematical induction, which is considered in Section 9.5.

Theorem 9.1   *The nth term in the sequence defined by*

$$s_1 = a, \qquad s_{n+1} = s_n + d,$$

where n is a positive integer and a and d are real numbers, is

$$s_n = a + (n - 1)d. \qquad (2)$$

E X A M P L E 5   Find the general term and the one-hundredth term in the arithmetic progression.

(a) $s_1 = 4, \quad d = 5$ (b) $s_1 = -3, \quad d = 4$

Solution   (a) $s_n = 4 + (n - 1)5 = 5n - 1$ (b) $s_n = -3 + (n - 1)4 = 4n - 7$

$s_{100} = 5(100) - 1 = 499$ $s_{100} = 4(100) - 7 = 393$

E X A M P L E 6   Find the general term and the fourteenth term in the arithmetic progression -6, $-1, \ldots .$

Solution   We first use the first two terms to obtain the common difference

$$d = -1 - (-6) = 5.$$

We then use Equation (2) to obtain

$$s_n = -6 + (n - 1)5 = 5n - 11.$$

Thus,

$$s_{14} = 5(14) - 11 = 59. \qquad ■$$

Geometric Progressions

In Definition 9.2, each term of the sequence, after the first term, is defined by its relation to previous terms. Such a formulation is called a **recursive** definition. The second kind of sequence we consider can also be defined recursively.

Definition 9.3

*A **geometric progression** is a sequence defined by equations of the form*

$$s_1 = a, \qquad s_{n+1} = rs_n,$$

where n is a positive integer and a and r are real numbers with $r \neq 0$.

Thus, $3, 9, 27, 81, \ldots$ is a **geometric progression** in which each term except the first is obtained by multiplying the preceding term by 3. Since the effect of multiplying the terms in this way is to produce a fixed ratio between any two successive terms, the multiplier, r, is called the **common ratio**.

E X A M P L E 7   Find the first four terms in the geometric progression.

(a) $s_1 = -3, \quad r = 2$ (b) $s_1 = 2, \quad r = -1$

Solution   (a) $s_1 = -3$ (b) $s_1 = 2$

$\qquad s_2 = rs_1 = -6 \qquad\qquad s_2 = rs_1 = -2$

$\qquad s_3 = rs_2 = -12 \qquad\qquad s_3 = rs_2 = 2$

$\qquad s_4 = rs_3 = -24 \qquad\qquad s_4 = rs_3 = -2$

$\qquad -3, -6, -12, -24 \qquad\quad 2, -2, 2, -2$

E X A M P L E 8   Find the next four terms in the geometric progression.

(a) $3, 6, \ldots$ (b) $x, 2, \ldots$

Solution   We first find the common ratio and then continue the sequence.

(a) $r = \dfrac{6}{3} = 2;$ (b) $r = \dfrac{2}{x} \quad (x \neq 0);$

$\qquad 6(2) = 12 \qquad\qquad 2\left(\dfrac{2}{x}\right) = \dfrac{4}{x}$

$\qquad 12(2) = 24 \qquad\qquad \left(\dfrac{4}{x}\right)\left(\dfrac{2}{x}\right) = \dfrac{8}{x^2}$

$\qquad 24(2) = 48 \qquad\qquad \left(\dfrac{8}{x^2}\right)\left(\dfrac{2}{x}\right) = \dfrac{16}{x^3}$

$\qquad 48(2) = 96 \qquad\qquad \left(\dfrac{16}{x^3}\right)\left(\dfrac{2}{x}\right) = \dfrac{32}{x^4}$

$\qquad 12, 24, 48, 96 \qquad\qquad \dfrac{4}{x}, \dfrac{8}{x^2}, \dfrac{16}{x^3}, \dfrac{32}{x^4}$ ■

To obtain the general term of a geometric progression with $s_1 = a$, observe that the sequence progresses from term to term by multiplying by the common ratio. Thus, any term is obtained by multiplying by the common ratio an appropriate number of times. The second term requires one factor of the common ratio, the third requires two factors, and, in general, the nth term, s_n, requires $n - 1$ factors of the common ratio multiplied by a. This indicates that the following theorem is true. The proof by mathematical induction is deferred until Section 9.5.

Theorem 9.2   *The nth term in the sequence defined by*

$$s_1 = a, \qquad s_{n+1} = rs_n,$$

where n is a positive integer and a and r are nonzero real numbers, is

$$s_n = ar^{n-1}. \tag{3}$$

EXAMPLE 9   Find the general term and the ninth term in the geometric progression.

(a) $s_1 = 3, \quad r = 2$ (b) $s_1 = 1, \quad r = -3$

Solution   (a) $s_n = 3(2)^{n-1}$ (b) $s_n = 1(-3)^{n-1}$

$\qquad\qquad s_9 = 3(2^8) = 768$ $s_9 = (-3)^8 = 6561$

EXAMPLE 10   Find the general term and the ninth term in the geometric progression $-24, 12, \ldots$.

Solution   We first use the first two terms to obtain the common ratio

$$r = \frac{12}{-24} = -\frac{1}{2}.$$

We then use Equation (3) to obtain

$$s_n = -24\left(-\frac{1}{2}\right)^{n-1}.$$

Thus,

$$s_9 = -24\left(-\frac{1}{2}\right)^8 = -\frac{3}{32}. \qquad ■$$

EXERCISE 9.1

A

■ Find the first four terms in the sequence with the general term as given. See Examples 1 and 2.

1. $s_n = n - 5$

2. $s_n = 2n - 3$

3. $s_n = \dfrac{n^2 - 2}{2}$

4. $s_n = \dfrac{3}{n^2 + 1}$

5. $s_n = 1 + \dfrac{1}{n}$

6. $s_n = \dfrac{n}{2n - 1}$

7. $s_n = \dfrac{n(n - 1)}{2}$

8. $s_n = \dfrac{5}{n(n + 1)}$

9. $s_n = (-1)^n$

10. $s_n = (-1)^{n+1}$

11. $s_n = \dfrac{(-1)^n(n - 2)}{n}$

12. $s_n = (-1)^{n-1}3^{n+1}$

■ Find the next three terms in each arithmetic progression. See Examples 3 and 4.

13. $3, 7, \ldots$

14. $-6, -1, \ldots$

15. $-4, -1, \ldots$

16. $-8, 5, \ldots$

17. $x, x + 1, \ldots$

18. $a, a + 5, \ldots$

19. $2x + 1, 2x + 4, \ldots$

20. $3a, 5a, \ldots$

■ Find the next four terms in each geometric progression. See Examples 7 and 8.

21. $2, 8, \ldots$

22. $4, 8, \ldots$

23. $\dfrac{2}{3}, \dfrac{4}{3}, \ldots$

24. $\dfrac{1}{2}, -\dfrac{3}{2}, \ldots$

25. $\dfrac{a}{x}, -1, \ldots$

26. $\dfrac{a}{b}, \dfrac{a}{bc}, \ldots$

27. $-4, x, \ldots$

28. $-x, x^2, \ldots$

■ For Exercises 29–34, see Examples 5 and 6.

29. Find the general term and the seventh term in the arithmetic progression $7, 11, \ldots$.

30. Find the general term and the twelfth term in the arithmetic progression $2, \dfrac{5}{2}, \ldots$.

31. Find the general term and the twentieth term in the arithmetic progression $3, -2, \ldots$.

32. Find the general term and the ninth term in the arithmetic progression $\dfrac{3}{4}, 2, \ldots$.

33. Find the general term and the eighth term in the arithmetic progression $x, 5x, \ldots$.

34. Find the general term and the twelfth term in the arithmetic progression $2a, -2a, \ldots$.

■ For Exercises 35–38, see Examples 9 and 10.

35. Find the general term and the sixth term in the geometric progression $48, 96, \ldots$.

36. Find the general term and the eighth term in the geometric progression $-3, \dfrac{3}{2}, \ldots$.

37. Find the general term and the seventh term in the geometric progression $-\dfrac{1}{3}, 1, \ldots$.

38. Find the general term and the ninth term in the geometric progression $-81, 27, \ldots$.

B

39. Find the general term and the fifth term in the geometric progression $x, x^3, \ldots$.

40. Find the general term and the tenth term in the geometric progression $x, x(1 + y), \ldots$.

41. Find the general term and the sixth term in the arithmetic progression $x, y, \ldots$.

42. Find the general term and the tenth term in the arithmetic progression $x, x^2, \ldots$.

43. Find the general term and the eighth term in the geometric progression $x, y, \ldots$.

44. Find the general term and the tenth term in the geometric progression $x, x + 2, \ldots$.

45. If the third term in an arithmetic progression is 7 and the eighth term is 17, find the common difference. What are the first and the twentieth terms?

46. If the fifth term of an arithmetic progression is -16 and the twentieth term is -46, what is the twelfth term?

47. Which term in the arithmetic progression $4, 1, \ldots$ is -77?

48. What is the twelfth term in an arithmetic progression in which the second term is x and the third term is y?

49. Find the first term in a geometric progression with fifth term 48 and ratio 2.

50. Find two different values for x so that $-\dfrac{3}{2}, x, -\dfrac{8}{27}$ will be in geometric progression.

9.2

SERIES

Associated with any sequence is a *series*.

Definition 9.4   *A **series** is the indicated sum of the terms in a sequence.*

We shall ordinarily denote a series of n terms by S_n. For example, with the finite sequence

$$4, 7, 10, \ldots, 3n + 1,$$

for a given positive integer n, there is associated the finite series

$$S_n = 4 + 7 + 10 + \cdots + (3n + 1).$$

Similarly, with the finite sequence

$$x, x^2, x^3, x^4, \ldots, x^n,$$

there is associated the finite series

$$S_n = x + x^2 + x^3 + x^4 + \cdots + x^n.$$

Since the terms in the series are the same as those in the sequence, we can refer to the first term or the second term or the general term of a series in the same manner as we do for a sequence.

EXAMPLE 1   Write the series S_j for the sequence $s_n = 2n$.

Solution $S_j = s_1 + s_2 + \cdots + s_j = 2 + 4 + \cdots + 2j$ ■

Sigma Notation

A series for which the general term is known can be represented by using a very convenient **notation** called **sigma**, or **summation**. The Greek letter Σ (sigma) is used to denote a sum. For example,

$$S_n = 4 + 7 + 10 + \cdots + (3n + 1)$$

can be written

$$S_n = \sum_{j=1}^{n} (3j + 1),$$

where we understand that S_n is the series having terms that are obtained by replacing j in the expression $3j + 1$ with the numbers $1, 2, 3, \ldots, n$, successively. Similarly,

$$S = \sum_{j=3}^{6} j^2$$

appears in expanded form as

$$S = 3^2 + 4^2 + 5^2 + 6^2,$$

where the first value for j is 3 and the last is 6.

The variable used in conjunction with summation notation is called the **index of summation**, and the set of integers over which we sum (in the preceding series, $\{3, 4, 5, 6\}$) is called the **range of summation**.

EXAMPLE 2   Write each series in expanded form.

(a) $\displaystyle\sum_{j=3}^{7} 3^{-j}$ (b) $\displaystyle\sum_{j=0}^{3} (2j + 1)$

Solution   (a) $\displaystyle\sum_{j=3}^{7} 3^{-j} = 3^{-3} + 3^{-4} + 3^{-5} + 3^{-6} + 3^{-7}$

$$= \frac{1}{27} + \frac{1}{81} + \frac{1}{243} + \frac{1}{729} + \frac{1}{2187}$$

(b) $\displaystyle\sum_{j=0}^{3} (2j + 1) = (2 \cdot 0 + 1) + (2 \cdot 1 + 1) + (2 \cdot 2 + 1) + (2 \cdot 3 + 1)$

$$= 1 + 3 + 5 + 7$$

E X A M P L E 3   Write each series using sigma notation with the range of summation starting with 1.

(a) $5 + 8 + 11 + 14$ (b) $x^2 + x^4 + x^6$

Solution   In each case we first find an expression for the general term.

(a) $\quad 5 = 3(1) + 2$ (b) $\quad x^2 = x^{2 \cdot 1}$

$\quad\quad 8 = 3(2) + 2$ $\quad\quad x^4 = x^{2 \cdot 2}$

$\quad\quad 11 = 3(3) + 2$ $\quad\quad x^6 = x^{2 \cdot 3}$

$\quad\quad 14 = 3(4) + 2$

general term: $3j + 2$ general term: x^{2j}

We then write the series using sigma notation.

(a) $\displaystyle\sum_{j=1}^{4} (3j + 2)$ (b) $\displaystyle\sum_{j=1}^{3} x^{2j}$ ■

Notation for an Infinite Sum

To indicate that a series has an infinite number of terms, we cannot use the notation S_n for the sum, because there is no value to substitute for n. We therefore adopt notation such as

$$S_\infty = \sum_{j=1}^{\infty} \frac{1}{2^j} \tag{1}$$

to indicate that there is no last term in a series. In expanded form, the infinite series (1) is given by

$$S_\infty = \frac{1}{2} + \frac{1}{4} + \frac{1}{8} + \cdots.$$

The meaning of such an infinite sum will be discussed in Section 9.3.

E X A M P L E 4   Write each series in expanded form.

(a) $\displaystyle\sum_{k=1}^{\infty} (-1)^k 2^{k+1}$ (b) $\displaystyle\sum_{j=3}^{\infty} 4^{-j}$

Solution   (a) $\displaystyle\sum_{k=1}^{\infty} (-1)^k 2^{k+1} = (-1)^1 2^{1+1} + (-1)^2 2^{2+1} + (-1)^3 2^{3+1} + \cdots$

$$= -4 + 8 - 16 + \cdots$$

(b) $\displaystyle\sum_{j=3}^{\infty} 4^{-j} = 4^{-3} + 4^{-4} + 4^{-5} + \cdots$

$$= \frac{1}{64} + \frac{1}{256} + \frac{1}{1024} + \cdots$$

EXAMPLE 5   Write each series in sigma notation with the range of summation starting at 1.

(a) $\dfrac{1}{2} + \dfrac{2}{3} + \dfrac{3}{4} + \cdots$ (b) $\dfrac{2}{3} - \dfrac{4}{9} + \dfrac{8}{27} - \cdots$

Solution   In each case we find an expression for the general term.

(a) $\dfrac{1}{2} = \dfrac{1}{1+1}$ (b) $\dfrac{2}{3} = (-1)^{1+1}\left(\dfrac{2}{3}\right)^1$

$\dfrac{2}{3} = \dfrac{2}{2+1}$ $-\dfrac{4}{9} = (-1)^{2+1}\left(\dfrac{2}{3}\right)^2$

$\dfrac{3}{4} = \dfrac{3}{3+1}$ $\dfrac{8}{27} = (-1)^{3+1}\left(\dfrac{2}{3}\right)^3$

general term: $\dfrac{j}{j+1}$ general term: $(-1)^{j+1}\left(\dfrac{2}{3}\right)^j$

We then write the series using sigma notation, keeping in mind that the series is infinite.

(a) $\displaystyle\sum_{j=1}^{\infty} \frac{j}{j+1}$ (b) $\displaystyle\sum_{j=1}^{\infty} (-1)^{j+1}\left(\frac{2}{3}\right)^j$ ■

Basic Algebraic Properties   Since $\displaystyle\sum_{j=1}^{n} s_j$ represents a sum, we can use the basic properties of real numbers to obtain the following properties.

$$\sum_{j=1}^{n} (a_j + b_j) = \left(\sum_{j=1}^{n} a_j\right) + \left(\sum_{j=1}^{n} b_j\right) \tag{2}$$

$$\sum_{j=1}^{n} k a_j = k \sum_{j=1}^{n} a_j \tag{3}$$

$$\sum_{j=1}^{n} k = nk \tag{4}$$

Equation (4) follows from the fact that

$$\sum_{j=1}^{n} k = \underbrace{k + k + \cdots + k}_{n \text{ terms}} = nk.$$

EXAMPLE 6 Assume that $\sum_{j=1}^{n} a_j = 55$ and $\sum_{j=1}^{n} b_j = 75$. Find $\sum_{j=1}^{n} (2a_j - 3b_j)$.

Solution From Equations (2) and (3), we have

$$\sum_{j=1}^{n} (2a_j - 3b_j) = \sum_{j=1}^{n} 2a_j + \sum_{j=1}^{n} (-3b_j)$$

$$= 2 \sum_{j=1}^{n} a_j - 3 \sum_{j=1}^{n} b_j$$

$$= 2(55) - 3(75) = -115. \qquad \blacksquare$$

Sum of the First n Terms of an Arithmetic Progression

Consider the series S_n of the first n terms in the general arithmetic progression

$$S_n = \qquad a \qquad + \qquad (a + d) \qquad + \cdots + [a + (n - 1)d] \qquad (5)$$
$$= [s_n - (n - 1)d] + [s_n - (n - 2)d] + \cdots + \qquad s_n,$$

and then consider the same series written as

$$S_n = s_n + (s_n - d) + (s_n - 2d) + \cdots + [s_n - (n - 1)d], \qquad (6)$$

where the terms are displayed in reverse order. Adding Equations (5) and (6) term-by-term, we have

$$S_n + S_n = (a + s_n) + (a + s_n) + (a + s_n) + \cdots + (a + s_n),$$

where the term $(a + s_n)$ occurs n times. Then

$$2S_n = n(a + s_n),$$

$$S_n = \frac{n}{2}(a + s_n). \qquad (7)$$

If Equation (7) is rewritten as

$$S_n = n\left(\frac{a + s_n}{2}\right),$$

we observe that the sum is given by the product of the number of terms in the series and the average of the first and last terms. The validity of Equation (7) can be established by mathematical induction and is deferred until Section 9.5.

An alternative form for Equation (7) is obtained by substituting $a + (n - 1)d$ for s_n in Equation (7) to obtain

$$S_n = \frac{n}{2}(a + [a + (n - 1)d]),$$

$$S_n = \frac{n}{2}[2a + (n - 1)d], \tag{7'}$$

where the sum is now expressed in terms of a, n, and d.

Equation (7') can be expressed using sigma notation as

$$\sum_{j=1}^{n} (a + (j - 1)d) = \frac{n}{2}[2a + (n - 1)d]. \tag{8}$$

EXAMPLE 7   Compute $1 + 3 + \cdots + 99$, where the general term of the series is $2j - 1$.

Solution   The sequence $1, 3, \ldots$ is an arithmetic progression with $s_1 = 1$, $d = 2$. We have $s_n = 99$, $a = 1$, and $d = 2$. Thus, from Equation (2) of Section 9.1, we have

$$99 = 1 + (n - 1)2.$$

Solving for n, we obtain $n = 50$. Now we use Equation (8) to compute

$$S_{50} = \sum_{j=1}^{50} (2j - 1) = \frac{50}{2}[2a + (50 - 1)d]$$

$$= \frac{50}{2}[2(1) + (49)(2)] = 2500.$$

Equation (7) could also have been used to find S_{50}.   ■

Sum of the First n Terms of a Geometric Progression

To find an explicit representation for the sum of a given number of terms in a geometric progression in terms of a, r, and n, we employ a device that is somewhat similar to the one used in finding the sum in an arithmetic progression. Consider the geometric series (9) containing n terms and the series (10) obtained by multiplying both members of series (9) by r:

$$S_n = a + ar + ar^2 + ar^3 + \cdots + ar^{n-1}, \tag{9}$$

$$rS_n = \quad ar + ar^2 + ar^3 + \cdots + ar^{n-1} + ar^n. \tag{10}$$

When we subtract series (10) from series (9), all terms in the right-hand members except the first term in series (9) and the last term in series (10) vanish, yielding

$$S_n - rS_n = a - ar^n.$$

Factoring S_n from the left-hand member gives

$$(1 - r)S_n = a - ar^n,$$

$$S_n = \frac{a - ar^n}{1 - r} \qquad (r \neq 1) \tag{11}$$

and we have a formula for the sum of the first n terms in a geometric progression. Establishing the validity of Equation (11), which can be accomplished by mathematical induction, is deferred until Section 9.5.

Equation (11) can be expressed using sigma notation as

$$\sum_{j=0}^{n} ar^j = \frac{a - ar^n}{1 - r} \qquad (r \neq 1). \tag{12}$$

EXAMPLE 8   Find the sum of the first ten terms in each geometric progression.

(a) $a = 2, \quad r = \frac{1}{2}$

(b) $a = \frac{1}{2}, \quad r = -2$

Solution   (a) $S_{10} = \dfrac{a - ar^{10}}{1 - r}$

$$= \frac{2 - 2(\frac{1}{2})^{10}}{1 - (\frac{1}{2})} = \frac{1023}{256}$$

(b) $S_{10} = \dfrac{a - ar^{10}}{1 - r}$

$$= \frac{(\frac{1}{2}) - (\frac{1}{2})(-2)^{10}}{1 - (-2)} = \frac{1023}{-6} \qquad ∎$$

An alternative expression for Equation (11) can be obtained by first writing ar^n as $r(ar^{n-1})$, which yields

$$S_n = \frac{a - r(ar^{n-1})}{1 - r},$$

and then, since $s_n = ar^{n-1}$, expressing this as

$$S_n = \frac{a - rs_n}{1 - r}, \tag{13}$$

where the sum is now given in terms of a, s_n, and r.

EXAMPLE 9 Find the sum of the terms in the geometric progression up to the nth term.

(a) $a = 2, \quad r = \dfrac{1}{2}, \quad s_n = \dfrac{1}{8}$ (b) $a = \dfrac{1}{2}, \quad r = -3, \quad s_n = \dfrac{81}{2}$

Solution (a) $S_n = \dfrac{a - rs_n}{1 - r}$ (b) $S_n = \dfrac{a - rs_n}{1 - r}$

$$= \dfrac{2 - (\frac{1}{2})(\frac{1}{8})}{1 - (\frac{1}{2})} = \dfrac{31}{8}$$ $$= \dfrac{(\frac{1}{2}) - (-3)(\frac{81}{2})}{1 - (-3)} = \dfrac{61}{2}$$ ■

EXERCISE 9.2

A

■ *Write each series in expanded form. See Examples 2 and 4.*

1. $\displaystyle\sum_{j=1}^{4} j^2$

2. $\displaystyle\sum_{j=1}^{4} (3j - 2)$

3. $\displaystyle\sum_{j=1}^{3} \dfrac{(-1)^j}{2^j}$

4. $\displaystyle\sum_{j=3}^{5} \dfrac{(-1)^{j+1}}{j - 2}$

5. $\displaystyle\sum_{j=-3}^{4} (2j + 1)$

6. $\displaystyle\sum_{j=-2}^{2} j^3$

7. $\displaystyle\sum_{j=0}^{\infty} \dfrac{1}{2^j}$

8. $\displaystyle\sum_{j=0}^{\infty} \dfrac{j}{1 + j}$

■ *Write each series in sigma notation with the range of summation starting at 1. See Examples 3 and 5.*

9. $x + x^3 + x^5 + x^7$

10. $x^3 + x^5 + x^7 + x^9 + x^{11}$

11. $1 + 4 + 9 + 16 + 25$

12. $\dfrac{1}{3} + \dfrac{1}{9} + \dfrac{1}{27} + \dfrac{1}{81}$

13. $1 \cdot 2 + 2 \cdot 3 + 3 \cdot 4 + 4 \cdot 5 + \cdots$

14. $\dfrac{1}{2} + \dfrac{2}{3} + \dfrac{3}{4} + \dfrac{4}{5} + \cdots$

15. $\dfrac{2}{1} + \dfrac{3}{2} + \dfrac{4}{3} + \dfrac{5}{4} + \cdots$

16. $\dfrac{1}{1} + \dfrac{2}{3} + \dfrac{3}{5} + \dfrac{4}{7} + \cdots$

■ *Find the value of the indicated sum. Assume*

$$\sum_{j=1}^{100} a_j = 30 \quad and \quad \sum_{j=1}^{100} b_j = -10.$$

See Example 6.

17. $\displaystyle\sum_{j=1}^{100} (a_j + j)$

18. $\displaystyle\sum_{j=1}^{100} (b_j + j)$

19. $\displaystyle\sum_{j=1}^{100} (3a_j + b_j)$

20. $\displaystyle\sum_{j=1}^{100} \left(a_j + \dfrac{1}{2}b_j\right)$

21. $\displaystyle\sum_{j=1}^{100} (2a_j - 3b_j)$

22. $\displaystyle\sum_{j=1}^{100} (-a_j + 2b_j)$

■ *Find each of the following sums. See Example 7.*

23. $\displaystyle\sum_{j=1}^{7} (2j + 1)$ **24.** $\displaystyle\sum_{j=1}^{21} (3j - 2)$ **25.** $\displaystyle\sum_{j=3}^{15} (7j - 1)$ **26.** $\displaystyle\sum_{j=10}^{20} (2j - 3)$

27. $\displaystyle\sum_{k=1}^{8} \left(\frac{1}{2}k - 3\right)$ **28.** $\displaystyle\sum_{k=1}^{100} k$ **29.** $\displaystyle\sum_{k=1}^{100} (2k - 1)$ **30.** $\displaystyle\sum_{k=1}^{100} (10k - 2)$

■ *For Exercises 31–38, see Examples 8 and 9.*

31. $\displaystyle\sum_{j=1}^{6} 3^{j}$ **32.** $\displaystyle\sum_{j=1}^{4} 2^{j}$ **33.** $\displaystyle\sum_{k=3}^{7} \left(\frac{1}{2}\right)^{k-2}$ **34.** $\displaystyle\sum_{j=3}^{12} 2^{j-5}$

35. $\displaystyle\sum_{j=1}^{6} \left(\frac{1}{3}\right)^{j+1}$ **36.** $\displaystyle\sum_{j=1}^{4} 2^{j+3}$ **37.** $\displaystyle\sum_{j=4}^{8} (-2)^{j+4}$ **38.** $\displaystyle\sum_{j=5}^{10} \left(-\frac{1}{2}\right)^{j+2}$

B

39. Find the sum of all even integers n, for $13 < n < 29$.

40. Find the sum of all integral multiples of 7 between 8 and 110.

41. Find the sum of the powers of 2 between 5 and 2000.

42. Find the sum of the powers of $\frac{1}{2}$ between $\frac{1}{300}$ and 1.

43. How many bricks will there be in a wall one brick in thickness if there are 27 bricks in the bottom row, 25 in the second row, and so forth, to the top row, which has one brick?

44. If there are a total of 256 bricks in a wall and they are arranged in the manner of those in Exercise 43, how many bricks are there in the tenth row from the bottom?

45. If $10 is deposited each month in an account that draws 12% annual interest compounded monthly, how much will be in the account at the end of 10 years?

46. How much has to be deposited monthly in an account earning 9% annual interest compounded monthly to guarantee that the account will be worth $1 million at the end of 30 years?

47. Find $\displaystyle\sum_{j=1}^{n} \left(\frac{1}{2}\right)^{j}$ for $n = 2, 3, 4,$ and 5. What value do you think $\displaystyle\sum_{j=1}^{n} \left(\frac{1}{2}\right)^{j}$ approximates as n becomes greater and greater?

48. Find p if $\displaystyle\sum_{j=1}^{5} pj = 14$.

49. Find p and q if $\displaystyle\sum_{j=1}^{4} (pj + q) = 28$ and $\displaystyle\sum_{j=2}^{5} (pj + q) = 44$.

50. Consider $S_n = \displaystyle\sum_{j=1}^{n} f(j)$. Explain why this equation defines a sequence function. What is the variable denoting an element in the domain? The range?

9.3

LIMITS OF SEQUENCES AND SERIES

A Sequence That Is Strictly Increasing but Bounded

Consider the sequence defined by

$$s_n = \frac{n}{n+1}. \tag{1}$$

If we write the range of Equation (1) in the form

$$\frac{1}{2}, \frac{2}{3}, \frac{3}{4}, \frac{4}{5}, \ldots, \frac{n}{n+1}, \ldots,$$

then it seems that the terms are getting close to 1 as n grows large. In fact, the value of $n/(n+1)$ is as close to 1 as we please if n is large enough.

EXAMPLE 1 The value of $n/(n+1)$ is within $\frac{1}{1000}$ of 1 if $n > 999$, because

$$1 - \frac{n}{n+1} = \frac{1}{n+1},$$

and we have $\dfrac{1}{n+1} < \dfrac{1}{1000}$, provided $n+1 > 1000$—that is, $n > 999$. ■

Limit of a Sequence

If it is true that the nth term in a sequence differs from the number L by as little as we please for all sufficiently large n, we say that **the sequence approaches the number L as a limit.** The symbolism

$$\lim_{n \to \infty} s_n = L$$

(read "the limit, as n increases without bound, of s_n is L") is used to denote this situation. A thorough discussion of the notion of a limit is included in courses in the calculus and will not be attempted here. A sequence in which the nth term approaches a number L as $n \to \infty$ is said to be a **convergent sequence,** and the sequence is said to **converge** to L.

EXAMPLE 2 (a) $\displaystyle\lim_{n \to \infty} \frac{n}{n+1} = 1$ (b) $\displaystyle\lim_{n \to \infty} \frac{1}{n} = 0$ ■

A slightly more formal definition of "$\displaystyle\lim_{n \to \infty} s_n = L$" is given in Definition 9.5.

Definition 9.5   *A sequence* $s_1, s_2, \ldots, s_n, \ldots$ ***converges*** *to the number L,*

$$\lim_{n \to \infty} s_n = L,$$

if and only if the absolute value of the difference between the nth term in the sequence and the number L is as small as we please for all sufficiently large n.

E X A M P L E   3   Find $\lim_{n \to \infty} \left(\dfrac{-1}{3} \right)^n$.

Solution   The sequence

$$-\frac{1}{3}, \frac{1}{9}, -\frac{1}{27}, \ldots, \frac{(-1)^n}{3^n}, \ldots$$

converges to 0 since the absolute value of the difference between $(-1)^n/3^n$ and 0 is as small as we please for sufficiently large n.   ■

Divergent Sequences   A sequence that does not converge is said to **diverge**.

E X A M P L E   4   Discuss $\lim_{n \to \infty} (-1)^{n+1} \dfrac{n}{n+1}$.

Solution   The sequence

$$\frac{1}{2}, -\frac{2}{3}, \frac{3}{4}, -\frac{4}{5}, \ldots, (-1)^{n+1} \frac{n}{n+1}, \ldots$$

does not converge. As n increases, the nth term oscillates back and forth from a neighborhood of $+1$ to a neighborhood of -1, and we cannot find a number L such that L differs from $s_n = (-1)^{n+1}n/(n+1)$ by a small amount *for all sufficiently large n.*   ■

A sequence with terms that increase without bound, such as

$$1, 2, 3, \ldots, n, \ldots,$$

also diverges.

Sequence of Partial Sums of a Series

For an infinite series,

$$S_\infty = \sum_{j=1}^{\infty} s_j,$$

we can consider the infinite **sequence of partial sums**:

$$S_1 = s_1$$
$$S_2 = s_1 + s_2$$
$$\vdots$$
$$S_n = s_1 + s_2 + \cdots + s_n$$
$$\vdots$$

Definition 9.6

An infinite series

$$S_\infty = \sum_{j=1}^{\infty} s_j$$

converges if and only if $S_1, S_2, \ldots, S_n, \ldots$, the corresponding sequence of partial sums, converges.

If the sequence of partial sums converges to the number L,

$$\lim_{n \to \infty} S_n = L,$$

then L is said to be the **sum** of the **infinite series**, and we write

$$S_\infty = \sum_{j=1}^{\infty} s_j = L.$$

If the sequence of partial sums diverges, then the **series** is said to **diverge**.

EXAMPLE 5 Determine whether the series converges or diverges.

(a) $\displaystyle\sum_{j=0}^{\infty} 2^j$ (b) $\displaystyle\sum_{j=1}^{\infty} \left(\frac{1}{j} - \frac{1}{j+1} \right)$

Solution (a) From Equation (6) in Section 9.2, the nth partial sum of the series is

$$S_n = 1 + 2^1 + 2^2 + \cdots + 2^n$$
$$= \frac{1 - 2^{n+1}}{1 - 2} = 2^{n+1} - 1.$$

Since the terms in this sequence increase without bound, it diverges. Thus, the infinite series $\displaystyle\sum_{j=0}^{\infty} 2^j$ diverges also.

(b)   The nth partial sum of this series is

$$S_n = \left(\frac{1}{1} - \frac{1}{2}\right) + \left(\frac{1}{2} - \frac{1}{3}\right) + \cdots + \left(\frac{1}{n-1} - \frac{1}{n}\right) + \left(\frac{1}{n} - \frac{1}{n+1}\right)$$

$$= 1 + \left(-\frac{1}{2} + \frac{1}{2}\right) + \left(-\frac{1}{3} + \frac{1}{3}\right) + \cdots + \left(-\frac{1}{n} + \frac{1}{n}\right) - \frac{1}{n+1}$$

$$= 1 - \frac{1}{n+1}.$$

Since this sequence converges to 1 as n grows without bound, the infinite series converges.   ■

In these examples we were able to obtain a formula for the nth partial sum of each series. In general, it is not possible to do this and indirect methods for determining whether or not a series converges must be used. Such methods are studied in the calculus.

Sum of an Infinite Geometric Progression

We recall from Section 9.2 that the sum of n terms (the nth partial sum) of a geometric progression is given, for $r \neq 1$, by

$$S_n = \frac{a - ar^n}{1 - r}. \tag{2}$$

If $|r| < 1$; that is, if $-1 < r < 1$, then $|r|^n$ becomes smaller and smaller for increasingly large n. For example, if $r = \frac{1}{2}$, then

$$r^2 = \frac{1}{4}, \qquad r^3 = \frac{1}{8}, \qquad r^4 = \frac{1}{16},$$

and so forth; and $(\frac{1}{2})^n$ is as small as we please if n is sufficiently large. Writing Equation (2) as

$$S_n = \frac{a}{1 - r}(1 - r^n), \tag{3}$$

we see that the value of the factor $(1 - r^n)$ is as close as we please to 1 provided $|r| < 1$ and n is taken large enough. Since this argument shows that the sequence of partial sums in Equation (3) converges to

$$\frac{a}{1 - r},$$

we have the following result.

Theorem 9.3   *The sum of an infinite geometric progression*

$$\sum_{j=0}^{\infty} ar^j = a + ar + ar^2 + \cdots + ar^n + \cdots,$$

with $|r| < 1$, is

$$S_\infty = \lim_{n \to \infty} S_n = \frac{a}{1 - r}.$$

EXAMPLE 6   Find the indicated sum.

(a) $\displaystyle\sum_{j=0}^{\infty} 2\left(\frac{1}{3}\right)^j$

(b) $\displaystyle\sum_{j=0}^{\infty} 3\left(-\frac{1}{2}\right)^j$

Solution   (a)   We use Theorem 9.3 with $a = 2$ and $r = \frac{1}{3}$. Since $|r| < 1$, we have

$$\sum_{j=0}^{\infty} 2\left(\frac{1}{3}\right)^j = \frac{2}{1 - \frac{1}{3}}$$

$$= 3.$$

(b)   We use Theorem 9.3 with $a = 3$ and $r = -\frac{1}{2}$. Since $|r| < 1$, we have

$$\sum_{j=0}^{\infty} 3\left(-\frac{1}{2}\right)^j = \frac{3}{1 - \left(-\frac{1}{2}\right)}$$

$$= 2. \quad ■$$

Repeating Decimals

An interesting application of the geometric series arises in connection with repeating decimals—that is, decimal numerals that, after a finite number of decimal places, have endlessly repeating groups of digits. For example,

$$0.2121\overline{21} \quad \text{and} \quad 0.138512512\overline{512}$$

are repeating decimals. The bar denotes that the numerals appearing under it are repeated endlessly. Consider the problem of expressing such a decimal fraction as the quotient of two integers.

EXAMPLE 7   Write the decimal $0.2121\overline{21}$ as the quotient of two positive integers.

Solution   The decimal can be written as

$$0.21 + 0.0021 + 0.000021 + \cdots, \tag{4}$$

which is a geometric series with $a = 0.21$ and ratio $r = 0.01$. Since the ratio is less than 1 in absolute value, we can use Theorem 9.3 to find the sum of the infinite series (4). Thus,

$$S_\infty = \frac{a}{1-r} = \frac{0.21}{1-0.01}$$

$$= \frac{21}{99} = \frac{7}{33},$$

and $0.21\overline{21} = \frac{7}{33}$.   ■

EXERCISE 9.3

A

■ *Discuss the limiting behavior of each expression as* $n \to \infty$. *See Examples 1–3.*

1. $\dfrac{1}{n}$

2. $1 + \dfrac{1}{n^2}$

3. $\dfrac{n+1}{n}$

4. $\dfrac{n+3}{n^2}$

5. $2n$

6. $(-1)^n$

7. $\dfrac{1}{2^n}$

8. $(-1)^n \dfrac{1}{n}$

■ *State which of the following sequences are convergent. See Examples 1–4.*

9. $\dfrac{1}{2}, \dfrac{1}{4}, \dfrac{1}{8}, \dfrac{1}{16}, \ldots, \dfrac{1}{2^n}, \ldots$

10. $2, \dfrac{3}{2}, \dfrac{4}{3}, \dfrac{5}{4}, \ldots, \dfrac{n+1}{n}, \ldots$

11. $1, 2, 3, 4, 5, \ldots, n, \ldots$

12. $2, 4, 6, 8, \ldots, 2n, \ldots$

13. $1, -\dfrac{1}{2}, \dfrac{1}{4}, -\dfrac{1}{8}, \ldots, (-1)^{n-1} \dfrac{1}{2^{n-1}}, \ldots$

14. $1, -1, 1, -1, \ldots, (-1)^{n+1}, \ldots$

■ *Find the sum of each infinite geometric series. If the series has no sum, so state. See Examples 5 and 6.*

15. $12 + 6 + \cdots$

16. $2 + 1 + \cdots$

17. $\dfrac{1}{36} + \dfrac{1}{30} + \cdots$

18. $\dfrac{1}{16} - \dfrac{1}{8} + \cdots$

19. $\displaystyle\sum_{j=1}^{\infty} \left(\dfrac{2}{3}\right)^j$

20. $\displaystyle\sum_{j=1}^{\infty} \left(-\dfrac{1}{4}\right)^j$

21. $\displaystyle\sum_{j=2}^{\infty} \left(\dfrac{2}{3}\right)^{j+1}$

22. $\displaystyle\sum_{j=-1}^{\infty} \left(\dfrac{-1}{3}\right)^{j+2}$

■ *Write each repeating decimal as the quotient of two positive integers. See Example 7.*

23. $0.3131\overline{31}$

24. $0.4545\overline{45}$

25. $2.4104\overline{10}$

26. $3.027\overline{027}$

27. $0.12888\overline{8}$

28. 0.83333

B

29. A force is applied to a particle moving in a straight line in such a fashion that each second it moves only one-half of the distance it moved the preceding second. If the particle moves 10 centimeters the first second, approximately how far will it move before coming to rest?

30. The arc length through which the bob on a pendulum moves is nine-tenths of its preceding arc length. Approximately how far will the bob move before coming to rest if the first arc length is 12 inches?

9.4

THE BINOMIAL THEOREM

Factorial Notation

There are situations in which it is necessary to write the product of several consecutive positive integers. To facilitate writing products of this type, we use a special symbol $n!$ (read "n factorial" or "factorial n"), which is defined by

$$n! = n(n - 1)(n - 2) \cdot \cdots \cdot (3)(2)(1).$$

Thus,

$$5! = 5 \cdot 4 \cdot 3 \cdot 2 \cdot 1 \quad \text{(read "five factorial")}$$

and

$$8! = 8 \cdot 7 \cdot 6 \cdot 5 \cdot 4 \cdot 3 \cdot 2 \cdot 1 \quad \text{(read "eight factorial")}.$$

EXAMPLE 1   Write each expression in expanded form and simplify.

(a) $\dfrac{7!}{4!}$ (b) $\dfrac{4!6!}{8!}$

Solution   (a) $\dfrac{7 \cdot 6 \cdot 5 \cdot 4!}{4!} = 210$ (b) $\dfrac{4 \cdot 3 \cdot 2 \cdot 1 \cdot 6!}{8 \cdot 7 \cdot 6!} = \dfrac{3}{7}$

EXAMPLE 2   Write each expression in expanded form showing the first three and the last three factors.

(a) $(2n + 1)!$ (b) $2[(n - 1)!]$

Solution   (a) $(2n + 1)(2n)(2n - 1) \cdot \cdots \cdot 3 \cdot 2 \cdot 1$

(b) $2(n - 1)(n - 2) \cdot \cdots \cdot 3 \cdot 2 \cdot 1$ ■

Division along with factorial notation can be used to represent products of consecutive positive integers, beginning with integers different from 1.

EXAMPLE 3 (a) $8 \cdot 7 \cdot 6 \cdot 5 = \dfrac{8 \cdot 7 \cdot 6 \cdot 5 \cdot (4 \cdot 3 \cdot 2 \cdot 1)}{4 \cdot 3 \cdot 2 \cdot 1} = \dfrac{8!}{4!}$

(b) $6 \cdot 5 \cdot 4 \cdot 3 = \dfrac{6 \cdot 5 \cdot 4 \cdot 3 \cdot (2 \cdot 1)}{2 \cdot 1} = \dfrac{6!}{2!}$

EXAMPLE 4 Simplify each expression.

(a) $\dfrac{(n-1)!}{(n-3)!}$ (b) $\dfrac{(n-1)!(2n)!}{2(n!)(2n-2)!}$

Solution (a) $\dfrac{(n-1)(n-2)(n-3)!}{(n-3)!}$ (b) $\dfrac{(n-1)!(2n)(2n-1)(2n-2)!}{2(n)(n-1)!(2n-2)!}$

$= (n-1)(n-2)$ $= 2n-1$ ■

Recursive Relationship Since

$$n! = n(n-1)(n-2)(n-3) \cdot \cdots \cdot 5 \cdot 4 \cdot 3 \cdot 2 \cdot 1$$

and

$$(n-1)! = (n-1)(n-2)(n-3) \cdot \cdots \cdot 5 \cdot 4 \cdot 3 \cdot 2 \cdot 1,$$

for $n > 1$, we can write the recursive relationship

$$n! = n \cdot (n-1)!. \tag{1}$$

For example,

$$7! = 6! \cdot 7,$$

$$27! = 26! \cdot 27,$$

$$(n+2)! = (n+1)! \cdot (n+2).$$

If Equation (1) is to hold also for $n = 1$, then we must have

$$1! = 1 \cdot (1-1)!$$

or

$$1! = 1 \cdot 0!.$$

Therefore, for consistency, we define 0! by

$$0! = 1.$$

Symbol $\binom{n}{r}$

A special case of the use of factorial notation occurs in the binomial expansion that follows. We make the following definition.

$$\binom{n}{r} = \frac{n!}{r!(n-r)!}. \tag{2}$$

The symbol $\binom{n}{r}$ is read "binomial coefficient n, r," or simply, "n, r." Some examples are

$$\binom{5}{3} = \frac{5!}{3!(5-3)!} = \frac{5!}{3!2!} = \frac{5 \cdot 4 \cdot 3!}{3! \cdot 2 \cdot 1} = 10,$$

$$\binom{5}{1} = \frac{5!}{1!(5-1)!} = \frac{5!}{4!} = \frac{5 \cdot 4!}{4!} = 5,$$

and

$$\binom{5}{0} = \frac{5!}{0!(5-0)!} = \frac{5!}{5!} = 1.$$

EXAMPLE 5 Write each expression in factorial notation and simplify.

(a) $\binom{6}{2}$ (b) $\binom{4}{4}$

Solution (a) $\binom{6}{2} = \frac{6!}{2!(6-2)!} = \frac{6!}{2!4!}$ (b) $\binom{4}{4} = \frac{4!}{4!(4-4)!}$

$= \frac{6 \cdot 5 \cdot 4!}{2 \cdot 1 \cdot 4!} = 15$ $= \frac{4!}{4!0!} = 1$ ■

Binomial Expansions

The series that is obtained by expanding a binomial of the form

$$(a+b)^n$$

is particularly useful in certain branches of mathematics. Starting with familiar examples, where n takes the values 1, 2, 3, 4, and 5 in turn, we can show by direct multiplication that

$$(a+b)^1 = a + b$$

$$(a+b)^2 = a^2 + 2ab + b^2$$

$$(a+b)^3 = a^3 + 3a^2b + 3ab^2 + b^3$$

$$(a+b)^4 = a^4 + 4a^3b + 6a^2b^2 + 4ab^3 + b^4$$

$$(a+b)^5 = a^5 + 5a^4b + 10a^3b^2 + 10a^2b^3 + 5ab^4 + b^5.$$

We observe that in each case:

1.   The first term is a^n.

2.   The variable factors of the second term are $a^{n-1}b^1$, and the coefficient is n, which can be written in the form

$$\frac{n}{1!}.$$

3.   The variable factors of the third term are $a^{n-2}b^2$, and the coefficient can be written in the form

$$\frac{n(n-1)}{2!}.$$

4.   The variable factors of the fourth term are $a^{n-3}b^3$, and the coefficient can be written in the form

$$\frac{n(n-1)(n-2)}{3!}.$$

The foregoing expansions suggest the following result, known as the **binomial theorem**. Its proof, which requires the use of mathematical induction, is omitted.

Theorem 9.4   *For each positive integer n,*

$$(a+b)^n = a^n + \frac{n}{1!}a^{n-1}b + \frac{n(n-1)}{2!}a^{n-2}b^2 + \frac{n(n-1)(n-2)}{3!}a^{n-3}b^3 + \cdots$$

$$+ \frac{n(n-1)(n-2)\cdots\cdots(n-r+2)}{(r-1)!}a^{n-r+1}b^{r-1} + \cdots + b^n, \qquad (3)$$

where r is the number of the term, counting from the left.

E X A M P L E   6   Write $(x-2)^4$ in expanded form.

Solution   Using Theorem 9.4 with $a = x$ and $b = -2$, we have

$$(x-2)^4 = x^4 + \frac{4}{1!}x^3(-2)^1 + \frac{4 \cdot 3}{2!}x^2(-2)^2 + \frac{4 \cdot 3 \cdot 2}{3!}x(-2)^3 + \frac{4 \cdot 3 \cdot 2 \cdot 1}{4!}(-2)^4$$

$$= x^4 - 8x^3 + 24x^2 - 32x + 16. \qquad ■$$

Binomial Coefficients   Observe that the coefficients of the terms in the binomial expansion (3) can be represented as follows.

1st term: $1 = \dfrac{n!}{0!\,n!} = \dbinom{n}{0}$

2nd term: $\dfrac{n}{1!} = \dfrac{n(n-1)!}{1!(n-1)!} = \dfrac{n!}{1!(n-1)!} = \dbinom{n}{1}$

3rd term: $\dfrac{n(n-1)}{2!} = \dfrac{n(n-1)(n-2)!}{2!(n-2)!} = \dfrac{n!}{2!(n-2)!} = \dbinom{n}{2}$

rth term: $\dfrac{n(n-1)(n-2)\cdot\cdots\cdot(n-r+2)}{(r-1)!}$

$$= \dfrac{n(n-1)(n-2)\cdot\cdots\cdot(n-r+2)(n-r+1)!}{(r-1)!(n-r+1)!}$$

$$= \dfrac{n!}{(r-1)!(n-r+1)!} = \dbinom{n}{r-1}$$

Hence, the binomial expansion (3) can be represented by

$$(a+b)^n = \dbinom{n}{0}a^n + \dbinom{n}{1}a^{n-1}b + \dbinom{n}{2}a^{n-2}b^2 + \dbinom{n}{3}a^{n-3}b^3 + \cdots$$
$$+ \dbinom{n}{r-1}a^{n-r+1}b^{r-1} + \cdots + \dbinom{n}{n}b^n. \qquad (4)$$

For example,

$$(x-2)^4 = \dbinom{4}{0}x^4 + \dbinom{4}{1}x^3(-2)^1 + \dbinom{4}{2}x^2(-2)^2 + \dbinom{4}{3}x(-2)^3 + \dbinom{4}{4}(-2)^4.$$

Using formula (2) for $\dbinom{n}{r}$ to find the coefficients, we obtain the same result that we obtained in Example 6.

$$(x-2)^4 = x^4 - 8x^3 + 24x^2 - 32x + 16.$$

rth Term in a Binomial Expansion

Note that the rth term from the left in a binomial expansion is given by

$$\dbinom{n}{r-1}a^{n-r+1}b^{r-1} = \dfrac{n!}{(r-1)!(n-r+1)!}a^{n-r+1}b^{r-1}$$
$$= \dfrac{n(n-1)(n-2)\cdot\cdots\cdot(n-r+2)}{(r-1)!}a^{n-r+1}b^{r-1}. \quad (5)$$

EXAMPLE 7   By using the left-hand member of expansion (5), the seventh term of $(x-2)^{10}$ is

$$\dbinom{10}{6}x^4(-2)^6 = \dfrac{10!}{6!\,4!}x^4(-2)^6 = \dfrac{10\cdot9\cdot8\cdot7\cdot6!}{6!\cdot4\cdot3\cdot2\cdot1}x^4(64)$$
$$= 13{,}440x^4.$$

By using the right-hand member of expansion (5), we also have

$$\frac{10 \cdot 9 \cdot 8 \cdot 7 \cdot 6 \cdot 5}{6 \cdot 5 \cdot 4 \cdot 3 \cdot 2 \cdot 1} x^4 (64) = 13,440 x^4. \qquad ■$$

E X E R C I S E 9.4

A

1. Write $(2n)!$ in expanded form for $n = 4$. **2.** Write $2(n!)$ in expanded form for $n = 4$.

3. Write $n[(n-1)!]$ in expanded form for $n = 6$. **4.** Write $2n[(2n-1)!]$ in expanded form for $n = 2$.

■ *Write in expanded form and simplify. See Example 1.*

5. $5!$ **6.** $7!$ **7.** $\dfrac{9!}{7!}$ **8.** $\dfrac{12!}{11!}$

9. $\dfrac{5!7!}{8!}$ **10.** $\dfrac{12!8!}{16!}$ **11.** $\dfrac{8!}{2!(8-2)!}$ **12.** $\dfrac{10!}{4!(10-4)!}$

■ *Write each product in factorial notation. See Example 3.*

13. $1 \cdot 2 \cdot 3$ **14.** $1 \cdot 2 \cdot 3 \cdot 4 \cdot 5$ **15.** $3 \cdot 4 \cdot 5 \cdot 6$

16. 7 **17.** $8 \cdot 7 \cdot 6$ **18.** $28 \cdot 27 \cdot 26 \cdot 25 \cdot 24$

■ *Write each expression in factorial notation and simplify. See Example 5.*

19. $\dbinom{6}{5}$ **20.** $\dbinom{4}{2}$ **21.** $\dbinom{3}{3}$ **22.** $\dbinom{5}{5}$

23. $\dbinom{7}{0}$ **24.** $\dbinom{2}{0}$ **25.** $\dbinom{5}{2}$ **26.** $\dbinom{5}{3}$

■ *Write each expression in expanded form, showing the first three factors and the last three factors. See Example 2.*

27. $n!$ **28.** $(n+4)!$ **29.** $(3n)!$

30. $3(n!)$ **31.** $(n-2)!$ **32.** $(3n-2)!$

■ *Simplify each expression. See Example 4.*

33. $\dfrac{(n+2)!}{n!}$ **34.** $\dfrac{(n+2)!}{(n-1)!}$ **35.** $\dfrac{(n+1)[(n+2)!]}{(n+3)!}$

36. $\dfrac{(2n+4)!}{(2n+2)!}$ **37.** $\dfrac{(2n)!(n-2)!}{4(2n-2)!(n)!}$ **38.** $\dfrac{(2n+1)!(2n-1)!}{[(2n)!]^2}$

■ *Expand. See Example 6.*

39. $(x+3)^5$ **40.** $(2x+y)^4$ **41.** $(x-3)^4$ **42.** $(2x-1)^5$

43. $\left(2x - \dfrac{y}{2}\right)^3$ **44.** $\left(\dfrac{x}{3} + 3\right)^5$ **45.** $\left(\dfrac{x}{2} + 2\right)^6$ **46.** $\left(\dfrac{2}{3} - a^2\right)^4$

■   *Write the first four terms in each expansion. Do not simplify the terms. See Example 6.*

47. $(x + y)^{20}$ **48.** $(x - y)^{15}$ **49.** $(a - 2b)^{12}$

50. $(2a - b)^{12}$ **51.** $(x - \sqrt{2})^{10}$ **52.** $\left(\dfrac{x}{2} + 2\right)^8$

■   *Find the specified term. See Example 7.*

53. $(a - b)^{15}$,   the sixth term   **54.** $(x + 2)^{12}$,   the fifth term
55. $(x - 2y)^{10}$,   the fifth term   **56.** $(a^3 - b)^9$,   the seventh term

B

■   *Find each power to the nearest hundredth. [Hint: Write the given power as $(1 + x)^n$ where $-1 < x < 1$, e.g., $(0.97)^7 = (1 - 0.03)^7$.]*

57. $(1.02)^{10}$ **58.** $(1.01)^{15}$ **59.** $(0.99)^8$ **60.** $(0.95)^8$

61. Given that the binomial formula holds for $(1 + x)^n$, where n is a negative integer:
   (a) Write the first four terms of $(1 + x)^{-1}$.
   (b) Find the first four terms of the quotient $1/(1 + x)$, by dividing 1 by $(1 + x)$. Compare the results of parts **(a)** and **(b)**.

62. Given that the binomial formula holds as an infinite "sum" for $(1 + x)^n$, where n is a rational number and $|x| < 1$, find the following to two decimal places.
   (a) $\sqrt{1.02}$ **(b)** $\sqrt{0.99}$

63. Show that $\dbinom{k}{r} + \dbinom{k}{r - 1} = \dbinom{k + 1}{r}$.

64. Show that $n\dbinom{n - 1}{k} = (n - k)\dbinom{n}{k}$.

9.5

MATHEMATICAL INDUCTION

The material in the present section depends on two special properties of the set of positive integers, $N = \{1, 2, 3, \ldots\}$:

   1.   1 is a positive integer.
   2.   If k is a positive integer, then $k + 1$ is a positive integer.

In addition, N *contains no elements that are not implied by properties* 1 *and* 2. These properties underlie the following theorem, called the **principle of mathematical induction**, which we state without proof.

Theorem 9.5

If a given sentence involving positive integers n is true for n = 1 and if its truth fo
n = k implies its truth for n = k + 1, then it is true for every positive integer n.

**Requirements
of a Proof by
Mathematical
Induction**

We can use Theorem 9.5 to prove a number of assertions. Although the technique
we use is called **proof by mathematical induction**, the argument we employ is
deductive, as have been all the other arguments in this book. Proofs by
mathematical induction require two things:

1. A demonstration that the assertion to be proved is true for the   e
   integer 1.

2. A demonstration that the truth of the assertion for a positive int   r k
   implies its truth for $k + 1$.

When these two demonstrations have been made, the principle of mathematica.
induction assures us that the assertion is true for every positive integer.
    This method of proof is often compared to lining up a row of dominoes, with
the assumption that whenever one domino is toppled, the one following will topple.
One then needs only to topple the first domino ($n = 1$) and the whole row behind it
will topple, as indicated in Figure 9.1.

FIGURE 9.1

EXAMPLE 1   Prove that the sum of the first n positive integers is $\dfrac{n(n + 1)}{2}$.

Solution   We wish to show that

$$1 + 2 + 3 + \cdots + n = \frac{n(n + 1)}{2}.$$

We must do two things:

1. We must first show that the assertion is true for $n = 1$, that is,

$$1 = \frac{1(1 + 1)}{2},$$

which is true.

2. We must next show that the truth of the assertion for $n = k$ implies its truth for $n = k + 1$. That is, we must show that the truth of

$$1 + 2 + 3 + \cdots + k = \frac{k(k + 1)}{2} \tag{1}$$

implies the truth of

$$1 + 2 + 3 + \cdots + k + (k + 1) = \frac{(k + 1)[(k + 1) + 1]}{2}. \tag{2}$$

Assuming the truth of Equation (1), we add $k + 1$ to each member of this equation, obtaining

$$1 + 2 + 3 + \cdots + k + (k + 1) = \frac{k(k + 1)}{2} + (k + 1)$$

$$= (k + 1)\left(\frac{k}{2} + 1\right)$$

$$= (k + 1)\left(\frac{k + 2}{2}\right),$$

from which we obtain

$$1 + 2 + 3 + \cdots + k + (k + 1) = \frac{(k + 1)[(k + 1) + 1]}{2},$$

which is Equation (2).

Thus, the second fact that is necessary for our proof is established. By the principle of mathematical induction, the assertion is true for every positive integer n.

EXAMPLE 2 Prove that for any arithmetic progression,

$$S_n = \frac{n}{2}[2a + (n - 1)d].$$

Solution For $n = 1$, the assertion is that

$$S_1 = \frac{1}{2}(2a + 0 \cdot d) = a,$$

which is true. We must next show that

$$S_k = \frac{k}{2}[2a + (k - 1)d]$$

implies that

$$S_{k+1} = \frac{k + 1}{2}[2a + (k + 1 - 1)d] = \frac{k + 1}{2}(2a + kd).$$

Now, since S_{k+1} is the sum of the first $k + 1$ terms, S_{k+1} is the sum of the first k terms and the $k + 1$st term. That is,

$$S_{k+1} = S_k + s_{k+1}.$$

Therefore,

$$S_{k+1} = \frac{k}{2}[2a + (k-1)d] + (a + kd)$$

$$= ka + a + \frac{k(k-1)}{2}d + kd$$

$$= (k+1)a + \frac{k(k+1)}{2}d = \frac{k+1}{2}(2a + kd).$$

Thus, by the principle of mathematical induction, the assertion is true for every positive integer n.   ■

E X E R C I S E   9.5

A

■ *By mathematical induction, prove the validity of the formulas in Exercises 1–14 for all positive integers.*

1. $\dfrac{1}{2} + \dfrac{2}{2} + \dfrac{3}{2} + \cdots + \dfrac{n}{2} = \dfrac{n(n+1)}{4}$

2. $1 + 3 + 5 + \cdots + (2n-1) = n^2$

3. $2 + 4 + 6 + \cdots + 2n = n(n+1)$

4. $2 + 6 + 10 + \cdots + (4n-2) = 2n^2$

5. $2 + 8 + 14 + \cdots + (6n-4) = n(3n-1)$

6. $1 + 5 + 9 + \cdots + (4n-3) = n(2n-1)$

7. $7 + 10 + 13 + \cdots + (3n+4) = \dfrac{n(3n+11)}{2}$

8. $6 + 11 + 16 + \cdots + (5n+1) = \dfrac{n(5n+7)}{2}$

9. $1^2 + 2^2 + 3^2 + \cdots + n^2 = \dfrac{n(n+1)(2n+1)}{6}$

10. $2 + 2^2 + 2^3 + \cdots + 2^n = 2^{n+1} - 2$

11. $1^3 + 3^3 + 5^3 + \cdots + (2n-1)^3 = n^2(2n^2 - 1)$

12. $\dfrac{1}{1 \cdot 2} + \dfrac{1}{2 \cdot 3} + \dfrac{1}{3 \cdot 4} + \cdots + \dfrac{1}{n(n+1)} = \dfrac{n}{n+1}$

13. $1 \cdot 2 + 2 \cdot 3 + 3 \cdot 4 + \cdots + n(n+1) = \dfrac{n(n+1)(n+2)}{3}$

14. $1 \cdot 4 + 2 \cdot 9 + 3 \cdot 16 + \cdots + n(n+1)^2 = \dfrac{1}{12}n(n+1)(n+2)(3n+5)$

B

15. Show that $n^3 + 2n$ is divisible by 3 for every positive integer n.

16. Show that $3^{2n} - 1$ is divisible by 8 for every positive integer n.

17. Show that if $2 + 4 + 6 + \cdots + 2n = n(n + 1) + 2$ is true for $n = k$, then it is true for $n = k + 1$. Is it true for every positive integer n?

18. Show that $n^3 + 11n = 6(n^2 + 1)$ is true for $n = 1, 2$, and 3. Is it true for every positive integer n?

19. Use mathematical induction to prove Theorem 9.1.

20. Use mathematical induction to prove Theorem 9.2.

21. Use mathematical induction to prove that for a geometric progression, $S_n = \dfrac{a - ar^n}{1 - r}$ for every positive integer n, provided $r \neq 1$.

22. Use mathematical induction to show that $n! \geq 2^n$ for every integer $n \geq 4$.

CHAPTER REVIEW

Key Words, Phrases, and Symbols

■ *Define or explain each of the following words, phrases, and symbols.*

1. sequence

2. terms of a sequence

3. arithmetic progression

4. geometric progression

5. finite series

6. sigma, or summation, notation

7. $\lim\limits_{n \to \infty} s_n = L$

8. convergent sequence

9. divergent sequence

10. partial sums of an infinite series

11. convergent series

12. divergent series

13. geometric series

14. $n!$

15. $\dbinom{n}{r}$

16. binomial theorem

17. principle of mathematical induction

Review Exercises

A

[9.1] ■ *Find the first three terms in the sequence with the general term as given.*

1. $s_n = n^2 + 1$

2. $s_n = \dfrac{1}{n + 1}$

■ *Find the next three terms in each arithmetic progression.*

3. $7, 10, \ldots$

4. $a, a - 2, \ldots$

■ *Find the next three terms in each geometric progression.*

5. $-2, 6, \ldots$

6. $\dfrac{2}{3}, 1, \ldots$

7. Find the general term and the seventh term in the arithmetic progression $-3, 2, \ldots$.

8. Find the general term and the fifth term in the geometric progression $-2, \dfrac{2}{3}, \ldots$.

9. If the fourth term in an arithmetic progression is 13 and the ninth term is 33, find the seventh term.

10. Which term in the geometric progression $-\dfrac{2}{9}, \dfrac{2}{3}, \ldots$ is 54?

[9.2] **11.** Write $\displaystyle\sum_{k=2}^{5} k(k-1)$ in expanded form.

12. Write $x^2 + x^3 + x^4 + \cdots$ in sigma notation.

13. Find the value of $\displaystyle\sum_{j=3}^{9} (3j - 1)$.

14. Find the value of $\displaystyle\sum_{j=1}^{5} \left(\dfrac{1}{3}\right)^j$.

[9.3] **15.** Specify the limit of $\dfrac{3n^2 - 1}{n^2}$ as $n \to \infty$.

16. Find the value of the geometric series $4 - 2 + 1 - \dfrac{1}{2} + \cdots$.

17. Find the value of $\displaystyle\sum_{i=1}^{\infty} \left(\dfrac{1}{3}\right)^i$.

18. Find a fraction equivalent to $0.44\overline{4}$.

[9.4] **19.** Write $n[(n-3)!]$ in expanded form for $n = 5$.

■ *Write each expression in expanded form and simplify.*

20. $\dfrac{8!3!}{7!}$

21. $\dbinom{7}{2}$

22. $\dfrac{(n-1)!}{n!(n+1)!}$

23. Find the first four terms in the binomial expansion of $(x - 2y)^{10}$.

24. Find the eighth term in the expansion of $(x - 2y)^{10}$.

[9.5] ■ *By mathematical induction, prove each formula for all positive integral values of n.*

25. $3 + 6 + 9 + \cdots + 3n = \dfrac{3n(n+1)}{2}$

26. $\dfrac{1}{2} + \dfrac{1}{4} + \dfrac{1}{8} + \cdots + \dfrac{1}{2^n} = 1 - \dfrac{1}{2^n}$

B

27. Find two values of x so that $\frac{3}{4}, x, \frac{16}{9}$ will be in geometric progression.

28. What is the tenth term in an arithmetic progression in which the third term is x and the fourth term is y?

29. Find p and q if $\sum\limits_{j=3}^{6} (pj + q) = 102$ and $\sum\limits_{j=1}^{4} (pj + q) = 62$.

30. Find the sum of the powers of $\frac{2}{3}$ between $\frac{1}{1000}$ and 1.

31. Use the binomial formula to find $(1.05)^{10}$ to the nearest hundredth.

32. Use the binomial formula as an infinite "sum" to find $\sqrt{0.98}$ to two decimal places.

33. Show that if $s_n = \dfrac{(2n + 1)}{n^2(n + 1)^2}$, then $S_n = 1 - \dfrac{1}{(n + 1)^2}$.

34. Show that for $n \geq 8$, $n! \geq 3^{n+1}$.

10 Counting and Probability

In this chapter we consider several important counting techniques, permutations and combinations. These notions are then applied to computing probabilities.

BASIC COUNTING PRINCIPLES; PERMUTATIONS

Associated with each finite set A is a nonnegative integer n, namely the number of elements in A. Hence, we have a function from the set of all finite sets to the set of nonnegative integers. The symbolism $n(A)$ is used to denote elements in the range of this set function n. For example, if

$$A = \{5, 7, 9\}, \qquad B = \left\{\frac{1}{2}, 0, 3, -5, 7\right\}, \qquad \text{and} \qquad C = \varnothing,$$

then

$$n(A) = 3, \qquad n(B) = 5, \qquad \text{and} \qquad n(C) = 0.$$

Counting Properties

All the sets with which we shall hereafter be concerned are assumed to be finite sets. We then have the following properties, called **counting properties**, for the function n.

$$\text{I.} \quad n(A \cup B) = n(A) + n(B), \qquad \text{if} \qquad A \cap B = \varnothing.$$

Thus, if A and B are disjoint sets, that is, if they have no elements in common, then the number of elements in their union is the sum of the number of elements in A and the number of elements in B.

EXAMPLE 1 Suppose there are five roads from town R to town S and two railroads from town R to town S. If A is the set of roads and B the set of railroads from R to S, then

$n(A) = 5$, $n(B) = 2$, and

$$n(A \cup B) = 5 + 2 = 7.$$

Thus, there are seven ways one can go from town R to town S by driving or riding on a train. ■

If A and B have elements in common, then to count the number of elements in $A \cup B$, we might add the number of elements in A to the number of elements in B. But since any elements in the intersection of A and B are counted twice in this process (once in A and once in B), we must subtract the number of such elements from the sum $n(A) + n(B)$ to obtain the number of elements in $A \cup B$, as suggested in Figure 10.1. This gives us the second counting property.

FIGURE 10.1

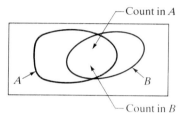

II. $n(A \cup B) = n(A) + n(B) - n(A \cap B),$ if $A \cap B \neq \varnothing.$

EXAMPLE 2 If $A = \{a, b, c\}$ and $B = \{c, d\}$, find $n(A \cup B)$. We first note that

$$A \cap B = \{c\}$$

and thus $n(A \cap B) = 1$. Using Property II, we have

$$\begin{aligned} n(A \cup B) &= n(A) + n(B) - n(A \cap B) \\ &= 3 + 2 - 1 \\ &= 4. \end{aligned}$$

EXAMPLE 3 Suppose there are 15 unrelated girls and 17 unrelated boys in a mathematics class, and suppose there are precisely 2 brother-sister pairs in the class. How many different families are represented by all members of the class?

Solution We let A denote the set of different families represented by the girls and B denote the set of different families represented by the boys. The number of different families represented by all the members of the class is $n(A \cup B)$. Notice that since there are 2 brother-sister pairs we have

$$n(A \cap B) = 2.$$

Thus, from Property II

$$n(A \cup B) = n(A) + n(B) - n(A \cap B)$$
$$= 15 + 17 - 2$$
$$= 30.$$

Therefore, 30 different families are represented by all the members of the class.   ■

Notice that Property I is a special case of Property II because if $A \cap B = \varnothing$ then $n(A \cap B) = 0$.

Definition 10.1   The **Cartesian product** of two sets A and B, denoted by $A \times B$, is the set of all ordered pairs (a, b) such that $a \in A$ and $b \in B$.

The following property relates $n(A \times B)$ to $n(A)$ and $n(B)$.

III. $n(A \times B) = n(A) \cdot n(B)$

This asserts that the number of elements in the Cartesian product of sets A and B is the product of the number of elements in A and the number of elements in B.

EXAMPLE 4   Suppose there are five roads from town R to town S (set A) and there are three roads from town S to town T (set B). How many ways can we drive from R to T via S?

Solution   For each element of A there are three elements of B, and the total possible ways one can drive from R to T via S is

$$n(A \times B) = n(A) \cdot n(B) = 5 \cdot 3 = 15.   ■$$

Property III can be stated in an equivalent way as follows. Suppose the first of two operations (for example, choosing a road from R to S) can be done in a ways. Suppose the second operation (choosing a road from S to T) can be done in b ways and each possible second operation can succeed every first operation. Then the two operations in sequence can be done in $a \cdot b$ ways. This formulation has the following very useful generalization.

IV.   Suppose the first of several operations can be one in a ways, the second in b ways, no matter what came first, the third in c ways, no matter what came prior, and so on. Then the number of ways the operation can be done in sequence is $a \cdot b \cdot c \cdots$.

EXAMPLE 5   Given the set of digits $A = \{1, 2, 3\}$, how many different three-digit numerals can be constructed from the members of A if no member is used more than once?

Solution   We apply the fourth counting property. The first digit can be any of the three numerals 1, 2, or 3. No matter what comes first, two choices remain for the second digit, and then one choice remains for the third digit. By the fourth counting property, the number of ways of choosing the three digits in order is $3 \cdot 2 \cdot 1 = 6$.

EXAMPLE 6   In how many ways can three members of a class be assigned a grade of A, B, C, or D?

Solution   We sometimes use a simple diagram, such as __, __, __, to designate a sequence. This is a helpful preliminary device. Since each of the students may receive any one of four different grades, our sequence is

$$\underline{4}, \underline{4}, \underline{4}.$$

From Property IV, there are $4 \cdot 4 \cdot 4$, or 64, possible ways the grades may be assigned.

EXAMPLE 7   In how many ways can three members of a class be assigned a grade of A, B, C, or D so that no two members receive the same grade?

Solution   Since the first student may receive any one of four different grades, the second student may then receive any one of three different grades, and the third student may then receive any one of two different grades, our sequence would appear as

$$\underline{4}, \underline{3}, \underline{2}.$$

From Property IV, there are $4 \cdot 3 \cdot 2$, or 24, possible ways the grades may be assigned.   ■

Permutations   Notice in the preceding examples that the order in which digits were listed and/or grades were assigned was important when counting the number of possible numerals and/or grade assignments.

Definition 10.2   A ***permutation*** of a set A is an ordering (first, second, etc.) of the members of A.

To count the number of permutations of a set with n members, note that we have n choices for the first member in the permutation, $n - 1$ for the second, and so on until only one remains for the last. By Property IV, the total number of possible permutations is

$$n(n - 1)(n - 2) \cdot \cdots \cdot 1 = n!$$

This proves the following theorem.

Theorem 10.1   *Let $P_{n,n}$ denote the number of distinct permutations of the members of a set A containing n members. Then*

$$P_{n,n} = n! \tag{1}$$

The symbol $P_{n,n}$ (or sometimes $_nP_n$ or P_n^n) is read "the number of permutations of n things taken n at a time."

E X A M P L E 8   In how many ways can nine men be assigned positions to form distinct baseball teams?

Solution   Let A denote the set of men, so that $n(A) = 9$. The total number of ways in which nine men can be assigned nine positions on a team, or, in other words, the number of possible permutations of the members of a nine-element set, is, by Equation (1),

$$P_{9,9} = 9! = 9 \cdot 8 \cdot 7 \cdot \cdots \cdot 1 = 362{,}880. \quad ■$$

In many applications there are more members from which to choose than there are places to fill, as the following theorem and example illustrate.

Theorem 10.2   *Let $P_{n,r}$ denote the number of permutations of the members, taken r at a time, of a set A containing n members; that is, let $P_{n,r}$ be the number of distinct orderings of r elements when there is a set A of n elements from which to choose. Then*

$$P_{n,r} = n(n-1)(n-2) \cdot \cdots \cdot (n-r+1). \tag{2}$$

The proof follows the proof of Theorem 10.1 except that only r choices are made, so the last choice involves $n - (r-1) = n - r + 1$ possibilities.

E X A M P L E 9   In how many ways can a basketball team be formed by assigning one player, from a set of ten players, to each of the five positions?

Solution   Let A denote the set of players, so that $n(A) = 10$. Then from Equation (2) and the fact that a basketball team consists of 5 players, we have

$$P_{10,5} = 10 \cdot 9 \cdot 8 \cdot \cdots \cdot (10 - 5 + 1)$$
$$= 10 \cdot 9 \cdot 8 \cdot 7 \cdot 6 = 30{,}240. \quad ■$$

An alternative expression for $P_{n,r}$ can be obtained by observing that

$$P_{n,r} = n(n-1)(n-2) \cdot \cdots \cdot (n-r+1)$$
$$= \frac{n(n-1)(n-2) \cdot \cdots \cdot (n-r+1)(n-r)!}{(n-r)!},$$

so that

$$P_{n,r} = \frac{n!}{(n-r)!}.$$ (3)

Distinguishable Permutations

The problem of finding the number of distinguishable permutations of n objects taken n at a time, if some of the objects are identical, requires a little more careful analysis. As an example, consider the number of permutations of the letters in the word *DIVISIBLE*. We can make a distinction between the three *I*'s by assigning subscripts to each so that we have nine distinct letters,

$$D, I_1, V, I_2, S, I_3, B, L, E.$$

The number of permutations of these nine letters is of course 9!. If the letters other than I_1, I_2, and I_3 are retained in the positions they occupy in a permutation of the preceding nine letters, I_1, I_2, and I_3 can be permuted among themselves 3! ways. Thus, if P is the number of *distinguishable* permutations of the letters

$$D, I, V, I, S, I, B, L, E,$$

then, since for each of these there are 3! ways in which the *I*'s can be permuted without otherwise changing the order of the other letters, it follows that

$$3! \cdot P = 9!,$$

from which

$$P = \frac{9!}{3!} = 60{,}480.$$

As another example, consider the letters in the word *MISSISSIPPI*. There would be 11! distinguishable permutations of the letters in this word if each letter were distinct. Note, however, that the letters *S* and *I* each appear four times and the letter *P* appears twice. Reasoning as we did in the previous example, we see that the number P of distinguishable permutations of the letters in *MISSISSIPPI* is given by

$$4!4!2!P = 11!,$$

from which

$$P = \frac{11!}{4!4!2!} = 34{,}650.$$

These two examples suggest the following theorem, which we state without proof.

Theorem 10.3

If there are n_1 identical objects of a first kind, n_2 of a second, . . . , and n_k of a kth, with $n_1 + n_2 + \cdots + n_k = n$, then the number of distinguishable permutations of the n objects is given by

$$P = \frac{n!}{n_1! n_2! \cdot \cdots \cdot n_k!}.$$

EXAMPLE 10   Find the number of distinguishable permutations of the letters in the word *CHATTANOOGA*.

Solution   There are 11 letters in *CHATTANOOGA*. The letter C, H, N, and G each appear only 1 time; the letters T and O each appear 2 times; and the letter A appears 3 times. Thus, from Theorem 10.3 the number of permutations of the letters in *CHATTANOOGA* is

$$P = \frac{11!}{1!1!1!1!2!2!3!} = 1{,}663{,}200. \qquad ■$$

Circular Permutations   An arrangement, ignoring rotations, of *n* objects around a circle is called a **circular permutation** of the objects. Notice that since we are ignoring rotations, the two circular permutations shown in Figure 10.2 are the same because one is obtained from the other by rotating the circle.

FIGURE 10.2

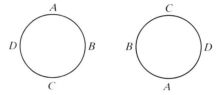

We count the number of circular permutations of *n* objects by noting that each object could be initially placed in any position without affecting the arrangement. Further, after the first object is placed, there are $(n-1)!$ ways to place the remaining $n-1$ objects relative to the first. Thus, we have the following theorem.

Theorem 10.4   *There are* $(n-1)!$ *distinct circular permutations of n objects.*

EXAMPLE 11   In how many ways can a committee of seven be seated around a circular table?

Solution   We want the number of circular permutations of the seven people. From Theorem 10.4, there are $(7-1)! = 720$ such seating arrangements.   ■

EXERCISE 10.1

A

■ *Given the following sets, find* $n\,(A \cap B)$, $n(A \cup B)$, *and* $n(A \times B)$. *See Examples 1–3.*

1. $A = \{d, e\}$, $B = \{e, f, g, h\}$ **2.** $A = \{e\}$, $B = \{a, b, c, d\}$

3. $A = \{1, 2, 3, 4\}, \quad B = \{3, 4, 5, 6\}$

4. $A = \{1, 2\}, \quad B = \{3, 4, 5\}$

5. $A = \{1, 2\}, \quad B = \{1, 2\}$

6. $A = \varnothing, \quad B = \{2, 3, 4\}$

■ *In each problem, a digit or letter may be used more than once unless stated otherwise. See Examples 4–9.*

7. How many different two-digit numerals can be formed from the digits 5 and 6?

8. How many different two-digit numerals can be formed from the digits 7, 8, and 9?

9. In how many different ways can four students be seated in a row?

10. In how many different ways can five students be seated in a row?

11. In how many different ways can four questions on a true–false test be answered?

12. In how many different ways can five questions on a true–false test be answered?

13. In how many ways can you write different three-digit numerals from $\{2, 3, 4, 5\}$?

14. How many different seven-digit telephone numbers can be formed from the set of digits $\{1, 2, 3, 4, 5, 6, 7, 8, 9, 0\}$?

15. In how many ways can you write different three-digit numerals, using $\{2, 3, 4, 5\}$, if no digit is to be used more than once in each numeral?

16. How many different seven-digit telephone numbers can be formed from the set of digits $\{1, 2, 3, 4, 5, 6, 7, 8, 9, 0\}$ if no digit is to be used more than once in any number?

17. How many three-letter sequences can be formed from $\{A, N, S, W, E, R\}$?

18. If no letter is to be used more than once in any permutation, how many different three-letter permutations can be formed from $\{A, N, S, W, E, R\}$?

19. How many four-digit numerals for odd integers can be formed from $\{1, 2, 3, 4, 5\}$?

20. How many four-digit numerals for even integers can be formed from $\{1, 2, 3, 4, 5\}$?

21. How many numerals for positive integers less than 500 can be formed from $\{3, 4, 5\}$?

22. How many numerals for positive odd integers less than 500 can be formed from $\{3, 4, 5\}$?

23. How many numerals for positive even integers less than 500 can be formed from $\{3, 4, 5\}$?

24. How many numerals for positive even integers between 400 and 500, inclusive, can be formed from $\{3, 4, 5\}$?

25. How many permutations of the elements in $\{P, R, I, M, E\}$ end in a vowel?

26. How many permutations of the elements in $\{P, R, O, D, U, C, T\}$ end in a vowel?

■ *For Exercises 27–30, see Example 10.*

27. Find the number of distinguishable permutations of the letters in the word *LIMIT*.

28. Find the number of distinguishable permutations of the letters in the word *COMBINATION*.

29. Find the number of distinguishable permutations of the letters in the word *COLORADO*.

30. Find the number of distinguishable permutations of the letters in the word *TALLAHASSEE*.

■ *For Exercises* 31 *and* 32, *see Example* 11.

31. In how many ways can five students be seated around a circular table?

32. In how many ways can six students be seated around a circular table?

B

■ *Show that each of the following is true.*

33. $P_{5,3} = 5(P_{4,2})$

34. $P_{5,r} = 5(P_{4,r-1})$

35. $P_{n,3} = n(P_{n-1,2})$

36. $P_{n,3} - P_{n,2} = (n-3)(P_{n,2})$

37. Solve for n: $P_{n,5} = 5(P_{n,4})$.

38. Solve for n: $P_{n,5} = 9(P_{n-1,4})$.

39. In how many ways can six students be seated around a circular table if a certain two must be seated together?

40. In how many ways can three different keys be arranged on a key ring? [*Hint:* Arrangements should be considered identical if one can be obtained from the other by turning the ring over.]

10.2

COMBINATIONS

An additional counting concept is needed before we turn our attention to probability—namely, finding the number of distinct r-element subsets of an n-element set with no reference to relative order of the elements in the subset. For example, five different cards can be arranged in 5! permutations, but to a poker player they represent the same hand. The set of five cards (with no reference to the arrangement of the cards) is called a *combination*.

Definition 10.3 *A subset of an n-element set A is called a **combination**.*

The counting of combinations is related to the counting of permutations. From Theorem 10.2, we know that the number of distinct permutations of n elements of a set A taken r at a time is given by

$$P_{n,r} = \frac{n!}{(n-r)!}.$$

With this in mind, consider the following result concerning the number $C_{n,r}$ of combinations of n things taken r at a time. The symbol $C_{n,r}$ is read "the number of combinations of n things taken r at a time."

Theorem 10.5   *The number $C_{n,r}$ of distinct combinations of the members, taken r at a time, of a set containing n members is given by*

$$C_{n,r} = \frac{P_{n,r}}{r!}. \tag{1}$$

To see that this is true, we observe that there are, by definition, $C_{n,r}$ r-element subsets of set A, where $n(A) = n$. Also, from Theorem 10.1, each of these subsets has $r!$ permutations of its members. Thus, there are $C_{n,r}r!$ permutations of n elements of A taken r at a time. Therefore,

$$P_{n,r} = C_{n,r}r!$$

from which we obtain

$$C_{n,r} = \frac{P_{n,r}}{r!}.$$

Thus, to find the number of r-element subsets of an n-element set A, we count the number of permutations of the elements of A taken r at a time and then divide by the number of possible permutations of an r-element set. This seems very much like counting a set of people by counting the number of arms and legs and dividing the result by 4, but this approach gives us a very useful expression for the number we seek, $C_{n,r}$. Since

$$P_{n,r} = n(n-1)(n-2)\cdots\cdots(n-r+1),$$

it follows that

$$C_{n,r} = \frac{P_{n,r}}{r!} = \frac{n(n-1)(n-2)\cdots\cdots(n-r+1)}{r!}. \tag{2}$$

EXAMPLE 1   In how many ways can a committee of five be selected from a set of twelve persons?

Solution   We want the number of 5-element subsets of a 12-element set. From Equation (2), we have

$$C_{12,5} = \frac{12 \cdot 11 \cdot 10 \cdot 9 \cdot 8}{5 \cdot 4 \cdot 3 \cdot 2 \cdot 1} = 792. \quad ■$$

By Equation (3) on page 418, we have the alternative expression

$$C_{n,r} = \frac{P_{n,r}}{r!} = \frac{n!}{r!(n-r)!}. \tag{3}$$

Notice that the value of $C_{n,r}$ given in the right-hand member of (3) is the same as the value of the binomial coefficient $\binom{n}{r}$ given in the right-hand member of (2) on page 401. Thus,

$$C_{n,r} = \binom{n}{r}.$$

EXAMPLE 2   How many different amounts of money can be formed from a penny, a nickel, a dime, and a quarter?

Solution   We want to find the total number of combinations that can be formed by taking the coins 1, 2, 3, and 4 at a time. By Equation (3) we have

$$C_{4,1} = \frac{4!}{1!3!} = 4, \qquad C_{4,2} = \frac{4!}{2!2!} = 6, \qquad C_{4,3} = \frac{4!}{3!1!} = 4, \qquad C_{4,4} = \frac{4!}{4!0!} = 1,$$

and the total number of combinations is 15. Clearly, each combination gives a different amount.   ■

Symmetry   Since each time a distinct set of r objects is chosen, a distinct set of $n - r$ objects remains unchosen, we have the following plausible assertion.

Theorem 10.6 $C_{n,r} = C_{n,n-r}$

To verify this, note that from Equation (3) we have

$$C_{n,n-r} = \frac{n!}{(n-r)![n-(n-r)]!} = \frac{n!}{(n-r)!r!} = C_{n,r}.$$

EXERCISE 10.2

A

■   *Solve. See Examples 1 and 2.*

1. How many different amounts of money can be formed from a penny, a nickel, and a dime?

2. How many different amounts of money can be formed from a penny, a nickel, a dime, a quarter, and a half-dollar?

3. How many different committees of four persons each can be chosen from a group of six persons?

4. How many different committees of four persons each can be chosen from a group of ten persons?

5. In how many different ways can a set of five cards be selected from a standard deck of 52 cards?

6. In how many different ways can a set of 13 cards be selected from a standard deck of 52 cards?

7. In how many different ways can a hand consisting of five spades, five hearts, and three diamonds be selected from a standard deck of 52 cards?

8. In how many different ways can a hand consisting of ten spades, one heart, one diamond, and one club be selected from a standard deck of 52 cards?

9. In how many different ways can a hand consisting of either five spades, five hearts, five diamonds, or five clubs be selected from a standard deck of 52 cards?

10. In how many different ways can a hand consisting of three aces and two cards that are not aces be selected from a standard deck of 52 cards?

11. A combination of three balls is picked at random from a box containing five red, four white, and three blue balls. In how many ways can the set chosen contain at least one white ball?

12. In Exercise 11, in how many ways can the set chosen contain at least one white and one blue ball?

13. A set of five distinct points lies on a circle. How many inscribed triangles can be drawn having all of their vertices in this set?

14. A set of ten distinct points lies on a circle. How many inscribed triangles can be drawn having all of their vertices in this set?

B

15. A set of ten distinct points lies on a circle. How many inscribed six-sided figures can be drawn having all their vertices in this set?

16. A set of ten distinct points lies on a circle. How many inscribed quadrilaterals can be drawn having all their vertices in this set?

17. Given $C_{n,7} = C_{n,5}$, find n.

18. Given $C_{n,3} = C_{50,47}$, find n.

10.3

PROBABILITY FUNCTIONS

Sample Spaces, Outcomes, and Events

When an experiment of some kind is undertaken, associated with the experiment is a set of possible results. For example, when a die is rolled, it will come to a stop with the number of spots on its upper face corresponding to one of the numerals 1, 2, 3, 4, 5, 6. This exhausts all possibilities. Of course, in the absence of chicanery, exactly which one of these random results will occur cannot be precisely specified in advance of the experiment. A question of great practical importance regarding the result of casting a die, and the result of any comparable experiment in which the outcome is uncertain, is "Can we assign some kind of measure to the degree of uncertainty involved?" The answer is "Yes," but before we assign a measure, let us define some necessary terms.

Definition 10.4   *The set of all possible results of an experiment is called the **sample space** for the experiment, and is commonly denoted by the letter S.*

Definition 10.5   *Each element of a sample space is called an **outcome**, or **sample point**.*

Definition 10.6   *Any subset of a sample space is called an **event**, and is commonly denoted by the letter E.*

The reason for the terminology in Definition 10.6 is that, in conducting an experiment, one may be interested in sets of outcomes rather than in individual outcomes. In the tossing of a die, for example, if the sample space is taken as $\{1, 2, 3, 4, 5, 6\}$, then the event that an outcome (a numeral) denotes an even integer is the set $\{2, 4, 6\}$, which is a subset of the sample space. The event that an outcome denotes an odd integer is the set $\{1, 3, 5\}$. These two events are complements of each other and are examples of **complementary events;** that is, two events whose intersection is $\varnothing$ and whose union is the entire sample space.

The likelihood of any outcome in a sample space is given by a **probability function.** Such a function assigns to each outcome a number called the probability of the outcome. Probabilities are numbers in $[0, 1]$, and the sum of the probabilities of all outcomes must be 1. If the probability of an outcome is 0.25, this means that the outcome should occur 1/4 or 25% of the time in the long run. Probabilities can be assigned by either a priori or a posteriori considerations, as follows.

A priori considerations involve physical, geometrical, and other inherent properties of the experiment in question. They involve no sampling of outcomes. One way to assign a probability to an event is simply to use the ratio of the number of outcomes of the event to the number of possible outcomes. Thus, when a die is cast and we admit as outcomes the die's stopping with any of its six different faces uppermost, then without making any trial throws we would assign the value 1/6 as the probability of each of the six possible outcomes.

In general, we have the following definition.

Definition 10.7   *If E is any subset containing $n(E)$ members (outcomes) of a sample space containing $n(S)$ equally likely outcomes, then the (a priori) **probability** of the occurrence of E, $P(E)$, is given by*

$$P(E) = \frac{n(E)}{n(S)}. \tag{1}$$

E X A M P L E   1   If two dice are cast, what is the (a priori) probability that the sum of the number of dots showing on the top faces of the dice is less than 6?

Solution   For our sample spaces, let us consider A the set of possible outcomes for one die and

B the set for the other. Then

$$n(A) = 6 \qquad \text{and} \qquad n(B) = 6.$$

The possible outcomes for both would be $S = A \times B$, the Cartesian product of A and B, and $n(A \times B) = n(S) = 36$. Each outcome here is an ordered pair (a, b), where a is the numeral on the upper face of the first die and b is that of the second die. The event we seek is $\{(a, b) \mid a + b < 6\}$. The lattice in the figure shows the situation schematically. Since

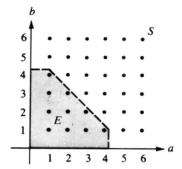

$$E = \{(1, 1), (1, 2), (1, 3), (1, 4), (2, 1),$$
$$(2, 2), (2, 3), (3, 1), (3, 2), (4, 1)\},$$

we have $n(E) = 10$, and

$$P(E) = \frac{n(E)}{n(S)} = \frac{10}{36} = \frac{5}{18}. \qquad ■$$

Care must be taken in interpreting the meaning of "the probability of an event." In the foregoing example, $\frac{5}{18}$ does not assure us, for instance, that E will occur 5 times out of 18 casts, or, indeed, that even one cast out of 18 will produce the event described. What it does imply, however, is that if you cast the dice a very great number of times, then you can *expect* the sum on the exposed faces to be less than 6 about $\frac{5}{18}$ of the time.

A *posteriori considerations* involve testing the experiment a certain number of times. Mortality tables give probability functions of this sort. Actually, Equation (1) can still be used in defining such probability functions, provided we interpret the function $n(E)$ as being the number of times the event E occurred in the test and $n(S)$ as being the total number of times the experiment was performed in the test.

Although diagrams are often useful in illustrating an event and a sample space, the number of outcomes $n(E)$ in the event and the number $n(S)$ of possible outcomes in the sample space ordinarily are computed directly.

EXAMPLE 2   If three marbles are drawn at random from an urn containing six white marbles and four blue marbles, what is the probability that the three marbles are all white?

Solution   Since the number of ways of drawing three white marbles from six white marbles is $C_{6,3}$ and the number of ways of drawing three marbles from the total number of marbles is $C_{10,3}$, we have

$$P(E) = \frac{n(E)}{n(S)} = \frac{C_{6,3}}{C_{10,3}} = \frac{\dfrac{6!}{3!3!}}{\dfrac{10!}{3!7!}} = \frac{1}{6}.$$ ■

EXERCISE 10.3

■ *Solve. See Examples 1 and 2.*

1. A die is cast. List the outcomes in the sample space. List the outcomes in the event E that the number on the upper face of the die is greater than 2. Determine $P(E)$.

2. A coin is tossed. List the outcomes in the sample space. List the outcomes in the event E that a head appears. Determine $P(E)$.

3. Two coins are tossed. List the outcomes in the sample space. List the outcomes in the event E that both coins show the same face. Determine $P(E)$.

4. A die is cast and a coin is tossed. List the outcomes in the sample space. List the outcomes in the event E that the coin shows a head and the die shows a numeral greater than 4. Determine $P(E)$.

■ *In Exercises 5–10, consider an experiment in which two dice are cast. Determine the probability that the specified event occurs.*

5. The sum of the numbers of dots shown is 7.

6. The sum of the numbers of dots shown is 8.

7. The sum of the numbers of dots shown is 3.

8. The sum of the numbers of dots shown is 12.

9. At least one of the numbers of dots shown is less than 3.

10. Both dice show the same number of dots.

■ *In Exercises 11–14, consider an experiment in which a single card is drawn at random from a standard deck of 52 cards. Determine the probability that the specified event occurs.*

11. The card is the king of hearts.   **12.** The card is an ace.

13. The card is a black face card.   **14.** The card is a 5, 6, 7, or 8.

■ *In Exercises 15–18, consider an experiment in which two cards are drawn at random from a standard deck of 52 cards. Determine the probability that the specified event occurs.*

15. Both cards are spades.   **16.** Both cards are red.

17. Both cards are face cards.   **18.** Both cards are aces.

■   *In Exercises 19–22, consider an experiment in which two marbles are drawn at random from an urn containing eight red, six blue, four green, and two white marbles. Determine the probability that the specified event occurs.*

19. Both marbles are white.   **20.** Both marbles are green.

21. Both marbles are blue.   **22.** Both marbles are red.

10.4

PROBABILITY OF THE UNION OF EVENTS

Probability of the Complement of an Event

The **complement** of a subset E of a set S is the set of all members of S that do not belong to E. If we denote the complement of an event E in a sample space S by E', then $P(E')$ denotes the probability of the occurrence of E'. Since an outcome in S must lie in either E or E', but not both, and $P(S) = 1$, it follows that

$$P(E) + P(E') = 1,$$

or

$$P(E') = 1 - P(E).$$

EXAMPLE 1   If two marbles are drawn at random from an urn containing six white, two red, and five green marbles, what is the probability that not both are white; that is, that at least one is not white?

Solution   Rather than consider all of the possible pairs of marbles in which not both are white, we simply compute the probability of the event E that they *are* both white and subtract the result from 1. We have

$$P(E) = \frac{C_{6,2}}{C_{13,2}} = \frac{\dfrac{6 \cdot 5}{1 \cdot 2}}{\dfrac{13 \cdot 12}{1 \cdot 2}}$$

$$= \frac{15}{78} = \frac{5}{26}.$$

Then

$$P(E') = 1 - \frac{5}{26} = \frac{21}{26},$$

and this is just the probability that not both marbles are white.   ■

Probability of the Union of Events

By the second counting property on page 414, if E_1 and E_2 are events in a sample space S, then

$$n(E_1 \cup E_2) = n(E_1) + n(E_2) - n(E_1 \cap E_2).$$

This leads directly to the following theorem.

Theorem 10.7   *If S is a sample space and E_1 and E_2 are any events in S, then*

$$P(E_1 \text{ or } E_2) = P(E_1 \cup E_2) = P(E_1) + P(E_2) - P(E_1 \cap E_2).$$

EXAMPLE 2   If two cards are drawn from a standard deck of 52 cards, what is the probability that either both are red or both are jacks?

Solution   The number of elements of the sample space is the number of ways (combinations) one can draw two cards from 52, which is $C_{52,2}$. Let E_1 be the event that both are red. Since there are 26 red cards in a deck, the number of outcomes (combinations) in the event E_1 is

$$n(E_1) = C_{26,2}.$$

Let E_2 be the event that both cards are jacks. Then, because there are four jacks,

$$n(E_2) = C_{4,2}.$$

Since there is only one red pair of jacks, $n(E_1 \cap E_2) = C_{2,2} = 1$. We then have

$$P(E_1 \cup E_2) = P(E_1) + P(E_2) - P(E_1 \cap E_2)$$

$$= \frac{C_{26,2}}{C_{52,2}} + \frac{C_{4,2}}{C_{52,2}} - \frac{1}{C_{52,2}}$$

$$= \frac{C_{26,2} + C_{4,2} - 1}{C_{52,2}}$$

$$= \frac{\dfrac{26 \cdot 25}{1 \cdot 2} + \dfrac{4 \cdot 3}{1 \cdot 2} - \dfrac{2 \cdot 1}{1 \cdot 2}}{\dfrac{52 \cdot 51}{1 \cdot 2}}$$

$$= \frac{325 + 6 - 1}{1326} = \frac{330}{1326} = \frac{55}{221}. \qquad ■$$

Probability of the Union of Disjoint Events

Of course, if E_1 and E_2 are *disjoint*, then $E_1 \cap E_2 = \varnothing$, so that $P(E_1 \cap E_2) = 0$. Then the equation in Theorem 10.7 reduces to

$$P(E_1 \cup E_2) = P(E_1) + P(E_2).$$

Disjoint events are said to be **mutually exclusive.**

EXAMPLE 3   A card is drawn at random from a standard deck of 52 cards. What is the probability that the card is either a face card (jack, queen, or king) or a 4?

Solution   Let E_1 be the event that the card is a 4. There are four such cards in a deck, so that $n(E_1) = C_{4,1} = 4$. Let E_2 be the event that the card is a jack, queen, or king. There are twelve such cards in a deck. Hence, $n(E_2) = C_{12,1} = 12$. Since the sample space is just the entire deck, $n(S) = C_{52,1} = 52$, and since E_1 and E_2 are mutually exclusive,

$$P(E_1 \cup E_2) = P(E_1) + P(E_2)$$
$$= \frac{4}{52} + \frac{12}{52} = \frac{16}{52} = \frac{4}{13}.$$

Therefore, the probability is $\dfrac{4}{13}$.   ■

EXERCISE 10.4

A

■   *For Exercises 1–18, see Examples 1–3.*

■   *Two dice are cast. Let E be the event that both dice show the same numeral. Let F be the event that the sum of the numbers thrown is greater than 8. Find each of the following probabilities.*

1. $P(E)$	**2.** $P(F)$	**3.** $P(E \cup F)$
4. $P(E')$	**5.** $P(F')$	**6.** $P(E' \cup F')$

■   *A box contains five red, four white, and three blue marbles. Two marbles are drawn from the box. Let RR be the event that both marbles are red, WW that both marbles are white, BB that both marbles are blue, and RW, RB, BW that a red and a white, a red and a blue, and a blue and a white are drawn, respectively. Find each probability.*

7. $P(RR)$	**8.** $P(BB)$	**9.** $P(WW)$
10. $P(RW)$	**11.** $P(RB)$	**12.** $P(BW)$

13. What is the probability that neither is white?

14. What is the probability that neither is blue?

15. What is the probability that at least one is red?

16. What is the probability that either one is red or else both are white?

17. What is the probability that by drawing a single card from a standard deck of 52 cards, one will get a 2, 3, or 4?

18. What is the probability that if two cards are drawn from a standard deck of 52 cards, they will be of the same suit? Different suits?

B

■ *If the probability of the event E that a person will receive k dollars is P(E), then the person's* **mathematical expectation relative to this event** *is kP(E).*

19. A lottery offers a prize of $50, and 70 tickets are sold. What is the mathematical expectation of a person who buys three tickets? If each ticket costs $1, is the person's expectation greater or less than the outlay?

20. The probability that a certain horse will win the Irish Sweepstakes is $\frac{2}{9}$. If you hold a ticket on this horse to pay $100,000 if he wins, what is your mathematical expectation?

■ *If E_1, E_2, E_3, etc., are mutually exclusive events and the return to a person is k_1 if E_1 occurs, k_2 if E_2 occurs, etc., then the person's mathematical expectation is $\sum_{i=1}^{n} k_i P(E_i)$.*

21. One coin is selected at random from a collection containing a penny, a nickel, and a dime. What is the expectation?

22. One coin is selected at random from a collection containing a dime, a quarter, and a half-dollar. What is the expectation?

23. Three $1 bills and four $5 bills are hidden from view. What is the expectation on a single selection?

24. Three $1 bills, four $5 bills, and one $10 bill are hidden from view. What is the expectation on a single draw?

10.5

PROBABILITY OF THE INTERSECTION OF EVENTS

Independent Events

In some experiments, we may be interested in events that are not dependent on each other, in the sense that the occurrence of one has no effect on the probability of the occurrence or nonoccurrence of the other.

Consider an experiment in which two cards are drawn at random, one after the other, from a deck of ten cards, six of which are red and four blue. We can inquire into the probability that the first card drawn is red and the second blue. The simplest such situation would be one in which the first card is drawn, observed, and returned to the deck, which is then shuffled thoroughly before the second card is drawn. In this case, the sample space would consist of a set of ordered pairs $\langle x, y \rangle$, where x is the result of the first draw and y the result of the second draw. Since there are ten possibilities in each case, the sample space consists of $10 \times 10 = 100$ ordered pairs. In the Cartesian graph of Figure 10.3 on page 432, r_i and b_i are used to designate the drawing of red and blue cards, respectively.

FIGURE 10.3

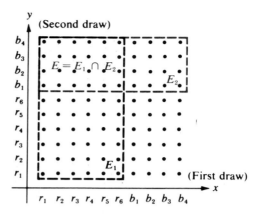

The events E_1 that the first card drawn is a red card and E_2 that the second card drawn is a blue card are outlined in Figure 10.3. The event E that both E_1 and E_2 occur is the intersection of E_1 and E_2; that is, $E = E_1 \cap E_2$. By inspection,

$$P(E_1) = \frac{60}{100} = \frac{3}{5}, \qquad P(E_2) = \frac{40}{100} = \frac{2}{5},$$

and

$$P(E) = P(E_1 \cap E_2) = \frac{24}{100} = \frac{6}{25}.$$

Moreover, in this experiment, it is evident that

$$P(E) = P(E_1 \cap E_2) = P(E_1) \cdot P(E_2).$$

Definition 10.8   *If E_1 and E_2 are events in a sample space and if*

$$P(E_1 \cap E_2) = P(E_1) \cdot P(E_2),$$

*then E_1 and E_2 are **independent events**. If two events are not independent, then they are said to be **dependent**.*

Dependent Events

Now consider the same experiment, except that this time the first card is not returned to the deck before the second is taken. Then there will be ten possible first draws, but only nine possible second draws. The sample space will therefore contain 10×9 ordered pairs, such that no ordered pair with first and second components the same remains in the set.

Figure 10.4 on page 433 shows a graph of the sample space, which is the same as that in Figure 10.3 except that one diagonal is missing. Again, the figure shows E_1 and E_2, the events that a red and blue are obtained on the first and second draw, respectively. The event that both occur is $E = E_1 \cap E_2$, which is also shown in the figure.

FIGURE 10.4

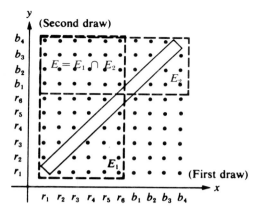

EXAMPLE I   Are events E_1 and E_2 described above independent?

Solution   By inspection,

$$P(E_1) = \frac{54}{90} = \frac{3}{5}, \qquad P(E_2) = \frac{36}{90} = \frac{2}{5},$$

and

$$P(E) = P(E_1 \cap E_2) = \frac{24}{60} = \frac{4}{15}.$$

This time,

$$P(E_1 \cap E_2) \neq P(E_1) \cdot P(E_2),$$

so the events are dependent.   ■

Conditional Probability

Suppose that in the experiment described in Example 1 we wanted to know the probability of the second draw producing a blue card, given that the first draw produced a red card. We would then be interested in the probability of the occurrence of the event E_2 given the occurrence of the event E_1. This is called the **conditional probability of E_2 given E_1** and is denoted $P(E_2 \mid E_1)$.

To compute $P(E_2 \mid E_1)$, observe that since we are assuming E_1 has occurred, we do not consider outcomes outside of E_1 to be possible. Thus,

$$P(E_2 \mid E_1) = \frac{n(E_1 \cap E_2)}{n(E_1)}, \qquad n(E_1) \neq 0. \tag{1}$$

To state this result in terms of probabilities of E_1, E_2, and $E_1 \cap E_2$, we divide the numerator and denominator of the right-hand member of Equation (1) by $n(S)$

to obtain

$$P(E_2 \mid E_1) = \frac{n(E_1 \cap E_2)/n(S)}{n(E_1)/n(S)}$$

$$= \frac{P(E_1 \cap E_2)}{P(E_1)}, \qquad P(E_1) \neq 0. \tag{2}$$

EXAMPLE 2   Find the conditional probability of drawing an ace on the second draw from a standard deck of 52 cards, given that the first draw produced a king, which was not replaced.

Solution   Let E_1 be the event that the first draw produces a king and E_2 be the event that the second draw produces an ace. Then

$$P(E_1 \cap E_2) = \frac{16}{(52)(51)}, \qquad P(E_1) = \frac{4}{52},$$

and

$$P(E_2 \mid E_1) = \frac{\dfrac{16}{(52)(51)}}{\dfrac{4}{52}} = \frac{4}{51}. \qquad ■$$

Conditional Probability and Independence

If E_1 and E_2 are independent, $P(E_1) \neq 0$, then by Definition 10.8 and Equation (2),

$$P(E_2 \mid E_1) = \frac{P(E_1 \cap E_2)}{P(E_1)} = \frac{P(E_1) \cdot P(E_2)}{P(E_1)} = P(E_2).$$

EXAMPLE 3   In the experiment described on page 431, the probability that the second card is blue, given that the first is red, is

$$P(E_2 \mid E_1) = \frac{P(E_1 \cap E_2)}{P(E_1)} = \frac{\dfrac{6}{25}}{\dfrac{3}{5}} = \frac{2}{5} = P(E_2),$$

provided the first card is returned to the deck before the second is drawn. Thus, if the first card is returned to the deck, then the two events are independent. But in the case where the first card is not returned, we have

$$P(E_2 \mid E_1) = \frac{P(E_1 \cap E_2)}{P(E_1)} = \frac{\frac{4}{15}}{\frac{3}{5}} = \frac{4}{9} \neq P(E_2),$$

and the events are not independent.   ■

EXERCISE 10.5

A

■ *Solve. See Examples* 1–3.

1. A bag contains four red and ten blue marbles. If two marbles are drawn in succession and if the first is not replaced, what is the probability that the first is red and the second is blue? Are the two draws independent events?

2. A red die and a green die are cast. What is the probability of obtaining a sum greater than 9, given that the green die shows 4?

3. A bag contains four white and six red marbles. Two marbles are drawn from the bag and replaced, and two more marbles are then drawn from the bag. What is the probability of drawing:

(a) Two red marbles on the first draw and two white ones on the second draw?

(b) A total of two white marbles?

(c) Four white marbles?

(d) Four red marbles?

4. In Exercise 3, what is the probability of drawing:

(a) Exactly three white marbles in the two draws?

(b) At least three white marbles in the two draws?

(c) Exactly two red marbles in the two draws?

(d) At least two red marbles in the two draws?

5. A red die and a green die are cast. Let E_1 be the event that at least one die shows 3 and E_2 be the event that the sum of the two numbers thrown is 8.

(a) Find $P(E_1)$.   (b) Find $P(E_2)$.

(c) Find $P(E_2 \mid E_1)$.   (d) Are E_1 and E_2 independent?

6. In Exercise 5, let E_1 be the event that neither die shows a result larger than 4 and let E_2 be the event that the dice do not show the same number.

(a) Find $P(E_1)$.   (b) Find $P(E_2)$

(c) Find $P(E_2 \mid E_1)$ (d) Are E_1 and E_2 independent?

7. A coin is tossed three consecutive times. What is the probability that:

(a) The second toss is a head?   (b) The third toss is a head?

(c) Both the second and third tosses are heads?   (d) The first and third tosses are heads?

(e) The first and third tosses are heads but the second is not?

8. In Exercise 7, state whether each of the following pairs of events is independent.

 (a) (a) and (b) (b) (a) and (c) (c) (a) and (d)

 (d) (a) and (e) (e) (d) and (e)

9. The probability that A will pass a course in college algebra is $\frac{5}{6}$, that B will pass is $\frac{3}{4}$, and that C will pass is $\frac{2}{3}$. What is the probability of each of the following events?

 (a) At least one of the three will pass. (b) At least A and C will pass.

 (c) A and C will pass but B will not. (d) At least two of the three will pass.

10. In Exercise 9, what is the probability of (c), given the occurrence of (a)? Are the events (a) and (c) independent?

11. A day is selected at random in some fashion such that any day of the week is an equally likely choice. Let the probability be $\frac{1}{30}$ that a day selected at random will be a rainy day.

 (a) What is the probability that a rainy Wednesday will be selected?

 (b) What is the probability that a dry Thursday will be selected?

 (c) What is the probability that either Monday, Tuesday, or Wednesday will be selected and that it will not rain that day?

 (d) Let E_1 be the selection of Sunday, and let E_2 be the event that it does not rain on the day selected. What is the conditional probability of E_2, given the occurrence of E_1? Are E_1 and E_2 independent?

12. Two identical urns contain marbles. One urn contains all red marbles, while half of the contents of the other urn is white and half is red. If a marble is drawn at random from one of the urns and found to be red, what is the probability that it was drawn from the urn containing only red marbles?

13. Of two dice, one is normal, but the other has three faces showing 4 spots and three faces showing 3 spots. If one of the dice is chosen at random and a 4 is thrown with it, what is the probability that the normal die was chosen?

14. Argue that if E_1 and E_2 are mutually exclusive events with nonzero probabilities, then E_1 and E_2 are not independent.

CHAPTER REVIEW

Key Words, Phrases, and Symbols

■ *Define or explain each of the following words, phrases, and symbols.*

1. $n(A)$, A is a finite set

2. Cartesian product of two sets

3. permutation of a set

4. $P_{n,r}$

5. combination

6. $C_{n,r}$

7. sample space

8. outcome (or sample point)

9. event

10. complementary events

11. probability function

12. (a priori) probability of an event

13. mutually exclusive events

14. independent events

15. dependent events

16. conditional probability of E_2 given E_1

Review Exercises

A

[10.1]   **1.** How many different two-digit numerals can be formed from $\{6, 7, 8, 9\}$?

2. How many different ways can six questions on a true–false test be answered?

3. How many four-digit numerals for positive odd integers can be formed from $\{3, 4, 5, 6\}$?

4. How many distinguishable permutations can be formed using the letters in the word *TENNIS*?

[10.2]   **5.** How many different committees of five persons can be formed from a group of twelve persons?

6. In how many different ways can a hand consisting of three spades, five hearts, four diamonds, and one club be selected from a standard deck of 52 cards?

7. A box contains four red, six white, and two blue marbles. In how many ways can you select three marbles from the box if at least one of the marbles chosen is white?

8. How many six-sided figures can be drawn whose vertices are members of a set of nine fixed points on the circle?

[10.3]   ■   *In Exercises 9–15, consider an experiment in which two cards are drawn at random from a deck of 52 cards. What is the probability that:*

9. Both cards are clubs?

10. Both cards are tens?

11. Neither card is a face card?

12. Both cards are black?

[10.4]   **13.** One card is red and one is black?

14. One card is a face card and the other is not?

15. Both cards are face cards, or one card is a ten and the other an ace?

16. What is the probability of randomly drawing one red and one blue marble from a box containing four red and six blue marbles?

[10.5]   ■   *In Exercises 17–20, consider an experiment in which two cards are drawn at random from a deck of 52 cards. What is the probability that:*

17. One card is black and the other is the ace of spades?

18. Both cards are red and one (but not the other) is a face card?

19. Both cards are red and at least one is a face card?

20. Neither card is a face card and at least one is red?

B

21. In how many ways can six students be seated at a circular table if a certain two must not be seated together?

22. In how many ways can a committee of eight be seated around a circular table if the chairperson and the vice-chairperson are to sit directly opposite one another?

23. Solve $2\dbinom{n}{n-2} = 56$ for n.   Solve $3\dbinom{n-1}{n-3} = 108$ for n.

25. A certain sweepstakes has three types of prizes. There is a super-grand prize of $5,000,000, a grand prize of $500,000, and a first prize of $100,000. The probability of winning the super-grand prize is $1/150,000,000$, the probability of winning the grand prize is $1/10,000,000$, and the probability of winning first prize is $1/1,000,000$. If it costs 22 cents (the price of a stamp) to enter the sweepstakes, what is your expected return to the nearest penny? Should you enter?

26. It costs $1 to blindly draw a ball at random from balls marked 1¢, 5¢, 25¢, 50¢, $1, and $4. You will be paid back the amount shown on the ball. Should you pay your dollar and take your chances? What if the $4 ball is replaced by a ball marked $4.50?

11 Conic Sections

In previous chapters we introduced some basic notions of analytic geometry that related ordered pairs of real numbers with points in a plane. In this chapter we shall continue our study of analytic geometry with particular emphasis on graphing conic sections.

RELATIONS WHOSE GRAPHS ARE GIVEN

Let us begin our discussion by describing a method of finding an equation or inequality for the coordinates of the points on a graph when a geometric description of the graph is given. Any set of points in the plane can be called a *graph* or a **locus** (plural **loci**). The words *graph* and *locus* are ordinarily used, however, to refer to the set of points meeting one or more predesignated conditions. The following examples illustrate ways of finding equations (or inequalities) for given loci.

EXAMPLE 1 Find an equation of the locus of points at a distance of 7 units from the point $P_1(3, -2)$.

Solution We first make a sketch showing a representative point $P(x, y)$ in the locus. By the distance formula, the distance d from P_1 to P is

$$d = \sqrt{(x - 3)^2 + (y + 2)^2}.$$

Since this distance is given to be 7, we have

$$\sqrt{(x - 3)^2 + (y + 2)^2} = 7.$$

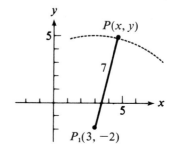

Squaring both members produces

$$(x - 3)^2 + (y + 2)^2 = 49,$$

$$x^2 - 6x + 9 + y^2 + 4y + 4 = 49,$$

or, finally,

$$x^2 + y^2 - 6x + 4y - 36 = 0. \tag{1}$$

Conversely, since the steps are reversible, any point whose coordinates satisfy (1) is on the graph. Hence, (1) is an equation of the locus.   ■

In a similar way, by replacing the sign $=$ with $\leq$ in Example 1, it could be shown that the set of points in the plane at distance not more than 7 from P_1 is the graph of the relation

$$x^2 + y^2 - 6x + 4y - 36 \leq 0.$$

EXAMPLE 2   Find an equation of the locus of all points that are twice as far from the point $(3, -2)$ as they are from $(-2, 1)$.

Solution   We first make a sketch. We then note that, by the distance formula, any point $P(x, y)$ in the plane located at a distance d_1 from $(-2, 1)$ has coordinates that satisfy

$$d_1 = \sqrt{(x + 2)^2 + (y - 1)^2},$$

and any point $P(x, y)$ at a distance d_2 from $(3, -2)$ has coordinates that satisfy

$$d_2 = \sqrt{(x - 3)^2 + (y + 2)^2}.$$

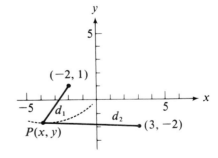

We now observe that, in order for $P(x, y)$ to lie in the described locus, it is necessary and sufficient that

$$d_2 = 2d_1,$$

or

$$\sqrt{(x - 3)^2 + (y + 2)^2} = 2\sqrt{(x + 2)^2 + (y - 1)^2}.$$

Then, equating squares of both members, we have

$$(x - 3)^2 + (y + 2)^2 = 4[(x + 2)^2 + (y - 1)^2],$$

$$x^2 - 6x + 9 + y^2 + 4y + 4 = 4x^2 + 16x + 16 + 4y^2 - 8y + 4,$$

from which we obtain the simplified equation

$$3x^2 + 3y^2 + 22x - 12y + 7 = 0. \qquad (2)$$

Again the steps are reversible, so that (2) is an equation of the given locus.   ■

Loci as Paths   Another way of considering a locus intuitively is to think of it as a path that has been traced in the plane by a moving point (or particle).

E X A M P L E 3   Find an equation of the locus of a point that moves in the plane so that it is always at the same distance from the line with equation $x = 3$ as it is from the point $(8, 0)$.

Solution   We first make a sketch as shown. We let $P(x, y)$ be any point on the locus and let P_1 be the point with coordinates $(3, y)$. Since segment P_1P is parallel to the x-axis, the distance d_1 from $P_1(3, y)$ to $P(x, y)$ is given by

$$d_1 = |x - 3|,$$

while d_2, the distance from $P_2(8, 0)$ to $P(x, y)$, is given by

$$d_2 = \sqrt{(x - 8)^2 + (y - 0)^2}.$$

The condition imposed on $P(x, y)$, is

$$d_1 = d_2.$$

Thus, we have

$$|x - 3| = \sqrt{(x - 8)^2 + (y - 0)^2}.$$

Squaring each member, we obtain

$$(x - 3)^2 = (x - 8)^2 + y^2,$$
$$x^2 - 6x + 9 = x^2 - 16x + 64 + y^2,$$

from which we have the simplified equation

$$y^2 - 10x + 55 = 0. \qquad (3)$$

Once more, the steps are reversible, so that (3) is an equation of the given locus.   ■

EXERCISE 11.1

A

■ *Find an equation of the locus of a point that moves in the plane in such a way that it meets the given condition. See Examples 1–3.*

1. It is a distance of 5 units from the point $(4, 2)$.

2. It is a distance of 3 units from the point $(-1, 3)$.

3. It is equidistant from $(3, 2)$ and $(5, 0)$.

4. It is equidistant from $(-3, -4)$ and $(1, 2)$.

5. Its distance from $(1, -4)$ is twice its distance from $(6, -1)$.

6. Its distance from $(5, 3)$ is three times its distance from $(5, 9)$.

7. Its distance from the line with equation $x = -2$ is equal to its distance from $(2, 0)$.

8. Its distance from the line with equation $y = 4$ is equal to its distance from the origin.

9. It is equidistant from the x-axis and the point $(-3, -2)$.

10. It is equidistant from the y-axis and the point $(4, 3)$.

11. The sum of its distances from $(2, 0)$ and $(-2, 0)$ is 8.

12. The sum of its distances from $(0, 4)$ and $(0, -4)$ is 12.

13. The difference of its distances from $(2, 0)$ and $(-2, 0)$ is 2.

14. The difference of its distances from $(0, 4)$ and $(0, -4)$ is 6.

15. Its distance from $(0, p)$ equals its distance from the x-axis.

16. Its distance from $(p, 0)$ equals its distance from the y-axis.

B

17. Show that the perpendicular distance d from the point $P(x_0, y_0)$ to the line $y = kx$, $k \neq 0$, is given by

$$d = \frac{|y_0 - kx_0|}{\sqrt{k^2 + 1}}.$$

[*Hint:* Find an equation of the line perpendicular to the given line through the point P and then use the two equations to find the point P' of the intersection of the two lines.]

18. Find an equation of the locus of a point that moves in such a way that it is equidistant from the line $y = 4x$ and the point $(5, 1)$. [*Hint:* Use the result of Exercise 17.]

19. Find an equation of the locus of a point that moves in such a way that its distance to the line $y = 2x$ is always the same as the sum of its distance to $(1, 3)$ and its distance to $(-1, 1)$.

20. Find a formula for the perpendicular distance from $P(x_0, y_0)$ to the line $Ax + By = 0$, where $A^2 + B^2 \neq 0$.

11.2

PARABOLAS

Here and in Sections 11.3 and 11.4 we use the methods that were illustrated in Section 11.1 to derive equations of special curves that occur frequently in applications. The first of these curves is the *parabola*.

Definition 11.1   A *parabola* is the set of points in the plane that are equidistant from a given point F and a given line l in the plane.

The given point F is called the **focus** of the parabola, and the given line l is called the **directrix** of the parabola.

Equation of a Parabola

To derive an equation of a parabola, let us make some simplifying assumptions about its location in the plane. First, let us assume that the focus $F(0, p)$ is on the positive y-axis and that the directrix is the line $y = -p$ (see Figure 11.1). By the distance formula, any point $P(x, y)$ that lies at a distance d_1 from F has coordinates that satisfy

FIGURE 11.1

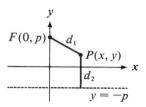

$$d_1 = \sqrt{(x - 0)^2 + (y - p)^2}$$

Also, any point $P(x, y)$ that lies at a distance d_2 from the line $y = -p$ has coordinates that satisfy

$$d_2 = |y + p|.$$

In order for a point P to lie on the parabola, by Definition 11.1 it must satisfy

$$d_1 = d_2.$$

Thus, we have

$$\sqrt{(x - 0)^2 + (y - p)^2} = |y + p|.$$

Squaring both sides of this equation, we obtain

$$x^2 + (y - p)^2 = |y + p|^2.$$

This is equivalent to

$$x^2 + y^2 - 2py + p^2 = y^2 + 2py + p^2,$$

which upon simplification yields

$$x^2 = 4py. \tag{1}$$

These steps are reversible, and thus any point whose coordinates satisfy Equation (1) lies on the parabola. Hence, Equation (1) is an equation of the parabola. The graph of Equation (1) is shown in Figure 11.2.

FIGURE 11.2

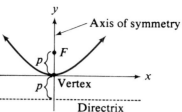

Axis of Symmetry, Vertex

The line that passes through the focus and is perpendicular to the directrix is called the **axis of symmetry** of the parabola. The midpoint of the segment of the axis of symmetry joining the focus and the directrix is called the **vertex** of the parabola (see Figure 11.2).

E X A M P L E I   Find an equation of the parabola with focus $F(0, 4)$ and directrix $y = -4$.

Solution   Since the focus is on the positive y-axis and the directrix is parallel to the x-axis, we can use Equation (1) with $p = 4$. Thus, the desired equation is

$$x^2 = 4(4)y = 16y. \qquad ■$$

Standard Position

The simplifying assumptions that were made in the preceding derivation were the following:

1. The focus of the parabola is on the positive y-axis.
2. The directrix of the parabola is parallel to the x-axis.
3. The focus and directrix are on opposite sides of the x-axis and equidistant from it.

Similar assumptions are made for three other cases. The equations and the corresponding graphs are shown in Figure 11.3. In each case, assume $p > 0$. The derivations are similar to the preceding one and are left as exercises.

FIGURE 11.3

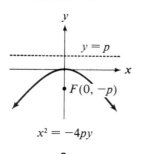

$$x^2 = -4py$$

a

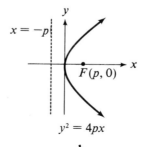

$$y^2 = 4px$$

b

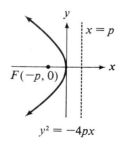

$$y^2 = -4px$$

c

The parabolas in Figures 11.2 and 11.3 are in **standard position**, since in each case the vertex is at the origin and the axis of symmetry coincides with one of the coordinate axes. Equation (1) and the equations in Figure 11.3 are the **standard forms** of the equation of a parabola in standard position. Note that Equation (1) and the equation in Figure 11.3a define quadratic functions. These functions were considered in Section 4.4.

EXAMPLE 2   Find the focus and an equation of the directrix, and sketch the graph of the parabola with equation $y^2 = -12x$.

Solution   This equation is of the form $y^2 = -4px$ with $-12 = -4p$. Thus, $p = 3$, the focus of the parabola is $F(-3, 0)$, and the directrix is

$$x = 3.$$

The graph is shown in the figure.

The parabola is one of a group of curves that are called conic sections. **Conic sections** result from the intersection of a plane with a right circular cone. In the case of the parabola, the plane is parallel to one of the elements of the cone, as illustrated by Figure 11.4.

FIGURE 11.4

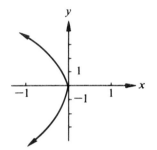

Parabola

EXERCISE 11.2

A

■   *Find an equation of the parabola in standard position satisfying the given condition or conditions and sketch the graph. Let F and d denote the focus and the directrix, respectively. See Example 1.*

1. $F(0, 4)$ **2.** $d: y = 3$ **3.** $d: x = 2$

4. $F(4, 0)$ **5.** $F(2, 0)$ **6.** $d: y = -1$

7. The directrix is parallel to the x-axis, the vertex is at the origin, and the parabola contains the point $(2, 4)$.

8. The directrix is parallel to the y-axis, the vertex is at the origin, and the parabola contains the point $(1, 3)$.

9. The directrix is parallel to the y-axis, the vertex is at the origin, and the parabola contains the point $(-1, 1)$.

10. The directrix is parallel to the x-axis, the vertex is at the origin, and the parabola contains the point $(2, 8)$.

■ *Find the focus and the directrix of the given parabola. See Example 2.*

11. $x^2 = 2y$ **12.** $y^2 = 4x$ **13.** $y^2 = -16x$ **14.** $x^2 = -18y$
15. $y^2 = 12x$ **16.** $4x^2 = 3y$ **17.** $y^2 + 4x = 0$ **18.** $4y^2 - 2x = 0$
19. $2y^2 - 3x = 0$ **20.** $2x^2 = 8y$

B

■ *Derive an equation of the parabola satisfying the given condition.*

21. Vertex $(0, 0)$, focus $(0, -p)$ $(p > 0)$ **22.** Vertex $(0, 0)$, focus $(p, 0)$ $(p > 0)$
23. Vertex $(0, 0)$, focus $(-p, 0)$ $(p > 0)$

24. The **latus rectum** of a parabola is the line segment that is perpendicular to the axis of symmetry, passing through the focus, and with endpoints on the parabola. Show that the length of the latus rectum of any parabola in standard position is $4p$.

11.3

CIRCLES AND ELLIPSES

Like the parabola, the *circle* and the *ellipse* are conic sections. A circle is the intersection of a right circular cone with a plane perpendicular to the axis of the cone (see Figure 11.5a). An ellipse is the intersection of a right circular cone with a plane that is not perpendicular to the axis of the cone but that intersects only one nappe of the cone (see Figure 11.5b).

FIGURE 11.5

a
Circle

b
Ellipse

Circle

The circle is defined as follows.

Definition 11.2

A *circle* is the set of all points in the plane at a given distance from a fixed point C in the plane.

The fixed point C in Definition 11.2 is called the **center** of the circle. The common distance of all points on the circle from the center is called the **radius** of the circle.

To derive an equation of a circle with the center at (0, 0), we employ the techniques we used in Section 11.1. If r is the radius, then by the distance formula, in order for a point $P(x, y)$ to lie on the circle (see Figure 11.6), we must have

FIGURE 11.6

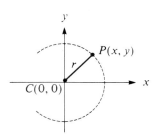

$$\sqrt{x^2 + y^2} = r. \qquad (1)$$

Squaring both sides of Equation (1) yields

$$x^2 + y^2 = r^2. \qquad (2)$$

These steps are reversible. Thus, any point whose coordinates satisfy Equation (2) will lie on the circle. Equation (2) is the **standard form** of the equation of a circle with center at the origin and radius r.

EXAMPLE 1   Show that $2x^2 + 2y^2 = 18$ is the equation of a circle and give the radius.

Solution   We divide both members of the given equation by 2 to obtain

$$x^2 + y^2 = 9.$$

This is an equation of a circle with $r^2 = 9$. Thus, the radius is $r = \sqrt{9} = 3$.

EXAMPLE 2   Find an equation of the circle with center (0, 0) and passing through the point (1, 4).

Solution   Since we know that the center of the circle is the origin, we need only compute the radius in order to use Equation (2). The radius is the distance from the center to any point on the circle. Thus, we have

$$r = \sqrt{(1 - 0)^2 + (4 - 0)^2} = \sqrt{17}.$$

Hence, the desired equation is

$$x^2 + y^2 = 17. \qquad ■$$

Ellipse

The ellipse is defined as follows.

Definition 11.3

*An **ellipse** is the set of all points in the plane, the sum of whose distances to two fixed points F_1 and F_2 in the plane is constant.*

The points F_1 and F_2 are called the **foci** of the ellipse. The midpoint of the line segment joining the foci is called the **center** of the ellipse.

Equation of an Ellipse

In order to simplify the derivation of the equation of the ellipse, we make the following assumptions:

1. The line passing through the foci is the x-axis.

2. The foci are the points $F_1(-c, 0)$ and $F_2(c, 0)$, $c > 0$, and the constant sum of the distances from a point on the ellipse to the foci is $2a$, where $a > 0$.

In order for a point $P(x, y)$ to be on the ellipse, it must satisfy

$$d_1 + d_2 = 2a, \qquad (3)$$

FIGURE 11.7

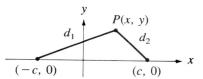

where d_1 and d_2 are the distances from P to F_1 and F_2, respectively (see Figure 11.7). Thus,

$$\sqrt{(x + c)^2 + (y - 0)^2} + \sqrt{(x - c)^2 + (y - 0)^2} = 2a. \qquad (4)$$

Rearranging terms in Equation (4), we obtain

$$\sqrt{(x - c)^2 + y^2} = 2a - \sqrt{(x + c)^2 + y^2}; \qquad (5)$$

and squaring both sides of Equation (5), we find

$$(x - c)^2 + y^2 = 4a^2 - 4a\sqrt{(x + c)^2 + y^2} + (x + c)^2 + y^2.$$

Performing the squaring operations and simplifying yields

$$cx + a^2 = a\sqrt{(x + c)^2 + y^2}. \qquad (6)$$

Squaring both sides of Equation (6), we obtain

$$c^2x^2 + 2ca^2x + a^4 = a^2[(x + c)^2 + y^2],$$

which upon simplification yields

$$(a^2 - c^2)x^2 + a^2y^2 = a^2(a^2 - c^2). \qquad (7)$$

Since $a^2 - c^2 > 0$ (see Exercise 24), we can simplify the notation by letting

$$b^2 = a^2 - c^2. \tag{8}$$

Then dividing both sides of Equation (7) by a^2b^2, we obtain

$$\frac{x^2}{a^2} + \frac{y^2}{b^2} = 1. \tag{9}$$

The steps in this derivation are reversible. Thus, any point whose coordinates satisfy Equation (9) is on the ellipse. Hence, (9) is an equation of the ellipse. The graph of the ellipse is shown in Figure 11.8a.

FIGURE 11.8

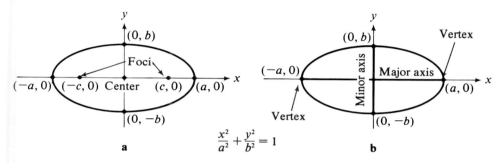

EXAMPLE 3 State whether $x^2 + 3y^2 = 9$ is the equation of a circle or an ellipse, and give the value of r or the values of a and b.

Solution We first divide both sides of the given equation by 9 to obtain

$$\frac{x^2}{9} + \frac{y^2}{3} = 1.$$

This is an equation of an ellipse with $a^2 = 9$ and $b^2 = 3$. Thus, $a = 3$ and $b = \sqrt{3}$. ■

Axes and Vertices

The line segment that passes through the foci with endpoints on the ellipse is called the **major axis** of the ellipse. The line segment passing through the center of the ellipse, perpendicular to the major axis and with endpoints on the ellipse, is called the **minor axis**. The endpoints of the major axis are called the **vertices** of the ellipse (see Figure 11.8b).

EXAMPLE 4 Find an equation of the ellipse with foci $F_1(-4, 0)$ and $F_2(4, 0)$ and vertices $V_1(-5, 0)$ and $V_2(5, 0)$, and sketch the graph.

Solution Since the vertices are at $(5, 0)$ and $(-5, 0)$, we have $a = 5$, and since the foci are at $(4, 0)$ and $(-4, 0)$, $c = 4$. Thus, from Equation (8),

$$b^2 = 5^2 - 4^2 = 9,$$

and, from Equation (9), the equation is

$$\frac{x^2}{25} + \frac{y^2}{9} = 1. \quad ■$$

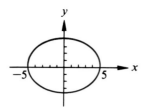

In the preceding derivation, it was assumed that the major axis lay along the x-axis. If the major axis lies along the y-axis, the derivation is essentially the same, and the equation takes the form

$$\frac{y^2}{a^2} + \frac{x^2}{b^2} = 1. \qquad (10)$$

FIGURE 11.9

$$\frac{y^2}{a^2} + \frac{x^2}{b^2} = 1$$

The graph of this equation is given in Figure 11.9.

An ellipse that is centered at the origin with axes lying along the coordinate axes is said to be in **standard position**. Equations (9) and (10) are the **standard forms** of the equation of an ellipse in standard position.

In the derivations of Equations (9) and (10), we let $b^2 = a^2 - c^2$. Thus,

$$a^2 = b^2 + c^2,$$

and since $c^2 > 0$, it follows that $a^2 > b^2$.

E X A M P L E 5 Find the foci and vertices of the ellipse with the equation

$$\frac{y^2}{9} + \frac{x^2}{4} = 1.$$

Solution Since $a^2 > b^2$, we conclude that $a^2 = 9$ and $b^2 = 4$. From Equation (8), we have

$$c^2 = a^2 - b^2$$
$$= 9 - 4 = 5.$$

Hence, $c = \sqrt{5}$.

Since the equation is of the form

$$\frac{y^2}{a^2} + \frac{x^2}{b^2} = 1,$$

the major axis of the ellipse lies along the y-axis. Hence, the foci and vertices lie on the y-axis. The foci are $F_1(0, -\sqrt{5})$ and $F_2(0, \sqrt{5})$, and the vertices are $V_1(0, -3)$ and $V_2(0, 3)$. ■

EXERCISE 11.3

A

■ *State whether the graph of the given equation is a circle or an ellipse and give the value of r or the values of a and b. See Examples 1 and 3.*

1. $2x^2 + 2y^2 = 4$ **2.** $3x^2 + y^2 = 3$ **3.** $\frac{x^2}{3} + \frac{y^2}{2} = 4$ **4.** $\frac{x^2}{2} + \frac{y^2}{2} = 1$

■ *Find an equation of the circle or ellipse in standard position satisfying the given condition, and sketch the graph. Assume that F_1, F_2 denote foci in the case of an ellipse and that r denotes the radius in the case of a circle. See Examples 2 and 4.*

5. Circle centered at the origin with radius 5.

6. Circle centered at the origin and containing the point $(-1, -3)$.

7. Ellipse with $F_1(0, -1)$ and $b = 2$.   **8.** Ellipse with vertices at $(0, \pm 4)$ and $b = 1$.

9. Circle centered at the origin with radius 4.   **10.** Circle centered at the origin with radius $\frac{1}{2}$.

11. Ellipse with $F_2(0, 2)$ and a vertex at $(0, 4)$.   **12.** Ellipse with $F_1(-4, 0)$ and $b = 4$.

13. Ellipse with major axis of length 4 along the x-axis and minor axis of length 1.

14. Ellipse with $F_2(3, 0)$ and major axis of length 12.

■ *Find the foci and vertices of the given ellipse. See Example 5.*

15. $\frac{x^2}{9} + \frac{y^2}{4} = 1$ **16.** $\frac{x^2}{16} + \frac{y^2}{25} = 1$ **17.** $\frac{x^2}{144} + \frac{y^2}{16} = 1$

18. $\frac{x^2}{9} + \frac{y^2}{16} = 1$ **19.** $3y^2 + 4x^2 = 12$ **20.** $2y^2 + x^2 = 1$

B

21. Derive Equation (10) on page 452.

22. Assume that $a = b$ in Equation (10). What figure does the resulting equation represent?

23. What effect does decreasing the value of c (that is, moving the foci closer to the origin) have on the shape of the ellipse? How does this relate to Exercise 22?

24. Show that $a^2 - c^2$ in Equation (7) is positive.

25. Show that if the major axis of an ellipse in standard position lies along the x-axis and if $2a$ is the constant sum of the distances from the foci to the points on the ellipse, then the endpoints of the major axis are $(\pm a, 0)$.

26. Show that the endpoints of the minor axis of the ellipse in Exercise 25 are $(0, \pm b)$, where $b^2 = a^2 - c^2$ and the foci are $(\pm c, 0)$.

11.4

HYPERBOLAS

The figure that is obtained by intersecting a right circular cone with a plane cutting both nappes (see Figure 11.10) is called a *hyperbola* and is defined as follows.

Definition 11.4   A ***hyperbola*** *is the set of all points in the plane such that the absolute value of the difference of the distances to two fixed points F_1 and F_2 in the plane is constant.*

FIGURE 11.10

Hyperbola

The two fixed points F_1 and F_2 are called the **foci** of the hyperbola. The midpoint of the line segment joining the foci of a hyperbola is called the **center** of the hyperbola.

Equation of a Hyperbola

In order to simplify the derivation of an equation for a hyperbola, let us make the following assumptions:

1. The line segment joining the foci lies along the x-axis.
2. The foci are $F_1(-c, 0)$ and $F_2(c, 0)$, $c > 0$, and the absolute value of the difference of the distances from the foci to the points on the ellipse is $2a$, $a > 0$.

By Definition 11.4, any point $P(x, y)$ on the hyperbola must have coordinates that satisfy

$$|d_2 - d_1| = 2a, \tag{1}$$

where d_1 and d_2 are the distances from P to F_1 and F_2, respectively (see Figure 11.11). Equation (1) can be rewritten equivalently as

FIGURE 11.11

$$d_2 - d_1 = \pm 2a. \tag{2}$$

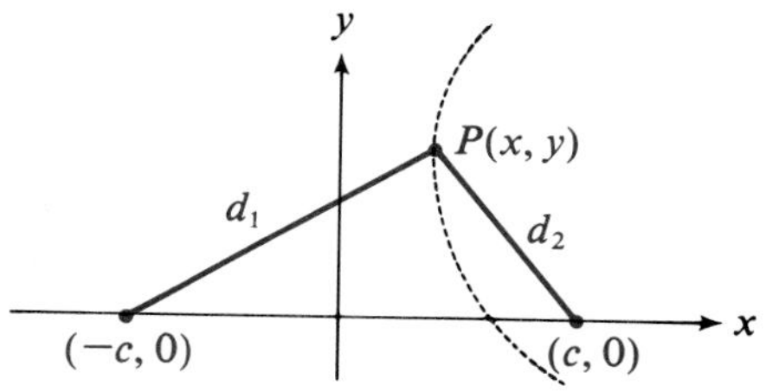

Thus, since

$$d_1 = \sqrt{(x + c)^2 + y^2} \quad \text{and} \quad d_2 = \sqrt{(x - c)^2 + y^2},$$

Equation (2) is equivalent to

$$\sqrt{(x - c)^2 + y^2} = \pm 2a + \sqrt{(x + c)^2 + y^2}. \tag{3}$$

Squaring both sides of Equation (3), we obtain

$$(x - c)^2 + y^2 = 4a^2 \pm 4a\sqrt{(x + c)^2 + y^2} + (x + c)^2 + y^2,$$

which upon simplification yields

$$cx + a^2 = \pm a\sqrt{(x + c)^2 + y^2}. \tag{4}$$

Then, squaring both sides of Equation (4), we have

$$c^2x^2 + 2a^2cx + a^4 = a^2[(x + c)^2 + y^2],$$

which is equivalent to

$$(c^2 - a^2)x^2 - a^2y^2 = a^2(c^2 - a^2). \tag{5}$$

Since $c^2 - a^2 > 0$ (see Exercise 17), we let

$$b^2 = c^2 - a^2.$$

Dividing both sides of Equation (5) by a^2b^2 yields

$$\frac{x^2}{a^2} - \frac{y^2}{b^2} = 1. \tag{6}$$

The graph of this equation is shown in Figure 11.12a on page 456.

Vertices and Axes

The points where the hyperbola intersects the line joining the foci are called the **vertices** of the hyperbola, and the line segment joining the vertices is called the **transverse axis**. The line segment of length $2b$ through the center that is bisected by and perpendicular to the transverse axis is the **conjugate axis** (see Figure 11.12b).

FIGURE 11.12

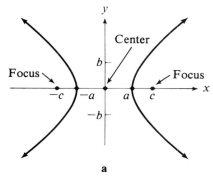

a

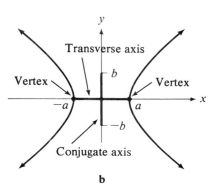

b

$$\frac{x^2}{a^2} - \frac{y^2}{b^2} = 1$$

EXAMPLE 1   Find an equation of the hyperbola with foci $F_1(-2, 0)$ and $F_2(2, 0)$ and with vertices $V_1(-1, 0)$ and $V_2(1, 0)$. Sketch the graph.

Solution   Since the foci are at $(\pm 2, 0)$, we have $c = 2$, and since the vertices are at $(\pm 1, 0)$, we have $a = 1$. Since

$$b^2 = c^2 - a^2,$$
$$b^2 = 4 - 1$$
$$= 3.$$

Since the transverse axis lies along the x-axis, the equation of the hyperbola is

$$\frac{x^2}{1} - \frac{y^2}{3} = 1.$$

The graph is shown in the figure.   ■

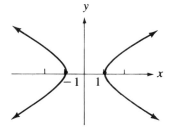

In the derivation of Equation (6), it was assumed that the transverse axis lay along the x-axis. Precisely the same steps can be used to derive an equation of a hyperbola with transverse axis along the y-axis (see Exercise 18). The resulting equation is

$$\frac{y^2}{a^2} - \frac{x^2}{b^2} = 1. \qquad (7)$$

The graph of Equation (7) is given in Figure 11.13.

FIGURE 11.13

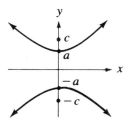

EXAMPLE 2   Find the foci and vertices of the hyperbola with equation

$$\frac{y^2}{4} - \frac{x^2}{9} = 1.$$

Solution   The equation has the same form as Equation (7). Hence, $a^2 = 4$ and $b^2 = 9$. Since $c^2 = a^2 + b^2$, we have

$$c^2 = 4 + 9 = 13.$$

Hence, $c = \sqrt{13}$.

Since the transverse axis of the hyperbola lies along the y-axis, the foci are $(0, \pm c)$ and the vertices are $(0, \pm a)$. Thus, the foci are $F_1(0, -\sqrt{13})$ and $F_2(0, \sqrt{13})$, and the vertices are $V_1(0, -2)$ and $V_2(0, 2)$. ■

A hyperbola centered at the origin with axes lying along the coordinate axes is said to be in **standard position**. Equations (6) and (7) are the **standard forms** of the equation of a hyperbola in standard position.

Asymptotes   It was observed in Section 5.5 that some curves approach asymptotes. It can be shown that each hyperbola has two asymptotes, and if the hyperbola is in standard position, the equations of the asymptotes can be obtained by setting the right-hand member of Equation (6) or (7) equal to 0. It follows that:

1. $y = \pm \dfrac{b}{a} x$ are asymptotes if Equation (6) is an equation of the hyperbola (see Figure 11.14a).

2. $y = \pm \dfrac{a}{b} x$ are asymptotes if Equation (7) is an equation of the hyperbola (see Figure 11.14b).

FIGURE 11.14

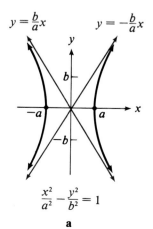

$$\frac{x^2}{a^2} - \frac{y^2}{b^2} = 1$$

a

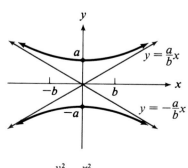

$$\frac{y^2}{a^2} - \frac{x^2}{b^2} = 1$$

b

EXAMPLE 3   Find equations of the asymptotes of the given hyperbola.

(a) $\dfrac{x^2}{16} - \dfrac{y^2}{4} = 1$ (b) $\dfrac{y^2}{4} - \dfrac{x^2}{36} = 1$

Solution   (a) This equation has the same form as Equation (6). Thus, the asymptotes have equations

(b) This equation has the same form as Equation (7). Thus, the asymptotes have equations

$$y = \pm \frac{2}{4}x = \pm \frac{1}{2}x.$$

$$y = \pm \frac{2}{6}x = \pm \frac{1}{3}x. \quad ■$$

EXERCISE 11.4

A

■ *Find an equation and sketch the graph of the hyperbola in standard position with the given property. The letters F and V stand for focus and vertex, respectively. See Example 1.*

1. $F_1(0, 3)$, $b = 2$

2. $F_2(4, 0)$, $a = 2$

3. $F_1(0, 4)$, $V_2(0, -2)$

4. $F_1\left(0, \dfrac{1}{2}\right)$, $V_1\left(0, \dfrac{1}{4}\right)$

5. $a = 4$, $F_1(5, 0)$

6. $b = 2$, $V_1(1, 0)$

7. $F_2(0, 1)$, asymptotes $y = \pm \dfrac{1}{2}x$

8. $V_1(6, 0)$, asymptotes $y = \pm x$

■ *Find the foci, the vertices, and the equations of the asymptotes of each of the following hyperbolas. See Examples 2 and 3.*

9. $\dfrac{x^2}{9} - \dfrac{y^2}{16} = 1$

10. $\dfrac{y^2}{4} - \dfrac{x^2}{9} = 1$

11. $x^2 - \dfrac{y^2}{2} = 1$

12. $\dfrac{x^2}{4} - \dfrac{y^2}{2} = 1$

13. $4y^2 - 2x^2 = 16$

14. $9x^2 - 4y^2 = 1$

15. $25y^2 - 16x^2 = 1$

16. $25y^2 - 9x^2 = 225$

B

17. Show that $c^2 - a^2 > 0$ in Equation (5).

18. Derive Equation (7).

19. Find an equation of the hyperbola with $a = \sqrt{2}$ and with foci $F_1(\sqrt{2}, \sqrt{2})$ and $F_2(-\sqrt{2}, -\sqrt{2})$. Sketch the graph.

20. Give an intuitive argument for the fact that the hyperbola

$$\frac{x^2}{a^2} - \frac{y^2}{b^2} = 1$$

has asymptotes $y = \pm\dfrac{b}{a}x$. [*Hint:* Solve for y in terms of x, and then factor x^2 from the quantity under the radical.]

11.5

MORE ABOUT GRAPHS OF QUADRATIC EQUATIONS

In Sections 11.2, 11.3, and 11.4 we derived equations of the conic sections of the form

$$y = ax^2, \qquad a \neq 0,$$

or

$$x = ay^2, \qquad a \neq 0,$$

or

$$Ax^2 + By^2 = C, \qquad A^2 + B^2 \neq 0.$$

These forms are frequently encountered in mathematics, and you should be able to sketch their graphs quickly. The following summary should be used as an aid.

1. A quadratic equation of the form

$$y = ax^2, \qquad a \neq 0,$$

has a graph that is a **parabola,** opening upward if $a > 0$ and downward if $a < 0$. Similarly, any equation of the form

$$x = ay^2, \qquad a \neq 0,$$

has a graph that is a **parabola,** opening to the right if $a > 0$ and to the left if $a < 0$.

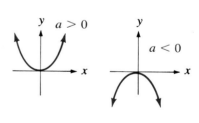

2. A quadratic equation of the form

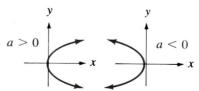

$$Ax^2 + By^2 = C,$$
$$A^2 + B^2 \neq 0,$$

has a graph that is:

a.   A **circle** if $A = B$ and A, B, and C have like signs.

b.   An **ellipse** if $A \neq B$ and A, B, and C have like signs.

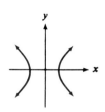

$$A < B, \ C > 0 \qquad A > B, \ C > 0$$

c.   A **hyperbola** if A and B are opposite in sign and $C \neq 0$.

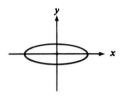

$$A < 0, \ B, \ C > 0 \qquad A, \ C > 0, \ B < 0$$

d.   **Two distinct lines** through the origin if A and B are opposite in sign and $C = 0$ (see Exercise 25).

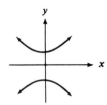

e.   **Two distinct parallel lines** if one of A and $B = 0$ and the other has the same sign as C.

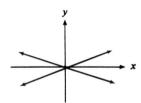

f.   **Two coincident parallel lines** (one line) through the origin if either $A = 0$ or $B = 0$ but not both, and also $C = 0$ (see Exercise 27).

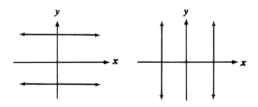
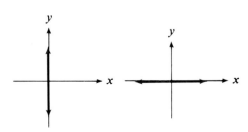

g. A **point** if A and B are both greater than 0 or both less than 0 and $C = 0$.

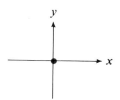

h. The **null set**, $\varnothing$, if A and B are both greater than or equal to 0 and $C < 0$, or if both A and B are less than or equal to 0 and $C > 0$ (see Exercise 28).

Note that we did not consider the special cases 2d to 2h in the preceding sections of this chapter. These cases are called the **degenerate cases** of the conic sections.

EXAMPLE I Name and sketch the graph of the equation

$$4x^2 - 36 = -9y^2.$$

Solution We first rewrite the given equation in standard form to obtain

$$4x^2 + 9y^2 = 36,$$

which is in the form $Ax^2 + By^2 = C$ with $A = 4$, $B = 9$, and $C = 36$. Since $A \neq B$ and A, B, and C are all of the same sign, the graph is an ellipse. The x-intercepts are -3 and 3, and the y-intercepts are -2 and 2. The graph is shown to the right. ■

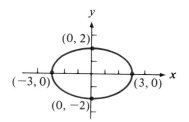

EXERCISE 11.5

A

■ *Name and sketch the graph of each of the following equations. Specify the vertices (or intercepts), and for the hyperbolas give equations of the asymptotes. See Example 1.*

1. $y = 4x^2$ **2.** $y = 9x^2$ **3.** $y = -8x^2$ **4.** $y = -3x^2$

5. $x^2 = 16y$ **6.** $x^2 = 4y$ **7.** $x^2 = -y$ **8.** $x^2 = -4y$

9. $x^2 + y^2 = 49$ **10.** $x^2 + y^2 = 64$ **11.** $4x^2 + 25y^2 = 100$ **12.** $x^2 + 2y^2 = 8$

13. $4x^2 = 4y^2$ **14.** $x^2 - 9y^2 = 0$ **15.** $x^2 = 9 + y^2$ **16.** $x^2 = 2y^2 + 8$

17. $4x^2 + 4y^2 = 1$ **18.** $9x^2 + 9y^2 = 2$ **19.** $3x^2 - 12 = -4y^2$ **20.** $12 - 3y^2 = 4x^2$

21. $4x^2 - y^2 = 0$ **22.** $4x^2 + y^2 = 0$ **23.** $y^2 = 0$ **24.** $x^2 = 4$

B

25. Discuss the graph of an equation of the form $Ax^2 - By^2 = 0$, $A, B > 0$.

26. Discuss the graph of an equation of the form $Ax^2 + By^2 = 0$, $A, B > 0$.

27. Discuss the graph of an equation of the form $Ax^2 = 0$, $A \neq 0$.

28. Explain why the graph of $x^2 + y^2 = -1$ is the null set. Generalize from the result, and discuss the graph of any equation of the form

$$Ax^2 + By^2 = C, \qquad A^2 + B^2 \neq 0, \quad A, B \geq 0, \quad C < 0.$$

11.6

TRANSLATION OF AXES

The equations of conic sections that were discussed in Sections 11.2, 11.3, and 11.4 were developed for the conic sections in standard position. By using a process called **translation of axes**, we can discuss conic sections with axes that are parallel to, but distinct from, the coordinate axes. Figure 11.15 shows a point P in the plane together with two sets of coordinate axes, an xy-system and an $x'y'$-system, whose corresponding axes are parallel. Each set of axes can be viewed as the result of "sliding" the other set across the plane while the axes remain parallel to their original position, suggesting the terminology "translation of axes."

FIGURE 11.15

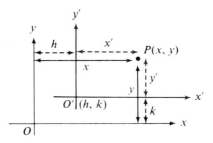

If the origin O' of the $x'y'$-system, shown in Figure 11.15, corresponds to the point (h, k) of the xy-system, then the $x'y'$-coordinates of the point P in the plane are related to the xy-coordinates of P by the equations

$$x = x' + h,$$
$$y = y' + k. \tag{1}$$

These equations can be solved for x' and y' to obtain the equivalent equations

$$x' = x - h,$$
$$y' = y - k. \tag{2}$$

EXAMPLE 1   The origin in an xy-system of coordinates is translated into the point $(4, -3)$, which becomes the origin in the $x'y'$-system. Under this translation:

(a)   What are the $x'y'$-coordinates of the point whose xy-coordinates are $(5, 8)$?

(b)   What is the $x'y'$-equation for $x + 5y = 3$?

Solution   (a)   Using Equations (2) with $h = 4$ and $k = -3$, we have

$$x' = x - 4,$$
$$y' = y + 3,$$

from which, for $(x, y) = (5, 8)$, we obtain

$$x' = 5 - 4 = 1,$$
$$y' = 8 + 3 = 11.$$

Therefore, the $x'y'$-coordinates are $(1, 11)$.

(b)   Using Equations (1) with $h = 4$ and $k = -3$, we have

$$x = x' + 4,$$
$$y = y' - 3.$$

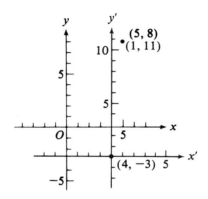

Replacing x and y in $x + 5y = 3$ with the right-hand members of these latter equations, we have

$$(x' + 4) + 5(y' - 3) = 3,$$

from which we obtain the desired equation,

$$x' + 5y' = 14.   ■$$

Translated Conic Sections

Now consider a parabola with vertex at the point $P(h, k)$ in an xy-coordinate system and with its axis of symmetry parallel to the x-axis as shown in Figure 11.16. If an $x'y'$-coordinate system had origin at P, then the $x'y'$-equation of the parabola would have the form

$$y'^2 = 4px'$$

As before, the constant p represents the distance between the focus and the vertex. By Equations (2), the xy-equation

FIGURE 11.16

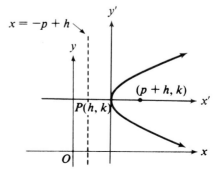

for the parabola would then have the form

$$(y - k)^2 = 4p(x - h). \tag{3}$$

Since in Figure 11.16, $p > 0$, the coordinates of the focus must then be $(p + h, k)$, while an equation for the directrix is $x = -p + h$.

Similar considerations lead to the conclusions given in Table 11.1 regarding parabolas, circles, ellipses, and hyperbolas with center (or vertex for parabolas) at the point (h, k) in the xy-plane. The data in the table can be used to help find an equation when certain information about a graph is given.

TABLE 11.1

Curve	xy-Equation	Major Axis Parallel to	Foci	Vertices
Parabola	$(y - k)^2 = 4p(x - h)$ $(y - k)^2 = -4p(x - h)$	x-axis	$(h + p, k)$ $(h - p, k)$	(h, k)
	$(x - h)^2 = 4p(y - k)$ $(x - h)^2 = -4p(y - k)$	y-axis	$(h, k + p)$ $(h, k - p)$	(h, k)
Circle	$(x - h)^2 + (y - k)^2 = r^2$	—	—	—
Ellipse $c^2 = a^2 - b^2$	$\dfrac{(x - h)^2}{a^2} + \dfrac{(y - k)^2}{b^2} = 1$	x-axis	$(h \pm c, k)$	$(h \pm a, k)$
	$\dfrac{(y - k)^2}{a^2} + \dfrac{(x - h)^2}{b^2} = 1$	y-axis	$(h, k \pm c)$	$(h, k \pm a)$
Hyperbola $c^2 = a^2 + b^2$	$\dfrac{(x - h)^2}{a^2} - \dfrac{(y - k)^2}{b^2} = 1$	x-axis	$(h \pm c, k)$	$(h \pm a, k)$
	$\dfrac{(y - k)^2}{a^2} - \dfrac{(x - h)^2}{b^2} = 1$	y-axis	$(h, k \pm c)$	$(h, k \pm a)$

EXAMPLE 2   Find an equation for the ellipse with major axis of length 12 and foci at $(-1, -1)$ and $(-1, 7)$. Sketch the curve.

Solution   The equal x-coordinates of the foci tell us that the major axis of the ellipse is parallel to the y-axis. Since the center is the midpoint of the segment with the foci as endpoints, the coordinates of the center are

$$\left(-1, \frac{7 - 1}{2}\right) = (-1, 3),$$

we see that c, the distance from the center to each focus, is 4. Because the major axis has length 12, it follows that a, the distance from the center to each vertex, is 6. Hence, $a^2 = 36$. From the fact that $b^2 = a^2 - c^2$, we have

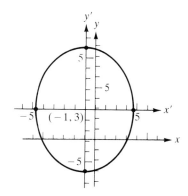

$$b^2 = 36 - 16 = 20.$$

The desired equation, then, is

$$\frac{(y-3)^2}{36} + \frac{(x+1)^2}{20} = 1.$$

Using $a = 6$ and $b = 2\sqrt{5}$, and starting with the center at $(-1, 3)$, we sketch the curve. ■

We can use the equations of translation (2) on page 462 together with the process of completing the square to identify and sketch the graph of an equation of the form

$$Ax^2 + By^2 + Dx + Ey + F = 0 \qquad (A \neq 0, B \neq 0). \qquad (4)$$

We do this by first completing the square in x and y in Equation (4) to obtain an equation of the form

$$A(x-h)^2 + B(y-k)^2 = C.$$

Then using Equations (2) and substituting x' for $x - h$ and y' for $y - k$, we have

$$Ax'^2 + By'^2 = C.$$

The graph of this equation is one of those discussed in Section 11.5. In standard form, we would have

$$\frac{x'^2}{C/A} + \frac{y'^2}{C/B} = 1 \qquad (A, B, \text{ and } C \neq 0).$$

If in Equation (4) either $A = 0$ or $B = 0$ but not both, then we can use translation of axes to transform Equation (4) into the standard form of the equation of a parabola.

EXAMPLE 3   Find an equation in standard form in the $x'y'$-system for the graph of

$$8x^2 + 4y^2 + 24x - 4y + 1 = 0.$$

Solution   We first rewrite the given equation equivalently as

$$8(x^2 + 3x \qquad) + 4(y^2 - y \qquad) = -1$$

and complete the square in x and y to obtain

$$8\left(x^2 + 3x + \frac{9}{4}\right) + 4\left(y^2 - y + \frac{1}{4}\right) = -1 + 18 + 1,$$

$$8\left(x + \frac{3}{2}\right)^2 + 4\left(y - \frac{1}{2}\right)^2 = 18.$$

If now we take $h = -\frac{3}{2}$ and $k = \frac{1}{2}$ in Equations (1), we find upon substituting $(x' - \frac{3}{2})$ for x and $(y' + \frac{1}{2})$ for y and dividing both members by 2 that this latter equation is equivalent to

$$4x'^2 + 2y'^2 = 9.$$

In standard form, we have

$$\frac{y'^2}{\frac{9}{2}} + \frac{x'^2}{\frac{9}{4}} = 1,$$

the equation of an ellipse, with $a = 3/\sqrt{2}$ and $b = 3/2$.   ■

It is sometimes convenient to use an auxiliary coordinate system as an aid in sketching graphs. For example, to graph

$$8x^2 + 4y^2 + 24x - 4y + 1 = 0$$

in Example 3, we can use the $x'y'$-coordinate system with origin at $(-\frac{3}{2}, \frac{1}{2})$ in the xy-system and then graph

$$\frac{x'^2}{\frac{9}{4}} + \frac{y'^2}{\frac{9}{2}} = 1.$$

FIGURE 11.17

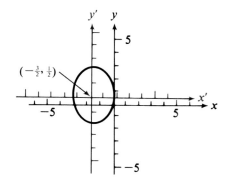

First, we sketch the $x'y'$-axes and identify the x'- and y'-intercepts, $\pm 3/2$ and $\pm 3/\sqrt{2}$. Then we complete the graph as shown in Figure 11.17.

EXERCISE 11.6

A

■   *For Exercises 1–8, assume that the origin in an xy-system of coordinates* is translated into the first point given, which then becomes the origin of an x'y'-system. See Example 1.

1. $(4, 4)$; find the x'y'-coordinates of the point whose xy-coordinates are $(-3, 1)$.
2. $(-3, 6)$; find the x'y'-coordinates of the point whose xy-coordinates are $(7, 0)$.
3. $(-5, -2)$; find the x'y'-coordinates of the point whose xy-coordinates are $(0, -3)$.
4. $(7, 0)$; find the x'y'-coordinates of the point whose xy-coordinates are $(0, 7)$.
5. $(3, 4)$; find the x'y'-equation for $x - 2y = 8$.
6. $(-1, 6)$; find the x'y'-equation for $3x + y = 1$.
7. $(2, -5)$; find the x'y'-equation for $x^2 - 2y^2 = 6$.
8. $(-6, -1)$; find the x'y'-equation for $x^2 + 2y = -2$.

■   Find an equation in the xy-system for each curve. Sketch the curve. See Example 2.

9. Circle with center at $(-5, 1)$ and radius 3.
10. Ellipse with major axis of length 10 and foci at $(3, 7)$ and $(3, -1)$.
11. Parabola with vertex at $(3, 3)$ and directrix with equation $y = 1$.
12. Hyperbola with foci at $(-1, 2)$ and $(-7, 2)$ and vertex at $(-3, 2)$.
13. Ellipse with vertices at $(3, 2)$ and $(-7, 2)$ and length of minor axis 6.
14. Parabola with focus at $(3, 2)$ and vertex at $(3, 4)$.
15. Hyperbola with center at $(3, 2)$, one focus at $(8, 2)$, and one vertex at $(0, 2)$.
16. Hyperbola with center $(-1, 5)$, length of transverse axis 8, and length of conjugate axis 6.

■   Find an equation in the x'y'-system and in standard form for the graph of the given equation in x and y. Use the auxiliary x'y'-system to graph the equation. See Example 3.

17. $y^2 - 4y = 12x - 52$
18. $x^2 + 4x + 8y = 4$
19. $x^2 - 6x + 3 = -y^2 - y$
20. $x^2 + y^2 = 6x - y + 3$
21. $3x^2 + 2y^2 + 12x - 4y = -2$
22. $16x^2 + 25y^2 - 32x + 100y - 284 = 0$
23. $9x^2 - 16y^2 - 90x + 64y + 17 = 0$
24. $9x^2 - 4y^2 = -61 - 54x - 16y$

B

■   Find conditions on the coefficients of the quadratic equation

$$Ax^2 + By^2 + Cx + Dy + F = 0$$

so that the graph of the equation is the given curve.

25. Parabola with the axis of symmetry parallel to the y-axis.
26. Ellipse
27. Circle
28. Hyperbola

CHAPTER REVIEW

Key Words and Phrases

■ *Define or explain each of the following words and phrases.*

  1. locus of points
  2. parabola
  3. focus of a parabola
  4. directrix of a parabola
  5. axis of symmetry of a parabola
  6. vertex of a parabola
  7. standard position of a parabola
  8. conic section
  9. circle
  10. center of a circle
  11. radius of a circle
  12. ellipse
  13. foci of an ellipse

  14. center of an ellipse
  15. major and minor axes of an ellipse
  16. vertices of an ellipse
  17. standard position of an ellipse
  18. hyperbola
  19. foci of a hyperbola
  20. center of a hyperbola
  21. vertices of a hyperbola
  22. transverse axis of a hyperbola
  23. conjugate axis of a hyperbola
  24. standard position of a hyperbola
  25. translation of axes

Review Exercises

[11.1] **1.** Find an equation of the locus of a point in the plane that moves so that its distance from $(2, -1)$ is twice its distance from $(0, 4)$.

  2. Find an equation of the locus of a point in the plane that moves so that its distance from $(3, 4)$ is 5.

  3. Find an equation of the locus of a point in the plane that moves so that its distance from the x-axis is 3.

  4. Find an equation of the locus of a point in the plane that moves so that the sum of its distances from $(-2, 6)$ and $(0, 4)$ is 8.

[11.2] ■ *Find an equation of the parabola in standard position satisfying the given condition, and sketch the graph.*

  5. Focus $(0, 6)$ **6.** Directrix $x = -3$ **7.** Directrix $y = 2$ **8.** Focus $(-4, 0)$

  ■ *Find the focus and the directrix of the given parabola.*

  9. $y^2 = \dfrac{1}{2}x$ **10.** $x^2 = -y$ **11.** $x^2 - 20y = 0$ **12.** $y^2 = -x$

[11.3] ■ *Find an equation and sketch the graph of the circle or ellipse in standard position satisfying the given condition.*

  13. Center $(0, 0)$, $r = 16$ **14.** Focus $(0, 4)$, $a = 8$

  15. Vertex $(9, 0)$, $b = 3$ **16.** Focus $\left(-\dfrac{1}{2}, 0\right)$, $b = 1$

■ *Find the foci and vertices of the given ellipse.*

17. $\dfrac{x^2}{4} + \dfrac{y^2}{16} = 1$

18. $\dfrac{x^2}{8} + y^2 = 1$

19. $\dfrac{x^2}{9} + \dfrac{y^2}{8} = 1$

20. $\dfrac{y^2}{9} + \dfrac{x^2}{4} = 1$

[11.4] ■ *Find an equation and sketch the graph of the hyperbola in standard position satisfying the given condition.*

21. Focus $(0, 2)$, $b = 1$

22. Focus $(0, 4)$,   vertex $(0, 2)$

23. Vertex $(1, 0)$,   asymptotes $y = \pm 2x$

24. Focus $(2, 0)$, $a = 1$

■ *Find the foci, vertices, and asymptotes of the given hyperbola.*

25. $\dfrac{x^2}{2} - y^2 = 1$

26. $y^2 - x^2 = 1$

27. $\dfrac{y^2}{4} - \dfrac{x^2}{2} = 1$

28. $\dfrac{x^2}{10} - \dfrac{y^2}{4} = 1$

[11.5] ■ *Name and sketch the graph of each of the following equations.*

29. $x^2 = 12y$

30. $x^2 = -4y$

31. $y^2 = 16x$

32. $y^2 = 8x$

33. $\dfrac{x^2}{2} - y^2 = 1$

34. $\dfrac{x^2}{2} + \dfrac{y^2}{4} = 3$

35. $x^2 - y^2 = 0$

36. $y^2 = -x^2$

[11.6] ■ *For Exercises 37 and 38, assume that the origin in an xy-system of coordinates is translated into the first point given, which then becomes the origin of an x′y′-system.*

37. $(2, -1)$;   find the $x'y'$-coordinates of the point whose xy-coordinates are $(-5, 3)$.

38. $(-3, 0)$;   find the $x'y'$-equation for $2x^2 - y = 1$.

■ *Find an equation in the xy-system for each curve. Sketch the curve.*

39. Circle with center at $(1, -3)$ and radius 4.

40. Parabola with focus at $(5, 0)$ and vertex at $(7, 0)$.

■ *Find an equation in standard form for the graph of the given equation in x and y.*

41. $x^2 - 2x = 6y + 14$

42. $2x^2 + y^2 - x + 4y - 5 = 0.$

43. Use an auxiliary system to graph the equation in Exercise 41.

44. Use an auxiliary system to graph the equation in Exercise 42.

A Using Logarithm Tables

For $b > 0$ and $b \neq 1$, values for b^x and $\log_b x$ $(x > 0)$ can be obtained by using prepared tables in conjunction with the laws of logarithms. Because we are familiar with the number 10 as the base of our numeration system, we shall first give our attention to logarithms and powers to the base 10. This was the base first used in the invention of logarithms to perform computations in astronomy and navigation.

Common Logarithms

The $\log_{10} x$ is the exponent that must be placed on 10 so that the resulting power is x (see Section 6.2, Equation (2) on page 260). Values for $\log_{10} x$ are sometimes called **common logarithms**.

Values for $\log_{10} 10^k$, $k \in J$

Some values of $\log_{10} x$ can be obtained simply by considering the definition of a logarithm, while other values require tables. Let us first consider values of $\log_{10} x$ for all values of x that are integral powers of 10. These can be obtained by inspection.

$$\text{Since} \quad 10^3 = 1000, \quad \log_{10} 1000 = 3.$$

$$\text{Since} \quad 10^2 = 100, \quad \log_{10} 100 = 2.$$

$$\text{Since} \quad 10^1 = 10, \quad \log_{10} 10 = 1.$$

$$\text{Since} \quad 10^0 = 1, \quad \log_{10} 1 = 0.$$

$$\text{Since} \quad 10^{-1} = 0.1, \quad \log_{10} 0.1 = -1.$$

$$\text{Since} \quad 10^{-2} = 0.01, \quad \log_{10} 0.01 = -2.$$

$$\text{Since} \quad 10^{-3} = 0.001, \quad \log_{10} 0.001 = -3.$$

Notice that the logarithm of a power of 10 is simply the exponent on the base 10. For example,

$$\log_{10} 100 = \log_{10} 10^2 = 2,$$

$$\log_{10} 0.01 = \log_{10} 10^{-2} = -2,$$

and so on.

EXAMPLE 1   Find $\log_{10} 1000$ by inspection.

Solution   We note that $1000 = 10^3$ and by definition $\log_{10} 10^3$ is the exponent on 10, so that the power equals 10^3. Hence,

$$\log_{10} 10^3 = 3. \qquad ■$$

Values for $\log_{10} x$, $1 < x < 10$

Table I in Appendix B gives values for $\log_{10} x$ for $1 < x < 10$. Consider the excerpt from the table shown below. Each number in the column headed x represents the first two digits of the numeral for x, while each of the other column-head numbers represents the third digit of the numeral for x. The number located at the intersection of a row and a column is the logarithm of x. For example, to find $\log_{10} 4.25$, we look at the intersection of the row containing 4.2 under x and the column containing 5. Thus,

$$\log_{10} 4.25 = 0.6284.$$

x	0	1	2	3	4	5	6	7	8	9
3.8	0.5798	0.5809	0.5821	0.5832	0.5843	0.5855	0.5866	0.5877	0.5888	0.5899
3.9	0.5911	0.5922	0.5933	0.5944	0.5955	0.5966	0.5977	0.5988	0.5999	0.6010
4.0	0.6021	0.6031	0.6042	0.6053	0.6064	0.6075	0.6085	0.6096	0.6107	0.6117
4.1	0.6128	0.6138	0.6149	0.6160	0.6170	0.6180	0.6191	0.6201	0.6212	0.6222
4.2	0.6232	0.6243	0.6253	0.6263	0.6274	0.6284	0.6294	0.6304	0.6314	0.6325
4.3	0.6335	0.6345	0.6355	0.6365	0.6375	0.6385	0.6395	0.6405	0.6415	0.6425
4.4	0.6435	0.6444	0.6454	0.6464	0.6474	0.6484	0.6493	0.6503	0.6513	0.6522
4.5	0.6532	0.6542	0.6551	0.6561	0.6571	0.6580	0.6590	0.6599	0.6609	0.6618
4.6	0.6628	0.6637	0.6646	0.6656	0.6665	0.6675	0.6684	0.6693	0.6702	0.6712

Similarly,

$$\log_{10} 4.02 = 0.6042,$$

$$\log_{10} 4.49 = 0.6522,$$

and so on.

Although the values in the tables are rational-number approximations of irrational numbers, we shall follow customary usage and write $=$ instead of $\approx$.

Values for $\log_{10} x$, $x > 10$

Now suppose we wish to find $\log_{10} x$ for values of x outside the range of the table—that is, for $x > 10$ or $0 < x < 1$ (see Figure A.1). This can be done quite readily by first representing the number in scientific notation and then applying the first law of logarithms.

FIGURE A.1

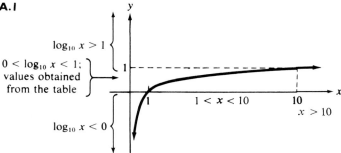

EXAMPLE 2 (a) $\log_{10} 42.5$
$= \log_{10}(4.25 \times 10^1)$
$= \log_{10} 4.25 + \log_{10} 10^1$
$= 0.6284 + 1$
$= 1.6284$

(b) $\log_{10} 425$
$= \log_{10}(4.25 \times 10^2)$
$= \log_{10} 4.25 + \log_{10} 10^2$
$= 0.6284 + 2$
$= 2.6284$

(c) $\log_{10} 4250$
$= \log_{10}(4.25 \times 10^3)$
$= \log_{10} 4.25 + \log_{10} 10^3$
$= 0.6284 + 3$
$= 3.6284$

(d) $\log_{10} 42,500$
$= \log_{10}(4.25 \times 10^4)$
$= \log_{10} 4.25 + \log_{10} 10^4$
$= 0.6284 + 4$
$= 4.6284$ ■

Observe that the decimal portion of the logarithms in each part of Example 2 is always 0.6284 and *the integral portion is the exponent on 10 when the number is written in scientific notation.*

This process can be reduced to a mechanical one by considering $\log_{10} x$ to consist of two parts, an *integral part* (called the **characteristic**) and a *nonnegative decimal fraction part* (called the **mantissa**). Thus, the table of values for $\log_{10} x$ for $1 < x < 10$ can be looked upon as a table of mantissas for $\log_{10} x$ for all $x > 0$.

In each part of Example 2, where $x > 10$, the mantissa in each case is 0.6284 and the characteristics are 1, 2, 3, and 4, respectively.

EXAMPLE 3 Find $\log_{10} 16.8$.

Solution We first represent 16.8 in scientific notation to obtain

$$16.8 = 1.68 \times 10^1.$$

Next, we use 1.68 to determine the mantissa from Table I in Appendix B. The mantissa is

$$0.2253.$$

Finally, we add the characteristic 1 as determined by the exponent on the base 10 to obtain

$$\log_{10} 16.8 = 0.2253 + 1$$
$$= 1.2253. \quad ■$$

Values for $\log_{10} x$, $0 < x < 1$

Now consider an example of the form $\log_{10} x$ for $0 < x < 1$. To find $\log_{10} 0.00425$, we write

$$\log_{10} 0.00425 = \log_{10}(4.25 \times 10^{-3})$$
$$= \log_{10} 4.25 + \log_{10} 10^{-3}.$$

We find from Table I that $\log_{10} 4.25 = 0.6284$. Upon adding 0.6284 to the characteristic -3, we obtain

$$\log_{10} 0.00425 = 0.6284 + (-3)$$
$$= -2.3716,$$

where the decimal part of the logarithm is no longer 0.6284. The decimal part is -0.3716, a negative number.

   If we want to use Table I, which contains only positive entries, it is customary to write the logarithm in a form in which the decimal part is positive. In Example 3, we write

$$\log_{10} 0.00425 = 0.6284 - 3,$$

where the decimal part is positive. Because -3 can be written $1 - 4$, $2 - 5$, $3 - 6$, $7 - 10$, and so on, the forms $1.6284 - 4$, $2.6284 - 5$, $3.6284 - 6$, $7.6284 - 10$, and so on, are equally valid representations of the desired logarithm. It will sometimes be convenient to use these alternative forms when we use the tables.

EXAMPLE 4   Find $\log_{10} 0.043$.

Solution   We first represent 0.043 in scientific notation to obtain

$$0.043 = 4.3 \times 10^{-2}.$$

Next, we use 4.3 to determine the mantissa from Table I. The mantissa is

$$0.6335.$$

Finally, we add the characteristic -2 as determined by the exponent on the base 10 to obtain

$$\log_{10} 0.043 = 0.6335 - 2$$
$$= -1.3665.$$

EXAMPLE 5 (a) $\log_{10} 0.294$

$$= \log_{10}(2.94 \times 10^{-1})$$
$$= \log_{10} 2.94 + \log_{10} 10^{-1}$$
$$= 0.4683 - 1$$
$$= -0.5317$$

(b) $\log_{10} 0.00294$

$$= \log_{10}(2.94 \times 10^{-3})$$
$$= \log_{10} 2.94 + \log_{10} 10^{-3}$$
$$= 0.4683 - 3$$
$$= -2.5164 \qquad ■$$

Antilog$_{10}$ N

Given a value for an exponent, $\log_{10} x$, we can use Table I (Appendix B) to find the power x by reversing the process described to find the logarithm of a number. In this case, the power x is sometimes called the **antilogarithm** of $\log_{10} x$. For example, if

$$\log_{10} x = 0.4409,$$

then

$$x = \text{antilog}_{10} 0.4409,$$

which can be obtained by locating 0.4409 in the body of Table I and observing that

$$\text{antilog}_{10} 0.4409 = 2.76.$$

If the $\log_{10} x$ is greater than 1, it can first be written as the sum of a positive decimal (the mantissa) and a positive integer (the characteristic). Antilog$_{10} x$ can then be written as the product of a number between 1 and 10 and a power of 10.

EXAMPLE 6 If $\log_{10} x = 2.4409$, then

$$x = \text{antilog}_{10} 2.4409$$
$$= \text{antilog}_{10}(0.4409 + 2)$$
$$= 2.76 \times 10^2$$
$$= 276. \qquad ■$$

If the decimal part of $\log_{10} x$ is negative and we wish to use Table I to obtain x, we cannot use the table directly. However, we can first write $\log_{10} x$ equivalently with a positive decimal part. For example, to find

$$\text{antilog}_{10}(-0.4522) \qquad \text{or} \qquad \text{antilog}_{10}(-2.4522),$$

we can first add $(+1 - 1)$ to write -0.4522 as

$$-0.4522 + 1 - 1 = 0.5478 - 1$$

and add $(+3 - 3)$ to write -2.4522 as

$$-2.4522 + 3 - 3 = 0.5478 - 3,$$

and then use the table. Thus,

$$\text{antilog}_{10}(-0.4522) = \text{antilog}_{10}(0.5478 - 1)$$

$$= 3.53 \times 10^{-1} = 0.353,$$

and

$$\text{antilog}_{10}(-2.4522) = \text{antilog}_{10}(0.5478 - 3)$$

$$= 3.53 \times 10^{-3} = 0.00353.$$

EXAMPLE 7   Use Table I to find the value of x.

(a) $\log_{10} x = -0.7292$ (b) $\log_{10} x = -1.4634$

Solution   (a) $x = \text{antilog}_{10}(-0.7292)$

$= \text{antilog}_{10}(-0.7292 + 1 - 1)$

$= \text{antilog}_{10}(0.2708 - 1)$

$= 1.87 \times 10^{-1} = 0.187$

(b) $x = \text{antilog}_{10}(-1.4634)$

$= \text{antilog}_{10}(-1.4634 + 2 - 2)$

$= \text{antilog}_{10}(0.5366 - 2)$

$= 3.44 \times 10^{-2} = 0.0344$ ▪

In Example 7, the mantissas, 0.2708 and 0.5366, were listed in Table I. If we seek the common logarithm of a number that is not an entry in the table (for example, $\log_{10} 23.42$) or if we seek x when $\log_{10} x$ is not an entry in the table, we shall simply use the entry in the table that is closest to the value we seek.

Powers to the Base 10   By the definition of a logarithm,

$$P = 10^E$$

can be written in logarithmic form as

$$\log_{10} P = E,$$

from which we see that the power P is the antilogarithm of the exponent E,

$$P = \text{antilog}_{10} E.$$

Since $P = 10^E$,

$$10^E = \text{antilog}_{10} E$$

and we can obtain a power 10^E simply by finding the antilogarithm of the exponent E.

EXAMPLE 8 Compute each power.

(a) $10^{0.2148}$ (b) $10^{-1.6345}$

Solution (a) $10^{0.2148}$

$= \text{antilog}_{10} 0.2148$

$= 1.64$

(b) $10^{-1.6345}$

$= \text{antilog}_{10}(-1.6345)$

$= \text{antilog}_{10}(\underbrace{-1.6345 + 2} - 2)$

$= \text{antilog}_{10}(0.3655 - 2)$

$= 2.32 \times 10^{-2}$ ■

Powers to the Base e

The number $e \approx 2.718281828$* is an irrational number that has applications in business, biological and physical sciences, and engineering. Because of its importance, special tables have been prepared for both e^x and $\log_e x$.

Table II in Appendix B gives approximations for e^x and e^{-x} for $0 \le x \le 1.00$ in 0.01 intervals and for $1.00 < x \le 10.00$ in 0.1 intervals.

EXAMPLE 9 Using Table II:

(a) $e^{2.4} = 11.023$ (b) $e^{-4.7} = 0.0091$ ■

Although we can obtain values for e^x and e^{-x} outside the interval 0 to 10 by using Table II along with the first law of exponents, at this time the function values in the table over this interval will be adequate for our work.

Natural Logarithms, ln x

The most-used values for $\log_e x$ $(x > 0)$ are printed in Table III in Appendix B. These values, like those for e^x, e^{-x}, and $\log_{10} x$, are *approximations* accurate to the number of decimals shown. The symbol $\log_e x$ is often written as **ln x** and read as

---

* The number e is discussed in more detail in Section 6.5.

"**natural logarithm of x.**" Unlike Table I for common logarithms, which provides only the decimal part of $\log_{10} x$, the table for natural logarithms gives the entire value for $\ln x$, both the integral and decimal portions.

EXAMPLE 10   Using Table III:

(a) $\ln 6.6 = 1.8871$ (b) $\ln 0.7 = -0.3567$ ■

If we seek a value of e^x or $\ln x$ for a value of x between two entries in the tables, we shall use the value in Table III that is closest to x.

Antilogarithms   Given an exponent $\ln x$, we can obtain the power x by using the definition of a logarithm and Table II.

EXAMPLE 11   (a) $\ln x = 1.3$ (b) $\ln x = -0.47$

Solution   (a) $\ln x = 1.3$ is equivalent to   (b) $\ln x = -0.47$ is equivalent to

$$x = e^{1.3}$$ $$x = e^{-0.47}$$
$$= 3.6693$$ $$= 0.6250 \quad ■$$

As noted, a power to the base b is called the antilog$_b$ of the exponent. Thus, in part (a) of Example 11,

$$x = e^{1.3} = \text{antilog}_e 1.3$$
$$= 3.6693.$$

EXERCISE A

A

■   *Find each logarithm by inspection. See Example 1.*

1. $\log_{10} 10^2$
2. $\log_{10} 10^4$
3. $\log_{10} 10^{-4}$
4. $\log_{10} 10^{-6}$
5. $\log_{10} 10^0$
6. $\log_{10} 10^n$

■   *Find an approximation for each logarithm using Table 1. See Examples 2–5.*

7. $\log_{10} 6.73$

8. $\log_{10} 891$

9. $\log_{10} 83.7$

10. $\log_{10} 21.4$

11. $\log_{10} 317$

12. $\log_{10} 219$

13. $\log_{10} 0.813$

14. $\log_{10} 0.00214$

15. $\log_{10} 0.08$

16. $\log_{10} 0.000413$

17. $\log_{10}(2.48 \times 10^2)$

18. $\log_{10}(5.39 \times 10^{-3})$

■   *Solve for x using Table I. See Examples 6 and 7.*

19. $\log_{10} x = 0.6128$

20. $\log_{10} x = 0.2504$

21. $\log_{10} x = 1.5647$

22. $\log_{10} x = 3.9258$

23. $\log_{10} x = 0.8075 - 2$

24. $\log_{10} x = 0.9722 - 3$

25. $\log_{10} x = 7.8562 - 10$

26. $\log_{10} x = 1.8155 - 4$

27. $\log_{10} x = -0.5272$

28. $\log_{10} x = -0.4123$

29. $\log_{10} x = -1.2984$

30. $\log_{10} x = -1.0545$

31. $\log_{10} x = -2.6882$

32. $\log_{10} x = -2.0670$

■   *Compute each power. See Example 8.*

33. $10^{0.8762}$

34. $10^{1.6405}$

35. $10^{2.8943}$

36. $10^{4.3766}$

37. $10^{-1.4473}$

38. $10^{-2.0958}$

■   *Find each power. See Example 9.*

39. $e^{0.43}$

40. $e^{0.62}$

41. $e^{-0.57}$

42. $e^{-0.08}$

43. $e^{1.5}$

44. $e^{2.6}$

45. $e^{-2.4}$

46. $e^{-1.2}$

■   *Find each logarithm. See Example 10.*

47. $\ln 3.9$

48. $\ln 6.3$

49. $\ln 16$

50. $\ln 55$

51. $\ln 0.4$

52. $\ln 0.7$

■   *Find each value of x. See Example 11.*

53. $\ln x = 0.16$

54. $\ln x = 0.25$

55. $\ln x = 1.8$

56. $\ln x = 2.4$

57. $\ln x = 4.5$

58. $\ln x = 6.0$

B   Tables

TABLE I   COMMON LOGARITHMS

x	0	1	2	3	4	5	6	7	8	9
1.0	0.0000	0.0043	0.0086	0.0128	0.0170	0.0212	0.0253	0.0294	0.0334	0.0374
1.1	0.0414	0.0453	0.0492	0.0531	0.0569	0.0607	0.0645	0.0682	0.0719	0.0755
1.2	0.0792	0.0828	0.0864	0.0899	0.0934	0.0969	0.1004	0.1038	0.1072	0.1106
1.3	0.1139	0.1173	0.1206	0.1239	0.1271	0.1303	0.1335	0.1367	0.1399	0.1430
1.4	0.1461	0.1492	0.1523	0.1553	0.1584	0.1614	0.1644	0.1673	0.1703	0.1732
1.5	0.1761	0.1790	0.1818	0.1847	0.1875	0.1903	0.1931	0.1959	0.1987	0.2014
1.6	0.2041	0.2068	0.2095	0.2122	0.2148	0.2175	0.2201	0.2227	0.2253	0.2279
1.7	0.2304	0.2330	0.2355	0.2380	0.2405	0.2430	0.2455	0.2480	0.2504	0.2529
1.8	0.2553	0.2577	0.2601	0.2625	0.2648	0.2672	0.2695	0.2718	0.2742	0.2765
1.9	0.2788	0.2810	0.2833	0.2856	0.2878	0.2900	0.2923	0.2945	0.2967	0.2989
2.0	0.3010	0.3032	0.3054	0.3075	0.3096	0.3118	0.3139	0.3160	0.3181	0.3201
2.1	0.3222	0.3243	0.3263	0.3284	0.3304	0.3324	0.3345	0.3365	0.3385	0.3404
2.2	0.3424	0.3444	0.3464	0.3483	0.3502	0.3522	0.3541	0.3560	0.3579	0.3598
2.3	0.3617	0.3636	0.3655	0.3674	0.3692	0.3711	0.3729	0.3747	0.3766	0.3784
2.4	0.3802	0.3820	0.3838	0.3856	0.3874	0.3892	0.3909	0.3927	0.3945	0.3962
2.5	0.3979	0.3997	0.4014	0.4031	0.4048	0.4065	0.4082	0.4099	0.4116	0.4133
2.6	0.4150	0.4166	0.4183	0.4200	0.4216	0.4232	0.4249	0.4265	0.4281	0.4298
2.7	0.4314	0.4330	0.4346	0.4362	0.4378	0.4393	0.4409	0.4425	0.4440	0.4456
2.8	0.4472	0.4487	0.4502	0.4518	0.4533	0.4548	0.4564	0.4579	0.4594	0.4609
2.9	0.4624	0.4639	0.4654	0.4669	0.4683	0.4698	0.4713	0.4728	0.4742	0.4757
3.0	0.4771	0.4786	0.4800	0.4814	0.4829	0.4843	0.4857	0.4871	0.4886	0.4900
3.1	0.4914	0.4928	0.4942	0.4955	0.4969	0.4983	0.4997	0.5011	0.5024	0.5038
3.2	0.5051	0.5065	0.5079	0.5092	0.5105	0.5119	0.5132	0.5145	0.5159	0.5172
3.3	0.5185	0.5198	0.5211	0.5224	0.5237	0.5250	0.5263	0.5276	0.5289	0.5302
3.4	0.5315	0.5328	0.5340	0.5353	0.5366	0.5378	0.5391	0.5403	0.5416	0.5428
3.5	0.5441	0.5453	0.5465	0.5478	0.5490	0.5502	0.5514	0.5527	0.5539	0.5551
3.6	0.5563	0.5575	0.5587	0.5599	0.5611	0.5623	0.5635	0.5647	0.5658	0.5670
3.7	0.5682	0.5694	0.5705	0.5717	0.5729	0.5740	0.5752	0.5763	0.5775	0.5786
3.8	0.5798	0.5809	0.5821	0.5832	0.5843	0.5855	0.5866	0.5877	0.5888	0.5899
3.9	0.5911	0.5922	0.5933	0.5944	0.5955	0.5966	0.5977	0.5988	0.5999	0.6010
4.0	0.6021	0.6031	0.6042	0.6053	0.6064	0.6075	0.6085	0.6096	0.6107	0.6117
4.1	0.6128	0.6138	0.6149	0.6160	0.6170	0.6180	0.6191	0.6201	0.6212	0.6222
4.2	0.6232	0.6243	0.6253	0.6263	0.6274	0.6284	0.6294	0.6304	0.6314	0.6325
4.3	0.6335	0.6345	0.6355	0.6365	0.6375	0.6385	0.6395	0.6405	0.6415	0.6425
4.4	0.6435	0.6444	0.6454	0.6464	0.6474	0.6484	0.6493	0.6503	0.6513	0.6522
4.5	0.6532	0.6542	0.6551	0.6561	0.6571	0.6580	0.6590	0.6599	0.6609	0.6618
4.6	0.6628	0.6637	0.6646	0.6656	0.6665	0.6675	0.6684	0.6693	0.6702	0.6712
4.7	0.6721	0.6730	0.6739	0.6749	0.6758	0.6767	0.6776	0.6785	0.6794	0.6803
4.8	0.6812	0.6821	0.6830	0.6839	0.6848	0.6857	0.6866	0.6875	0.6884	0.6893
4.9	0.6902	0.6911	0.6920	0.6928	0.6937	0.6946	0.6955	0.6964	0.6972	0.6981
5.0	0.6990	0.6998	0.7007	0.7016	0.7024	0.7033	0.7042	0.7050	0.7059	0.7067
5.1	0.7076	0.7084	0.7093	0.7101	0.7110	0.7118	0.7126	0.7135	0.7143	0.7152
5.2	0.7160	0.7168	0.7177	0.7185	0.7193	0.7202	0.7210	0.7218	0.7226	0.7235
5.3	0.7243	0.7251	0.7259	0.7267	0.7275	0.7284	0.7292	0.7300	0.7308	0.7316
5.4	0.7324	0.7332	0.7340	0.7348	0.7356	0.7364	0.7372	0.7380	0.7388	0.7396
x	0	1	2	3	4	5	6	7	8	9

TABLE I (Continued)

x	0	1	2	3	4	5	6	7	8	9
5.5	0.7404	0.7412	0.7419	0.7427	0.7435	0.7443	0.7451	0.7459	0.7466	0.7474
5.6	0.7482	0.7490	0.7497	0.7505	0.7513	0.7520	0.7528	0.7536	0.7543	0.7551
5.7	0.7559	0.7566	0.7574	0.7582	0.7589	0.7597	0.7604	0.7612	0.7619	0.7627
5.8	0.7634	0.7642	0.7649	0.7657	0.7664	0.7672	0.7679	0.7686	0.7694	0.7701
5.9	0.7709	0.7716	0.7723	0.7731	0.7738	0.7745	0.7752	0.7760	0.7767	0.7774
6.0	0.7782	0.7789	0.7796	0.7803	0.7810	0.7818	0.7825	0.7832	0.7839	0.7846
6.1	0.7853	0.7860	0.7868	0.7875	0.7882	0.7889	0.7896	0.7903	0.7910	0.7917
6.2	0.7924	0.7931	0.7938	0.7945	0.7952	0.7959	0.7966	0.7973	0.7980	0.7987
6.3	0.7993	0.8000	0.8007	0.8014	0.8021	0.8028	0.8035	0.8041	0.8048	0.8055
6.4	0.8062	0.8069	0.8075	0.8082	0.8089	0.8096	0.8102	0.8109	0.8116	0.8122
6.5	0.8129	0.8136	0.8142	0.8149	0.8156	0.8162	0.8169	0.8176	0.8182	0.8189
6.6	0.8195	0.8202	0.8209	0.8215	0.8222	0.8228	0.8235	0.8241	0.8248	0.8254
6.7	0.8261	0.8267	0.8274	0.8280	0.8287	0.8293	0.8299	0.8306	0.8312	0.8319
6.8	0.8325	0.8331	0.8338	0.8344	0.8351	0.8357	0.8363	0.8370	0.8376	0.8382
6.9	0.8388	0.8395	0.8401	0.8407	0.8414	0.8420	0.8426	0.8432	0.8439	0.8445
7.0	0.8451	0.8457	0.8463	0.8470	0.8476	0.8482	0.8488	0.8494	0.8500	0.8506
7.1	0.8513	0.8519	0.8525	0.8531	0.8537	0.8543	0.8549	0.8555	0.8561	0.8567
7.2	0.8573	0.8579	0.8585	0.8591	0.8597	0.8603	0.8609	0.8615	0.8621	0.8627
7.3	0.8633	0.8639	0.8645	0.8651	0.8657	0.8663	0.8669	0.8675	0.8681	0.8686
7.4	0.8692	0.8698	0.8704	0.8710	0.8716	0.8722	0.8727	0.8733	0.8739	0.8745
7.5	0.8751	0.8756	0.8762	0.8768	0.8774	0.8779	0.8785	0.8791	0.8797	0.8802
7.6	0.8808	0.8814	0.8820	0.8825	0.8831	0.8837	0.8842	0.8848	0.8854	0.8859
7.7	0.8865	0.8871	0.8876	0.8882	0.8887	0.8893	0.8899	0.8904	0.8910	0.8915
7.8	0.8921	0.8927	0.8932	0.8938	0.8943	0.8949	0.8954	0.8960	0.8965	0.8971
7.9	0.8976	0.8982	0.8987	0.8993	0.8998	0.9004	0.9009	0.9015	0.9020	0.9025
8.0	0.9031	0.9036	0.9042	0.9047	0.9053	0.9058	0.9063	0.9069	0.9074	0.9079
8.1	0.9085	0.9090	0.9096	0.9101	0.9106	0.9112	0.9117	0.9122	0.9128	0.9133
8.2	0.9138	0.9143	0.9149	0.9154	0.9159	0.9165	0.9170	0.9175	0.9180	0.9186
8.3	0.9191	0.9196	0.9201	0.9206	0.9212	0.9217	0.9222	0.9227	0.9232	0.9238
8.4	0.9243	0.9248	0.9253	0.9258	0.9263	0.9269	0.9274	0.9279	0.9284	0.9289
8.5	0.9294	0.9299	0.9304	0.9309	0.9315	0.9320	0.9325	0.9330	0.9335	0.9340
8.6	0.9345	0.9350	0.9355	0.9360	0.9365	0.9370	0.9375	0.9380	0.9385	0.9390
8.7	0.9395	0.9400	0.9405	0.9410	0.9415	0.9420	0.9425	0.9430	0.9435	0.9440
8.8	0.9445	0.9450	0.9455	0.9460	0.9465	0.9469	0.9474	0.9479	0.9484	0.9489
8.9	0.9494	0.9499	0.9504	0.9509	0.9513	0.9518	0.9523	0.9528	0.9533	0.9538
9.0	0.9542	0.9547	0.9552	0.9557	0.9562	0.9566	0.9571	0.9576	0.9581	0.9586
9.1	0.9590	0.9595	0.9600	0.9605	0.9609	0.9614	0.9619	0.9624	0.9628	0.9633
9.2	0.9638	0.9643	0.9647	0.9652	0.9657	0.9661	0.9666	0.9671	0.9675	0.9680
9.3	0.9685	0.9689	0.9694	0.9699	0.9703	0.9708	0.9713	0.9717	0.9722	0.9727
9.4	0.9731	0.9736	0.9741	0.9745	0.9750	0.9754	0.9759	0.9763	0.9768	0.9773
9.5	0.9777	0.9782	0.9786	0.9791	0.9795	0.9800	0.9805	0.9809	0.9814	0.9818
9.6	0.9823	0.9827	0.9832	0.9836	0.9841	0.9845	0.9850	0.9854	0.9859	0.9863
9.7	0.9868	0.9872	0.9877	0.9881	0.9886	0.9890	0.9894	0.9899	0.9903	0.9908
9.8	0.9912	0.9917	0.9921	0.9926	0.9930	0.9934	0.9939	0.9943	0.9948	0.9952
9.9	0.9956	0.9961	0.9965	0.9969	0.9974	0.9978	0.9983	0.9987	0.9991	0.9996
x	0	1	2	3	4	5	6	7	8	9

TABLE II
EXPONENTIAL FUNCTIONS, BASE e

x	e^x	e^{-x}	x	e^x	e^{-x}
0.00	1.0000	1.0000	1.5	4.4817	0.2231
0.01	1.0101	0.9901	1.6	4.9530	0.2019
0.02	1.0202	0.9802	1.7	5.4739	0.1827
0.03	1.0305	0.9705	1.8	6.0496	0.1653
0.04	1.0408	0.9608	1.9	6.6859	0.1496
0.05	1.0513	0.9512	2.0	7.3891	0.1353
0.06	1.0618	0.9418	2.1	8.1662	0.1225
0.07	1.0725	0.9324	2.2	9.0250	0.1108
0.08	1.0833	0.9331	2.3	9.9742	0.1003
0.09	1.0942	0.9139	2.4	11.023	0.0907
0.10	1.1052	0.9048	2.5	12.182	0.0821
0.11	1.1163	0.8958	2.6	13.464	0.0743
0.12	1.1275	0.8869	2.7	14.880	0.0672
0.13	1.1388	0.8781	2.8	16.445	0.0608
0.14	1.1503	0.8694	2.9	18.174	0.0550
0.15	1.1618	0.8607	3.0	20.086	0.0498
0.16	1.1735	0.8521	3.1	22.198	0.0450
0.17	1.1853	0.8437	3.2	24.533	0.0408
0.18	1.1972	0.8353	3.3	27.113	0.0369
0.19	1.2092	0.8270	3.4	29.964	0.0334
0.20	1.2214	0.8187	3.5	33.115	0.0302
0.21	1.2337	0.8106	3.6	36.598	0.0273
0.22	1.2461	0.8025	3.7	40.447	0.0247
0.23	1.2586	0.7945	3.8	44.701	0.0224
0.24	1.2712	0.7866	3.9	49.402	0.0202
0.25	1.2840	0.7788	4.0	54.598	0.0183
0.30	1.3499	0.7408	4.1	60.340	0.0166
0.35	1.4191	0.7047	4.2	66.686	0.0150
0.40	1.4918	0.6703	4.3	73.700	0.0136
0.45	1.5683	0.6376	4.4	81.451	0.0123
0.50	1.6487	0.6065	4.5	90.017	0.0111
0.55	1.7333	0.5769	4.6	99.484	0.0101
0.60	1.8221	0.5488	4.7	109.95	0.0091
0.65	1.9155	0.5220	4.8	121.51	0.0082
0.70	2.0138	0.4966	4.9	134.29	0.0074
0.75	2.1170	0.4724	5.0	148.41	0.0067
0.80	2.2255	0.4493	5.5	244.69	0.0041
0.85	2.3396	0.4274	6.0	403.43	0.0025
0.90	2.4596	0.4066	6.5	665.14	0.0015
0.95	2.5857	0.3867	7.0	1096.6	0.0009
1.0	2.7183	0.3679	7.5	1808.0	0.0006
1.1	3.0042	0.3329	8.0	2981.0	0.0003
1.2	3.3201	0.3012	8.5	4914.8	0.0002
1.3	3.6693	0.2725	9.0	8103.1	0.0001
1.4	4.0552	0.2466	10.0	22026	0.00005

TABLE III
NATURAL LOGARITHMS

n	$\log_e n$	n	$\log_e n$	n	$\log_e n$
	*	4.5	1.5041	9.0	2.1972
0.1	7.6974	4.6	1.5261	9.1	2.2083
0.2	8.3906	4.7	1.5476	9.2	2.2192
0.3	8.7960	4.8	1.5686	9.3	2.2300
0.4	9.0837	4.9	1.5892	9.4	2.2407
0.5	9.3069	5.0	1.6094	9.5	2.2513
0.6	9.4892	5.1	1.6292	9.6	2.2618
0.7	9.6433	5.2	1.6487	9.7	2.2721
0.8	9.7769	5.3	1.6677	9.8	2.2824
0.9	9.8946	5.4	1.6864	9.9	2.2925
1.0	0.0000	5.5	1.7047	10	2.3026
1.1	0.0953	5.6	1.7228	11	2.3979
1.2	0.1823	5.7	1.7405	12	2.4849
1.3	0.2624	5.8	1.7579	13	2.5649
1.4	0.3365	5.9	1.7750	14	2.6391
1.5	0.4055	6.0	1.7918	15	2.7081
1.6	0.4700	6.1	1.8083	16	2.7726
1.7	0.5306	6.2	1.8245	17	2.8332
1.8	0.5878	6.3	1.8405	18	2.8904
1.9	0.6419	6.4	1.8563	19	2.9444
2.0	0.6931	6.5	1.8718	20	2.9957
2.1	0.7419	6.6	1.8871	25	3.2189
2.2	0.7885	6.7	1.9021	30	3.4012
2.3	0.8329	6.8	1.9169	35	3.5553
2.4	0.8755	6.9	1.9315	40	3.6889
2.5	0.9163	7.0	1.9459	45	3.8067
2.6	0.9555	7.1	1.9601	50	3.9120
2.7	0.9933	7.2	1.9741	55	4.0073
2.8	1.0296	7.3	1.9879	60	4.0943
2.9	1.0647	7.4	2.0015	65	4.1744
3.0	1.0986	7.5	2.0149	70	4.2485
3.1	1.1314	7.6	2.0281	75	4.3175
3.2	1.1632	7.7	2.0412	80	4.3820
3.3	1.1939	7.8	2.0541	85	4.4427
3.4	1.2238	7.9	2.0669	90	4.4998
3.5	1.2528	8.0	2.0794	100	4.6052
3.6	1.2809	8.1	2.0919	110	4.7005
3.7	1.3083	8.2	2.1041	120	4.7875
3.8	1.3350	8.3	2.1163	130	4.8676
3.9	1.3610	8.4	2.1282	140	4.9416
4.0	1.3863	8.5	2.1401	150	5.0106
4.1	1.4110	8.6	2.1518	160	5.0752
4.2	1.4351	8.7	2.1633	170	5.1358
4.3	1.4586	8.8	2.1748	180	5.1930
4.4	1.4816	8.9	2.1861	190	5.2470

* Subtract 10 for $n < 1$. Thus, $\log_e 0.1 = 7.6974 - 10 = -2.3026$.

Reference Outline

[1.1] If n is a positive integer and a is a real number, then

$$a^n = \underbrace{a \cdot \cdots \cdot a}_{n \text{ factors}}.$$

The standard form of a polynomial of degree n in x is

$$a_n x^n + a_{n-1} x^{n-1} + \cdots + a_1 x + a_0 \qquad (a_n \neq 0).$$

[1.2] If m and n are positive integers and a and b are any real numbers, then

    I. $a^m \cdot a^n = a^{m+n},$ **II.** $(a^m)^n = a^{mn},$ **III.** $(ab)^n = a^n b^n.$

[1.3] Special products:

$$(x + a)(x + b) = x^2 + (a + b)x + ab$$
$$(x + a)^2 = x^2 + 2ax + a^2$$
$$(x + a)(x - a) = x^2 - a^2$$
$$(ax + b)(cx + d) = acx^2 + (bc + ad)x + bd$$
$$(x + a)^3 = x^3 + 3x^2 a + 3a^2 x + a^3$$
$$(x + a)(x^2 - ax + a^2) = x^3 + a^3$$
$$(x - a)(x^2 + ax + a^2) = x^3 - a^3$$

[1.4] If a and b are nonzero real numbers and m and n are positive integers with $m > n$, then

    I. $\dfrac{a^m}{a^n} = a^{m-n},$ **II.** $\left(\dfrac{a}{b}\right)^m = \dfrac{a^m}{b^m}.$

[2.1] If a is a nonzero real number, then

$$a^0 = 1.$$

If a is a nonzero real number and n is a positive integer, then

$$a^{-n} = \frac{1}{a^n}.$$

If a and b are any nonzero real numbers and m and n are integers, then

    I. $a^m \cdot a^n = a^{m+n}$,   IV. $(ab)^n = a^n b^n$,

    II. $\dfrac{a^m}{a^n} = a^{m-n}$,   V. $\left(\dfrac{a}{b}\right)^n = \dfrac{a^n}{b^n}$.

    III. $(a^m)^n = a^{mn}$,

[2.2] An nth root of a is any number b where $b^n = a$.

If a is a real number and n is a positive integer, then:

    I.   If n is odd, then $a^{1/n}$ is the nth root of a.
    II.   If n is even and $a > 0$, then $a^{1/n}$ is the positive nth root of a.
    III.   If $a = 0$, then $a^{1/n} = 0$.
    IV.   If n is even and $a < 0$, then $a^{1/n}$ is undefined.

If $a^{1/n}$ is a real number and m and n are integers $n > 0$, then

$$a^{m/n} = (a^{1/n})^m.$$

[2.3] If n is a natural number and $a^{1/n}$ is defined, then

$$\sqrt[n]{a} = a^{1/n}.$$

For integers m, n, and c, with $n > 0$, $c > 0$, and for real values of a and b for which all the radical expressions denote real numbers,

    IA. $\sqrt[n]{a^n} = a$ (n an odd positive integer),

    IB. $\sqrt[n]{a^n} = |a|$ (n an even integer),

    II. $\sqrt[n]{a^m} = (\sqrt[n]{a})^m$,

    III. $\sqrt[n]{a} \cdot \sqrt[n]{b} = \sqrt[n]{ab}$,

    IV. $\dfrac{\sqrt[n]{a}}{\sqrt[n]{b}} = \sqrt[n]{\dfrac{a}{b}}$,

    V. $\sqrt[cn]{a^{cm}} = \sqrt[n]{a^m}$.

[2.5] The set of complex members is the set

$$C = \{a + bi \mid a, b \text{ are real numbers}\}.$$

Let $z_1 = a + bi$ and $z_2 = c + di$. Then

$$z_1 = z_2 \qquad \text{if and only if} \qquad a = c \text{ and } b = d,$$
$$z_1 + z_2 = (a + c) + (b + d)i,$$
$$z_1 z_2 = (ac - bd) + (ad + bc)i,$$
$$\bar{z}_1 = a - bi,$$
$$z_1 - z_2 = (a - c) + (b - d)i,$$
$$\frac{z_1}{z_2} = \frac{z_1 \bar{z}_2}{z_2 \bar{z}_2}.$$

[3.1] The following operations are used to generate equivalent equations.

  I.   Any expression may be added to both sides of an equation.
  II. Both sides of an equation may be multiplied by any expression that does not represent zero.

[3.2] The following steps are frequently helpful in constructing models.

  1. Represent the unknown quantities by using word phrases.
  2. Represent each unknown quantity in terms of symbols.
  3. Where applicable, draw a sketch and label all known quantities in terms of symbols.
  4. Find a quantity that can be represented in two different ways and write these representations as an equation (mathematical model). Summarizing given information in a table is sometimes helpful.

[3.3] If $P(x)$ and $Q(x)$ are expressions in x, then $P(x)Q(x) = 0$ if and only if $P(x) = 0$ or $Q(x) = 0$ (or both) and both expressions are defined.

The procedure for completing the square on a quadratic equation in standard form is as follows.

  1. Multiply both sides of the equation by the reciprocal of the leading coefficient.
  2. Add the negative of the constant term to both sides of the equation.
  3. Add the square of one-half times the coefficient of x to both sides of the equation.
  4. Write the left-hand member of the equation as the square of a binomial.

[3.4] ***Quadratic formula*** If a, b, and c are real numbers $(a \neq 0)$, then the solutions to $ax^2 + bx + c = 0$ are

$$x = \frac{-b \pm \sqrt{b^2 - 4ac}}{2a}.$$

[3.6] The solution set of $P(x) = Q(x)$ is a subset of but not necessarily equal to the solution set of $[P(x)]^n = [Q(x)]^n$, for each positive integer n.

[3.7] The following properties are used to generate equivalent inequalities.

    I. The addition of the same expression to each member of an inequality produces an equivalent inequality in the same sense.

    II. If each member of an inequality is multiplied by the same expression representing a positive number, the result is an equivalent inequality.

    III. If each member of an inequality is multiplied by the same expression representing a negative number and the direction of the inequality is reversed, the result is an equivalent inequality.

[3.9]

    I. $|x| = a$ is equivalent to $x = a$ or $x = -a$.

    II. $|x| < a$ is equivalent to $-a < x$ and $x < a$,   i.e., $-a < x < a$.

    III. $|x| > a$ is equivalent to $x < -a$ or $x > a$.

    IV. $|x| \leq a$ is equivalent to $-a \leq x$ and $x \leq a$,   i.e., $-a \leq x \leq a$.

    V. $|x| \geq a$ is equivalent to $x \leq -a$ or $x \geq a$.

[4.1] A relation is a set of ordered pairs.

A function is a relation in which each element of the domain is paired with one and only one element of the range.

If f and g are functions, the composition of f and g is

$$(f \circ g)(x) = f(g(x)).$$

[4.2] The distance between two points $P_1(x_1, y_1)$ and $P_2(x_2, y_2)$ is given by

$$P_1 P_2 = \sqrt{(x_2 - x_1)^2 + (y_2 - y_1)^2}.$$

The slope of the line segment joining the points $P_1(x_1, y_1)$ and $P_2(x_2, y_2)$ is

$$m = \frac{y_2 - y_1}{x_2 - x_1},$$

$$x_2 \neq x_1.$$

[4.3] Forms for linear equations:

$$y - y_1 = m(x - x_1) \qquad \text{Point-slope form}$$

$$y - y_1 = \frac{y_2 - y_1}{x_2 - x_1}(x - x_1) \qquad \text{Two-point form}$$

$$y = mx + b \qquad \text{Slope-intercept form}$$

$$\frac{x}{a} + \frac{y}{b} = 1 \qquad \text{Intercept form}$$

Two nonvertical lines are parallel if and only if they have the same slope.

Two nonvertical lines with slopes m_1 and m_2 are perpendicular if and only if $m_1 \cdot m_2 = -1$.

[4.4] When graphing a quadratic function, $y = ax^2 + bx + c$, use the following steps.

1. Find and graph the intercepts, if possible.
2. Find and graph the axis of symmetry $(x = -b/2a)$ and the vertex.
3. Note whether the parabola opens up $(a > 0)$ or whether it opens down $(a < 0)$.
4. If there are too few points obtained in Steps 1 and 2 to get an accurate picture of the graph, graph a few more selected points that will be helpful.
5. Sketch the curve through the points obtained above.

[4.6] For every one-to-one function f, the function f^{-1} ("f inverse") is defined by

$$y = f^{-1}(x) \qquad \text{if and only if} \qquad x = f(y).$$

The procedure for finding f^{-1} given $y = f(x)$ is:

1. Interchange the symbols x and y in $y = f(x)$ to write $x = f(y)$.
2. Solve $x = f(y)$ for y to obtain

$$y = f^{-1}(x).$$

[5.1] ***Remainder theorem*** If $P(x)$ is a polynomial and c is any real number, then there is a unique polynomial $Q(x)$ such that

$$P(x) = (x - c)Q(x) + P(c).$$

Factor theorem If $P(x)$ is a polynomial with real-number coefficients and $P(c) = 0$, then $(x - c)$ is a factor of $P(x)$.

[5.2] Let $P(x)$ be a polynomial with real coefficients.

    I.   If $r_2 \geq 0$ and if the third row in the synthetic division of $P(x)$ by $x - r_2$ has only nonnegative values, then $P(x) = 0$ has no solutions greater than r_2.

    II. If $r_1 \leq 0$ and if the numbers in the third row of the synthetic division of $P(x)$ by $x - r_1$ alternate in sign (zero being considered $+$ or $-$ as needed), then $P(x) = 0$ has no solution less than r_1.

Descartes' rule of signs   If $P(x)$ is a polynomial with real coefficients, then:

    I.   The number of positive solutions of $P(x) = 0$ is either equal to the number of variations in sign in $P(x)$ or less than that number by an even integer.

    II. The number of negative solutions of $P(x) = 0$ is either equal to the number of variations in sign in $P(-x)$ or less than that number by an even integer.

If $P(z)$ is a polynomial with real-number coefficients and $P(z) = 0$ for some complex number z, then $P(\bar{z}) = 0$.

Every polynomial function of degree $n \geq 1$ over the complex numbers has at least one real or complex zero.

If $P(x)$ is a polynomial of degree $n \geq 1$ over the complex numbers, then $P(x)$ can be expressed as a product of a constant and n linear factors of the form $(x - x_j)$.

[5.3] Let $P(x)$ be a polynomial with real coefficients. If $x_1 < x_2$, and if $P(x_1)$ and $P(x_2)$ have different signs, then there is at least one value c between x_1 and x_2 such that $P(c) = 0$.

To approximate a zero of $P(x)$ to within 10^{-k}, complete the following steps k times, eliminating Step 4 the kth time.

    1. Use Theorem 5.9 to determine consecutive integers a and b with a zero between them.

    2. Subdivide the interval $[a, b]$ into ten equal subintervals.

    3. Use Theorem 5.9 to find one of these subintervals that contains a zero of $P(x)$.

    4. Start over at Step 2, with the endpoints of the subinterval determined in Step 3 taking the place of a and b.

If the rational number p/q, in lowest terms, is a solution of

$$P(x) = a_n x^n + a_{n-1} x^{n-1} + \cdots + a_0 = 0,$$

where each a_j is an integer, then p is an integer factor of a_0 and q is an integer factor of a_n.

[5.4] If $P(x) = a_n x^n + a_{n-1} x^{n-1} + \cdots + a_0$ is a real polynomial function of degree n, then the graph of

$$P(x) = a_n x^n + a_{n-1} x^{n-1} + \cdots + a_0, \qquad a_n \neq 0,$$

is a smooth curve that has at most $n - 1$ turning points.

The first components of the turning points of the graph of

$$P(x) = a_n x^n + a_{n-1} x^{n-1} + \cdots + a_1 x + a_0$$

are solutions of the equation

$$P'(x) = n a_n x^{n-1} + (n-1) a_{n-1} x^{n-2} + \cdots + 2 a_2 x + a_1 = 0.$$

[5.5] The graph of the rational function defined by $y = P(x)/Q(x)$ has the line $x = a$ as a vertical asymptote if $Q(a) = 0$ and $P(a) \neq 0$.

The graph of the rational function defined by

$$y = \frac{a_n x^n + a_{n-1} x^{n-1} + \cdots + a_0}{b_m x^m + b_{m-1} x^{m-1} + \cdots + b_0},$$

where $a_n, b_m \neq 0$ and n, m are nonnegative integers, has:

I. A horizontal asymptote at $y = 0$ if $n < m$,

II. A horizontal asymptote at $y = a_n/b_m$ if $n = m$,

III. No horizontal asymptotes if $n > m$.

[6.2] For any $b > 0$, $b \neq 1$, the domain of $g(x) = \log_b x$ is the set of positive real numbers and the range of g is the set of all real numbers.

For any $b > 0$, $b \neq 1$,

$$\log_b 1 = 0$$

[6.3] Assume x_1 and x_2 are values for which all expressions are defined. Then

I. $\log_b(x_1 x_2) = \log_b x_1 + \log_b x_2$,

II. $\log_b \dfrac{x_2}{x_1} = \log_b x_2 - \log_b x_1$,

III. $\log_b(x_1)^m = m \log_b x_1$.

[6.5] For large values of n,

$$e \approx \left(1 + \frac{1}{n}\right)^n.$$

[7.1] Two systems of equations are equivalent if they have the same solution set.

Any ordered pair that satisfies both the equations

$$f(x, y) = 0 \quad \text{and} \quad g(x, y) = 0$$

will also satisfy the equation

$$a \cdot f(x, y) + b \cdot g(x, y) = 0$$

for all real numbers a and b.

[7.2] If any equation in the system

$$f(x, y, z) = 0, \quad g(x, y, z) = 0, \quad \text{and} \quad h(x, y, z) = 0$$

is replaced by a linear combination, with nonzero coefficients, of itself and any one of the other equations in the system, then the result is an equivalent system.

[8.1] A matrix is a rectangular array of numbers.

The order, or dimension, of a matrix is the ordered pair of integers having as its first component the number of rows and as its second component the number of columns of the matrix.

Two matrices are equal if and only if both matrices are of the same order and the corresponding entries are equal.

The transpose of a matrix A, denoted by A^t, is the matrix in which the rows are the columns of A and the columns are the rows of A.

The sum of two matrices of the same order is the matrix of the same order whose entries are the sums of the corresponding entries of the two given matrices.

The zero matrix is the matrix with each entry equal to 0.

The negative of a matrix is the matrix of the same order with entries that are the negative of the corresponding entries in the given matrix.

If A, B, and C are $m \times n$ matrices with real-number entries, then:

    I. $(A + B)_{m \times n}$ is a matrix with real-number entries.

    II. $(A + B) + C = A + (B + C)$.

    III. The matrix $0_{m \times n}$ has the property that for every matrix $A_{m \times n}$,

$$A + 0 = A \quad \text{and} \quad 0 + A = A.$$

    IV. For every matrix $A_{m \times n}$, the matrix $-A_{m \times n}$ has the property that

$$A + (-A) = 0 \quad \text{and} \quad (-A) + A = 0.$$

    V. $A + B = B + A$.

[8.2] The product of a real number c and an $m \times n$ matrix A with entries a_{ij} is the matrix cA with corresponding entries ca_{ij}, where $i = 1, 2, 3, \ldots, m$ and $j = 1, 2, 3, \ldots, n$.

If A and B are $m \times n$ matrices and c and d are real numbers, then:

I.	cA is an $m \times n$ matrix,	V.	$1A = A$,
II.	$c(dA) = (cd)A$,	VI.	$(-1)A = -A$,
III.	$(c + d)A = cA + dA$,	VII.	$0A = 0$,
IV.	$c(A + B) = cA + cB$,	VIII.	$c0 = 0$.

For $A = [a_1 \cdots a_p]$ and $B = \begin{bmatrix} b_1 \\ \vdots \\ b_p \end{bmatrix}$, the product is

$$AB = a_1 b_1 + a_2 b_2 + \cdots + a_p b_p.$$

The product of matrices $A_{m \times p}$ and $B_{p \times n}$ is the $m \times n$ matrix whose i, jth entry is the product of the ith row of A and the jth column of B.

If A, B, and C are $n \times n$ square matrices, then

$$(AB)C = A(BC).$$

If A, B, and C are $n \times n$ square matrices, then

$$A(B + C) = AB + AC \quad \text{and} \quad (B + C)A = BA + CA.$$

If $I_{n \times n}$ denotes the $n \times n$ matrix with 1's along the principal diagonal and zero for every other entry, then for each matrix $A_{n \times n}$,

$$A_{n \times n} I_{n \times n} = I_{n \times n} A_{n \times n} = A_{n \times n}.$$

Furthermore, $I_{n \times n}$ is the unique matrix having this property for all matrices $A_{n \times n}$.

If A and B are $n \times n$ square matrices and a is a real number, then

$$a(AB) = (aA)B = A(aB).$$

[8.3] An elementary transformation of a matrix A is one of the following three operations upon the rows of the matrix.

1. Multiplication of the entries of any row of A by any nonzero real number k.

2. Interchanging of any two rows of A.

3. Multiplication of the entries of any row of A by a real number k and add to the corresponding entries of any other row.

Two matrices are row-equivalent if one is obtained from the other by a finite sequence of elementary row transformations.

In a linear system of the form

$$a_{11}x + a_{12}y + a_{13}z = c_1$$

$$a_{21}x + a_{22}y + a_{23}z = c_2$$

$$a_{31}x + a_{32}y + a_{33}z = c_3,$$

the matrices

$$\begin{bmatrix} a_{11} & a_{12} & a_{13} \\ a_{21} & a_{22} & a_{23} \\ a_{31} & a_{32} & a_{33} \end{bmatrix} \quad \text{and} \quad \left[\begin{array}{ccc|c} a_{11} & a_{12} & a_{13} & c_1 \\ a_{21} & a_{22} & a_{23} & c_2 \\ a_{31} & a_{32} & a_{33} & c_3 \end{array}\right]$$

are called the coefficient matrix and augmented matrix, respectively.

[8.4] The determinant of the matrix

$$\begin{bmatrix} a_{11} & a_{12} \\ a_{21} & a_{22} \end{bmatrix}$$

is the number $a_{11}a_{22} - a_{12}a_{21}$.

The minor M_{ij} of a_{ij} is the determinant of the $(n - 1) \times (n - 1)$ matrix obtained by deleting the ith row and the jth column of the matrix A.

The cofactor A_{ij} of the entry a_{ij} is

$$A_{ij} = (-1)^{i+j}M_{ij}.$$

The determinant of the square matrix

$$\begin{bmatrix} a_{11} & a_{12} & \cdots & a_{1n} \\ a_{21} & a_{22} & \cdots & a_{2n} \\ \vdots & \vdots & & \vdots \\ a_{n1} & a_{n2} & \cdots & a_{nn} \end{bmatrix}$$

is the sum of the n products formed by multiplying each entry in any single row (or any single column) by its cofactor.

[8.5] If each entry in any row (or any column) of a determinant is 0, then the determinant is equal to 0.

If any two rows (or any two columns) of a determinant are interchanged, then the resulting determinant is the negative of the original determinant.

If two rows (or two columns) in a determinant have corresponding entries that are equal, the determinant is equal to 0.

If each entry in one row (or column) of a determinant is multiplied by k, the determinant is multiplied by k.

If each entry in one row (or column) of a determinant is multiplied by a real number k and the resulting product is added to the corresponding entry in another row (or column, respectively) in the determinant, the resulting determinant is equal to the original determinant.

[8.6] For a given square matrix A of order n, if there is a square matrix A^{-1} of order n such that

$$AA^{-1} = A^{-1}A = I$$

where I is the $n \times n$ identity matrix, then A^{-1} is the multiplicative inverse of A. If

$$A = \begin{bmatrix} a_{11} & a_{12} & \cdots & a_{1n} \\ a_{21} & a_{22} & \cdots & a_{2n} \\ \vdots & \vdots & & \vdots \\ a_{n1} & a_{n2} & \cdots & a_{nn} \end{bmatrix}$$

and if $\delta(A) \neq 0$, then

$$A^{-1} = \frac{1}{\delta(A)} \begin{bmatrix} A_{11} & A_{21} & \cdots & A_{n1} \\ A_{12} & A_{22} & \cdots & A_{n2} \\ \vdots & \vdots & & \vdots \\ A_{1n} & A_{2n} & \cdots & A_{nn} \end{bmatrix}$$

where A_{ij} is the cofactor of a_{ij} in A. If $\delta(A) = 0$, then A has no inverse.

If A is an $n \times n$ nonsingular matrix and if $[A \mid I]$ is the $n \times 2n$ matrix obtained by adjoining the $n \times n$ identity matrix to A, then

$$[A \mid I] \sim [I \mid A^{-1}].$$

[8.8] *Cramer's rule:*

$$x = \frac{\delta(A_x)}{\delta(A)}, \qquad y = \frac{\delta(A_y)}{\delta(A)}, \qquad \text{and} \qquad z = \frac{\delta(A_z)}{\delta(A)}, \qquad \delta(A) \neq 0$$

[9.1] A sequence is a function having as its domain the set of positive integers.

An arithmetic progression is a sequence defined by equations of the form

$$s_1 = a, \qquad s_{n+1} = s_n + d,$$

where n is a positive integer and a and d are real numbers.

The nth term in the arithmetic progression with first term a and common difference d is

$$s_n = a + (n - 1)d.$$

A geometric progression is a sequence defined by equations of the form

$$s_1 = a, \qquad s_{n+1} = rs_n,$$

where n is a positive integer and a and r are nonzero real numbers.

The nth term in the geometric progression with first term a and common ratio $r \neq 0$ is

$$s_n = ar^{n-1}.$$

[9.2] A series is the indicated sum of the terms in a sequence.

The sum of the first n terms in an arithmetic progression is

$$S_n = \frac{n}{2}(a + s_n), \qquad \text{or} \qquad S_n = \frac{n}{2}[2a + (n - 1)d].$$

The sum of the first n terms in a geometric progression is

$$S_n = \frac{a - rs_n}{1 - r}, \qquad \text{or} \qquad S_n = \frac{a - ar^n}{1 - r} \qquad (r \neq 1).$$

[9.3] A sequence $s_1, s_2, \ldots, s_n, \ldots$ converges to the number L,

$$\lim_{n \to \infty} s_n = L,$$

if and only if the absolute value of the difference between the nth term in the sequence and the number L is as small as we please for all sufficiently large n.

An infinite series

$$S_\infty = \sum_{j=1}^{\infty} s_j$$

converges if and only if $S_1, S_2, \ldots, S_n, \ldots$, the corresponding sequence of partial sums, converges.

The sum of an infinite geometric progression, $a + ar + ar^2 + \cdots + ar^n + \cdots$, with $|r| < 1$, is

$$S_\infty = \lim_{n \to \infty} S_n = \frac{a}{1 - r}.$$

[9.4]
$$n! = n(n-1)(n-2)\cdots(3)(2)(1), \qquad n \text{ a positive integer}$$
$$0! = 1$$
$$\binom{n}{r} = \frac{n!}{r!(n-r)!}$$

For each positive integer n,

$$(a+b)^n = a^n + \frac{n}{1!}a^{n-1}b + \frac{n(n-1)}{2!}a^{n-2}b^2 + \frac{n(n-1)(n-2)}{3!}a^{n-3}b^3 + \cdots$$
$$+ \frac{n(n-1)(n-2)\cdots(n-r+2)}{(r-1)!}a^{n-r+1}b^{r-1} + \cdots + b^n,$$

where r is the number of the term. Alternatively,

$$(a+b)^n = \binom{n}{0}a^n + \binom{n}{1}a^{n-1}b + \binom{n}{2}a^{n-2}b^2 + \binom{n}{3}a^{n-3}b^3 + \cdots$$
$$+ \binom{n}{r-1}a^{n-r+1}b^{r-1} + \cdots + \binom{n}{n}b^n.$$

The rth term from the left in a binomial expansion is given by

$$\binom{n}{r-1}a^{n-r+1}b^{r-1} = \frac{n!}{(r-1)!(n-r+1)!}a^{n-r+1}b^{r-1}$$
$$= \frac{n(n-1)(n-2)\cdots(n-r+2)}{(r-1)!}a^{n-r+1}b^{r-1}.$$

[9.5] If a given statement involving n is true for $n = 1$ and if its truth for $n = k$ implies its truth for $n = k + 1$, then it is true for every positive integer n.

[10.1] Counting properties:

    I. $n(A \cup B) = n(A) + n(B)$ if $A \cap B = \emptyset$
    II. $n(A \cup B) = n(A) + n(B) - n(A \cap B)$ if $A \cap B \neq \emptyset$
    III. $n(A \times B) = n(A) \cdot n(B)$
    IV. Suppose the first of several operations can be done in a ways, the second in b ways, no matter what came first, the third in c ways, no matter what came prior, and so on. Then the number of ways the operation can be done in sequence is $a \cdot b \cdot c \cdots$.

A permutation of a set A is an ordering of the members of A.

Let $P_{n,n}$ denote the number of distinct permutations of a set A, where $n(A) = n$. Then

$$P_{n,n} = n!.$$

Let $P_{n,r}$ denote the number of permutations of the members, taken r at a time, of a set A containing n members; that is, let $P_{n,r}$ be the number of distinct orderings of r elements when there is a set A of n elements from which to choose. Then

$$P_{n,r} = n(n-1)(n-2) \cdots \cdots [n-(r-1)]$$
$$= n(n-1)(n-2) \cdots \cdots (n-r+1)$$
$$= \frac{n!}{(n-r)!} \cdot$$

If there are n_1 identical objects of a first kind, n_2 of a second kind, and so on, with $n_1 + n_2 + \cdots + n_k = n$, then the number of distinguishable permutations of the n objects is given by

$$\frac{n!}{n_1! n_2! \cdots \cdots n_k!} \cdot$$

[10.2]   A subset of an n-element set A is called a combination.

Let $C_{n,r}$ denote the number of distinct combinations of the members, taken r at a time, of a set containing n members. Then

$$C_{n,r} = \binom{n}{r} = \frac{P_{n,r}}{r!} = \frac{n!}{r!(n-r)!},$$
$$\binom{n}{r} = \binom{n}{n-r}.$$

[10.3]   The set of all possible results of an experiment is called the sample space for the experiment.

Each element of a sample space is called an outcome, or sample point.

Any subset of a sample space is called an event.

If E is an event with $n(E)$ members of a sample space containing $n(S)$ equally likely outcomes, then the (a priori) probability of the occurrence of the event E is

$$P(E) = \frac{n(E)}{n(S)} \cdot$$

[10.4] $P(E') = 1 - P(E)$

If S is a sample space and E_1 and E_2 are any events in S, then

$$P(E_1 \text{ or } E_2) = P(E_1 \cup E_2) = P(E_1) + P(E_2) - P(E_1 \cap E_2),$$
$$P(E_1 \text{ or } E_2) = P(E_1 \cup E_2) = P(E_1) + P(E_2) \qquad \text{if} \qquad E_1 \cap E_2 = \varnothing.$$

[10.5]   The events E_1 and E_2 are independent if

$$P(E_1 \cap E_2) = P(E_1) \cdot P(E_2).$$

The conditional probability of E_2 given E_1 is

$$P(E_2|E_1) = \frac{P(E_1 \cap E_2)}{P(E_1)}, \qquad P(E_1) \neq 0.$$

If E_1 and E_2 are independent and $P(E_1) \neq 0$, then

$$P(E_2|E_1) = P(E_2).$$

[11.2] A parabola is the set of points in the plane that are equidistant from a given point F and a given line l in the plane.

The standard forms of the equation of a parabola in standard position are as follows.

Equation	Axis of Symmetry	Focus
$x^2 = 4py$	$x = 0$	$(0, p)$
$x^2 = -4py$	$x = 0$	$(0, -p)$
$y^2 = 4px$	$y = 0$	$(p, 0)$
$y^2 = -4px$	$y = 0$	$(-p, 0)$

[11.3] A circle is the set of all points in the plane at a given distance from a fixed point C in the plane.

The standard form of the equation of a circle of radius r that is centered at $(0, 0)$ is

$$x^2 + y^2 = r^2.$$

An ellipse is the set of all points in the plane, the sum of whose distances to two fixed points F_1 and F_2 in the plane is constant.

The standard forms of the equation of an ellipse in standard position that is centered at $(0, 0)$ are:

Equation	Vertices	Foci
$\dfrac{x^2}{a^2} + \dfrac{y^2}{b^2} = 1$	$(\pm a, 0)$	$(\pm c, 0)$
$\dfrac{y^2}{a^2} + \dfrac{x^2}{b^2} = 1$	$(0, \pm a)$	$(0, \pm c)$

where $c^2 = a^2 - b^2$.

[11.4] A hyperbola is the set of all points in the plane such that the absolute value of the difference of the distances to two fixed points F_1 and F_2 is constant.

The standard forms of the equation of a hyperbola in standard position that is centered at $(0, 0)$ are:

Equation	Vertices	Foci	Asymptotes
$\dfrac{x^2}{a^2} - \dfrac{y^2}{b^2} = 1$	$(\pm a, 0)$	$(\pm c, 0)$	$y = \pm \dfrac{b}{a} x$
$\dfrac{y^2}{a^2} - \dfrac{x^2}{b^2} = 1$	$(0, \pm a)$	$(0, \pm c)$	$y = \pm \dfrac{a}{b} x$

where $c^2 = a^2 + b^2$.

[11.6] If the origin in the $x'y'$-coordinate system has xy-coordinates (h, k), then the xy-coordinates and $x'y'$-coordinates of a point are related by the equations

$$x' = x - h \qquad y' = y - k.$$

Answers to Odd-Numbered Exercises

Have you obtained your *Student Guide to College Alegbra, Seventh Edition?* It can help you with this course by acting as:

1. a tutor for those sections of the course with which you have difficulty,
2. an aid in catching up with work covered during an absence from class,
3. a vehicle for reviewing for examinations.

You can get a copy of the *Student Guide to College Algebra, Seventh Edition,* at your college bookstore. If your bookstore doesn't have it in stock, please ask the bookstore manager to order you a copy.

Exercise 1.1 (Page 6)

1. $-3x + 4$ 3. $-2x^2 - 6x - 15$ 5. $-x^3 - x^2 + x - 1$ 7. $-2x^2y^2z^2 + x^2z^2 - x^2y^2 - 4xyz$
9. $x^2y^2 + y^2z^2 - xyz - 4xy + 2xz + yz$ 11. $2x^2 + 10x$ 13. $-2x^2 - 6x - 12$
15. $-x^3 + x^2 + 3x - 1$ 17. $-5x^3 + 3x^2 + x + 1$ 19. $x + 5$ 21. $5x - 8$ 23. $-2x + 1$
25. $3x^2 - 4x$ 27. $x^3 - x^2 - 4x + 5$ 29. $x^3 - x^2 + 2x - 1$ 31. $3x^2 + x + 2$
33. $-2x^2 + 6x - 4$ 35. $-x^2 + 3x - 4$ 37. $2x^2 - x - 1$ 39. $2x^2 - x - 1$ 41. $-6x^2 + 8x - 6$
43. $-1, -1, 1, 0$ 45. $0, 2, -10, 2$ 47. $-1, 0, 1, 2$ 49. $11, 1$ 51. $9, 1$ 53. 5 55. -8
57. -9 59. $n, n, n - 2$, no degree

Exercise 1.2 (Page 11)

1. $-2x^3y^5$ 3. $-6x^7y^8$ 5. $-24x^6y^6$ 7. $3x^{2n+6}$ 9. $-2x^{2n+2}y^{2n+1}$ 11. $x^{4n}y^{5n}$
13. $10x + 15$ 15. $2x^2 + 6x$ 17. $x^5 - x^4 - x^3$ 19. $x^5 - x^4 + x^3$ 21. $x^2 + x - 6$
23. $3x^2 + 5x - 2$ 25. $6x^2 - 9x - 6$ 27. $-x^2 + 1$ 29. $-4x^2 - 4x + 3$ 31. $2x^3 + 5x^2 + 5x + 6$
33. $2x^3 - 3x^2 + 1$ 35. $x^4 + 2x^3 + 2x^2 + x$ 37. $x^4 + 2x^2 + x + 2$ 39. $x^3 + 3x^2 + 2x$
41. $x^3 + 8$ 43. $x^3 + x^2 + 2x + 2$ 45. $-2x^5 + 5x^4 + 7x^3 - 23x^2 + 4x + 12$
47. $-3x^4 - 7x^3 + 12x^2 + 20x - 16$ 49. $2c^2 + c + 2, 2c^2 + 4ch + 2h^2 + c + h + 2, 4ch + 2h^2 + h$
51. $x^2 + 4x + 3, x^4 - 1, x^4 - 2x^2 + 1$ 53. $x^4 + 2x^3 - x^2 - 2x$ 55. $x^4 + 4x^3 - 8x + 3$
57. $m + n$ 59. $2n, kn$

A–1

Exercise 1.3 (Page 18)

1. $(x + 1)(x + 2)$ **3.** $(x - 4)(x - 2)$ **5.** $(x + 5)(x - 1)$ **7.** $(x - 4)(x + 2)$ **9.** $(x + 1)^2$

11. $(x - 2)^2$ **13.** $(2x + 3)^2$ **15.** $(x - 1)(x + 1)$ **17.** $(2x - 3)(2x + 3)$ **19.** $(5 - 2x)(5 + 2x)$

21. $(3x + 1)(x + 1)$ **23.** $(3x - 2)(x - 1)$ **25.** $(3x + 2)(2x + 3)$ **27.** $(2x - 3)(3x + 4)$

29. $(x + 1)^3$ **31.** $(x - 1)^3$ **33.** $(2x - 1)^3$ **35.** $(x - 2)(x^2 + 2x + 4)$ **37.** $(2x - 1)(4x^2 + 2x + 1)$

39. $(3 - 2x)(9 + 6x + 4x^2)$ **41.** $(x + 3)(x^2 - 3x + 9)$ **43.** $(3x + 1)(9x^2 - 3x + 1)$

45. $(5x + 3)(25x^2 - 15x + 9)$ **47.** $(x - y)(x - 3)$ **49.** $(x - 1)(x + 1)(y + 2)$

51. $(x - 2)(x + 2)(x + 3)$ **53.** $x^2(2x - 1)$ **55.** $2x^2(4x^4 + 1)$ **57.** $2x^2y(xy - 4 - 16y)$

59. $(x - 4)(x + 4)$ **61.** $3x^2(y - 3)(y + 3)$ **63.** $2(x - 3)^2$ **65.** $4y^3(x + 1)^2$

67. $x(xyz + 10yz + 25)$ **69.** $(x + 6)(x - 3)$ **71.** $(x + 3)(x + 5)$ **73.** $(2x + 1)(x + 1)$

75. $(3x + 1)(x + 2)$ **77.** $(2z + 1)(z - 2)$ **79.** $(3y - 1)(2y - 3)$ **81.** $(8a + 1)(2a + 1)$

83. $(4x + 2)(3x - 1)$ **85.** $3(x - 1)(x + 5)$ **87.** $a(b + 7)(b - 2)$ **89.** $(b + 3)(a - 1)$

91. $2(x - 2)(y + 3)$ **93.** $(x - y^2)(x^2 - 2y)$ **95.** $(x - 2y + 5)(x + 2y + 5)$

97. $(y - 2x + 1)(y + 2x - 1)$ **99.** $(3x + 1 - 2y)(3x + 1 + 2y)$ **101.** $(2x + 1)^3$ **103.** $(2x - 3)^3$

105. $(5x - 2)^3$ **107.** $(3 - x)(9 + 3x + x^2)$ **109.** $8(2x + y)(4x^2 - 2xy + y^2)$

111. $(1 + (x - 1))(1 - (x - 1) + (x - 1)^2)$ **113.** $((x^3 + 8) + 1)((x^3 + 8)^2 - (x^3 + 8) + 1)$

115. $(x^2 + y)(x + 3)$ **117.** $(x + 6)(x + 5)$ **119.** $(x^2 + 1)(y^2 + 3)$ **121.** $x(x^2 + 1)(x - 1)(x + 1)$

123. $(x^2 - xy + y^2)(x^2 + xy + y^2)$ **125.** $(x^2 - x + 1)(x^2 + x + 1)$ **127.** $(3x^2 - 2x + 1)(3x^2 + 2x + 1)$

Exercise 1.4 (Page 23)

1. $6y^3z^2$ **3.** $32x^3y$ **5.** $2x^2 + 4x$ **7.** $3x - 2y + 3xy$ **9.** $x - 2 + \dfrac{4}{x}$ **11.** $4x + \dfrac{-2x + 6}{x^2}$

13. $2x^3 + x + \dfrac{x - 4}{2x^2}$ **15.** $x + y + \dfrac{1}{x}$ **17.** $xy^2 + 1 + \dfrac{x + y}{x^2y}$ **19.** $3y + \dfrac{y^2 + y - x}{xy^2}$ **21.** $x + 1$

23. $2x^2 + 3x + \dfrac{7}{2} + \dfrac{\frac{33}{2}}{2x - 3}$ **25.** $x^4 + 4x^3 + 18x^2 + 72x + 289 + \dfrac{1152}{x - 4}$ **27.** $x^2 + 3 + \dfrac{4x - 4}{x^3 - x}$

29. $x^4 - x^3 + x^2 - x + 1$ **31.** $2x^2 + \dfrac{3}{2}x - \dfrac{\frac{1}{2}x}{2x^2 + 1}$ **33.** $x + \dfrac{x - 1}{x^3 + x - 1}$ **35.** $x^2 - 1 + \dfrac{x^2 + x + 1}{x^3 + x}$

37. $x^2 - 2x + 3 + \dfrac{-4x - 3}{x^2 + 2x + 1}$ **39.** $x^2 - \dfrac{1}{3}x + \dfrac{1}{9} + \dfrac{-\frac{1}{9}}{3x + 1}$ **41.** $x^2 + 1 + \dfrac{4}{x^2 + 1}$

43. $3(x^2 + 9) - \dfrac{x}{(x^2 + 9)^2}$ **45.** $1 + \dfrac{-(x + 1)}{(x + 1)^2 - 1}$ **47.** $\dfrac{1}{2}(x^2 + 1) + \dfrac{3}{4} + \dfrac{-\frac{7}{4}}{2(x^2 + 1) - 1}$

Exercise 1.5 (Page 30)

1. $4xy^3$ **3.** $x + 1$ **5.** $2x$ **7.** $a - b$ **9.** $b^2(1 - ab)$ **11.** $x - 1$ **13.** $\dfrac{x - 4}{x - 1}$ **15.** $\dfrac{1}{x + 3}$

17. $\dfrac{x^2 - 1}{2}$ **19.** $-(a^2 + x^2)$ **21.** $\dfrac{x^2 - 7}{x}$ **23.** $\dfrac{(1 + a^2)(1 - a)}{(1 + a)^2}$ **25.** $x^4 + x^2 + 1$ **27.** $\dfrac{y^2 + 3}{x - y}$

29. $\dfrac{x^2 - xy + y^2}{x - y}$ **31.** 12 **33.** a^3c **35.** x^2 **37.** $3(x + y)$ **39.** 1890 **41.** $x(x - 1)(x + 1)$

43. $x^3(x - 1)(x + 4)$ **45.** $(x - 1)(x + 1)(x^2 + 1)$ **47.** $\dfrac{x^2 - x + 1}{x^2(x - 1)}$ **49.** $\dfrac{y^2 - 3y - 2}{4y^2}$ **51.** $\dfrac{2x}{x^2 - 1}$

53. $\dfrac{x^2 - 10x - 3}{(x + 2)(x - 1)}$ **55.** $\dfrac{-5}{x + 3}$ **57.** $\dfrac{x - 1}{x - 2}$ **59.** $\dfrac{-w^2 + 3w + 1}{(w - 1)(w + 1)(w - 2)}$ **61.** $\dfrac{x^2 - 7}{(x - 3)^2(x - 2)}$

63. $\dfrac{-7x + 9}{(x + 1)(x - 3)^2}$ **65.** $\dfrac{x^2 + 2x - 2}{x^2 - 1}$ **67.** $\dfrac{z^3 - z - 1}{z^2}$ **69.** $\dfrac{x^2 - 2}{x^2 - 1}$ **71.** $\dfrac{4}{(x + 1)(x + 3)}$

73. $\dfrac{x^2 + a^2}{x^2 - a^2}$ **75.** $\dfrac{8(3x - 4)}{x^3}$ **77.** $\dfrac{(x + 1)(x - 1)}{4}$ **79.** $\dfrac{(4x - 1)(2x + 1)}{x - 1}$ **81.** $\dfrac{x^3 + 3x^2 + 2x + 2}{(x^3 - 1)(x^2 + 1)}$

83. $\dfrac{x^3 - 2x^2 + 2x}{(x - 1)(x^2 + x + 1)(x^2 - x + 1)}$ **85.** $\dfrac{2(y - x)}{(x + y)(x - 2)(y - 2)}$ **87.** $\dfrac{y^2 + xy + y}{(x^2 + y^2)(x + y)(x - y)}$

89. $\dfrac{3(x^2 - y^2)}{(2x - 3z)(x - 2y)(2x - y)}$

Exercise 1.6 (Page 36)

1. $\dfrac{a^2y^2z^2b}{x^4}$ **3.** $\dfrac{xac^2}{yzb}$ **5.** $\dfrac{20zb}{9y}$ **7.** $\dfrac{y^2a^2bc^2}{x^2z}$ **9.** $\dfrac{x}{x + 1}$ **11.** $(x - 1)(x + 1)$ **13.** $\dfrac{x}{2(x + 2)^2}$

15. $\dfrac{(x + 1)(x - 2)}{(x + 2)(x - 1)}$ **17.** $\dfrac{b^4}{a^2c}$ **19.** $\dfrac{(x - y)y^2}{x^2}$ **21.** $\dfrac{a^2b^2}{a - b}$ **23.** $\dfrac{z^2 - 1}{z^2}$ **25.** $\dfrac{1}{x + 5}$ **27.** $\dfrac{x - 2}{x + 3}$

29. $\dfrac{x^5 - 13x^3 + 5x^2 + 36x}{(x^2 - 9)(x - 4)}$ **31.** $\dfrac{a^3}{b^3}$ **33.** $\dfrac{x^5}{y^5}$ **35.** $\dfrac{xy}{ab}$ **37.** $\dfrac{1}{(x + 1)(y - 3)^2}$ **39.** $\dfrac{(x - 1)^2}{(x + 1)(x - 2)}$

41. $\dfrac{x}{x - 1}$ **43.** $\dfrac{2x + 2}{2x + 1}$ **45.** $\dfrac{2x}{2x + 1}$ **47.** $\dfrac{x^2 + 2x}{x^2 - 1}$ **49.** $\dfrac{a^2 + 8a + 27}{a^2 + 10a + 41}$ **51.** $\dfrac{x - 3}{x + 3}$ **53.** $\dfrac{2y - x}{4y - x}$

55. $\dfrac{xy^2 - yx^2}{x^2 + 2y^2}$ **57.** $\dfrac{-2x^3}{(x^2 - 1)(x^2 + 1)}$ **59.** $\dfrac{x(2x - 1)(2x + 1)}{-(x + 2)(x - 2)}$ **61.** $\frac{22}{7}$, 3.1428 **63.** $\frac{355}{113}$, 3.1415929

65. $\dfrac{17x + 7}{12x + 5}$ **67.** $\dfrac{201x + 38}{72x + 17}$

Chapter 1 Review (Page 39)

1. $2x^2 - 2x - 1$ **2.** $6x^2$ **3.** 13 **4.** -3 **5.** $6x^2 + 10x + 4$ **6.** $3x^3 + 7x^2 + 3x + 2$

7. $(y - 5)(y - 3)$ **8.** $x(x + 2)^2$ **9.** $(x^2 + 4)(x - 2)(x + 2)$ **10.** $(z - 4)(z^2 + 4z + 16)$ **11.** $6xy^2z$

12. $6y^2 + 7y + 1$ **13.** $2x - 5$ **14.** $3x + 2$ **15.** $x + \dfrac{3}{x - 1}$ **16.** $x^2 + \dfrac{1}{x - 2}$ **17.** $2y + 11z$

18. $x + 7$ **19.** $\dfrac{x + 7}{x + 3}$ **20.** $\dfrac{x + 2}{x - 2}$ **21.** $\dfrac{7x + 3}{12}$ **22.** $\dfrac{6x}{x^2 + x - 2}$ **23.** $\dfrac{x^2 - 2x + 1}{x + 1}$

24. $\dfrac{y^2 - 4y + 8}{y^2 - y - 2}$ **25.** $\dfrac{x^4z}{y^2}$ **26.** $\dfrac{x^2 + 8x + 7}{x + 5}$ **27.** $\dfrac{x^2 + x - 2}{x + 3}$ **28.** $\dfrac{x - 3}{x + 1}$

29. $(x - y)(x + y)(x^2 - xy + y^2)(x^2 + xy + y^2)$ **30.** $(x^2 - 2)(x - 1)(x + 1)$

31. $(x - 1)(x + 1)(x^2 + 1)(x^4 + 1)(x^8 + 1)$ **32.** $(x - 1)(x^2 + x + 1)(x^6 + x^3 + 1)(x^{18} + x^9 + 1)$

33. $\dfrac{2x^3 - 3x^2 - x - 1}{(x - 1)(x + 1)^2(x - 4)}$ **34.** $\dfrac{-x^7 - x^6 - x^5 - x^4 - x^3}{(x^4 + 1)(x^2 + 1)(x + 1)(x^2 + x + 1)(x - 1)}$

35. $\dfrac{(x - 1)^3(x + 1)(x^2 + 2x + 4)}{-2(x - 2)}$ **36.** $\dfrac{x(x^4 - 8x^2 + 16)}{x^2 + 4}$ **37.** $\frac{127}{24}$, 5.2917 **38.** $\dfrac{5x + 3}{3x + 2}$

Exercise 2.1 (Page 48)

1. 1 **3.** $\frac{1}{3}$ **5.** $-\frac{1}{64}$ **7.** 36 **9.** $\frac{3}{4}$ **11.** $\frac{27}{8}$ **13.** $\frac{1}{x^3}$ **15.** $\frac{1}{y^2x^4}$ **17.** x^6y^3 **19.** $\frac{x^{10}}{32y^5}$

21. $\frac{4x}{y^2}$ **23.** $\frac{z^7}{x^5y^2}$ **25.** y^2 **27.** $\frac{x^9}{y^{17}z^{10}}$ **29.** x^5y^3 **31.** $\frac{1}{x^9y^4}$ **33.** $\frac{y^{13}z^9}{x^5}$ **35.** $1-y$

37. $xy^{-1}-y^{-1}$ **39.** $1+x^{-1}y^3$ **41.** $2+x^{-2}y^2+x^2y^{-2}$ **43.** y^3-3x^4 **45.** $x-3y$

47. y^3z^6+x **49.** $\frac{1-xy}{2y}$ **51.** $\frac{2}{y^2-x^2}$ **53.** $\frac{xy}{x+y}$ **55.** 1.23456×10^5 **57.** -1.945×10^3

59. 4.321×10^{-2} **61.** -1.0010001×10^{-2} **63.** $428{,}000{,}000$ **65.** $-5{,}642{,}000$ **67.** 0.000563

69. -0.0000000455 **71.** 8.6×10^{-1} **73.** 6.7×10^{11} **75.** 10^7 **77.** 1.99 **79.** 7.28×10^2

81. 5.38×10^{-1} **83.** a. 6.70×10^7 b. 1.61×10^{10} c. 1.13×10^{11} d. 5.84×10^{12}

85. 6.45×10^{-4} sec **87.** 3.16×10^{21} lb **89.** 8×10^7 **91.** x^{2n-3} **93.** y^2 **95.** y^{5n+1}

97. x^{2n-2} **99.** $x^{n+1}y^{n+3}$ **101.** $\frac{x^{n-1}}{y}$

Exercise 2.2 (Page 56)

1. 4 **3.** $\frac{1}{27}$ **5.** -9 **7.** 4 **9.** $\frac{1}{4}$ **11.** $\frac{27}{8}$ **13.** $x^{11/6}$ **15.** $y^{10/3}$ **17.** $y^{2/9}$ **19.** $64x^{9/4}$

21. $x^{10}y^2$ **23.** $\frac{25x^6y^2}{9}$ **25.** $\frac{1}{x^{9/2}y^{16/3}}$ **27.** $\frac{x^{8/3}}{y^{29/6}}$ **29.** $\frac{1}{y}+\frac{2}{x^{1/2}y^{1/2}}+\frac{1}{x}$ **31.** $x-y$

33. $x^{3/2}-x^2$ **35.** $x^2+x^{3/2}$ **37.** $y^{-1/3}-1$ **39.** $x-y$ **41.** $(x+y)-(x+y)^{3/2}$ **43.** $x^{4/3}+x$

45. $x^{1/2}-1$ **47.** $1-x^2$ **49.** $1+(x+1)^{-1}$ **51.** 1.59 **53.** 0.48 **55.** 0.40 **57.** 0.83

59. $x^{1/n^2}yz^n$ **61.** $x^{3n/2}y^{3n+3}$ **63.** a^{5n-3} **65.** 9 **67.** 8 **69.** $3|x|$ **71.** $\frac{1}{x^2(1-x)^{1/2}}$

73. 1.41×10^4 **75.** 5.09×10^{-7} **77.** 4.72×10^5 **79.** -1.53×10^3

Exercise 2.3 (Page 61)

1. $27^{1/3}=3^1$ **3.** $8^{2/3}=4^1$ **5.** $16^{1/2}=4^1$ **7.** $2x^{4/3}$ **9.** $x^{1/2}y^{3/4}$ **11.** $(x+y)^{3/2}$

13. $(x^2+1)^{1/3}$ **15.** $(x^3+y^3)^{1/3}$ **17.** $(x^2+1)^{3/2}$ **19.** $[-(x^2-1)]^{5/4}$ **21.** $\sqrt{5}$ **23.** $\sqrt[3]{x^3}$

25. $2\sqrt[3]{x}$ **27.** $3\sqrt{y}$ **29.** $x\sqrt[3]{y}$ **31.** $\sqrt[3]{xy}$ **33.** $\sqrt[4]{x-y}$ **35.** $\sqrt{x^2+y^2}$ **37.** $\sqrt[3]{(x^3+y^3)^4}$

39. $\frac{1}{\sqrt[4]{(x^2-y^2)^3}}$ **41.** $x^3\sqrt{x}$ **43.** $-3x\sqrt[3]{x}$ **45.** $x\sqrt{y}$ **47.** $4x^4$ **49.** x **51.** $2x\sqrt[3]{y}$ **53.** $\sqrt[3]{9}$

55. $2\sqrt{x}$ **57.** $\sqrt{2x}$ **59.** $(x-1)^2$ **61.** $\frac{\sqrt{2}}{2}$ **63.** $\frac{2\sqrt{2x}}{x}$ **65.** $\frac{\sqrt[3]{2}}{2}$ **67.** $\frac{\sqrt[3]{y}}{y}$ **69.** $\frac{\sqrt{x^2+y^2}}{x^2+y^2}$

71. $\frac{1}{\sqrt{2}}$ **73.** $\frac{x}{y\sqrt[3]{x^2}}$ **75.** $\frac{y}{x\sqrt[3]{x^2y^2}}$ **77.** $\frac{x^2+y^2}{\sqrt{x^2+y^2}}$ **79.** 5.44 cps **81.** 16.33 cps

83. $36{,}732.6$ ft/sec **85.** 258.6 days **87.** 31 years **89.** $\frac{\sqrt[4]{(x-1)^2(x+1)^3}}{x+1}$ **91.** $\frac{\sqrt[3]{x^2(y+1)^2}}{x^2}$

93. No, because if $\sqrt{a^2+b^2}=a+b$ then $a^2+b^2=(a+b)^2$, or $2ab=0$. This is not true for every $a\geq0$, $b\geq0$.

Exercise 2.4 (Page 65)

1. $8\sqrt{5}$ 3. $\sqrt{3}$ 5. $-2\sqrt{3}$ 7. $15\sqrt{2}$ 9. $-3\sqrt[3]{2}$ 11. $\sqrt[4]{2}$ 13. $6 - 3\sqrt{6}$ 15. 4

17. $-5 - \sqrt{7}$ 19. $-\sqrt{2}\,x$ 21. $2x + 5 - 7\sqrt{x}$ 23. $5 - x$ 25. $\dfrac{1 - \sqrt{7}}{6}$ 27. $\dfrac{4x - x\sqrt{x}}{16 - x}$

29. $\dfrac{x\sqrt{2} - 2\sqrt{2}}{x - 4}$ 31. $\dfrac{(x + 1)(\sqrt{x} - 1)}{x - 1}$ 33. $\dfrac{(2x - 1)(\sqrt{x} + 1)}{x - 1}$ 35. $\dfrac{(x - 2)(8 + 2\sqrt{x})}{16 - x}$ 37. $\dfrac{2}{\sqrt{5} + 1}$

39. $\dfrac{1 - 4x}{x^2(1 - 2\sqrt{x})}$ 41. $\dfrac{3 - 2x}{2x + 1 + 2\sqrt{2x + 1}}$ 43. $\dfrac{x - 1}{(x + 1)(\sqrt{x} - 1)}$ 45. $\dfrac{2x - 1}{(3x - 1)(\sqrt{2x} + 1)}$

47. $\dfrac{9 - x}{(3x + 1)(9 + 3\sqrt{x})}$ 49. $\dfrac{x + y - 2\sqrt{xy}}{x - y}$ 51. $2x + 1 - 2\sqrt{x^2 + x}$

53. $\dfrac{(x\sqrt{x} - x\sqrt{y} + x)(x - y - 1 - 2\sqrt{y})}{(x - y - 1)^2 - 4y}$ 55. $\dfrac{x - 2}{x - 1 + \sqrt{x - 1}}$ 57. $\dfrac{x - y}{x + y - 2\sqrt{xy}}$

59. $\dfrac{(x - y - 1)^2 - 4y}{(x\sqrt{x} - x\sqrt{y} + x)(x - y - 1 - 2\sqrt{y})}$

Exercise 2.5 (Page 71)

1. $4i$ 3. $3\sqrt{3}i$ 5. $12\sqrt{2}i$ 7. $6\sqrt{7}i$ 9. $3 - 2i$ 11. $1 + 25\sqrt{2}i$ 13. $3 + 3i$ 15. $5 + i$

17. $1 - i$ 19. $3 - 2i$ 21. $2 - 16i$ 23. $7 + 26i$ 25. $-1 - 18i$ 27. $-21 + 20i$ 29. 17

31. $-4 + 7i$ 33. -15 35. $5 + 5i$ 37. $-\frac{1}{8}i$ 39. $\frac{1}{4} - \frac{3}{4}i$ 41. $\frac{3}{2} - \frac{3}{2}i$ 43. $1 - i$

45. $\frac{12}{13} - \frac{5}{13}i$ 47. $\frac{9}{13} + \frac{9}{13}i$ 49. $\frac{1}{5} + \frac{2}{5}i$ 51. $\frac{18}{5} - \frac{1}{5}i$ 53. $\frac{-7}{5} + \frac{4}{5}i$ 55. $4 + 2i$ 57. $15 + 3i$

59. $-\frac{3}{2}i$ 61. $\frac{3}{5} - \frac{4}{5}i$ 63. -1 65. $2i\sqrt{5}$ 67. a. -1 b. 1 c. $-i$ d. -1 69. $5 + 4i$

71. Let $z_1 = a_1 + b_1 i$ and $z_2 = a_2 + b_2 i$. Then

$$\overline{z_1 \pm z_2} = \overline{(a_1 \pm a_2) + (b_1 \pm b_2)i}$$
$$= (a_1 \pm a_2) - (b_1 \pm b_2)i$$
$$= (a_1 - b_1 i) \pm (a_2 - b_2 i)$$
$$= \overline{z_1} \pm \overline{z_2}.$$

73. Repeated applications of Exercise 72 with $z = z_1 = z_2$ yields this result.

75. Let $z = a + bi$, then $\bar{z} = a - bi$ and

$$z + \bar{z} = (a + bi) + (a - bi)$$
$$= 2a + 0i$$
$$= 2a.$$

Chapter 2 Review (Page 72)

1. $x^{12}y^8$ 2. $\dfrac{y^3}{x^3}$ 3. $\dfrac{y}{x^2}$ 4. x^3 5. $\dfrac{x^2 + 1}{x^3}$ 6. $xy^2 + x^2$ 7. 3.51×10^4 8. 1.8×10^{-4}

9. $314{,}000{,}000$ 10. 0.00000675 11. $x^{7/6}$ 12. $\dfrac{y}{x^{1/2}}$ 13. $x^4 y^8$ 14. $\dfrac{1}{xy}$ 15. $y^{4/3} + y$

16. $4y^2 - y$ **17.** $1 + x^{3/4}$ **18.** $x^{4/3} - 1$ **19.** $2x^2y\sqrt{xy}$ **20.** $6x\sqrt{y}$ **21.** $x\sqrt{3}$ **22.** $3x$

23. $\sqrt{5}$ **24.** $\sqrt{3xy}$ **25.** $\dfrac{3 + \sqrt{x}}{9 - x}$ **26.** $\dfrac{2x + 1 + 3\sqrt{x}}{4x - 1}$ **27.** $\dfrac{\sqrt[3]{x^2}}{x}$ **28.** $\dfrac{2\sqrt[3]{x} - 1}{x - 1}$

29. $\dfrac{x - 1}{2(\sqrt{x} - 1)}$ **30.** $\dfrac{x - 1}{x + 2 + 3\sqrt{x}}$ **31.** $\dfrac{x}{(x + 1)\sqrt[3]{x^2}}$ **32.** $\dfrac{x + 1}{x\sqrt[3]{x} + 1}$ **33.** $3 - 3i$ **34.** $-9i\sqrt{2}$

35. $5 + i$ **36.** $-2 - 3i$ **37.** 5 **38.** $\frac{2}{5} + \frac{4}{5}i$ **39.** $-6 + 3i$ **40.** $-\frac{3}{4} - \frac{1}{2}i$ **41.** 4×10^8

42. 5×10^6 **43.** $\dfrac{\sqrt{x + y} - 1}{x + y - 1}$ **44.** $\dfrac{\sqrt{x + y} + 1}{1 - x - y}$ **45.** $\dfrac{1}{\sqrt{x + h} + \sqrt{x}}$ **46.** $\dfrac{-1}{\sqrt{x - h} + \sqrt{x}}$

47. 16 **48.** -8 **49.** i **50.** -1

Exercise 3.1 (Page 81)

1. $\{\frac{3}{2}\}$ **3.** $\{\frac{1}{2}\}$ **5.** $\{-1\}$ **7.** $\{-11\}$ **9.** $\{-5\}$ **11.** $\{3\}$ **13.** $\{\frac{5}{4}\}$ **15.** $\{-3\}$ **17.** $\{6\}$

19. $\{-\frac{1}{3}\}$ **21.** $\{\frac{1}{2}\}$ **23.** The empty set **25.** Any x; $x \neq 1, -1$ **27.** $\{0\}$ **29.** $r = \dfrac{S}{2\pi h}$; $h \neq 0$

31. $k = v - gt$ **33.** $c = \dfrac{2A - bh}{h}$; $h \neq 0$ **35.** $n = \dfrac{l - a + d}{d}$; $d \neq 0$ **37.** $r^2 = \dfrac{\pi R^2 h - V}{\pi h}$; $h \neq 0$

39. $y' = -\dfrac{x}{y}$; $y \neq 0$ **41.** $y' = \dfrac{1}{x}$; $x \neq 0, y \neq 0$ **43.** $y' = \dfrac{1 + 3x}{x^2 - 2y^3}$; $x^2 \neq 2y^3$

45. $x_1 = \dfrac{x_4}{x_2 - 2x_3}$; $x_2 \neq 2x_3$ **47.** $37°C$ **49.** $62.6°F$ **51.** 153 ft/sec **53.** 464 ft

55. $21,000$ rpm **57.** 2.5 in. **59.** 32-g weight is 6 cm from fulcrum, 24-g weight is 8 cm from fulcrum

61. 5.5 feet from the man **63.** $120 **65.** 12 amperes **67.** 120 ohms **69.** 30 years

71. 750 kilograms **73.** $\frac{3}{4}$ **75.** $\frac{1}{2}$ **77.** $\frac{1}{3}$ **79.** -1

Exercise 3.2 (Page 90)

1. a. $c + (c - 15) = 75$, where c is the cost of a Touch-Tone phone
  b. $45 for Touch-Tone, $30 for rotary

3. a. $10c + 25(c + 24) = 1510$, where c is the cost of a business calculator
  b. $26 for business model, $50 for scientific model

5. a. $w + (w - 122) = 584$, where w is the number of votes for the winner
  b. 353 votes for the winner, 231 votes for the loser

7. a. $600 = \frac{3}{4}n$, where n is the number of applicants **b.** 800 applicants

9. a. $80F + 64(42 - F) = 2880$, where F is the number of first-class passengers
  b. 12 in first class, 30 in tourist

11. a. $0.10(1 - 0.28)g = 45$, where g is the gross earnings **b.** $625

13. a. $0.10b + (8000 - b)(0.12) = 844$, where b is the amount in the bank
  b. $5800 in the bank, $2200 in the savings and loan

15. a. $3000(0.08) + 0.13s = (3000 + s)(0.10)$, where s is the amount in stocks **b.** $2000

17. a. $x + 2x + (x + 2x + 12°) = 180°$, where x is the measure of the smallest angle **b.** $28°, 56°, 96°$

19. a. $(x - 70) + (x - 50) + x = 180$, where x is the measure of the largest angle **b.** $100°, 30°, 50°$

21. a. $(2x + 3) + (2x + 3) + x = 86$, where x is the length of the base **b.** 16 cm, 35 cm, 35 cm

23. a. $(x-2)^2 = x^2 - 20$, where x is the length of the side of the original square   b.   6 cm

25. a. $0.32x + 0.48(25) = (x + 25)(0.42)$, where x is the amount of 32% silver alloy   b.   15 pounds

27. a. $\frac{22}{24}x + 1(\frac{14}{24}) = (x + 1)\frac{18}{24}$, where x is the amount of 22K gold   b.   1 ounce

29. a. $50(0.50) = (x + 50)(0.12)$, where x is the amount of water added   b. $158\frac{1}{3}$ gallons

31. a. $(20 - x)(0.16) = 20(0.12)$, where x is the amount of water added   b.   5 gallons

33. a. $\dfrac{t}{30} + \dfrac{t}{45} = 1$, where t is time in hours   b.   18 hours

35. a. $6\left(\dfrac{1}{10} + \dfrac{1}{t}\right) = 1$, where t is time in hours   b.   15 hours

37. a. $10 - x = \dfrac{10}{7}(6)$, where x is the distance from Jack to Georgia at the finish   b.   1.43 kilometers

39. a. $\dfrac{r}{1260} = \dfrac{r - 120}{420}$, where r is the rate of the airplane   b.   180 mph for the airplane, 60 mph for the car

41. a. $20t + 5 = 30t$, where t is the time (in hours) elapsed after the second ship leaves   b. $\frac{1}{2}$ hour

Exercise 3.3 (Page 97)

1. $\{-1, 2\}$ 3. $\{-8, 0\}$ 5. $\{1\}$ 7. $\{-\frac{1}{2}, -\frac{1}{3}\}$ 9. $\{-1, \frac{3}{2}\}$ 11. $\{-\frac{4}{3}, -\frac{1}{2}\}$ 13. $\{-6, \frac{1}{2}\}$

15. $\{-\frac{5}{2}, -2\}$ 17. $\{-6, -1\}$ 19. $\{-4, 4\}$ 21. $\{-3, 3\}$ 23. $\{-3i, 3i\}$ 25. $\{-1, 5\}$

27. $\{-\frac{3}{2}, \frac{1}{2}\}$ 29. $\{2 + i, 2 - i\}$ 31. $\left\{\dfrac{1}{2} - \dfrac{\sqrt{3}}{2}, \dfrac{1}{2} + \dfrac{\sqrt{3}}{2}\right\}$ 33. $\{\frac{3}{10}, \frac{7}{10}\}$ 35. $\{2 - \frac{1}{2}i, 2 + \frac{1}{2}i\}$

37. $\{-2, 1\}$ 39. $\{-4, -3\}$ 41. $\{2, -\frac{1}{3}\}$ 43. $\left\{\dfrac{1}{2} + \dfrac{\sqrt{17}}{2}, \dfrac{1}{2} - \dfrac{\sqrt{17}}{2}\right\}$ 45. $\left\{\dfrac{3}{2} + \dfrac{i\sqrt{3}}{2}, \dfrac{3}{2} - \dfrac{i\sqrt{3}}{2}\right\}$

47. $\{-3 + 2i, -3 - 2i\}$ 49. $x = 0; x = y^2$ 51. $y = 0$, x any real number; $x = \dfrac{\pm 1}{\sqrt{y}}$ 53. $x = y \pm 1$

55. $x = -\frac{2}{3}y \pm 1$ 57. $(x + 1)^2 = -c + 1$ 59. $\left(x + \dfrac{b}{2}\right)^2 = -1 + \dfrac{b^2}{4}$ 61. $\left(x + \dfrac{1}{a}\right)^2 = \dfrac{4}{a} + \dfrac{1}{a^2}$

63. $(x + 1)^2 + (y + 1)^2 = 2$ 65. $(x + 2)^2 + 3(y + 1)^2 = 6$

Exercise 3.4 (Page 101)

1. $\{-4, -3\}$ 3. $\{\frac{3}{4}, 2\}$ 5. $\{-1, \frac{1}{6}\}$ 7. $\{-1 - \sqrt{7}, -1 + \sqrt{7}\}$ 9. $\left\{\dfrac{-3}{2} + \dfrac{\sqrt{5}}{2}, -\dfrac{3}{2} - \dfrac{\sqrt{5}}{2}\right\}$

11. $\{4 + \sqrt{2}i, 4 - \sqrt{2}i\}$ 13. $\{\frac{3}{5} + \frac{1}{5}i, \frac{3}{5} - \frac{1}{5}i\}$ 15. $\left\{-\dfrac{3}{2} + \dfrac{\sqrt{15}}{2}i, -\dfrac{3}{2} - \dfrac{\sqrt{15}}{2}i\right\}$

17. $\left\{\dfrac{3}{4} - \dfrac{\sqrt{15}}{4}i, \dfrac{3}{4} + \dfrac{\sqrt{15}}{4}i\right\}$ 19. $\{-2, \frac{1}{2}\}$ 21. $\{1 - \sqrt{3}, 1 + \sqrt{3}\}$ 23. $\left\{1 - \dfrac{\sqrt{38}}{2}i, 1 + \dfrac{\sqrt{38}}{2}i\right\}$

25. $\{-2 - \sqrt{5}, -2 + \sqrt{5}\}$ 27. $\{-1 - \sqrt{6}, -1 + \sqrt{6}\}$ 29. Two real roots   31. Two imaginary roots

33. One real root of multiplicity two   35. $y = \dfrac{-1 \pm \sqrt{1 + 4x}}{2x}$ 37. $y = \dfrac{-1 \pm \sqrt{1 - 2x^2}}{-x^2}$

39. $y = -x$ 41. $y = \dfrac{x \pm \sqrt{5}|x|}{-2}$ 43. $\frac{1}{2}$ second, $\frac{7}{2}$ seconds   45. $\frac{5}{2}$ seconds, $5\sqrt{6}/4$ seconds

47. 4.57 seconds, 7.32 seconds   49. 15 sides   51. 5 ohms   53. 2.17 ohms

55. If $ax^2 + bx + c = 0$ and $a \neq 0$, then completing the square, we have

$$x^2 + \frac{b}{a}x + \frac{b^2}{4a^2} = \frac{-c}{a} + \frac{b^2}{4a^2}$$

$$\left(x + \frac{b}{2a}\right)^2 = \frac{b^2 - 4ac}{4a^2}.$$

$$x + \frac{b}{2a} = \frac{\pm\sqrt{b^2 - 4ac}}{2a}$$

$$x = \frac{-b \pm \sqrt{b^2 - 4ac}}{2a}.$$

57. $-2, 2$ **59.** $\frac{1}{16}$ **61.** $0, 8$ **63.** 1

Exercise 3.5 (Page 108)

1. a. $2n(2n + 2) = 168$ **b.** $12, 14$ **3. a.** $(2n + 1)(2n + 3) = 323$ **b.** $17, 19$

5. a. $\dfrac{1}{n} + \dfrac{1}{n + 1} = \dfrac{17}{72}$ **b.** $8, 9$ **7. a.** $\dfrac{1}{n} - \dfrac{1}{n + 1} = \dfrac{1}{20}$ **b.** $4, 5$

9. a. $w(w + 2) = 63$, where w is the width   **b.** 9 meters × 7 meters

11. a. $w(4w - 1) = 33$, where w is the width   **b.** 11 meters × 3 meters

13. a. $\dfrac{b(b + 25)}{2} = 75$, where b is the length of the base

  b. Base is 5 centimeters, altitude is 30 centimeters.

15. a. $(2w - 4)(w - 4) = 48$, where w is the width of the paper   **b.** 8 inches × 16 inches

17. a. $(11 - 2x)(14 - 2x) = 108$, where x is the width of the margin   **b.** 1 inch

19. a. $(18 + 2x)(12 + 2x) - (12)(18) = 216$, where x is the width of the border   **b.** 3 meters

21. a. $(6 + x)^2\pi = 64\pi$, where x is the increase in the radius   **b.** 2 yards

23. a. $(50 + 2x)(25 + x) = (50)(25) + 318$, where x is the width of the border   **b.** 3 meters

25. a. $x(1000 - 2x) = 120{,}000$, where x is the length of the shorter side

  b. 200 feet by 600 feet or 300 feet by 400 feet

27. a. $\pi\left(\dfrac{x}{2}\right)^2 + 4x = 19$, where x is the width of the rectangular portion of the table   **b.** 3 feet

29. a. $\left(\dfrac{50{,}000}{n} + 10\right)(n - 5) = 50{,}000 + 9700$, where n is the number of barrels purchased

  b. 1000 barrels

31. a. $\left(\dfrac{1500}{x} + 0.10\right)(x + 50) = 1500 + 285$, where x is the number of pounds purchased   **b.** 300 pounds

33. a. $\left(\dfrac{x}{4}\right)^2 = \left(\dfrac{1 - x}{6}\right)^2 (2)$, where x is the length of the shorter piece   **b.** 0.49 foot

35. a. $\dfrac{1}{t - \frac{3}{2}} + \dfrac{1}{t} = 1$, where t is the time it takes B to load 1 pile of gravel   **b.** 3 hours

37. a. $\dfrac{30}{18 - r} = \dfrac{63}{18 + r} - 1$, where r is the rate of the current   **b.** 3 mph

39. **a.** $\dfrac{10}{r} + \dfrac{10}{r+10} = \dfrac{5}{6}$, where r is the rate of the train taking the woman to work

b. 20 mph to work, 30 mph home from work

Exercise 3.6 (Page 117)

1. $\{25\}$ **3.** $\{6\}$ **5.** $\{2\}$ **7.** $\{6\}$ **9.** $\{-1\}$ **11.** $\{1\}$ **13.** No solution   **15.** $\{5\}$ **17.** $\{\frac{1}{16}\}$

19. $A = \pi r^2$ **21.** $y = \dfrac{1}{x^3}$ **23.** $y = \pm\sqrt{a^2 - x^2}$ **25.** $V = \pi h(R^2 - r^2)$ **27.** $\{1\}$

29. $\{216, -27\}$ **31.** $\{\frac{1}{9}, -\frac{1}{2}\}$ **33.** $\{-\frac{6}{7}, -\frac{5}{4}\}$ **35.** $\{16, 256\}$ **37.** $\{87\}$ **39.** 1 centimeter

41. $\dfrac{2}{\sqrt{5}}$ hours   **43.** $\frac{13}{8}$ miles   **45.** 44 feet   **47.** $\{16\}$ **49.** $\{\frac{1}{9}\}$ **51.** $\{1, -1\}$ **53.** $\{64\}$

55. No solution   **57.** $\left\{\dfrac{2}{\sqrt{3}}\right\}$ **59.** $\left\{\dfrac{4}{\sqrt{3}}\right\}$

Exercise 3.7 (Page 124)

1. $[0, 1]$ **3.** $(-2, 5)$ **5.** $[2, 7)$ **7.** $[-5, \infty)$ **9.** $(-\infty, -1]$

11. $(-6, +\infty)$ **13.** $(-\infty, 3]$ **15.** $[-6, +\infty)$ **17.** $[\frac{5}{2}, +\infty)$

19. $(-\infty, -1)$ **21.** $[-2, +\infty)$ **23.** $[-3, +\infty)$ **25.** $(-\infty, \frac{8}{3})$

27. $[-\frac{6}{5}, +\infty)$ **29.** $[-6, 1]$ **31.** $(-\frac{22}{3}, -4]$ **33.** $[-25, -16]$

35. Greater than 15 days   **37.** At least \$3600   **39.** At least 94%   **41.** Between 14°F and 68°F

43. $(\frac{1}{4}(5 - \varepsilon), +\infty)$ **45.** $(\frac{1}{3}(-2 - \varepsilon), \frac{1}{3}(\varepsilon - 2))$ **47.** $(\frac{1}{3}(7 - \varepsilon), \frac{1}{3}(7 + \varepsilon))$ **49.** $(-\infty, 1]$

51. $(-\infty, \frac{1}{8}]$ **53.** $(-\infty, \frac{13}{6}]$ **55.** $(\frac{9}{4}, \infty)$ **57.** $(-\infty, -\frac{25}{8})$ **59.** $(-\frac{1}{3}, \infty)$

Exercise 3.8 (Page 131)

1. $(-\infty, -2) \cup (3, \infty)$ **3.** $[-\frac{8}{3}, \frac{3}{2}]$ **5.** $(-\infty, -3) \cup (4, \infty)$ **7.** $(-\infty, -1) \cup (3, \infty)$

9. $(-\infty, -2) \cup (1, \infty)$ **11.** $[-3, \frac{1}{3}]$ **13.** $(-\infty, -\frac{3}{2}] \cup [\frac{1}{4}, \infty)$ **15.** Empty set

17. $(-\infty, 2 - \sqrt{3}] \cup [2 + \sqrt{3}, \infty)$ **19.** Empty set   **21.** $(-2, \infty)$ **23.** $(-\infty, -\frac{1}{2}) \cup (0, \infty)$

25. $(-\infty, -\frac{3}{2}) \cup [0, \infty)$ **27.** $(-\infty, 0) \cup [\frac{1}{3}, \infty)$ **29.** $(-\infty, -3) \cup (-1, \infty)$ **31.** $(-2, 2)$

33. $(-1, 1) \cup (3, +\infty)$ **35.** $(-\infty, -1) \cup (0, 2)$ **37.** $[-3, -1] \cup [1, +\infty)$

39. $(-\infty, -4) \cup (-2, 0) \cup (2, +\infty)$ **41.** $(-\infty, -\sqrt{3}) \cup (-1, 1) \cup (\sqrt{3}, \infty)$ **43.** $(1, 2)$

45. $(-\infty, -1) \cup (1, \infty)$ **47.** $(-1, 1) \cup (2, \infty)$

49. Between 1 and 1.5 seconds and between 2 and 2.5 seconds after the ball is thrown   **51.** Between 5 and 9 feet

53. Between 0.98 feet and 0.99 feet   **55.** $(-\infty, -2) \cup (2, \infty)$ **57.** $(-\infty, 0) \cup (1, \infty)$

59. $(-\infty, -\frac{24}{13}) \cup (0, \infty)$

Exercise 3.9 (Page 136)

1. $\{-2, 6\}$ 3. $\{-11, -1\}$ 5. $\{1, 4\}$ 7. $\{-4, 4\}$ 9. $\{-5, 0\}$ 11. $\{-2, -\sqrt[3]{4}\}$

13. $[-1, 1]$ 15. $(-4, 0)$ 17. $(-\frac{5}{6}, -\frac{1}{6})$ 19. $(-\infty, -8) \cup (0, \infty)$ 21. $(-\infty, \frac{1}{3}) \cup (1, \infty)$

23. $[-9, 1]$ 25. $[0, 3]$ 27. $(-\infty, -\frac{3}{2}] \cup [\frac{5}{2}, +\infty)$ 29. $(-\infty, -\frac{2}{3}] \cup [\frac{4}{3}, +\infty)$ 31. $|x| \le 3$

33. $|x - 2| < 4$ 35. $|x + 5| \le 3$ 37. $|x + \frac{3}{2}| < \frac{7}{2}$ 39. $\{-2, -\frac{4}{3}\}$ 41. $\{1, -1\}$

43. $(-1, 0) \cup (0, 1)$ 45. $(-\infty, 2 - 2\sqrt{2}] \cup \{2\} \cup [2 + 2\sqrt{2}, +\infty)$ 47. $(-\infty, -\frac{2}{3}] \cup [-\frac{3}{7}, +\infty)$

49. $|x - 4| < 5$ 51. $|x^2 - 16| \le 0.1$ 53. $|(x - 2) - L| < 10^{-4}$ 55. $|(x^2 + 1) - L| < \varepsilon$

57. $|x - y| > 4$ 59. $|x - 4| \ge 10^{-3}$ 61. $|P(x) - L| < \varepsilon$

Chapter 3 Review (Page 137)

1. $\{-\frac{23}{2}\}$ 2. $\{3\}$ 3. $\{-\frac{5}{2}\}$ 4. $\{\frac{3}{5}\}$ 5. $y = \frac{x}{4}$ 6. $x = 4y$ 7. $\frac{5}{6}$ mile   8. 75 gallons

9. $\{3, -2\}$ 10. $\{-7, 3\}$ 11. $\{\sqrt{5}, -\sqrt{5}\}$ 12. $\{5 + \sqrt{2}, 5 - \sqrt{2}\}$ 13. $(x - 2)^2 = 3$

14. $(x + \frac{3}{2})^2 = \frac{1}{4}$ 15. $\left\{\dfrac{-5}{2} + \dfrac{\sqrt{33}}{2}, \dfrac{-5}{2} - \dfrac{\sqrt{33}}{2}\right\}$ 16. $\{\frac{1}{2}, -1\}$ 17. $\left\{\dfrac{1 \pm \sqrt{15}}{4}\right\}$ 18. $\{-1 \pm \sqrt{5}\}$

19. $\left\{\dfrac{3 \pm \sqrt{9 - 4k}}{2k}\right\}$ 20. $\left\{\dfrac{-k \pm \sqrt{k^2 + 16}}{2}\right\}$ 21. 12 yd × 24 yd   22. 2 inches   23. $\{4, 9\}$

24. $\{8\}$ 25. $\{1, 4\}$ 26. $\{\frac{1}{7}, -\frac{1}{6}\}$ 27. $[32, +\infty)$ 28. $(-\infty, -\frac{6}{5})$ 29. $(-5, 2)$ 30. $(\frac{1}{3}, 1)$

31. $\{-2, \frac{4}{3}\}$ 32. $\{-1, \frac{7}{3}\}$ 33. $\{0, 4\}$ 34. $\left\{0, -1, -\dfrac{1}{2} - \dfrac{\sqrt{17}}{2}, -\dfrac{1}{2} + \dfrac{\sqrt{17}}{2}\right\}$ 35. $[-1, 0]$

36. $(-\infty, -\frac{1}{2}] \cup [\frac{7}{2}, +\infty)$ 37. $(-\sqrt{3}, -1) \cup (1, \sqrt{3})$ 38. $(-\infty, -\frac{5}{3}) \cup (-\frac{1}{3}, +\infty)$

39. $|x + 1| \le 4$ 40. $|x + 2| < 4$ 41. $y' = \dfrac{y}{y + 2x}$ 42. $y' = \dfrac{-2xy}{y^2 - 3xy^2}$ 43. $k = 0$

44. $k \ne 0$ 45. No real solutions   46. $\left\{\dfrac{27}{2} - \dfrac{\sqrt{245}}{2}, \dfrac{27}{2} + \dfrac{\sqrt{245}}{2}\right\}$ 47. $\{1\}$ 48. $\{-1, 1\}$

49. $(-3, -2) \cup (1, \infty)$ 50. $(-1, \frac{1}{2}] \cup [4, \infty)$ 51. $[-3, 1]$ 52. $(-\infty, -4] \cup [-\frac{4}{3}, \infty)$

Exercise 4.1 (Page 149)

1. a. Domain: $\{2, 5, 7\}$ b. A function   3. a. Domain: $\{2, 3\}$ b. Not a function

5. a. Domain: $\{5, 6, 7\}$ b. A function   7. a. Domain: $\{-1, 1\}$ b. Not a function

9. $(0, -\frac{4}{3}), (1, -\frac{2}{3}), (2, 0), (-3, -\frac{10}{3}), (\frac{2}{3}, -\frac{8}{9})$ 11. $(0, \sqrt{15}), (1, \sqrt{18}), (2, \sqrt{21}), (-3, \sqrt{6}), (\frac{2}{3}, \sqrt{17})$

13. Set of all real numbers   15. Set of all real numbers   17. Set of all real numbers $\ne 2$

19. Set of all real numbers ≥ -4 21. $\{x \mid -3 \le x \le 3\}$ 23. $\{x \mid x \in R, x \ne 1 \text{ and } x \ne -1\}$

25. Function   27. Function   29. Function   31. Not a function   33. Not a function

35. Function   37. Function   39. Not a function   41. Function   43. -3 45. 5 47. $-2a + 2$

49. 4 51. 0 53. $\dfrac{a^2 + 1}{a^2}$ 55. -1 57. $a^4 - 8a^3 + 22a^2 - 24a + 9$ 59. $-\frac{1}{3}$ 61. $\sqrt{3}$

63. 5 65. $2h$ 67. $-3h$ 69. $-2ah - h^2$ 71. $h^2 + 2ah - 3h$ 73. 3 75. $\frac{11}{6}$ 77. 1

79. a. 13 b. 49 81. a. 5 b. 3 83. a. $\dfrac{1}{a}$ b. $\dfrac{1}{a}$ 85. $2a + h - 1$ 87. $\dfrac{1}{\sqrt{a + h} + \sqrt{a}}$

89. $h, h \ne 0$

Exercise 4.2 (Page 158)

1. x-intercept is 2; y-intercept is 3   **3.** x-intercept is $-\frac{1}{2}$; y-intercept is 1   **5.** Both intercepts are 0.

7. x-intercept is $\frac{1}{2}$; y-intercept is 3   **9.** x-intercept is $\frac{3}{5}$; y-intercept is $-\frac{4}{5}$

11. x-intercept is $\dfrac{1}{2\sqrt{2}}$; y-intercept is $-\dfrac{1}{2\sqrt{5}}$

13.   **15.**   **17.**

19. 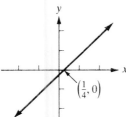   **21.**   **23.**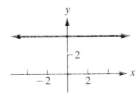

25. 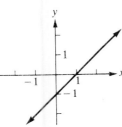   **27.**   **29.**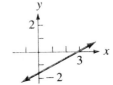

31. Distance, 5; slope, $\frac{4}{3}$ **33.** Distance, 13; slope, $\frac{12}{5}$ **35.** Distance, $\sqrt{2}$; slope, 1

37. Distance, $3\sqrt{5}$; slope, $\frac{1}{2}$ **39.** Distance, 5; slope, 0   **41.** Distance, 10; slope, undefined

43. Distance, 4; slope, 0   **45.** Distance, 4; slope, undefined   **47.** Yes   **49.** Yes   **51.** No

53. l_2 has smallest slope, l_3 has largest slope   **55.** l_1 has smallest slope, l_2 has largest slope

57. l_3 has smallest slope, l_1 has largest slope

59.   **61.**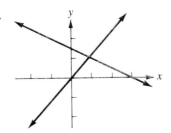

(1, 1); the coordinates satisfy both equations

63. 3; slope of the line segment joining $(a, f(a))$ and $(a + h, f(a + h))$

65. 2; slope of the line segment joining $(a, f(a))$ and $(a + h, f(a + h))$

67. $\frac{2}{3}$; slope of the line segment joining $(a, f(a))$ and $(a + h, f(a + h))$

69. 0; slope of the line segment joining $(a, f(a))$ and $(a + h, f(a + h))$

Exercise 4.3 (Page 166)

1. $4x - y - 7 = 0$ **3.** $x - 2y - 1 = 0$ **5.** $y + 3 = 0$ **7.** $x + 1 = 0$ **9.** $\sqrt{2}x - y + 3 - \sqrt{2} = 0$

11. $2x - y + 2 - 2\sqrt{3} = 0$ **13.** $x + 2y = 0$ **15.** $-x + 3y - 5 = 0$ **17.** $y - 1 = 0$

19. $y = -x + 3$; slope, -1; intercept, 3 **21.** $y = -\frac{3}{2}x + \frac{1}{2}$; slope, $-\frac{3}{2}$; intercept, $\frac{1}{2}$

23. $y = \frac{1}{3}x - \frac{2}{3}$; slope, $\frac{1}{3}$; intercept, $-\frac{2}{3}$ **25.** $y = \frac{4}{3}x - \frac{7}{3}$; slope, $\frac{4}{3}$; intercept, $-\frac{7}{3}$

27. $y = 6$; slope, 0; intercept, 6 **29.** $x = 5$; slope, undefined; no y-intercept **31.** $3x + 2y - 6 = 0$

33. $5x + 2y + 10 = 0$ **35.** $6x - 2y + 3 = 0$ **37.** **a.** $3x + y = 16$ **b.** $x - 3y = 2$

39. **a.** $3x - 2y = 0$ **b.** $2x + 3y = 0$ **41.** **a.** $2x + 3y = -7$ **b.** $3x - 2y = -14$

43. **a.** $x + 2y - 5 = 0$ **b.** $y - 2x = 0$ **45.** $x - y = 0$ **47.** $7x - 3y = -8$

49. $ax - by = \frac{1}{2}(a^2 - b^2)$ **51.** $y - x = 0$

53. Slope of $\overline{AB}$: $m_1 = -1$; slope of $\overline{AC}$: $m_2 = 1$; $m_1 m_2 = -1$; thus, by Theorem 4.3, $\overline{AB} \perp \overline{AC}$. Hence, $\triangle ABC$ is a right triangle.

55. $v = -32t$ **57.** $C = 900n + 100,000$ **59.** $F = \frac{9}{5}C + 32$ **61.** $D = -55P + 20,000$

63. From geometry, if L_1 is parallel to L_2, then $\angle P_3 P_1 Q_1$ is congruent to $\angle P_4 P_2 Q_2$. Also, $\angle P_1 Q_1 P_3$ is congruent to $\angle P_2 Q_2 P_4$, since each is a right angle. Hence, $\triangle P_1 Q_1 P_3$ is similar to $\triangle P_2 Q_2 P_4$. Because the lengths of corresponding sides of similar triangles are proportional, it follows that $\dfrac{Q_1 P_3}{P_1 Q_1} = \dfrac{Q_2 P_4}{P_2 Q_2}$.

But $\dfrac{Q_1 P_3}{P_1 Q_1} = \dfrac{y_3}{x_3 - x_1} = m_1$ and $\dfrac{Q_2 P_4}{P_2 Q_2} = \dfrac{y_4}{x_4 - x_2} = m_2$. Hence, $m_1 = m_2$.

65. $\dfrac{y_1 - y_2}{x_2 - x_1} = -\dfrac{y_2 - y_1}{x_2 - x_1} = -m_2$; $\dfrac{x_2 - x_1}{y_3 - y_1} = \dfrac{1}{m_1}$. Hence, $-m_2 = \dfrac{1}{m_1}$, or $m_1 m_2 = -1$.

67. From geometry, $\triangle ABC$ is similar to $\triangle AME$. Thus, $\dfrac{AM}{AE} = \dfrac{AB}{AC}$ or $\dfrac{AM}{AB} = \dfrac{AE}{AC}$. But since M is the midpoint of line AB, we know $\dfrac{AM}{AB} = \dfrac{1}{2}$. Thus, $\dfrac{1}{2} = \dfrac{AE}{AC}$ or $AC = 2AE$. Similarly, $BC = 2BD$.

Exercise 4.4 (Page 175)

1. **a.** $1, 2$ **b.** 2 **c.** $x = \frac{3}{2}$ **d.** $(\frac{3}{2}, -\frac{1}{4})$ **3.** **a.** $-2, -\frac{1}{2}$ **b.** -2 **c.** $x = -\frac{5}{4}$ **d.** $(-\frac{5}{4}, \frac{9}{8})$

5. **a.** 1 **b.** 1 **c.** $x = 1$ **d.** $(1, 0)$ **7.** **a.** $\frac{1}{2}$ **b.** -1 **c.** $x = \frac{1}{2}$ **d.** $(\frac{1}{2}, 0)$

9. **a.** None **b.** 1 **c.** $x = 0$ **d.** $(0, 1)$ **11.** **a.** None **b.** -9 **c.** $x = 0$ **d.** $(0, -9)$

13. $y, 4$; $x, 1,$ and 4; $(\frac{5}{2}, -\frac{9}{4})$ **15.** $y, -7$; $x, -1,$ and 7; $(3, -16)$

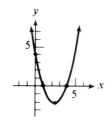

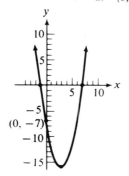

17. $y, -3$; $x, \frac{1}{2}$, and -3; $(-\frac{5}{4}, -\frac{98}{16})$

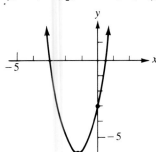

19. $y, -6$; $x, 2$, and 3; $(\frac{5}{2}, \frac{1}{4})$

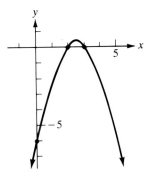

21. $y, -4$; $x, 1$, and 4; $(\frac{5}{2}, \frac{9}{4})$

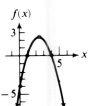

23. $y, 2$; no x; $(0, 2)$

25. $y, 2$; no x; $(0, 2)$

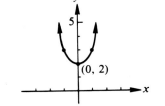

27. $y, 0$; $x, 0$; $(0, 0)$

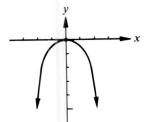

29. $y, -1$; $x, -1$; $(-1, 0)$

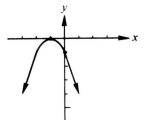

31. $y, 9$; $x, \frac{3}{2}$; $(\frac{3}{2}, 0)$

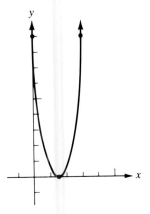

33. 4, slope of secant line

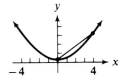

35. Varying k has the effect of translating the graph along the y-axis.

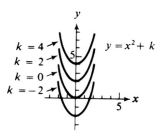

37. Varying k has the effect of translating the axis of symmetry along the x-axis and thus changes the vertex.

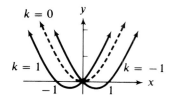

39. $4a + 2h + 1$; the slope of the line segment joining $(2, f(2))$ and $(2 + h, f(2 + h))$

41. $6a + 3h + 4$; the slope of the line segment joining $(0, f(0))$ and $(0 + h, f(0 + h))$

43. Completing the square on $y = ax^2 + bx + c$ and solving for y, we obtain $y = a\left(x + \dfrac{b}{2a}\right)^2 + k$ where $k = c - \dfrac{b^2}{4a}$. Note that if $x \neq -\dfrac{b}{2a}$, then $\left(x + \dfrac{b}{2a}\right)^2 > 0$. Thus, if $a > 0$, then $a\left(x + \dfrac{b}{2a}\right)^2 + k > k$ for any $x \neq -\dfrac{b}{2a}$. Hence, the minimum value of $ax^2 + bx + c$ is k, which is attained at $x = -\dfrac{b}{2a}$.

45. $\frac{3}{8}$ second, $\frac{9}{4}$ feet   **47.** 2 hours, 54.08 miles

49. The equation $ax^2 + bx + c = 0$ has at least one real solution because $b^2 - 4ac \geq 0$. Thus, there is a real number x_0 so that $f(x_0) = 0$. Since $f(x) \geq 0$ (because it is an expression to an even power), we know 0 is the minimum value of f.

Exercise 4.5 (Page 182)

1. 1   **3.** 3   **5.** 8   **7.** 4

9.

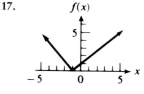

11.

13.

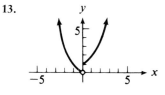

15.

17.

19.

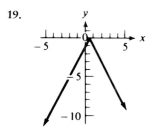

21.

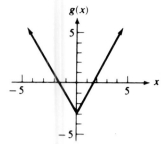

23.

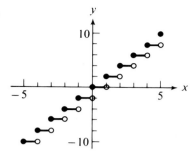

25.

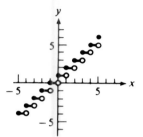

27.

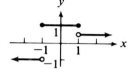

29.

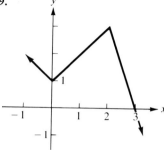

31.

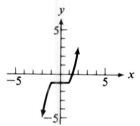

33.

35.

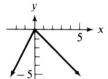

37.

39.

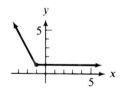

41.

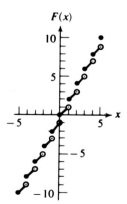

43.

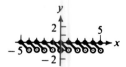

45. $3   **47.** Nothing   **49.** $10,000   **51.** $52,745   **53.** loses $2090   **55.** $8750

Exercise 4.6 (Page 189)

1. Yes   **3.** No   **5.** No   **7.** $\{(1, 4), (3, 2), (5, 1)\}$ **9.** $\{(2, -2), (0, 0), (-2, 2)\}$ **11.** $f^{-1}(x) = x$

13. $f^{-1}(x) = -\frac{1}{2}x + 2$ **15.** $f^{-1}(x) = -\frac{4}{3}x + 4$ **17.** $f^{-1}(x) = \sqrt[3]{x + 1}$ **19.** $f^{-1}(x) = \dfrac{1}{x}$

21. $f^{-1}(x) = -\sqrt{x}, x \geq 0$

23.

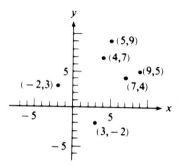

25.

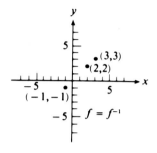

27.

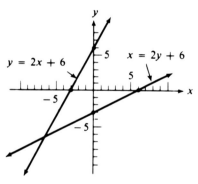

29.

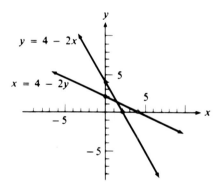

31.

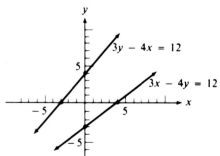

33.

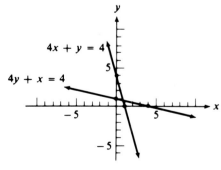

35.

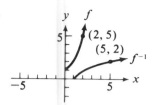

37.

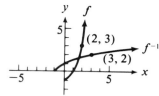

39.

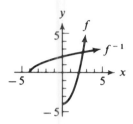

41.

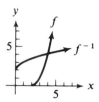

43. $f(f^{-1}(x)) = f(x + 1) = (x + 1) - 1 = x$
$f^{-1}(f(x)) = f^{-1}(x - 1) = (x - 1) + 1 = x$

45. $f(f^{-1}(x)) = f(\sqrt[3]{x}) = (\sqrt[3]{x})^3 = x$
$f^{-1}(f(x)) = f^{-1}(x^3) = \sqrt[3]{x^3} = x$

47. $f(f^{-1}(x)) = f\left(\dfrac{x + 1}{x}\right) = \dfrac{1}{\dfrac{x + 1}{x} - 1} = \dfrac{1}{\dfrac{1}{x}} = x; \quad f^{-1}(f(x)) = f^{-1}\left(\dfrac{1}{x - 1}\right) = \dfrac{\dfrac{1}{x - 1} + 1}{\dfrac{1}{x - 1}} = \dfrac{\dfrac{x}{x - 1}}{\dfrac{1}{x - 1}} = x$

49. Yes   **51.** No   **53.** Yes   **55.** 4   **57.** 3

59. The line joining the points (b, a) and (a, b) has slope -1 and hence is perpendicular to the line $y = x$. The midpoint of the line segment joining the points (a, b) and (b, a) is $\left(\dfrac{a + b}{2}, \dfrac{a + b}{2}\right)$, which is on the line $y = x$. Thus, the line $y = x$ is the perpendicular bisector of the line segment joining the points (b, a) and (a, b).

61. Yes   **63.** No   **65.** Yes

Exercise 4.7 (Page 194)

1.

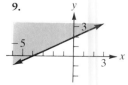

3.

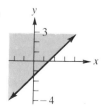

5.

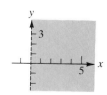

7.

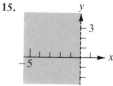

9.

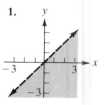

11.

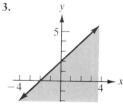

13.

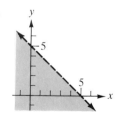

15.

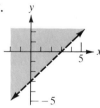

17.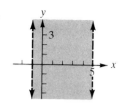

19.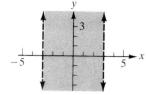

21.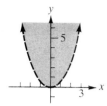

23.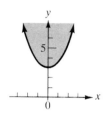

25.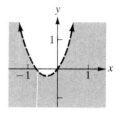

27.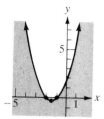

29.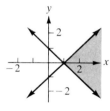

31.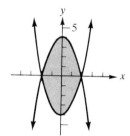

33.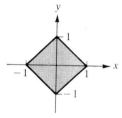

35.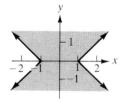

37.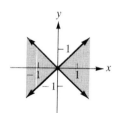

39.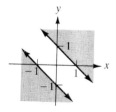

Exercise 4.8 (Page 201)

1. $s = 8t$, $t \geq 0$ **3.** $L = \dfrac{6000}{l}$, $l > 0$ **5.** $C = 5000 + 10n$, $n \geq 0$ **7.** $C = 30s$, $s \geq 0$

9. $C = 52.50s$, $s \geq 0$ **11.** $I = 0.27x$, $x \geq 0$ **13.** $y = 2n$, $n > 0$ **15.** $n = y$, $y > 0$

17. $A = \dfrac{l^2}{32}$, $l > 0$ **19.** $A = \dfrac{x^2}{4\pi} + \dfrac{(50 - x)^2}{4\pi}$, $0 < x < 25$

21. $N = 3(11 - 2x)(10)(8.5 - 2x)$, $0 < x < 4.25$ **23.** $A = (x - 2)^2$, $x > 2$

25. $C = \dfrac{1500}{w} + 15w$, $w > 0$ **27.** $C = 10[2x(50 + 2x) + 2x(50)] + 25(50)^2$, $x \geq 0$

29. $C = 30x + 50\sqrt{144 + (18 - x)^2}$ $0 \leq x \leq 18$ **31.** $r_u = \frac{5}{2}r_c$, $r_c > 0$ **33.** $r = 40\left(\dfrac{8 - t}{2t - 8}\right)$, $4 < t < 8$.

35. $P = \dfrac{9500}{x - 4} - \dfrac{8000}{x}$, $x > 4$ **37.** $P = \begin{cases} 10 & 0 \leq x \leq 3 \\ 10 - 0.30x & 3 < x \leq 10, \quad 0 \leq x \leq 10 \end{cases}$

Chapter 4 Review (Page 205)

1. $\{4, 2, 3\}$ 2. $\{1\}$ 3. $\{x \mid x \neq -4\}$ 4. $\{x \mid x \geq 6\}$ 5. 1   6. 0   7. 2   8. $\frac{1}{3}$ 9. $3h$

10. $-4h$ 11. $h^2 + 2ah - 2h$ 12. $-h^2 - 2ah + h$ 13. 12   14. $\frac{\sqrt{3}}{3} - 1$ 15. a. 0   b. -5

16. a. -1 b. 1

17. 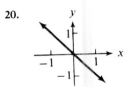   18. 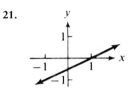   19.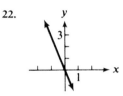

20. 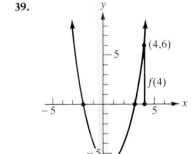   21. 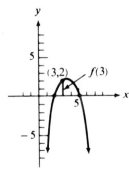   22.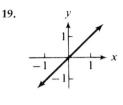

23. $\sqrt{41}, m = -\frac{4}{5}$ 24. $\sqrt{5}, m = -\frac{1}{2}$ 25. 6, slope is undefined   26. 12, $m = 0$

27. $-4x + y + 15 = 0$ 28. $2y - x - 12 = 0$ 29. $y + x - 1 = 0$ 30. $5x + 6y + 14 = 0$

31. $y = -4x + 6, m = -4, b = 6$ 32. $y = \frac{3}{2}x - 8, m = \frac{3}{2}, b = -8$ 33. $-\frac{2}{3}x + y - 2 = 0$

34. $y = 12x - 4$ 35. a. $y - 1 = \frac{3}{2}(x - 1)$ b. $y - 1 = -\frac{2}{3}(x - 1)$

36. a. $y - 1 = x + 1$ b. $y - 1 = -(x + 1)$

37. x-intercepts 3, -2;   axis of symmetry $x = \frac{1}{2}$;   vertex $(\frac{1}{2}, -\frac{25}{4})$

38. x-intercepts 5, 2;   axis of symmetry $x = \frac{7}{2}$;   vertex $(\frac{7}{2}, \frac{9}{4})$

39. 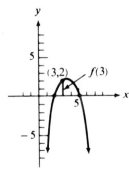   40. 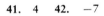   41. 4   42. -7

43.
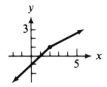

44.
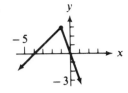

45.
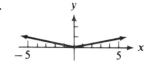

46.
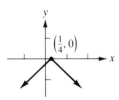

47.
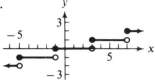

48.
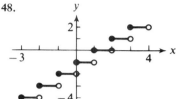

49. $f^{-1}(x) = \frac{1}{2}(x + 3)$ **50.** $f^{-1}(x) = -\sqrt{x - 9}; \quad x \geq 9$

51.
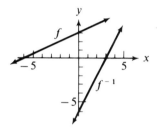

52.
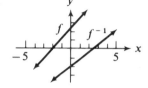

53.
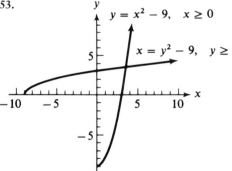

54.
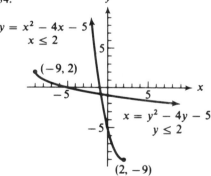

55. $f(f^{-1}(x)) = f(\frac{1}{2}(x + 1)) = 2[\frac{1}{2}(x + 1)] - 1 = x$

$f^{-1}(f(x)) = f^{-1}(2x - 1) = \frac{1}{2}[(2x - 1) + 1] = x$

56. $f(f^{-1}(x)) = f\left(\dfrac{1}{\sqrt[3]{x}}\right) = \dfrac{1}{(1/\sqrt[3]{x})^3} = \dfrac{1}{\dfrac{1}{x}} = x$

$f^{-1}(f(x)) = f^{-1}\left(\dfrac{1}{x^3}\right) = \dfrac{1}{1/\sqrt[3]{x^3}} = \dfrac{1}{\dfrac{1}{x}} = x$

57.

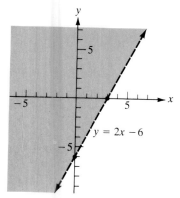

$y = 2x - 6$

58.

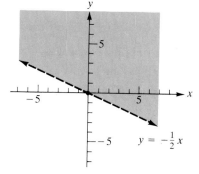

$y = -\frac{1}{2}x$

59.

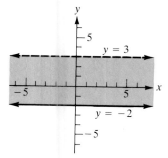

$y = 3$

$y = -2$

60.

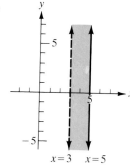

$x = 3$ $x = 5$

61.

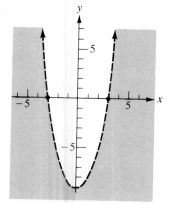

62.

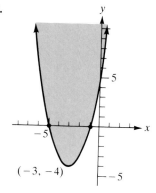

$(-3, -4)$

63. $y = 5x^2$ **64.** $r = \dfrac{32x^2}{9z^3}$ **65.** $n = \dfrac{9600}{d}$ **66.** $r = 24n$ **67.** $\dfrac{-2}{(a + h + 1)(a + 1)}$

68. $\dfrac{-2}{\sqrt{a + 1}\sqrt{a + h + 1}(\sqrt{a + 1} + \sqrt{a + h + 1})}$ **69.** $d = -80t + 640$ **70.** $d = \frac{1}{6}t$ **71.** 0   **72.** $\frac{9}{16}$

73. 3   **74.** 10

75.

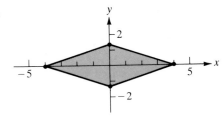

76.

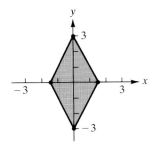

77. $d(t) = 100 + 270t, \ t \geq 0$ **78.** $d(t) = \sqrt{(100 + 150t)^2 + (120t)^2}, \ t \geq 0$

Exercise 5.1 (Page 215)

1. $x + 7$ **3.** $x - 5 - \dfrac{5}{x + 5}$ **5.** $2x^2 - 4x + 9$ **7.** $3x^3 + 7x^2 + 12x + 24 + \dfrac{54}{x - 2}$

9. $x^5 + x^4 + x^3 + x^2 + x + 1$ **11.** $x^6 - x^5 + x^4 - x^3 + x^2 - x + 1$

13. $x^3 - 3x^2 + 9x - 26 + \dfrac{64}{x + 3}$ **15.** $4x^4 + 13x^3 + 37x^2 + 112x + 336 + \dfrac{1008}{x - 3}$

17. $x^2 - 4x + 7 - \dfrac{8}{x + 1}$ **19.** $x^4 + 3x^2 + 3x - 6 + \dfrac{5}{x + 2}$ **21.** $3; 17; 47$ **23.** $56; 12; 326$

25. $-685; 95; 719$ **27.** $(x - 2)(x + 1)(x + 1)$ **29.** $(x + 1)(x + 3)(x - 2)$ **31.** $(2x - 1)(x - 2)(x + 2)$

33. $(x - 1)(x + 2)(2x + 1)(x - 3)$ **35.** $x(x + 2)(3x - 2)(2x - 1)(x - 1)$ **37.** $\dfrac{1}{2}x + 1 + \dfrac{3}{2x - 2}$

39. $x^3 - \dfrac{1}{2}x^2 - \dfrac{1}{2}x - \dfrac{1}{2x - 2}$ **41.** $\dfrac{1}{2}x^3 - \dfrac{3}{4}x^2 - \dfrac{3}{8}x + \dfrac{5}{16} + \dfrac{\frac{37}{16}}{2x - 1}$ **43.** $\dfrac{1}{3}x^2 + \dfrac{2}{9}x + \dfrac{13}{27} - \dfrac{\frac{28}{27}}{3x - 2}$

45. $-1 + i$ **47.** $-13 - 57i$ **49.** $10 - 2i$

Exercise 5.2 (Page 223)

1. Upper bound 3;   lower bound -4 **3.** Upper bound 4;   lower bound -3

5. Upper bound 2;   lower bound -3 **7.** Upper bound 1;   lower bound -2

9. Upper bound 6;   lower bound -4 **11.** Upper bound 2;   lower bound -2

13. 0 or 2 positive, 0 or 2 negative   **15.** 0 positive, 1 negative   **17.** 0 or 2 positive, 0 or 2 negative

19. $-2i$ **21.** $-i$ **23.** $i, \dfrac{1}{2}$ **25.** $i, -2i$ **27.** $3i, \dfrac{-3 + \sqrt{29}}{2}, \dfrac{-3 - \sqrt{29}}{2}$

29. $-2i$ (multiplicity two) and 2   **31.** $P(x) = (2x - 3)(x - 2 + 2i)(x - 2 - 2i)$

33. $P(x) = (x - 2 - 3i)(x - 2 + 3i)(x - i)(x + i)$ **35.** $P(x) = (x - 2i)^2(x + 2i)^2(x - 4)$

37. $P(x) = (x - i)(x + i)(x - 2i)(x + 2i)$ **39.** $4 + i, 1 + i, 1 - i$ **41.** $1 - i, x^3 - 2x + 4 = 0$

43. No, because Theorem 5.6 only applies to polynomials with real-number coefficients. For example, $P(x) = x - 1 - i$ has $1 + i$ as a zero but not $1 - i$.

45. $-x^3 + 6x^2 - 11x + 6$ **47.** $x^4 - 6x^3 + 18x^2 - 30x + 25$ **49.** $\dfrac{1}{2}x^3 - 2x^2 + \dfrac{1}{2}x + 3$

51. $\dfrac{1}{3}x^3 - \dfrac{4}{3}x^2 + 2x - \dfrac{4}{3}$ **53.** $-1, \dfrac{1 + i\sqrt{3}}{2}, \dfrac{1 - i\sqrt{3}}{2}$

55. Let $P(x) = 0$ be a polynomial equation with real coefficients of degree n, where n is odd; then, $P(x) = 0$ must have at least n zeros and, since $P(x)$ has only real coefficients, any complex zeros must occur as conjugate pairs; since n is odd, there must always be at least one real zero, then $P(x) = 0$ must have at least one real solution.

57. Let $P(x) = a_n x^n + \cdots + a_0$ and suppose $P(z) = 0$; then, by Exercises 71–73, Section 2.5, we can write

$$P(\bar{z}) = a_n \bar{z}^n + \cdots + a_0$$
$$= \overline{a_n z^n} + \cdots + \bar{a}_0 \quad (\text{since } a_i \in R)$$
$$= \overline{a_n z^n + \cdots + a_0}$$
$$= \overline{P(z)} = \bar{0} = 0.$$

59. Let $S(x) = P(x) - Q(x)$; then the degree of S is less than or equal to n. But there are more than n values of x for which $S(x) = 0$ [namely, the values where $P(x) = Q(x)$]. But the number of solutions of $S(x) = 0$ must be the same as the degree of S if the degree of S is positive. Since the number of solutions of $S(x) = 0$ is greater than the degree of S, it must be the case that the degree of S is 0 and $S(x)$ must be a constant. Since the equation $S(x) = 0$ has solutions, it must be the case that $P(x) - Q(x) = S(x) = 0$, or $P(x) = Q(x)$ for all x.

Exercise 5.3 (Page 230)

1. $f(0) = 1$ and $f(1) = -1$ **3.** $g(-3) = 10$ and $g(-2) = -33$

5. $P(-3) = 2$ and $P(-2) = -4$, $P(0) = -4$, and $P(1) = 2$ **7.** Zero between 1.7 and 1.8

9. Zero between -2.7 and -2.6 **11.** Zero between 1.2 and 1.3   **13.** Zero between -1.3 and -1.2

15. Zero between 1.58 and 1.59   **17.** Zero between -1.71 and -1.70 **19.** Zero between 1.62 and 1.63

21. Zero between -0.85 and -0.84 **23.** $\{4\}$ **25.** $\{-2, 1\}$ **27.** None   **29.** None   **31.** $\{-\frac{3}{2}\}$

33. $\{-\frac{7}{4}, 2\}$ **35.** $\{\frac{3}{2}\}$ **37.** $\{\frac{1}{2}, 2\}$ **39.** $\{-\frac{1}{3}, 1 + \sqrt{5}, 1 - \sqrt{5}\}$

41. $\left\{-1, \dfrac{1}{2}, \dfrac{-1 + \sqrt{3}}{2}, \dfrac{-1 - \sqrt{3}}{2}\right\}$

43. $\{\frac{3}{4}, -\frac{4}{3}, \sqrt{2}, -\sqrt{2}\}$ **45.** $\{0, \frac{1}{2}, \frac{2}{3}, \sqrt{2}, -\sqrt{2}\}$ **47.** Zero between -1.415 and -1.414

49. Zero between -0.895 and -0.894 **51.** Zero between 0.754 and 0.755

53. Zero between -1.452 and -1.451

55. Zeros between -0.6 and -0.5, between 1.9 and 2.0, and between 4.6 and 4.7

57. Zeros between -2.8 and -2.7, between -0.6 and -0.5, between 0.7 and 0.8, and between 1.2 and 1.3

59. $(2x + 3)(x - 1)(x + 1)$

61. Consider $x^2 - 3 = 0$. If the equation has real solutions, then they are either rational or irrational. If rational, by Theorem 5.10, then the only possibilities are ± 1 and ± 3; however, direct substitution shows none of these satisfies the equation. Now, $\sqrt{3}$ is a real number, and since $(\sqrt{3})^2 - 3 = 3 - 3 = 0$, $\sqrt{3}$ is a real solution of the equation, and since it is not among the rational solutions, it must be irrational.

63. By Theorem 5.10, the only possible rational roots of the equation $x^2 - p = 0$ are ± 1 and $\pm p$. However, since p is prime, $p \neq 1$, and thus $(\pm 1)^2 - p = 1 - p \neq 0$ and neither 1 nor -1 is a solution. If either p or $-p$ is a solution, then $p^2 - p = 0$, and $p = 0$ or $p = 1$. Since p is prime, $p \neq 0$ and $p \neq 1$. Thus, neither p nor $-p$ is a solution. Finally $\sqrt{p}$ is a real solution; since it is not rational, it must be irrational.

Exercise 5.4 (Page 237)

1. $(1, 15)$ is a local maximum; $(2, 14)$ is a local minimum.

3. $(-1, 0)$ and $(1, 0)$ are local minima; $(0, 1)$ is a local maximum.

5. $(0, 10)$ is neither a local maximum nor a local minimum; $(1, 8)$ is a local minimum.

7.

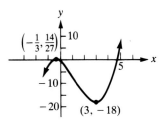

9.

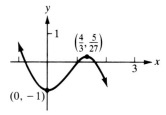

11.

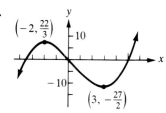

13.

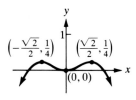

15.

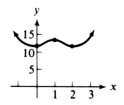

17.

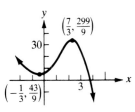

19.

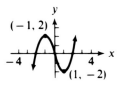

21.

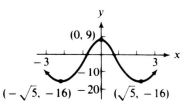

23.

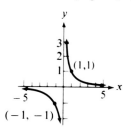

25.

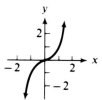

27. The vertex of the graph of $P(x) = ax^2 + bx + c$ is the only turning point. The only turning point occurs where $P'(x) = 2ax + b = 0$, or $x = \dfrac{-b}{2a}$. The y-coordinate is $P\left(-\dfrac{b}{2a}\right) = \dfrac{4ac - b^2}{4a}$.

Exercise 5.5 (Page 246)

1. $x = 3$ **3.** $x = 3;\quad x = -2$ **5.** $x = -1;\quad x = -4$ **7.** None   **9.** None

11. $x = -1$ **13.** $x = 1;\quad x = -1$ **15.** $x = 2;\quad x = -2;\quad y = 0$ **17.** $x = 4;\quad y = x + 4$

19. $x = 4;\quad x = -1;\quad y = 1$ **21.** No vertical asymptote; $\quad y = 1$

23. No vertical asymptote; $\quad y = x + 1$ **25.** No straight line asymptotes

27.

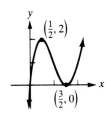

29.

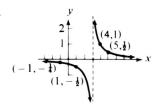

31.

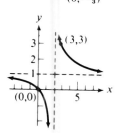

33.

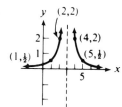

35.

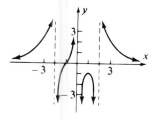

37.

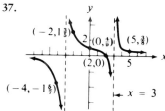

39.

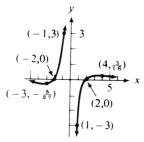

41.

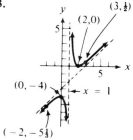

43.

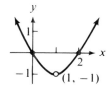

45.

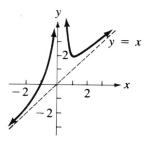

47.

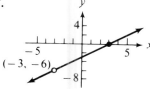

49.

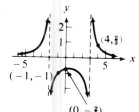

51.

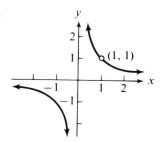

53. Behaves like $y = x^2$ as $|x|$ becomes large.

55. No, because if $x = a$ is a vertical asymptote, then the function is not defined at $x = a$; thus, there is no y so that the point (a, y) is on the graph.

57. $(-1, 1)$ **59.** $y = \dfrac{(x-1)(x-2)}{x^3}$

Chapter 5 Review (Page 248)

1. $x^3 + 2x^2 - x + 5$ **2.** $3x^2 - 4 + \dfrac{3}{x+2}$ **3.** $26, -4, -2$ **4.** $14, -1, -34$

5. $(x + 2)(2x - 3)(x - 1)$ 6. $(2x - 1)(3x + 2)(x + 2)$ 7. Upper bound 2;   lower bound -2

8. Upper bound 3;   lower bound -4 9. 1 positive, 0 or 2 negative   10. 0 or 2 positive, 0 or 2 negative

11. $\{2, 2i, -2i\}$ 12. $\{\frac{3}{2}, 2 + 2i, 2 - 2i\}$ 13. Zero between 3.4 and 3.5

14. Zero between 1.9 and 2.0   15. $\{4\}$ 16. $\{-\frac{1}{2}, 3\}$ 17. $\{-2, \sqrt{2}, -\sqrt{2}\}$

18. $\{\frac{1}{2}, -\frac{1}{3}, \sqrt{\frac{2}{3}}, -\sqrt{\frac{2}{3}}\}$ 19. $(1, 5)$ is a local maximum; $(2, 4)$ is a local minimum.

20. $(0, 0)$ is a local maximum; $(\frac{2}{3}, \frac{-4}{27})$ is a local minimum.

21.
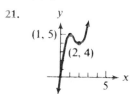

22.
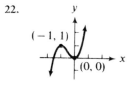

23.
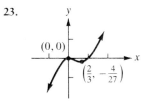

24.
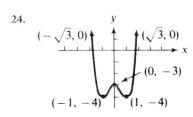

25. Vertical asymptote: $x = -2$
Horizontal asymptote: $y = 0$

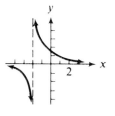

26. Vertical asymptotes: $x = -3$; $x = 4$
Horizontal asymptote: $y = 0$

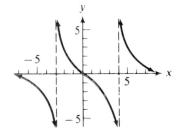

27. Vertical asymptote: $x = 3$
Horizontal asymptote: $y = 1$

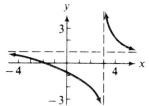

28. Vertical asymptotes: $x = 1$; $x = 2$
Horizontal asymptote: $y = 0$

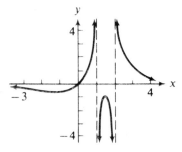

29. $1, -1, i, -i$

30. $2, -2, 2i, -2i$ **31.** Zero between -0.725 and -0.724 **32.** Zero between -0.475 and -0.474

33. Zeros between -0.6 and -0.5, between 0.6 and 0.7, and between 2.8 and 2.9

34. Zeros between -0.4 and -0.3, and between 1.3 and 1.4; zero at 0.5

35.

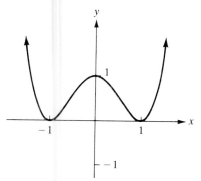

36.

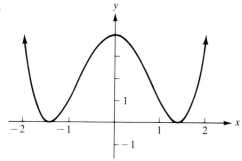

37.

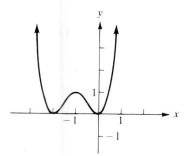

38.

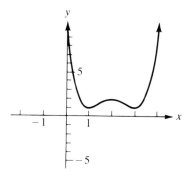

39.

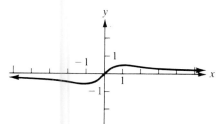

40.

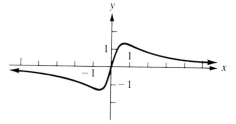

41.

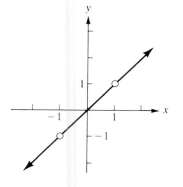

42.

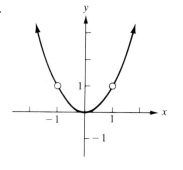

Exercise 6.1 (Page 256)

1. 1, 3 **3.** −1, −25 **5.** 8, $\frac{1}{8}$ **7.** $\frac{1}{100}$, 10 **9.** 0.3679, 2.7183 **11.** 0.0631, 3162.2777
13. 0.1003, 0.2466 **15.** 2.1544, 25.9546 **17.** 0.8669, 9.3565

19.

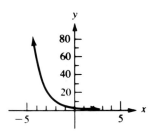

21.

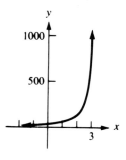

23.

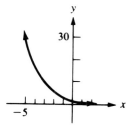

25.

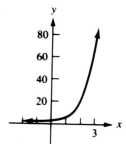

27.

29. {5} **31.** {−2} **33.** {−4} **35.** {$\frac{3}{4}$} **37.** 2 **39.** −2 **41.** 4 **43.** −4

45.

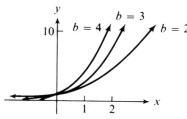

47.

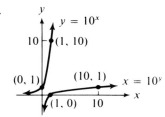

As $b > 1$ increases, the rate of increase of each curve becomes larger.

49.

51. 3.33×10^{-1}, 4.11×10^{-4}, 1.69×10^{-5}, 1.94×10^{-48}

53. 6.67×10^{-1}, 1.32×10^{-1}, 1.73×10^{-2}, 2.46×10^{-18}

55. 0.90000, 0.59049, 0.34868, 0.00003

57. **a.** 2.81×10^{-12}, 1.70×10^{-16}, 9.64×10^{-21}, 5.25×10^{-25}

    b. 7.58×10^{-8}, 1.09×10^{-11}, 1.21×10^{-15}, 1.13×10^{-19}

    c. $55.26, 0.04$, 1.88×10^{-5}, 5.29×10^{-9}

59. **a.** 2.71828, 29.68263, 2202.64658, 2.69×10^{41}

    b. 2.71828, 0.23746, 2.20265, 2.69×10^{35}

    c. 2.71828, 0.00002, 2.20×10^{-6}, 2.69×10^{23}

Exercise 6.2 (Page 262)

1. $2^6 = 64$ **3.** $3^2 = 9$ **5.** $(\frac{1}{3})^{-2} = 9$ **7.** $10^3 = 1000$ **9.** $10^{-1} = 0.1$ **11.** $e^2 = e^2$

13. $\log_4 16 = 2$ **15.** $\log_3 27 = 3$ **17.** $\log_{1/2} \frac{1}{4} = 2$ **19.** $\log_8 \frac{1}{2} = -\frac{1}{3}$ **21.** $\log_{10} 100 = 2$

23. $\log_{10}(0.1) = -1$ **25.** 2 **27.** 3 **29.** -1 **31.** 1 **33.** 2 **35.** -1 **37.** $\{2\}$ **39.** $\{2\}$

41. $\{64\}$ **43.** 0.1875 **45.** -0.2924 **47.** 1.3002 **49.** -0.6931

51. $y = \log_2 x$ **53.** $y = \log_2 4x$ **55.** $y = \log_2 (x + 3)$

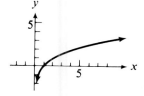 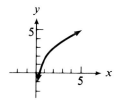 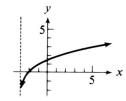

57. $y = \log_{10} x^3$

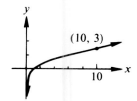

59. -0.006908, -0.000921, -0.000115, -0.000014

61. -0.000145, -0.000011, -0.000001, 0.000000

63. 0.0069, 1.38×10^{-5}, 2.07×10^{-8}, 2.76×10^{-11}, 3.45×10^{-14}

65. 1.2284, 0.43688, 0.11654, 0.02763, 0.00614

67. As $b > 1$ increases, $\log_b x$ increases more slowly.

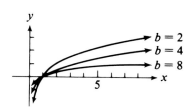

69. 5 **71.** 10 **73.** 22 **75.** 13

77. No horizontal asymptote. Since $f(x) = \ln x$ is increasing and there is no horizontal asymptote, we can make the value of $\ln x$ as large as desired by choosing sufficiently large x.

79. No horizontal asymptote

Exercise 6.3 (Page 267)

1. $\log_b x + \log_b y$ **3.** $\log_b x - \log_b y$ **5.** $5 \log_b x$ **7.** $\frac{1}{3} \log_b x$ **9.** $\frac{1}{3} \log_b x + \frac{2}{3} \log_b y - 3 \log_b z$

11. $\frac{3}{2} \log_b x + \frac{1}{2} \log_b y - \frac{1}{2} \log_b z - \frac{1}{2} \log_b 3$ **13.** 1.7917 **15.** -0.9163 **17.** 2.1972

19. 0.1116 **21.** 1.4067 **23.** -2.8133 **25.** $\log_b 2xy^3$ **27.** $\log_b x^{1/2} y^{2/3}$ **29.** $\log_b \dfrac{x^3 y}{z^2}$

31. $\log_b (x^6 y^{8/3})$ **33.** $\log_b (2^{1/3} x^{5/6} y^{1/2})$ **35.** $\log_b \left(\dfrac{x}{y}\right)^{\sqrt{2}-\sqrt{3}}$ **37.** $\frac{16}{3}$ **39.** ± 1 **41.** $\dfrac{10^2}{4} = 25$

43. $\pm\sqrt{1/50} \approx \pm 0.1414$ **45.** $\frac{1}{3} e^2 \approx 2.4630$ **47.** $\pm\sqrt{e^{-1}/2} \approx \pm 0.4289$ **49.** $\dfrac{10^5}{6} \approx 16{,}666.6667$

51. 500 **53.** 4 **55.** 3 **57.** 100 **59.** $74{,}621.5449$ **61.** $1{,}618.1780$

63. $2 \ln|x-1| + \frac{1}{2} \ln(x^2+2) - \ln(x^2-3)$ **65.** $\frac{1}{2} \ln|x-1| + \frac{1}{2} \ln|x+1| + \frac{1}{2} \ln(x^2+1) - \ln x - \ln 2$

67. $\frac{1}{2}[\ln(x^3-1) - \ln(x^2+1)]$ **69.** $\frac{1}{2}[\ln x - \frac{1}{2}\ln(x+1)]$

Exercise 6.4 (Page 273)

1. 8.7997 **3.** 220.2647 **5.** 1.0829 **7.** 0.6957 **9.** $\frac{1}{3}\log_{10} 6 = 0.2594$

11. $-\frac{1}{3} + \frac{1}{3}\log_{10} 9 = -0.0153$ **13.** $\pm\sqrt{\log_{10} 150} = \pm 1.4752$ **15.** $(\log_{10} 25)^2 = 1.9542$

17. $(\log_{10} 2)^2 - 1 = -0.9094$ **19.** $10 - 10\log_{10} 1.6429 = 7.8439$ **21.** $\frac{1}{3} \ln 5 = 0.5365$

23. $-\frac{1}{2} + \frac{1}{2} \ln 25 = 1.1094$ **25.** $\pm\sqrt{2 \ln 15} = \pm 2.3273$ **27.** $(\ln 20)^2 = 8.9744$

29. $-2 + (1 + \ln 2)^2 = 0.8667$ **31.** $-25 + 25 \ln 1.7857 = -10.5045$ **33.** $\dfrac{\log_{10} 7}{\log_{10} 2} = 2.8074$

35. $\dfrac{1}{2}\left(\dfrac{\log_{10} 3}{\log_{10} 7} + 1\right) = 0.7823$ **37.** $\pm\sqrt{\dfrac{\log_{10} 16}{\log_{10} 4}} = \pm 1.4142$ **39.** $\left(\dfrac{\log_{10} 8}{\log_{10} 4}\right)^2 = 2.25$

41. $\pm\sqrt{\dfrac{\log_{10} 14}{\log_{10} 6}} + 2 = \pm 1.8636$ **43.** $\pm\sqrt{\dfrac{\log_{10} 0.45}{\log_{10} 5}} + 3 = \pm 1.5824$ **45.** $t = \dfrac{1}{k}\log_{10} A$

47. $t = \dfrac{1}{k} \ln y$ **49.** $n = \dfrac{\log_{10} y}{\log_{10} x}$ **51.** $t = -\frac{1}{4}\ln[\frac{1}{3}(y+2)]$ **53.** $t = \frac{1}{2} - \frac{1}{2}\ln(-y - \frac{1}{2})$

55. $n = \dfrac{\log_{10} V - \log_{10} P}{\log_{10}(1+r)}$ **57.** $0.205672, \quad 0.045007, \quad 0.006884, \quad 0.000921$

59. $-0.124539, \quad -0.013468, \quad -0.001358, \quad -0.000136$ **61.** $k = \dfrac{\log_{10} V - \log_{10} P}{n \log_{10}[1 + (r/100n)]}$

63. $k = -\dfrac{1}{n}\left[\dfrac{\log_{10}(P/S)}{\log_{10}\left(1 + \dfrac{r}{n}\right)}\right]$

Exercise 6.5 (Page 283)

1. $16{,}200$ **3.** 0.58 **5.** 2003 **7.** 0.18 **9.** 548.8 grams   **11.** 13.86 seconds   **13.** $90.5°$

15. 1611.8 seconds   **17.** 380 elk   **19.** 0.063 **21.** 1832 **23.** $87{,}059$ **25.** 11 years   **27.** 7%

29. $\$12{,}391.14; \quad \$12{,}648.84$ **31.** A gives the best return by approximately 1.25 cents on the dollar.

33. 7.7 **35.** 2.5×10^{-6} **37.** -1573.3 joule   **39.** 4.50 **41.** $25{,}119$ **43.** 3.5

45. β increases by 3.01 db   **47.** 1000 **49.** 8.62% **51.** 7.06% **53.** $\$767.15$

55. $\$27{,}782$ **57.** $y = Q_M$

59. No. For large values of t, $\alpha t > 0$ and αt is large. Thus, $e^{\alpha t}$ continues to grow without bound. This model does not reflect any limit on growth. Thus, it is not a good model for a population that has a limited growth potential.

61. $\dfrac{\ln 2}{|\alpha|}$

Chapter 6 Review (Page 288)

1. 3.1623 **2.** 2.1544 **3.** 25.9546 **4.** 0.0185 **5.** 1.2840 **6.** 1.1814 **7.** 9.3565 **8.** 0.1069

9. 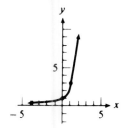   **10.**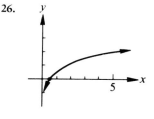

11. $\log_{16} \frac{1}{4} = -\frac{1}{2}$ **12.** $\log_7 343 = 3$ **13.** $2^3 = 8$ **14.** $10^{-4} = 0.0001$ **15.** $y = 4$ **16.** $x = 1000$
17. 1.6232 **18.** -2.5031 **19.** 2.8340 **20.** -1.3828 **21.** 1.9459 **22.** 3.1355 **23.** 5.4848
24. 6.2344

25. 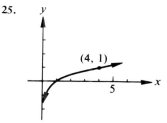   **26.**

27. $\frac{1}{3}\log_{10} x + \frac{2}{3}\log_{10} y$ **28.** $\log_{10} 2 + 3\log_{10} R - \frac{1}{2}\log_{10} P - \frac{1}{2}\log_{10} Q$ **29.** $\log_b \dfrac{x^2}{\sqrt[3]{y}}$

30. $\log_{10} \dfrac{\sqrt[3]{x^2 y}}{z^3}$ **31.** ± 5.2249 **32.** 0.7165 **33.** $\frac{1}{9}$ **34.** $\sqrt{11}$ **35.** $\dfrac{\log_{10} 8}{-2} = -0.4515$

36. $-2 - \log_{10} 25 = -3.3979$ **37.** $\dfrac{\ln 8}{4} = 0.5199$ **38.** $\frac{1}{2}(2 - \ln 16) = -0.3863$ **39.** $\dfrac{\log_{10} 2}{\log_{10} 3} = 0.6309$

40. $x = \dfrac{\log_{10} 80}{\log_{10} 3} - 1 = 2.9887$ **41.** $t = -\ln\left(\dfrac{y - A}{k}\right)$ **42.** $t = -c - \ln\left(\dfrac{y - A}{k}\right)$ **43.** 201,000

44. 0.7 second **45.** 497 **46.** 2 **47.** $5\frac{1}{2}\%$ **48.** $15\frac{1}{4}$ years **49.** 5.27 **50.** -2493.5 joule
51. 3,162,278 **52.** 3162 **53.** 1 **54.** 8 **55.** 150 **56.** 103 **57.** 225,097,187.2 **58.** 831,117.3
59. $y = 0$ is a horizontal asymptote because as x gets larger the value of $(0.95)^x$ gets very close to 0.
60. $y = 0$ is a horizontal asymptote because as $|x|$ gets larger with $x < 0$ the value of $(1.05)^x$ gets very close to 0.
61. \$634.84 **62.** A little more than 223 months **63.** $y = 100$
64. $y = 0$ for $|t|$ large, $t > 0$ and $y = 1$ for $|t|$ large, $t < 0$

Exercise 7.1 (Page 299)

1. $\{(3, 2)\}$ **3.** $\{(2, 1)\}$ **5.** $\{(-5, 4)\}$ **7.** $\{(0, \frac{3}{2})\}$ **9.** $\{(\frac{2}{3}, -1)\}$ **11.** $\{(1, 2)\}$

13. $\{(-\frac{19}{5}, -\frac{18}{5})\}$ **15.** $(x, -\frac{3}{2}x + 3), x \in R$ **17.** No solution   **19.** 7 quarters, 5 nickels

21. 32 pounds   **23.** $3040 at 11% and $4040 at 9%   **25.** 32 and 64 mph   **27.** $a = 1, b = -1$

29. $a = 0, b = 2$ **31.** $y = -\frac{10}{3}x + 2$ **33.** $\{(1, 1)\}$ **35.** $\{(0, 1)\}$

37. The left-hand members are independent if and only if both equations represent the same line. This is true if and only if the slopes and the y-intercepts are equal. This is the case if and only if

$$\frac{-a_1}{b_1} = \frac{-a_2}{b_2} \quad \text{and} \quad \frac{-c_1}{b_1} = \frac{-c_2}{b_2}.$$

But this is true if and only if

$$\frac{a_1}{a_2} = \frac{b_1}{b_2} = \frac{c_1}{c_2}.$$

39. $a = b = 0$

Exercise 7.2 (Page 307)

1. $\{(1, 2, -1)\}$ **3.** $\{(2, -2, 0)\}$ **5.** $\{(2, 2, 1)\}$ **7.** $\{(0, 1, 2)\}$ **9.** Dependent   **11.** $\{(4, -2, 2)\}$

13. 3, 6, 6   **15.** 60 nickels, 20 dimes, 5 quarters   **17.** $x = 40$ cm, $y = 60$ cm, $z = 55$ cm

19. $a = -6, b = -8, c = 0$ **21.** $a = 3, b = 1, c = -2$ **23.** $a = \frac{1}{4}, b = \frac{1}{2}, c = 0$ **25.** $\{(1, 1, 1)\}$

27. $\{(0, 1, 2)\}$ **29.** $(\frac{5}{3} - \frac{1}{3}z, -\frac{5}{3}z + \frac{1}{3}, z), z \in R; \{(\frac{5}{3}, \frac{1}{3}, 0), (2, 2, -1)\}$

Exercise 7.3 (Page 316)

1. $\{(-1, -4), (5, 20)\}$ **3.** $\{(2, 3), (3, 2)\}$ **5.** $\{(4, -3), (-3, 4)\}$

7. $\{(-1, 3), (-1, -3), (1, 3), (1, -3)\}$ **9.** $\{(-3, \sqrt{2}), (-3, -\sqrt{2}), (3, \sqrt{2}), (3, -\sqrt{2})\}$

11. $\{(\sqrt{3}, 4), (\sqrt{3}, -4), (-\sqrt{3}, 4), (-\sqrt{3}, -4)\}$ **13.** $\{(1, -2), (-1, 2), (2, -1), (-2, 1)\}$

15. $\{(3, 1), (-3, -1), (-2\sqrt{7}, \sqrt{7}), (2\sqrt{7}, -\sqrt{7})\}$ **17.** $\{(2\sqrt{\frac{1}{3}}, \sqrt{\frac{1}{3}}), (-2\sqrt{\frac{1}{3}}, -\sqrt{\frac{1}{3}}), (-6i, 4i), (6i, -4i)\}$

19. $\{(i\sqrt{7}, 4), (-i\sqrt{7}, 4)\}$ **21.** $\{(2i, -2i), (-2i, 2i), (2\sqrt{2}, \sqrt{2}), (-2\sqrt{2}, -\sqrt{2})\}$

23. $\{(-1, \frac{1}{10})\}$ **25.** $\{(1, 0), (10, -1)\}$ **27.** $\{(0, 2)\}$ **29.** $\{(1, 0), (\frac{3}{2}, \frac{1}{2})\}$

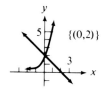

$\{(0,2)\}$

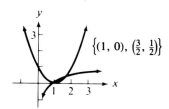

$\{(1, 0), (\frac{3}{2}, \frac{1}{2})\}$

31. $x = 2, y = 3$ **33.** $800 at 4%   **35.** $\{(2, 2), (-2, -2)\}$

Exercise 7.4 (Page 319)

1.

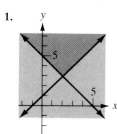

3.

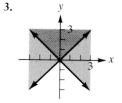

5.

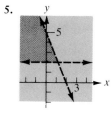

7.

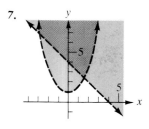

9.
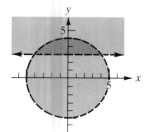

11.
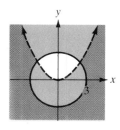

13.
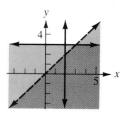

15.
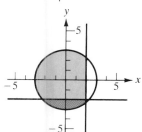

17.
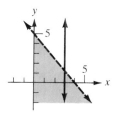

19.
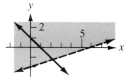

21.
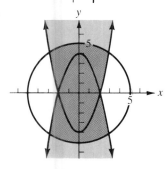

23.
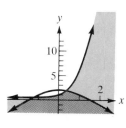

25.
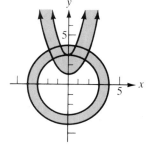

Chapter 7 Review (Page 320)

1. $\left\{\left(\frac{1}{2}, \frac{7}{2}\right)\right\}$ **2.** $\{(1, 2)\}$ **3.** $\left\{\left(\frac{37}{13}, -\frac{10}{13}\right)\right\}$ **4.** $\{(-2, 1)\}$ **5.** $(x, \frac{2}{3}x - \frac{1}{3}), x \in R$ **6.** No solution

7. $a = 2, b = 5$ **8.** $a = \frac{2}{3}, b = \frac{5}{3}$ **9.** $\{(2, 0, -1)\}$ **10.** $\{(2, 1, -1)\}$ **11.** $\{(2, -1, 3)\}$

12. $a = 2, b = -3, c = 4$ **13.** $\{(1, 2), (4, -13)\}$ **14.** $\{(1, 1), (-1, -1)\}$

15. $\{(2, 3), (2, -3), (-2, 3), (-2, -3)\}$

16. $\left\{\left(\frac{1}{\sqrt{5}}, -\frac{1}{\sqrt{5}}\right), \left(-\frac{1}{\sqrt{5}}, \frac{1}{\sqrt{5}}\right), \left(\frac{1}{\sqrt{11}}, \frac{-2}{\sqrt{11}}\right), \left(-\frac{1}{\sqrt{11}}, \frac{2}{\sqrt{11}}\right)\right\}$ **17.** $\{(-\frac{1}{3}, 10^{-2/3})\}$

18. $\{(\frac{1}{20}, -1)\}$ **19.** $\{(\frac{7}{8}, \frac{17}{8})\}$ **20.**

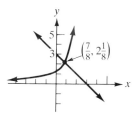

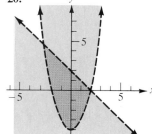

21.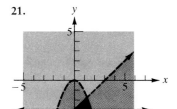

22.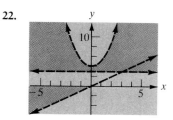

23. $\{(\frac{7}{16}, \frac{7}{6})\}$ **24.** $\{(-3, 2)\}$ **25.** $\{(1, -1, \frac{1}{2})\}$ **26.** $\{(\frac{1}{2}, -\frac{8}{3}, \frac{14}{3})\}$

27. $a^2 + 8b + 16 > 0,\quad a^2 + 8b + 16 = 0,\quad a^2 + 8b + 16 < 0$

28. $b^2 - a^2 > -1, a \neq \pm 1;\quad b^2 - a^2 < -1, a \neq \pm 1;\quad b^2 - a^2 < -1, a = \pm 1$

29.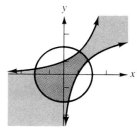

30.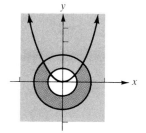

Exercise 8.1 (Page 329)

1. $2 \times 2, \begin{bmatrix} 6 & 2 \\ -1 & 3 \end{bmatrix}$ **3.** $2 \times 3, \begin{bmatrix} 2 & 1 \\ -7 & 4 \\ 3 & 0 \end{bmatrix}$ **5.** $3 \times 3, \begin{bmatrix} 2 & 4 & -2 \\ 3 & 0 & 3 \\ -1 & 1 & 1 \end{bmatrix}$ **7.** $2 \times 4, \begin{bmatrix} 4 & 2 \\ -3 & 1 \\ -1 & 1 \\ 0 & 6 \end{bmatrix}$

9. $\begin{bmatrix} 3 & 1 \\ 3 & 9 \end{bmatrix}$ **11.** $\begin{bmatrix} 9 & -1 & -1 \\ 2 & 3 & 6 \end{bmatrix}$ **13.** $\begin{bmatrix} -2 & -3 \\ 4 & 0 \end{bmatrix}$ **15.** $\begin{bmatrix} 2 & -9 & -13 \\ 7 & -4 & 1 \\ 9 & 0 & 0 \end{bmatrix}$ **17.** $\begin{bmatrix} 10 \\ 3 \\ -3 \end{bmatrix}$

19. $\begin{bmatrix} 2 & 3 & 4 \\ -1 & 6 & 2 \\ 1 & 0 & 3 \end{bmatrix}$ **21.** $\begin{bmatrix} 2 & -1 \\ 2 & 1 \end{bmatrix}$ **23.** $\begin{bmatrix} 1 & 0 \\ -5 & 3 \end{bmatrix}$ **25.** $\begin{bmatrix} 0 & 1 & 1 & 0 & 0 & 0 & 0 & 0 \\ 0 & 0 & 0 & 1 & 1 & 0 & 0 & 0 \\ 0 & 0 & 0 & 0 & 0 & 1 & 1 & 0 \\ 0 & 0 & 0 & 0 & 0 & 0 & 0 & 0 \\ 0 & 0 & 0 & 0 & 0 & 0 & 0 & 0 \\ 0 & 0 & 0 & 0 & 0 & 0 & 0 & 0 \\ 0 & 0 & 0 & 0 & 0 & 0 & 0 & 1 \\ 0 & 0 & 0 & 0 & 0 & 0 & 0 & 0 \end{bmatrix}$

27. $\begin{bmatrix} 1 & -2 \\ -2 & 1 \end{bmatrix}$

29. Let $A = \begin{bmatrix} a_{11} & a_{12} \\ a_{21} & a_{22} \end{bmatrix}$; then $A^t = \begin{bmatrix} a_{11} & a_{21} \\ a_{12} & a_{22} \end{bmatrix}$ and $[A^t]^t = \begin{bmatrix} a_{11} & a_{12} \\ a_{21} & a_{22} \end{bmatrix} = A$. This result holds for both $n \times n$ and $m \times n$ matrices.

Exercise 8.2 (Page 337)

1. $\begin{bmatrix} 2 & 2 \\ 0 & 4 \end{bmatrix}$ 3. $\begin{bmatrix} 0 & -5 & 5 \\ -15 & 5 & -10 \end{bmatrix}$ 5. $\begin{bmatrix} -1 & 1 \\ -1 & -1 \end{bmatrix}$ 7. $\begin{bmatrix} 1 & -1 \\ 1 & 0 \end{bmatrix}$ 9. 5 11. -1

13. $\begin{bmatrix} 1 & -13 \\ 4 & -7 \end{bmatrix}$ 15. $\begin{bmatrix} 30 & -39 \\ 29 & 14 \end{bmatrix}$ 17. $\begin{bmatrix} -5 & -1 \\ 8 & -1 \end{bmatrix}$ 19. $\begin{bmatrix} -1 & 0 & -2 \\ 1 & 2 & 8 \\ 0 & 1 & 3 \end{bmatrix}$ 21. $\begin{bmatrix} 1 & 0 & 0 \\ 0 & 1 & 0 \\ 0 & 0 & 1 \end{bmatrix}$

23. $\begin{bmatrix} 1 & 0 \\ -1 & 2 \end{bmatrix}$ 25. $\begin{bmatrix} 1 & -2 \\ 1 & 2 \end{bmatrix}$ 27. $\begin{bmatrix} -2 & 3 \\ 2 & -4 \end{bmatrix}$ 29. $\begin{bmatrix} 1 & 2 & 3 & 4 \\ 0 & 1 & 2 & 3 \\ 0 & 0 & 1 & 2 \\ 0 & 0 & 0 & 1 \end{bmatrix}$ 31. $\begin{bmatrix} 2 & 2 & 0 & 0 \\ 2 & 2 & 0 & 0 \\ 0 & 0 & 2 & 2 \\ 0 & 0 & 2 & 2 \end{bmatrix}$

33. $[2 \quad -4 \quad 4]$ 35. $[-\frac{2}{3} \quad -\frac{1}{2} \quad -\frac{7}{6}]$

37. Since $A + B = \begin{bmatrix} 0 & 2 \\ -1 & 3 \end{bmatrix}$, $A - B = \begin{bmatrix} -2 & 2 \\ 1 & -1 \end{bmatrix}$, $A^2 = \begin{bmatrix} 1 & 0 \\ 0 & 1 \end{bmatrix}$, $B^2 = \begin{bmatrix} 1 & 0 \\ -3 & 4 \end{bmatrix}$, and $AB = \begin{bmatrix} -3 & 4 \\ -1 & 2 \end{bmatrix}$, we have:

a. $(A + B)(A + B) = \begin{bmatrix} 0 & 2 \\ -1 & 3 \end{bmatrix} \cdot \begin{bmatrix} 0 & 2 \\ -1 & 3 \end{bmatrix} = \begin{bmatrix} -2 & 6 \\ -3 & 7 \end{bmatrix}$ and

$$A^2 + 2AB + B^2 = \begin{bmatrix} 1 & 0 \\ 0 & 1 \end{bmatrix} + 2\begin{bmatrix} -3 & 4 \\ -1 & 2 \end{bmatrix} + \begin{bmatrix} 1 & 0 \\ -3 & 4 \end{bmatrix} = \begin{bmatrix} -4 & 8 \\ -5 & 9 \end{bmatrix}.$$

Hence, $(A + B)(A + B) \neq A^2 + 2AB + B^2$.

b. $(A + B)(A - B) = \begin{bmatrix} 0 & 2 \\ -1 & 3 \end{bmatrix} \cdot \begin{bmatrix} -2 & 2 \\ 1 & -1 \end{bmatrix} = \begin{bmatrix} 2 & -2 \\ 5 & -5 \end{bmatrix}$ and

$$A^2 - B^2 = \begin{bmatrix} 1 & 0 \\ 0 & 1 \end{bmatrix} - \begin{bmatrix} 1 & 0 \\ -3 & 4 \end{bmatrix} = \begin{bmatrix} 0 & 0 \\ 3 & -3 \end{bmatrix}.$$

Hence, $(A + B)(A - B) \neq A^2 - B^2$.

39. $A_{2 \times 2} \cdot I_{2 \times 2} = \begin{bmatrix} a & b \\ c & d \end{bmatrix} \cdot \begin{bmatrix} 1 & 0 \\ 0 & 1 \end{bmatrix} = \begin{bmatrix} a \cdot 1 + b \cdot 0 & a \cdot 0 + b \cdot 1 \\ c \cdot 1 + d \cdot 0 & c \cdot 0 + d \cdot 1 \end{bmatrix}$

$= \begin{bmatrix} a & b \\ c & d \end{bmatrix} = A_{2 \times 2}$. Similarly, $I_{2 \times 2} \cdot A_{2 \times 2} = A_{2 \times 2}$.

41. $(A_{2 \times 2} \cdot B_{2 \times 2}) = \begin{bmatrix} a_{11}b_{11} + a_{12}b_{21} & a_{11}b_{12} + a_{12}b_{22} \\ a_{21}b_{11} + a_{22}b_{21} & a_{21}b_{12} + a_{22}b_{22} \end{bmatrix}$

Hence, $(A_{2 \times 2} \cdot B_{2 \times 2})^t = \begin{bmatrix} a_{11}b_{11} + a_{12}b_{21} & a_{21}b_{11} + a_{22}b_{21} \\ a_{11}b_{12} + a_{12}b_{22} & a_{21}b_{12} + a_{22}b_{22} \end{bmatrix}$.

By the commutative property of multiplication for real numbers,

$(A_{2 \times 2} \cdot B_{2 \times 2})^t = \begin{bmatrix} b_{11}a_{11} + b_{21}a_{12} & b_{11}a_{21} + b_{21}a_{22} \\ b_{12}a_{11} + b_{22}a_{12} & b_{12}a_{21} + b_{22}a_{22} \end{bmatrix} = \begin{bmatrix} b_{11} & b_{21} \\ b_{12} & b_{22} \end{bmatrix} \cdot \begin{bmatrix} a_{11} & a_{21} \\ a_{12} & a_{22} \end{bmatrix} = B_{2 \times 2}^t \cdot A_{2 \times 2}^t.$

Exercise 8.3 (Page 347)

1. $\{(2, -1)\}$ 3. $\{(5, 1)\}$ 5. $\{(8, 1)\}$ 7. $\{(-2, 2, 0)\}$ 9. $\{(\frac{5}{4}, \frac{5}{2}, -\frac{1}{2})\}$ 11. $\{(-\frac{77}{27}, -\frac{8}{27}, \frac{29}{27})\}$

13. $\{(1, 1 - z, z)\}$ **15.** $(3, 3, -1)$ **17.** No solution **19.** No solution **21.** $1, \frac{7}{5}, \frac{3}{5}$

23. 5 quarts of wine, $2\frac{1}{2}$ quarts of brandy, $2\frac{1}{2}$ quarts of fruit juice **25.** $a = -3, b = \frac{5}{3}, c = -\frac{2}{3}$

27. $\begin{bmatrix} k & 0 \\ 0 & 1 \end{bmatrix}\begin{bmatrix} a & b \\ c & d \end{bmatrix} = \begin{bmatrix} k \cdot a + 0 \cdot c & k \cdot b + 0 \cdot d \\ 0 \cdot a + 1 \cdot c & 0 \cdot b + 1 \cdot d \end{bmatrix} = \begin{bmatrix} ka & kb \\ c & d \end{bmatrix}$

29. $\begin{bmatrix} 1 & 0 \\ k & 1 \end{bmatrix}\begin{bmatrix} a & b \\ c & d \end{bmatrix} = \begin{bmatrix} 1 \cdot a + 0 \cdot c & 1 \cdot b + 0 \cdot d \\ k \cdot a + 1 \cdot c & k \cdot b + 1 \cdot d \end{bmatrix} = \begin{bmatrix} a & b \\ ka + c & kb + d \end{bmatrix}$

Exercise 8.4 (Page 352)

1. 3 **3.** 4 **5.** 0 **7.** -11 **9.** $A_{11} = \begin{vmatrix} 0 & 3 & -1 \\ 1 & 2 & 2 \\ -1 & 3 & 1 \end{vmatrix}$ **11.** $A_{23} = -\begin{vmatrix} 2 & 1 & 0 \\ -2 & 1 & 2 \\ 1 & -1 & 1 \end{vmatrix}$

13. $A_{31} = \begin{vmatrix} 1 & -2 & 0 \\ 0 & 3 & -1 \\ -1 & 3 & 1 \end{vmatrix}$ **15.** $A_{44} = \begin{vmatrix} 2 & 1 & -2 \\ 1 & 0 & 3 \\ -2 & 1 & 2 \end{vmatrix}$ **17.** 16 **19.** -4 **21.** 0 **23.** -24

25. 4 **27.** Let $A = \begin{vmatrix} x & y & 1 \\ x_1 & y_1 & 1 \\ x_2 & y_2 & 1 \end{vmatrix} = 0$. Expanding about the first row gives

$$\delta(A) = x\begin{vmatrix} y_1 & 1 \\ y_2 & 1 \end{vmatrix} - y\begin{vmatrix} x_1 & 1 \\ x_2 & 1 \end{vmatrix} + 1\begin{vmatrix} x_1 & y_1 \\ x_2 & y_2 \end{vmatrix} = 0.$$

Hence, $(y_1 - y_2)x + (x_2 - x_1)y + (x_1 y_2 - y_1 x_2) = 0$. Further, y_1, y_2, x_1, x_2 are real numbers; hence, there exist real numbers a, b, c such that $(y_1 - y_2) = a$, $(x_2 - x_1) = b$, and $(x_1 y_2 - y_1 x_2) = c$, with a and b not both 0 since $(x_1, y_1) \neq (x_2, y_2)$. Substituting yields $ax + by + c = 0$, which is the equation of a straight line. Since $ax_1 + by_1 + c = 0$ and $ax_2 + by_2 + c = 0$, the line passes through (x_1, y_1) and (x_2, y_2).

29. If the ith row of A is identically 0, then expanding about the ith row we have

$$\delta(A) = 0 \cdot A_{i1} + 0 \cdot A_{i2} + \cdots + 0 \cdot A_{in} = 0.$$

Similarly, if the ith column is identically 0, then $\delta(A) = 0$.

31. Let $A = \begin{bmatrix} a_{11} & a_{12} \\ a_{21} & a_{22} \end{bmatrix}$; by Definition 8.6 and the fact that $\delta(A) = a_{11}a_{22} - a_{12}a_{21}$, it follows that

$$aA = \begin{bmatrix} aa_{11} & aa_{12} \\ aa_{21} & aa_{22} \end{bmatrix} \text{ and } \delta(aA) = a^2 a_{11}a_{22} - a^2 a_{12}a_{21} = a^2(a_{11}a_{22} - a_{12}a_{21}) = a^2 \delta(A).$$

33. Let $A = \begin{bmatrix} a_{11} & a_{12} \\ a_{21} & a_{22} \end{bmatrix}$, $B = \begin{bmatrix} b_{11} & b_{12} \\ b_{21} & b_{22} \end{bmatrix}$; then $AB = \begin{bmatrix} a_{11}b_{11} + a_{12}b_{21} & a_{11}b_{12} + a_{12}b_{22} \\ a_{21}b_{11} + a_{22}b_{21} & a_{21}b_{12} + a_{22}b_{22} \end{bmatrix}$.

Thus, $\delta(AB) = (a_{11}b_{11} + a_{12}b_{21})(a_{21}b_{12} + b_{22}a_{22}) - (a_{11}b_{12} + a_{12}b_{22})(a_{21}b_{11} + a_{22}b_{21})$
$$= a_{11}b_{11}a_{21}b_{12} + a_{12}b_{21}a_{21}b_{12} + a_{11}b_{11}a_{22}b_{22} + a_{12}b_{21}a_{22}b_{22}$$
$$- a_{11}b_{12}a_{21}b_{11} - a_{12}b_{22}a_{21}b_{11} - a_{11}b_{12}a_{22}b_{21} - a_{12}b_{22}a_{22}b_{21}$$
$$= a_{11}b_{11}a_{22}b_{22} - a_{12}a_{21}b_{11}b_{22} - a_{11}a_{22}b_{21}b_{12} + a_{21}a_{12}b_{21}b_{12}$$
$$= b_{11}b_{22}(a_{11}a_{22} - a_{12}a_{21}) - b_{21}b_{12}(a_{11}a_{22} - a_{21}a_{12}) = (a_{11}a_{22} - a_{21}a_{12})(b_{11}b_{22} - b_{21}b_{12})$$
$$= \delta(A)\delta(B).$$

Exercise 8.5 (Page 356)

1. Theorem 8.7 **3.** Theorem 8.9 **5.** Theorem 8.10 **7.** Theorem 8.10 **9.** Theorem 8.11

11. Theorem 8.11   13. $\begin{vmatrix} 1 & 3 \\ 0 & -4 \end{vmatrix}$ 15. $\begin{vmatrix} 1 & -2 & 1 \\ 0 & 7 & 1 \\ 0 & 2 & 1 \end{vmatrix}$ 17. $\begin{vmatrix} 0 & 1 & -3 & -2 \\ 0 & 2 & 1 & 2 \\ 1 & 1 & 2 & 3 \\ 0 & 1 & 1 & 1 \end{vmatrix}$

19. $-1\begin{vmatrix} 2 & 1 \\ -1 & 2 \end{vmatrix} = -5$ 21. $\begin{vmatrix} -1 & -5 \\ 2 & -2 \end{vmatrix} = 12$ 23. $\begin{vmatrix} 4 & 4 \\ 3 & 7 \end{vmatrix} = 16$ 25. $\begin{vmatrix} 6 & 1 \\ 0 & 3 \end{vmatrix} = 18$

27. $-16\begin{vmatrix} 1 & 2 \\ 2 & 3 \end{vmatrix} = 16$ 29. $\begin{vmatrix} 4 & -4 \\ 3 & -9 \end{vmatrix} = -24$

31. Multiply Column 1 by $(-a)$ and add the result to Column 2; also, multiply Column 1 by $(-a^2)$ and add the result to Column 3, to obtain $\begin{vmatrix} 1 & a & a^2 \\ 1 & b & b^2 \\ 1 & c & c^2 \end{vmatrix} = \begin{vmatrix} 1 & 0 & 0 \\ 1 & b-a & b^2-a^2 \\ 1 & c-a & c^2-a^2 \end{vmatrix}$. Expand about the first row to obtain

$$1\begin{vmatrix} b-a & b^2-a^2 \\ c-a & c^2-a^2 \end{vmatrix} = (b-a)[c^2-a^2] - (c-a)[b^2-a^2]$$
$$= (b-a)[(c-a)(c+a)] - (c-a)[(b-a)(b+a)]$$
$$= -(a-b)[(c-a)(c+a)] + (c-a)[(a-b)(a+b)]$$
$$= (a-b)(c-a)[-(c+a) + (a+b)]$$
$$= (a-b)(c-a)(b-c) = (b-c)(c-a)(a-b).$$

33. If Rows i and j of the matrix A are identical, then interchanging the two rows results in the same matrix. Thus, by Theorem 8.8, $\delta(A) = -\delta(A)$ and hence, $\delta(A) = 0$. A similar argument shows $\delta(A) = 0$ if two columns are identical.

Exercise 8.6 (Page 364)

1. $\begin{bmatrix} 3 & -2 \\ -1 & 1 \end{bmatrix}$ 3. $\dfrac{1}{5}\begin{bmatrix} 1 & 3 \\ -1 & 2 \end{bmatrix}$ 5. $|A| = 0$; no inverse   7. $-1\begin{bmatrix} 4 & -7 \\ -3 & 5 \end{bmatrix}$ 9. $\dfrac{1}{2}\begin{bmatrix} -2 & -4 \\ 4 & 7 \end{bmatrix}$

11. $|A| = 0$; no inverse   13. $\dfrac{1}{6}\begin{bmatrix} 2 & 2 & -5 \\ -4 & 2 & 1 \\ 0 & 0 & 3 \end{bmatrix}$ 15. $\dfrac{1}{3}\begin{bmatrix} -2 & 3 & -1 \\ -1 & 0 & 1 \\ 6 & -6 & 3 \end{bmatrix}$ 17. $|A| = 0$; no inverse

19. $|A| = 0$; no inverse   21. $\begin{bmatrix} 2 & -3 & 11 \\ -2 & 4 & -13 \\ 1 & -2 & 7 \end{bmatrix}$ 23. $-1\begin{bmatrix} 0 & 0 & -1 \\ 0 & -1 & 0 \\ -1 & 0 & 0 \end{bmatrix}$

25. Let $A \cdot B = \begin{bmatrix} 2 & 3 \\ 1 & -1 \end{bmatrix} \cdot \begin{bmatrix} 0 & 1 \\ 3 & 1 \end{bmatrix}$;

$$A \cdot B = \begin{bmatrix} 9 & 5 \\ -3 & 0 \end{bmatrix} \text{ and } \delta(AB) = 15, \text{ so } (A \cdot B)^{-1} = \frac{1}{15}\begin{bmatrix} 0 & -5 \\ 3 & 9 \end{bmatrix}.$$

Also, since $\delta(A) = -5$ and $\delta(B) = -3$,

$$B^{-1} = -\frac{1}{3}\begin{bmatrix} 1 & -1 \\ -3 & 0 \end{bmatrix} \text{ and } A^{-1} = -\frac{1}{5}\begin{bmatrix} -1 & -3 \\ -1 & 2 \end{bmatrix}; B^{-1} \cdot A^{-1} = \frac{1}{15}\begin{bmatrix} 0 & -5 \\ 3 & 9 \end{bmatrix}.$$

Hence, $(A \cdot B)^{-1} = B^{-1} \cdot A^{-1}$.

27. Let $A = \begin{bmatrix} 3 & 0 & 1 \\ 2 & 1 & 0 \\ 0 & 1 & 2 \end{bmatrix}$; then $\delta(A) = 8$ and $A^{-1} = \dfrac{1}{8}\begin{bmatrix} 2 & 1 & -1 \\ -4 & 6 & 2 \\ 2 & -3 & 3 \end{bmatrix}$.

Let $B = \begin{bmatrix} 2 & 1 & 0 \\ 1 & 1 & 2 \\ 0 & 1 & 0 \end{bmatrix}$; then $\delta(B) = -\dfrac{1}{4}$ and $B^{-1} = -\dfrac{1}{4}\begin{bmatrix} -2 & 0 & 2 \\ 0 & 0 & -4 \\ 1 & -2 & 1 \end{bmatrix}$;

$A \cdot B = \begin{bmatrix} 6 & 4 & 0 \\ 5 & 3 & 2 \\ 1 & 3 & 2 \end{bmatrix}$, $\delta(A \cdot B) = -32$ and $(A \cdot B)^{-1} = -\dfrac{1}{32}\begin{bmatrix} 0 & -8 & 8 \\ -8 & 12 & -12 \\ 12 & -14 & -2 \end{bmatrix}$;

$B^{-1} \cdot A^{-1} = -\dfrac{1}{4}\begin{bmatrix} -2 & 0 & 2 \\ 0 & 0 & -4 \\ 1 & -2 & 1 \end{bmatrix} \cdot \dfrac{1}{8}\begin{bmatrix} 2 & 1 & -1 \\ -4 & 6 & 2 \\ 2 & -3 & 3 \end{bmatrix} = -\dfrac{1}{32}\begin{bmatrix} 0 & -8 & 8 \\ -8 & 12 & -12 \\ 12 & -14 & -2 \end{bmatrix}$.

Hence, $(A \cdot B)^{-1} = B^{-1} \cdot A^{-1}$.

29. Let $A = \begin{bmatrix} a_{11} & a_{12} \\ a_{21} & a_{22} \end{bmatrix}$; then $\delta(A) = (a_{11}a_{22} - a_{12}a_{21}) \neq 0$, since A is nonsingular. Hence, $\dfrac{1}{\delta(A)}$ is defined and A^{-1} exists.

$$A^{-1} = \frac{1}{\delta(A)}\begin{bmatrix} a_{22} & -a_{12} \\ -a_{21} & a_{11} \end{bmatrix} = \begin{bmatrix} \dfrac{a_{22}}{\delta(A)} & \dfrac{-a_{12}}{\delta(A)} \\ \dfrac{-a_{21}}{\delta(A)} & \dfrac{a_{11}}{\delta(A)} \end{bmatrix}$$

and $\delta(A^{-1}) = \dfrac{a_{22}a_{11}}{[\delta(A)]^2} - \dfrac{a_{12}a_{21}}{[\delta(A)]^2} = \dfrac{a_{22}a_{11} - a_{12}a_{21}}{[\delta(A)]^2} = \dfrac{\delta(A)}{[\delta(A)]^2} = \dfrac{1}{\delta(A)}$.

31. From the results of Exercise 8.4-33 and Exercise 29 above,

$$\delta[B^{-1}AB] = \delta[B^{-1}(AB)] = \delta(B^{-1}) \cdot \delta(AB) = \delta(B^{-1}) \cdot \delta(A) \cdot \delta(B) = \frac{1}{\delta(B)} \cdot \delta(A) \cdot \delta(B) = \delta(A).$$

Exercise 8.7 (Page 368)

1. $\{(1, 1)\}$ **3.** No solution   **5.** $\{(6, 4)\}$ **7.** $\{(-\frac{18}{7}, \frac{19}{7})\}$ **9.** $(x, -\frac{1}{3}x + \frac{2}{9})$, $x \in R$ **11.** $\{(2, 2)\}$
13. $\{(1, 1, 1)\}$ **15.** $\{(1, 1, 0)\}$ **17.** $\{(1, -2, 3)\}$ **19.** $\{(3, -1, -2)\}$
21. **a.** $\{(\frac{11}{6}, \frac{5}{3}, -\frac{1}{2})\}$ **b.** $\{(-\frac{13}{12}, -\frac{1}{6}, -\frac{3}{4})\}$ **c.** $\{(\frac{5}{6}, -\frac{1}{3}, \frac{1}{2})\}$

Exercise 8.8 (Page 372)

1. $\{(\frac{13}{5}, \frac{3}{5})\}$ **3.** $\{(\frac{22}{7}, \frac{20}{7})\}$ **5.** $\{(6, 4)\}$ **7.** Inconsistent   **9.** $\{(4, 1)\}$

11. $\left\{\left(\dfrac{1}{a + b}, \dfrac{1}{a + b}\right)\right\}$ $(a \neq -b)$ **13.** $\{(1, 1, 0)\}$ **15.** $\{(1, -2, 3)\}$ **17.** $\{(3, -1, -2)\}$

19. $\{(-\frac{1}{3}, -\frac{25}{24}, -\frac{5}{8})\}$ **21.** $\{(1, -\frac{1}{3}, \frac{1}{2})\}$ **23.** $\{(w, x, y, z)\} = \{(2, -1, 1, 0)\}$

25. $|A| = \begin{vmatrix} a_1 & b_1 \\ a_2 & b_2 \end{vmatrix}$ and $\delta(A) = a_1 b_2 - a_2 b_1$;

$|A_y| = \begin{vmatrix} a_1 & c_1 \\ a_2 & c_2 \end{vmatrix}$ and $\delta(A_y) = a_1 c_2 - a_2 c_1 = 0$, so $a_1 c_2 = a_2 c_1$;

$|A_x| = \begin{vmatrix} c_1 & b_1 \\ c_2 & b_2 \end{vmatrix}$ and $\delta(A_x) = b_2c_1 - b_1c_2 = 0$, so $b_1c_2 = b_2c_1$;

Hence, $\dfrac{a_1c_2}{b_1c_2} = \dfrac{a_2c_1}{b_2c_1}; \dfrac{a_1}{b_1} = \dfrac{a_2}{b_2}$ and $a_1b_2 = a_2b_1$, so $a_1b_2 - a_2b_1 = 0$; therefore, $\delta(A) = 0$.

Chapter 8 Review (Page 373)

1. $\begin{bmatrix} 1 & -1 \\ 1 & 1 \end{bmatrix}$ **2.** $\begin{bmatrix} 2 & 5 & -2 \\ 14 & -1 & 12 \end{bmatrix}$ **3.** $\begin{bmatrix} -2 & -3 & 1 \\ 3 & -7 & 6 \\ -5 & 2 & 5 \end{bmatrix}$ **4.** $\begin{bmatrix} -8 & -3 & -6 \\ 6 & -3 & 0 \\ -9 & 5 & 5 \end{bmatrix}$ **5.** $\begin{bmatrix} -21 & 7 \\ -14 & 0 \\ -7 & -7 \end{bmatrix}$

6. [13] **7.** $\begin{bmatrix} -13 & 3 \\ -19 & 27 \end{bmatrix}$ **8.** $\begin{bmatrix} 18 & 7 & 25 \\ 8 & -1 & 11 \\ 3 & 0 & 3 \end{bmatrix}$ **9.** $\{(2, -1)\}$ **10.** $\{(2, 1, 1)\}$ **11.** -3 **12.** 7

13. -3 **14.** 14 **15.** 0 **16.** -1 **17.** 15 **18.** -1578 **19.** $\dfrac{1}{34}\begin{bmatrix} -3 & 2 \\ 11 & 4 \end{bmatrix}$

20. $\dfrac{1}{6}\begin{bmatrix} 1 & 3 & -2 \\ -3 & -3 & 6 \\ 1 & -3 & 4 \end{bmatrix}$ **21.** $\{(-2, 1)\}$ **22.** $\{3, -1, 1)\}$

23. a. $\{(\frac{1}{2}, -\frac{1}{2})\}$ **b.** $\{(\frac{1}{4}, -\frac{7}{4})\}$ **c.** $\{(1, -2)\}$

24. a. $\{(1, 1, -1)\}$ **b.** $\{(1, 1, 2)\}$ **c.** $\{(0, -1, -2)\}$ **25.** $\{(-\frac{16}{7}, -\frac{13}{7})\}$ **26.** $\{(2, -1, 0)\}$

27. From Exercise 8.1-30 we have $(A + B)^t = A^t + B^t$. Since A and B are symmetric, $A^t = A$ and $B^t = B$. Thus, $(A + B)^t = A + B$ and $A + B$ is symmetric.

28. Let $A = \begin{bmatrix} a & b \\ c & d \end{bmatrix}$. Then $(rA)^t = \begin{bmatrix} ra & rb \\ rc & rd \end{bmatrix}^t = \begin{bmatrix} ra & rc \\ rb & rd \end{bmatrix} = rA^t = rA$. The last equality comes from the fact that A is symmetric.

29. If $AB = BA$, then AB is symmetric because

$$(AB)^t = B^tA^t \quad \text{(from Exercise 8.2-41)}$$
$$= BA \quad \text{(since } A \text{ and } B \text{ are symmetric)}$$
$$= AB \quad \text{(we assumed } AB = BA\text{).}$$

30. $(A^2)^t = (AA)^t = A^tA^t$ (from Exercise 8.2-41 with $B = A$). Thus, $(A^2)^t = A^tA^t = (A^t)^2$. Yes, the same result holds if A is an $n \times n$ matrix.

31. The matrix cA is obtained from A by multiplying each of the n rows of A by c. By Theorem 8.10, each of these results is a multiplication of $\delta(A)$ by one factor of c. Thus, $\delta(cA) = c^n\delta(A)$.

32. If A is nonsingular, then $\delta(A) \neq 0$. Thus, from Exercise 8.4-33 we have $\delta(A^2) = \delta(AA) = \delta(A)\delta(A) \neq 0$ (since $\delta(A) \neq 0$). Thus, A^2 is nonsingular.

Exercise 9.1 (Page 383)

1. $-4, -3, -2, -1$ **3.** $-\frac{1}{2}, 1, \frac{7}{2}, 7$ **5.** $2, \frac{3}{2}, \frac{4}{3}, \frac{5}{4}$ **7.** $0, 1, 3, 6$ **9.** $-1, 1, -1, 1$

11. $1, 0, -\frac{1}{3}, \frac{1}{2}$ **13.** $11, 15, 19$ **15.** $2, 5, 8$ **17.** $x + 2, x + 3, x + 4$ **19.** $2x + 7, 2x + 10, 2x + 13$

21. $32, 128, 512, 2048$ **23.** $\frac{8}{3}, \frac{16}{3}, \frac{32}{3}, \frac{64}{3}$ **25.** $\dfrac{x}{a}, -\dfrac{x^2}{a^2}, \dfrac{x^3}{a^3}, -\dfrac{x^4}{a^4}$ **27.** $\dfrac{x^2}{-4}, \dfrac{x^3}{16}, \dfrac{x^4}{-64}$

29. $4n + 3, 31$ **31.** $8 - 5n, -92$ **33.** $(4n + 1)x - 4x, 29x$ **35.** $48(2)^{n-1}, 1536$

37. $-\frac{1}{3}(-3)^{n-1}$, -243 **39.** x^{2n-1}, x^9 **41.** $x + (n-1)(y-x)$, $5y - 4x$ **43.** $\dfrac{y^{n-1}}{x^{n-2}}$, $\dfrac{y^7}{x^6}$

45. $2; 3, 41$ **47.** 28th **49.** 3

Exercise 9.2 (Page 391)

1. $1 + 4 + 9 + 16$ **3.** $-\frac{1}{2} + \frac{1}{4} - \frac{1}{8}$ **5.** $-5 - 3 - 1 + 1 + 3 + 5 + 7 + 9$ **7.** $1 + \frac{1}{2} + \frac{1}{4} + \cdots$

9. $\sum\limits_{j=1}^{4} x^{2j-1}$ **11.** $\sum\limits_{j=1}^{5} j^2$ **13.** $\sum\limits_{j=1}^{\infty} j(j+1)$ **15.** $\sum\limits_{j=1}^{\infty} \dfrac{j+1}{j}$ **17.** 5080 **19.** 80 **21.** 90

23. 63 **25.** 806 **27.** -6 **29.** 10,000 **31.** 1092 **33.** $\frac{31}{32}$ **35.** $\frac{364}{2187}$ **37.** 2816 **39.** 168

41. 2040 **43.** 196 **45.** \$2333.39 **47.** $\frac{3}{4}, \frac{7}{8}, \frac{15}{16}, \frac{31}{32}$; 1 **49.** $p = 4$, $q = -3$

Exercise 9.3 (Page 398)

1. $\lim\limits_{n \to \infty} s_n = 0$ **3.** $\lim\limits_{n \to \infty} s_n = 1$ **5.** $\lim\limits_{n \to \infty} s_n$ is undefined **7.** $\lim\limits_{n \to \infty} s_n = 0$

9. Convergent, $\lim\limits_{n \to \infty} \left| 0 - \dfrac{1}{2^n} \right| = 0$ **11.** Divergent, $\lim\limits_{n \to \infty} n$ is undefined

13. Convergent, $\lim\limits_{n \to \infty} \left| 0 - (-1)^{n+1} \dfrac{1}{2^{n-1}} \right| = 0$ **15.** 24 **17.** No sum **19.** 2 **21.** $\frac{8}{9}$ **23.** $\frac{31}{99}$

25. $2\frac{410}{999}$ **27.** $\frac{29}{225}$ **29.** 20 centimeters

Exercise 9.4 (Page 404)

1. $8 \cdot 7 \cdot 6 \cdot 5 \cdot 4 \cdot 3 \cdot 2 \cdot 1$ **3.** $6 \cdot 5 \cdot 4 \cdot 3 \cdot 2 \cdot 1$ **5.** $5 \cdot 4 \cdot 3 \cdot 2 \cdot 1 = 120$ **7.** $\dfrac{9 \cdot 8 \cdot 7!}{7!} = 72$

9. $\dfrac{5 \cdot 4 \cdot 3 \cdot 2 \cdot 1 \cdot 7!}{8 \cdot 7!} = 15$ **11.** $\dfrac{8 \cdot 7 \cdot 6!}{2 \cdot 1 \cdot 6!} = 28$ **13.** $3!$ **15.** $\dfrac{6!}{2!}$ **17.** $\dfrac{8!}{5!}$ **19.** $\dfrac{6!}{5!1!} = 6$

21. $\dfrac{3!}{3!0!} = 1$ **23.** $\dfrac{7!}{0!7!} = 1$ **25.** $\dfrac{5!}{2!3!} = 10$ **27.** $(n)(n-1)(n-2) \cdot \cdots \cdot 3 \cdot 2 \cdot 1$

29. $(3n)(3n-1)(3n-2) \cdot \cdots \cdot 3 \cdot 2 \cdot 1$ **31.** $(n-2)(n-3)(n-4) \cdot \cdots \cdot 3 \cdot 2 \cdot 1$ **33.** $(n+2)(n+1)$

35. $\dfrac{n+1}{n+3}$ **37.** $\dfrac{2n-1}{2n-2}$ **39.** $x^5 + 15x^4 + 90x^3 + 270x^2 + 405x + 243$

41. $x^4 - 12x^3 + 54x^2 - 108x + 81$ **43.** $8x^3 - 6x^2y + \dfrac{3}{2}xy^2 - \dfrac{1}{8}y^3$

45. $\dfrac{1}{64}x^6 + \dfrac{3}{8}x^5 + \dfrac{15}{4}x^4 + 20x^3 + 60x^2 + 96x + 64$

47. $x^{20} + 20x^{19}y + \dfrac{20 \cdot 19}{2!}x^{18}y^2 + \dfrac{20 \cdot 19 \cdot 18}{3!}x^{17}y^3$, or

$\dbinom{20}{0}x^{20} + \dbinom{20}{1}x^{19}y + \dbinom{20}{2}x^{18}y^2 + \dbinom{20}{3}x^{17}y^3$

49. $a^{12} + 12a^{11}(-2b) + \dfrac{12 \cdot 11}{2!}a^{10}(-2b)^2 + \dfrac{12 \cdot 11 \cdot 10}{3!}a^9(-2b)^3$, or

$\dbinom{12}{0}a^{12} + \dbinom{12}{1}a^{11}(-2b) + \dbinom{12}{2}a^{10}(-2b)^2 + \dbinom{12}{3}a^9(-2b)^3$

51. $x^{10} + 10x^9(-\sqrt{2}) + \dfrac{10 \cdot 9}{2!}x^8(-\sqrt{2})^2 + \dfrac{10 \cdot 9 \cdot 8}{3!}x^7(-\sqrt{2})^3$, or

$$\binom{10}{0}x^{10} + \binom{10}{1}x^9(-\sqrt{2}) + \binom{10}{2}x^8(-\sqrt{2})^2 + \binom{10}{3}x^7(-\sqrt{2})^3$$

53. $-3003a^{10}b^5$ **55.** $3360x^6y^4$ **57.** 1.22 **59.** 0.92

61. a. $1 - x + x^2 - x^3 + \cdots$ **b.** $1 - x + x^2 - x^3 + \cdots$

63.
$$\binom{k}{r} + \binom{k}{r-1} = \frac{k!}{r!(k-r)!} + \frac{k!}{(r-1)!(k-r+1)!}$$
$$= \frac{k!(k-r+1) + k!r}{r!(k-r+1)!} = \frac{k!(k+1)}{r!(k-r+1)!}$$
$$= \frac{(k+1)!}{r![(k+1)-r]!} = \binom{k+1}{r}$$

Exercise 9.5 (Page 408)

1. a. For $n = 1$, $\dfrac{n}{2} = \dfrac{1}{2}$; $\dfrac{n(n+1)}{4} = \dfrac{1(1+1)}{4} = \dfrac{1}{2}$.

b. For $n = k$, $\dfrac{1}{2} + \dfrac{2}{2} + \dfrac{3}{2} + \cdots + \dfrac{k}{2} = \dfrac{k(k+1)}{4}$ and $(k+1)$st term $= \dfrac{k+1}{2}$.

Hence, $\dfrac{1}{2} + \dfrac{2}{2} + \dfrac{3}{2} + \cdots + \dfrac{k}{2} + \dfrac{k+1}{2} = \dfrac{k(k+1)}{4} + \dfrac{k+1}{2} = \dfrac{k^2 + k + 2k + 2}{4} = \dfrac{k^2 + 3k + 2}{4} = \dfrac{(k+1)(k+2)}{4}$

3. a. For $n = 1$, $2n = 2(1) = 2$; $n(n+1) = 1(1+1) = 2$.

b. For $n = k$, $2 + 4 + 6 + \cdots + 2k = k(k+1)$ and $(k+1)$st term is $2(k+1)$.

Hence, $2 + 4 + 6 + \cdots + 2k + 2(k+1) = k(k+1) + 2(k+1)$
$$= (k+1)(k+2).$$

5. a. For $n = 1$, $6n - 4 = 6(1) - 4 = 2$; $n(3n-1) = 1[3(1)-1] = 2$.

b. For $n = k$, $2 + 8 + 14 + \cdots + (6k-4) = k(3k-1)$ and $(k+1)$st term is $6(k+1) - 4$.

Hence, $2 + 8 + 14 + \cdots + (6k-4) + [6(k+1) - 4] = k(3k-1) + 6k + 2 = 3k^2 + 5k + 2$
$$= (k+1)(3k+2) = (k+1)[3(k+1) - 1].$$

7. a. For $n = 1$, $3n + 4 = 3(1) + 4 = 7$; $\dfrac{n(3n+11)}{2} = \dfrac{1[3(1)+11]}{2} = 7$.

b. For $n = k$, $7 + 10 + 13 + \cdots + (3k+4) = \dfrac{k(3k+11)}{2}$ and $(k+1)$st term is $3(k+1) + 4$.

Hence, $7 + 10 + 13 + \cdots + (3k+4) + [3(k+1) + 4] = \dfrac{k(3k+11)}{2} + 3k + 7 = \dfrac{3k^2 + 17k + 14}{2}$
$$= \dfrac{(k+1)(3k+14)}{2} = \dfrac{(k+1)[3(k+1) + 11]}{2}.$$

9. a. For $n = 1$, $n^2 = 1^2 = 1$; $\dfrac{n(n+1)(2n+1)}{6} = \dfrac{1(2)(3)}{6} = 1$.

b. For $n = k$, $1^2 + 2^2 + 3^2 + \cdots + k^2 = \dfrac{k(k+1)(2k+1)}{6}$ and $(k+1)$st term is $(k+1)^2$.

Hence, $1^2 + 2^2 + 3^2 + \cdots + k^2 + (k+1)^2 = \dfrac{k(k+1)(2k+1)}{6} + (k+1)^2$
$$= \dfrac{k(k+1)(2k+1) + 6(k+1)^2}{6} = \dfrac{(k+1)[k(2k+1) + 6(k+1)]}{6} = \dfrac{(k+1)(2k^2 + 7k + 6)}{6}$$
$$= \dfrac{(k+1)(k+2)(2k+3)}{6} = \dfrac{(k+1)[(k+1) + 1][2(k+1) + 1]}{6}.$$

11. a. For $n = 1$, $(2n-1)^3 = (2-1)^3 = 1^3 = 1$; $n^2(2n^2 - 1) = 1(2-1) = 1(1) = 1$.

b. For $n = k$, $1^3 + 3^3 + 5^3 + \cdots + (2k-1)^3 = k^2(2k^2 - 1)$ and the $(k+1)$st term is $[2(k+1) - 1]^3 = (2k+1)^3$. Hence,

$$1^3 + 3^3 + 5^3 + \cdots + (2k-1)^3 + (2k+1)^3 = k^2(2k^2 - 1) + (2k+1)^3 = 2k^4 + 8k^3 + 11k^2 + 6k + 1.$$

By use of the factor theorem and synthetic division:

$$2k^4 + 8k^3 + 11k^2 + 6k + 1 = (k+1)(k+1)(2k^2 + 4k + 1).$$

Also, $2k^2 + 4k + 1 = 2(k^2 + 2k + 1) - 2 + 1 = 2(k+1)^2 - 1$.
Hence, $2k^4 + 8k^3 + 11k^2 + 6k + 1 = (k+1)^2[2(k+1)^2 - 1]$.

13. a. For $n = 1$, $n(n+1) = 1(2) = 2$; $\dfrac{n(n+1)(n+2)}{3} = \dfrac{1(2)(3)}{3} = 2$.

b. For $n = k$, $1 \cdot 2 + 2 \cdot 3 + 3 \cdot 4 + \cdots + k(k+1) = \dfrac{k(k+1)(k+2)}{3}$

and the $(k+1)$st term is $(k+1)[(k+1) + 1] = (k+1)(k+2)$. Hence,

$$1 \cdot 2 + 2 \cdot 3 + 3 \cdot 4 + \cdots + k(k+1) + [(k+1)(k+2)]$$

$$= \frac{k(k+1)(k+2)}{3} + (k+1)(k+2) = \frac{[k(k+1)(k+2)] + [3(k+1)(k+2)]}{3}$$

$$= \frac{(k+1)(k+2)(k+3)}{3} = \frac{(k+1)[(k+1) + 1][(k+1) + 2]}{3}.$$

15. a. For $n = 1$, $1^3 + 2 \cdot 1 = 3$;   3 is divisible by 3.

b. For $n = k$,   assume $k^3 + 2k$ is divisible by 3; for $n = k + 1$,

$$(k+1)^3 + 2(k+1) = k^3 + 3k^2 + 3k + 1 + 2k + 2$$

$$= (k^3 + 2k) + (3k^2 + 3k + 3).$$

Since by hypothesis $k^3 + 2k$ is divisible by 3, and since $3k^2 + 3k + 3$ is divisible by 3, $(k+1)^3 + 2(k+1)$ is divisible by 3.

17. For $n = k$, $2 + 4 + 6 + \cdots + 2k = k(k+1) + 2$ and $(k+1)$st term is $2(k+1)$. Hence,

$$2 + 4 + 6 + \cdots + 2k + 2(k+1) = k(k+1) + 2 + 2(k+1) = (k^2 + 3k + 2) + 2$$

$$= (k+1)(k+2) + 2 = (k+1)[(k+1) + 1] + 2.$$

However, for $n = 1$, $2n = 2(1) = 2$; $n(n+1) + 2 = 1(2) + 2 = 4$. Hence, it is not true for every $n \in N$.

19. Since $s_1 = a$ and $s_{n+1} = s_n + d$, it follows that

a. for $n = 1$, $s_n = s_1 = a = a + 0 \cdot d = a + (n-1)d$.

b. for $n = k$, $s_k = a + (k-1)d$; now

$$s_{k+1} = s_k + d = a + (k-1)d + d = a + kd = a + [(k+1) - 1]d.$$

21. Since $s_n = ar^{n-1}$ and $S_{n+1} = S_n + s_{n+1}$, it follows that

a. for $n = 1$, $S_1 = s_1 = a = a\dfrac{1-r}{1-r} = \dfrac{a - ar^1}{1-r}$.

b. for $n = k$, $S_k = \dfrac{a - ar^k}{1-r}$ and $s_{k+1} = ar^k$; now

$$S_{k+1} = S_k + s_{k+1} = \frac{a - ar^k}{1-r} + ar^k = \frac{a - ar^k + ar^k - ar^{k+1}}{1-r} = \frac{a - ar^{k+1}}{1-r}.$$

Chapter 9 Review (Page 409)

1. $2, 5, 10$ 2. $\frac{1}{2}, \frac{1}{3}, \frac{1}{4}$ 3. $13, 16, 19$ 4. $a - 4, a - 6, a - 8$ 5. $-18, 54, -162$ 6. $\frac{3}{2}, \frac{9}{4}, \frac{27}{8}$

7. $s_n = 5n - 8$; $s_7 = 27$ 8. $s_n = (-2)\left(\dfrac{-1}{3}\right)^{n-1}$; $s_5 = \dfrac{-2}{81}$ 9. 25 10. Sixth term

11. $2 + 6 + 12 + 20$ 12. $\displaystyle\sum_{k=1}^{\infty} x^{k+1}$ 13. 119 14. $\frac{121}{243}$ 15. 3 16. $\frac{8}{3}$ 17. $\frac{1}{2}$ 18. $\frac{4}{9}$

19. $5 \cdot (2 \cdot 1)$ 20. 48 21. 21 22. $\dfrac{1}{n(n+1)!}$ 23. $x^{10} - 20x^9 y + 180x^8 y^2 - 960x^7 y^3$

24. $-15{,}360 x^3 y^7$

25. Formula holds for 1. Assume formula holds for n; test for $n + 1$:

$$3 + 6 + 9 + \cdots + 3n + 3(n+1) = \frac{3n(n+1)}{2} + 3(n+1).$$

Right-hand member is equivalent to

$$\frac{3n^2 + 3n}{2} + \frac{6(n+1)}{2} = \frac{3n^2 + 9n + 6}{2} = \frac{3(n^2 + 3n + 2)}{2} = \frac{3(n+1)(n+2)}{2}$$

$$= \frac{3(n+1)[(n+1)+1]}{2}.$$

26. Formula holds for 1. Assume formula holds for n; test for $n + 1$:

$$\frac{1}{2} + \frac{1}{4} + \frac{1}{8} + \cdots + \frac{1}{2^n} + \frac{1}{2^{n+1}} = 1 - \frac{1}{2^n} + \frac{1}{2^{n+1}}.$$

27. $\dfrac{2}{\sqrt{3}}, \dfrac{-2}{\sqrt{3}}$ 28. $-6x + 7y$ 29. $p = 5, q = 3$ 30. $2 - 3\left(\dfrac{2}{3}\right)^{18} = 1.9980$ 31. 1.63 32. 0.99

33. For $n = 1$, $S_1 = s_1 = \dfrac{2(1) + 1}{(1)^2(1+1)^2}$ and $1 - \dfrac{1}{(1+1)^2} = \dfrac{3}{4}$. Therefore, the statement is true for $n = 1$. Assume

that $S_k = 1 - \dfrac{1}{(k+1)^2}$; then we have

$$S_{k+1} = S_k + s_{k+1} = 1 - \frac{1}{(k+1)^2} + \frac{2(k+1) + 1}{(k+1)^2(k+2)^2} = 1 + \frac{-(k+2)^2 + 2k + 3}{(k+1)^2(k+2)^2}$$

$$= 1 - \frac{-k^2 - 2k - 1}{(k+1)^2(k+2)^2} = 1 - \frac{1}{(k+2)^2} = 1 - \frac{1}{[(k+1)+1]^2}.$$

34. Since $8! = 40{,}320$ and $3^9 = 19{,}683$, the inequality is true for $n = 8$. Assume that $k! \geq 3^{k+1}$. Then

$$\begin{aligned}
(k+1)! = (k+1)(k!) &\geq (k+1)3^{k+1} \quad &&\text{(since } k! \geq 3^{k+1}) \\
&\geq 3 \cdot 3^{k+1} \quad &&\text{(since } k + 1 \geq 3) \\
&= 3^{(k+1)+1}.
\end{aligned}$$

Exercise 10.1 (Page 419)

1. $1, 5, 8$ 3. $2, 6, 16$ 5. $2, 2, 4$ 7. 4 9. 24 11. 16 13. 64 15. 24 17. 216

19. 375 21. 30 23. 10 25. 48 27. $\dfrac{5!}{2!}$, or 60 29. $\dfrac{8!}{3!}$, or 6720 31. 24

33. $P_{5,3} = \dfrac{5!}{2!} = \dfrac{5 \cdot 4!}{2!} = 5\left(\dfrac{4!}{2!}\right) = 5(P_{4,2})$ **35.** $P_{n,3} = \dfrac{n!}{(n-3)!} = \dfrac{n(n-1)!}{(n-3)!} = n\left[\dfrac{(n-1)!}{(n-3)!}\right] = n(P_{n-1,2})$

37. 9   **39.** 48

Exercise 10.2 (Page 423)

1. 7   **3.** 15   **5.** $\binom{52}{5}$ **7.** $\binom{13}{5} \cdot \binom{13}{5} \cdot \binom{13}{3}$ **9.** $4 \cdot \binom{13}{5}$ **11.** 164   **13.** 10   **15.** 210

17. 12

Exercise 10.3 (Page 427)

1. $\{1, 2, 3, 4, 5, 6\}$; $\{3, 4, 5, 6\}$; $\frac{2}{3}$ **3.** $\{(H, H), (H, T), (T, H), (T, T)\}$; $\{(H, H), (T, T)\}$; $\frac{1}{2}$ **5.** $\frac{1}{6}$

7. $\frac{1}{18}$ **9.** $\frac{5}{9}$ **11.** $\frac{1}{52}$ **13.** $\frac{3}{26}$ **15.** $\frac{1}{17}$ **17.** $\frac{11}{221}$ **19.** $\frac{1}{190}$ **21.** $\frac{3}{38}$

Exercise 10.4 (Page 430)

1. $\frac{1}{6}$ **3.** $\frac{7}{18}$ **5.** $\frac{13}{18}$ **7.** $\frac{5}{33}$ **9.** $\frac{1}{11}$ **11.** $\frac{5}{22}$ **13.** $\frac{14}{33}$ **15.** $\frac{15}{22}$ **17.** $\frac{3}{13}$ **19.** $2.14; less

21. 5.3 cents   **23.** $3.29

Exercise 10.5 (Page 435)

1. $\frac{20}{91}$; no   **3. a.** $\frac{2}{45}$ **b.** $\frac{28}{75}$ **c.** $\frac{4}{225}$ **d.** $\frac{1}{9}$ **5. a.** $\frac{11}{36}$ **b.** $\frac{5}{36}$ **c.** $\frac{2}{11}$ **d.** No

7. a. $\frac{1}{2}$ **b.** $\frac{1}{2}$ **c.** $\frac{1}{4}$ **d.** $\frac{1}{4}$ **e.** $\frac{1}{8}$ **9. a.** $\frac{71}{72}$ **b.** $\frac{5}{9}$ **c.** $\frac{5}{36}$ **d.** $\frac{61}{72}$.

11. a. $\frac{1}{210}$ **b.** $\frac{29}{210}$ **c.** $\frac{29}{70}$ **d.** $\frac{29}{30}$; yes   **13.** $\frac{1}{4}$

Chapter 10 Review (Page 436)

1. 16   **2.** 64   **3.** 128   **4.** 360   **5.** 792   **6.** 1,033,885,600   **7.** 200   **8.** 84   **9.** $\frac{1}{17}$ **10.** $\frac{1}{221}$

11. $\frac{10}{17}$ **12.** $\frac{25}{102}$ **13.** $\frac{26}{51}$ **14.** $\frac{80}{221}$ **15.** $\frac{41}{663}$ **16.** $\frac{12}{25}$ **17.** $\frac{25}{1326}$ **18.** $\frac{20}{221}$ **19.** $\frac{25}{221}$ **20.** $\frac{95}{663}$

21. 72   **22.** 1440   **23.** 8   **24.** 10   **25.** $-4\cancel{c}$; no

26. No, if the $4 is present; yes, if the $4.50 ball is present.

Exercise 11.1 (Page 444)

1. $x^2 + y^2 - 8x - 4y - 5 = 0$ **3.** $x - y - 3 = 0$ **5.** $3x^2 + 3y^2 - 46x + 131 = 0$ **7.** $y^2 - 8x = 0$

9. $x^2 + 6x + 4y + 13 = 0$ **11.** $3x^2 + 4y^2 - 48 = 0$ **13.** $3x^2 - y^2 - 3 = 0$ **15.** $x^2 - 2py + p^2 = 0$

17. The equation of the line perpendicular to the line $y = kx$ passing through the point $P(x_0, y_0)$ is

$(y - y_0) = -\dfrac{1}{k}(x - x_0)$. The point of intersection of these two lines is $Q\left(\dfrac{ky_0 + x_0}{k^2 + 1}, \dfrac{k^2 y_0 + kx_0}{k^2 + 1}\right)$.

$$d = \sqrt{\left(x_0 - \dfrac{ky_0 + x_0}{k^2 + 1}\right)^2 + \left(y_0 - \dfrac{k^2 y_0 + kx_0}{k^2 + 1}\right)^2} = \sqrt{\left(\dfrac{k^2 x_0 - ky_0}{k^2 + 1}\right)^2 + \left(\dfrac{y_0 - kx_0}{k^2 + 1}\right)^2}$$

$$= \sqrt{\dfrac{k^2(y_0 - kx_0)^2}{(k^2 + 1)^2} + \dfrac{(y_0 - kx_0)^2}{(k^2 + 1)^2}} = \sqrt{\dfrac{(y_0 - kx_0)^2}{k^2 + 1}} = \dfrac{|y_0 - kx_0|}{\sqrt{k^2 + 1}}.$$

19. $\dfrac{|y - 2x|}{\sqrt{5}} = \sqrt{(x - 1)^2 + (y - 3)^2} + \sqrt{(x + 1)^2 + (y - 1)^2}$

Exercise 11.2 (Page 447)

1.

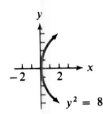

$x^2 = 16y$

3.

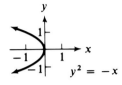

$y^2 = -8x$

5.

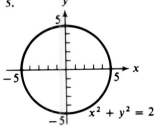

$y^2 = 8x$

7.

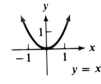

$y = x^2$

9.

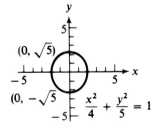

$y^2 = -x$

11. $F(0, \frac{1}{2})$, $d:y = -\frac{1}{2}$ **13.** $F(-4, 0)$, $d:x = 4$ **15.** $F(3, 0)$, $d:x = -3$

17. $F(-1, 0)$, $d:x = 1$ **19.** $F(\frac{3}{8}, 0)$, $d:x = -\frac{3}{8}$

21. The point $P(x, y)$ is on the parabola with vertex $(0, 0)$ and focus $(0, -p)$ if and only if $p - y = \sqrt{(x - 0)^2 + (y + p)^2}$. By the same procedure used in the text, this equation simplifies to $x^2 = -4py$.

23. The point $P(x, y)$ is on the parabola with vertex $(0, 0)$ and focus $(-p, 0)$ if and only if $p - x = \sqrt{(x + p)^2 + (y - 0)^2}$. By the same procedure used in the text, this equation simplifies to $y^2 = -4px$.

Exercise 11.3 (Page 453)

1. Circle, $r = \sqrt{2}$ **3.** Ellipse, $a = \sqrt{12}, b = \sqrt{8}$

5.

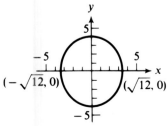

$x^2 + y^2 = 25$

7.

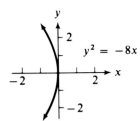

$(0, \sqrt{5})$

$(0, -\sqrt{5})$

$\dfrac{x^2}{4} + \dfrac{y^2}{5} = 1$

9.

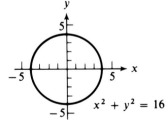

$x^2 + y^2 = 16$

11.

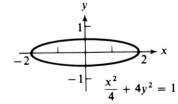

$(-\sqrt{12}, 0)$ $(\sqrt{12}, 0)$

$\dfrac{x^2}{12} + \dfrac{y^2}{16} = 1$

13.

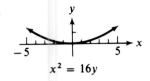

$\dfrac{x^2}{4} + 4y^2 = 1$

15. $F_1(-\sqrt{5}, 0)$, $F_2(\sqrt{5}, 0)$, $V_1(-3, 0)$, $V_2(3, 0)$

17. $F_1(-8\sqrt{2}, 0)$, $F_2(8\sqrt{2}, 0)$, $V_1(-12, 0)$, $V_2(12, 0)$

19. $F_1(0, -1)$, $F_2(0, 1)$, $V_1(0, -2)$, $V_2(0, 2)$

21. The steps in the derivation are the same as those in the derivation of Equation (9) except that instead of Equation (4), we start with $\sqrt{(x-0)^2 + (y-c)^2} + \sqrt{(x-0)^2 + (y+c)^2} = 2a$.

23. If $c \approx 0$, then $a \approx b$ and the ellipse is more nearly circular, as one would expect from Exercise 22.

25. The endpoints of the major axis are the x-intercepts of the curve $\dfrac{x^2}{a^2} + \dfrac{y^2}{b^2} = 1$. Setting $y = 0$, we have $\dfrac{x^2}{a^2} = 1$, or $x = \pm a$.

Exercise 11.4 (Page 458)

1.

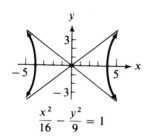

$$\frac{y^2}{5} - \frac{x^2}{4} = 1$$

3.

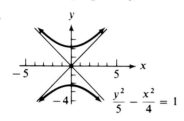

$$\frac{y^2}{4} - \frac{x^2}{12} = 1$$

5.

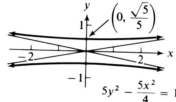

$$\frac{x^2}{16} - \frac{y^2}{9} = 1$$

7.

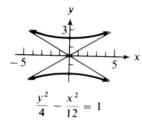

$$\left(0, \frac{\sqrt{5}}{5}\right)$$

$$5y^2 - \frac{5x^2}{4} = 1$$

9. $F_1(5, 0)$, $F_2(-5, 0)$, $V_1(3, 0)$, $V_2(-3, 0)$, $y = \pm\frac{4}{3}x$

11. $F_1(\sqrt{3}, 0)$, $F_2(-\sqrt{3}, 0)$, $V_1(1, 0)$, $V_2(-1, 0)$, $y = \pm\sqrt{2}x$

13. $F_1(0, 2\sqrt{3})$, $F_2(0, -2\sqrt{3})$, $V_1(0, 2)$, $V_2(0, -2)$, $y = \pm\dfrac{\sqrt{2}}{2}x$

15. $F_1\left(0, \dfrac{\sqrt{41}}{20}\right)$, $F_2\left(0, -\dfrac{\sqrt{41}}{20}\right)$, $V_1(0, \frac{1}{5})$, $V_2(0, -\frac{1}{5})$, $y = \pm\frac{4}{5}x$

17. Assume without loss of generality that $d_1 \geq d_2$ in Figure 11.11. Then $2a = |d_1 - d_2| = d_1 - d_2$. Now assume $c < a$, then $2c < 2a$. Therefore, $2c < d_1 - d_2$. But then $2c + d_2 < d_1$. This says that one side of $\triangle F_1 P F_2$ (namely $F_1 P$) has greater length than the sum of the lengths of the other two sides. This is a contradiction. Thus, $c \geq a$; and since both are positive we have $c^2 - a^2 \geq 0$.

19. The equation is $|\sqrt{(x + \sqrt{2})^2 + (y + \sqrt{2})^2} - \sqrt{(x - \sqrt{2})^2 + (y - \sqrt{2})^2}| = 2\sqrt{2}$, which simplifies to $xy = 1$.

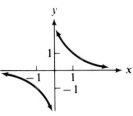

Exercise 11.5 (Page 461)

1. Parabola, intercepts $(0, 0)$

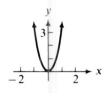

3. Parabola, intercepts $(0, 0)$

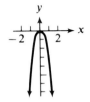

5. Parabola, intercepts $(0, 0)$

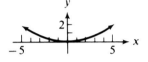

7. Parabola intercepts $(0, 0)$

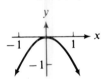

9. Circle, x-intercepts $(\pm 7, 0)$, y-intercepts $(0, \pm 7)$

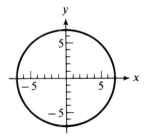

11. Ellipse, x-intercepts $(\pm 5, 0)$ y-intercepts $(0, \pm 2)$

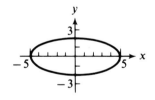

13. Two intersecting lines, intercepts $(0, 0)$

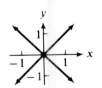

15. Hyperbola, x-intercepts $(\pm 3, 0)$, no y-intercepts

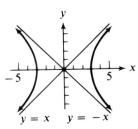

17. Circle, x-intercepts $\left(\pm \dfrac{1}{2}, 0\right)$, y-intercepts $\left(0, \pm \dfrac{1}{2}\right)$

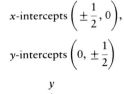

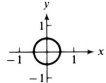

19. Ellipse,
x-intercepts $(\pm 2, 0)$,
y-intercepts $(0, \pm\sqrt{3})$

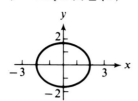

21. Two intersecting lines,
intercepts $(0, 0)$

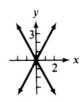

23. Straight line,
every real number is an
x-intercept,
y-intercept $(0, 0)$

25. Since A and B are positive, $Ax^2 - By^2 = 0$ is equivalent to $y = \pm\sqrt{A/B}\,x$. Thus, the graph of the relation is a pair of straight lines with slopes $\pm\sqrt{A/B}$ that intersect at $(0, 0)$.

27. Since $A \neq 0$, $Ax^2 = 0$ is equivalent to $x^2 = 0$, which is equivalent to $x = 0$. Thus, the graph of $Ax^2 = 0$ is a vertical straight line passing through the point $(0, 0)$.

Exercise 11.6 (Page 466)

1. $(-7, -3)$ **3.** $(5, -1)$ **5.** $x' - 2y' = 13$ **7.** $x'^2 - 2y'^2 + 4x' + 20y' - 52 = 0$

9. $(x + 5)^2 + (y - 1)^2 = 9$ **11.** $(x - 3)^2 = 8(y - 3)$

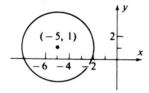

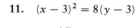

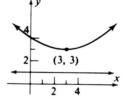

13. $\dfrac{(x + 2)^2}{25} + \dfrac{(y - 2)^2}{9} = 1$ **15.** $\dfrac{(x - 3)^2}{9} - \dfrac{(y - 2)^2}{16} = 1$

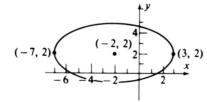

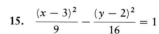

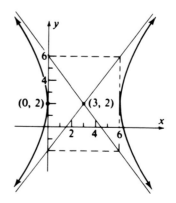

17.

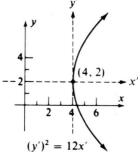

$(y')^2 = 12x'$

19.

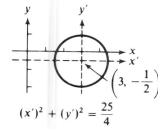

$$(x')^2 + (y')^2 = \frac{25}{4}$$

with point $\left(3, -\frac{1}{2}\right)$

21.

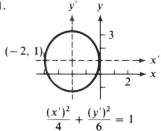

$(-2, 1)$

$$\frac{(x')^2}{4} + \frac{(y')^2}{6} = 1$$

23.

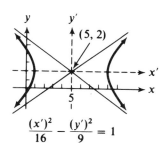

$(5, 2)$

$$\frac{(x')^2}{16} - \frac{(y')^2}{9} = 1$$

25. $A \neq 0$, $B = 0$, $D \neq 0$ **27.** $A \neq 0$, $B \neq 0$, $A = B$, $\dfrac{C^2}{4A} + \dfrac{D^2}{4B} - F > 0$

Chapter 11 Review (Page 468)

1. $3x^2 + 3y^2 + 4x - 34y + 59 = 0$ **2.** $(x - 3)^2 + (y - 4)^2 = 25$ **3.** $|y| = 3$

4. $\sqrt{(x + 2)^2 + (y - 6)^2} + \sqrt{x^2 + (y - 4)^2} = 8$

5.

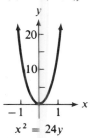

$x^2 = 24y$

6.

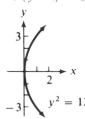

$y^2 = 12x$

7.

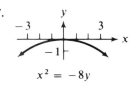

$x^2 = -8y$

8.

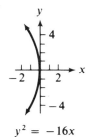

$y^2 = -16x$

9. $F\left(\frac{1}{8}, 0\right)$, $x = -\frac{1}{8}$ **10.** $F\left(0, -\frac{1}{4}\right)$, $y = \frac{1}{4}$ **11.** $F(0, 5)$, $y = -5$ **12.** $F\left(-\frac{1}{4}, 0\right)$, $x = \frac{1}{4}$

13.

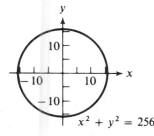

$x^2 + y^2 = 256$

14.

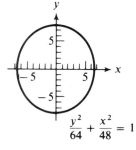

$$\frac{y^2}{64} + \frac{x^2}{48} = 1$$

15.

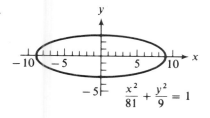

$$\frac{x^2}{81} + \frac{y^2}{9} = 1$$

16.

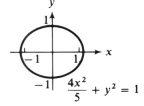

$$\frac{4x^2}{5} + y^2 = 1$$

17. $F_1(0, -2\sqrt{3})$, $F_2(0, 2\sqrt{3})$, $V_1(0, -4)$, $V_2(0, 4)$

18. $F_1(-\sqrt{7}, 0)$, $F_2(\sqrt{7}, 0)$, $V_1(-\sqrt{8}, 0)$, $V_2(\sqrt{8}, 0)$

19. $F_1(-1, 0)$, $F_2(1, 0)$, $V_1(-3, 0)$, $V_2(3, 0)$

20. $F_1(0, -\sqrt{5})$, $F_2(0, \sqrt{5})$, $V_1(0, -3)$, $V_2(0, 3)$

21.

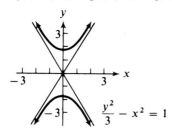

$$\frac{y^2}{3} - x^2 = 1$$

22.

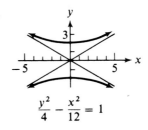

$$\frac{y^2}{4} - \frac{x^2}{12} = 1$$

23.

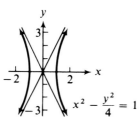

$$x^2 - \frac{y^2}{4} = 1$$

24.

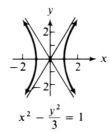

$$x^2 - \frac{y^2}{3} = 1$$

25. $F_1(\sqrt{3}, 0)$, $F_2(-\sqrt{3}, 0)$, $V_1(\sqrt{2}, 0)$, $V_2(-\sqrt{2}, 0)$, $y = \pm\dfrac{\sqrt{2}}{2}x$

26. $F_1(0, \sqrt{2})$, $F_2(0, -\sqrt{2})$, $V_1(0, 1)$, $V_2(0, -1)$, $y = \pm x$

27. $F_1(0, \sqrt{6})$, $F_2(0, -\sqrt{6})$, $V_1(0, 2)$, $V_2(0, -2)$, $y = \pm\sqrt{2}x$

28. $F_1(\sqrt{14}, 0)$, $F_2(-\sqrt{14}, 0)$, $V_1(\sqrt{10}, 0)$, $V_2(-\sqrt{10}, 0)$, $y = \pm\dfrac{\sqrt{10}}{5}x$

29. Parabola

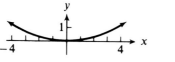

30. Parabola

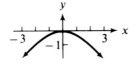

31. Parabola

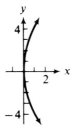

32. Parabola

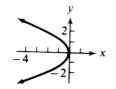

33. Hyperbola

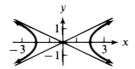

34. Ellipse

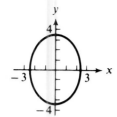

35. Two intersecting lines

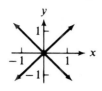

36. Single point

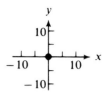

37. $(-7, 4)$ **38.** $2x'^2 - 12x' - y' + 17 = 0$

39. $(x - 1)^2 + (y + 3)^2 = 16$

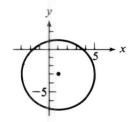

40. $y^2 = -8(x - 7)$

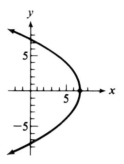

41. $x'^2 = 6y'$ **42.** $2x'^2 + y'^2 = \frac{73}{8}$

43.

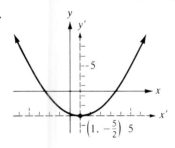

44.

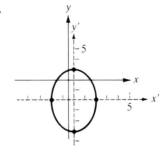

Exercise A (Page 478)

1. 2 **3.** -4 **5.** 0 **7.** 0.8280 **9.** 1.9227 **11.** 2.5011 **13.** $0.9101 - 1$ **15.** $0.9031 - 2$
17. 2.3945 **19.** 4.10 **21.** 36.7 **23.** 0.0642 **25.** 0.00718 **27.** 0.297 **29.** 0.0503
31. 0.00205 **33.** 7.52 **35.** 784 **37.** 0.0357 **39.** 1.5373 **41.** 0.5655 **43.** 4.4817
45. 0.0907 **47.** 1.3610 **49.** 2.7726 **51.** $0.0837 - 1$ or -0.9163 **53.** 1.1735
55. 6.0496 **57.** 90.017

INDEX